W0079229

Measures of
Complexity
and Chaos

NATO ASI Series

Advanced Science Institutes Series

A series presenting the results of activities sponsored by the NATO Science Committee, which aims at the dissemination of advanced scientific and technological knowledge, with a view to strengthening links between scientific communities.

The series is published by an international board of publishers in conjunction with the NATO Scientific Affairs Division

A	**Life Sciences**	Plenum Publishing Corporation
B	**Physics**	New York and London
C	**Mathematical and Physical Sciences**	Kluwer Academic Publishers Dordrecht, Boston, and London
D	**Behavioral and Social Sciences**	
E	**Applied Sciences**	
F	**Computer and Systems Sciences**	Springer-Verlag
G	**Ecological Sciences**	Berlin, Heidelberg, New York, London,
H	**Cell Biology**	Paris, and Tokyo

Recent Volumes in this Series

Volume 202—Point and Extended Defects in Semiconductors
edited by G. Benedek, A. Cavallini, and W. Schröter

Volume 203—Evaluation of Advanced Semiconductor Materials
by Electron Microscopy
edited by David Cherns

Volume 204—Techniques and Concepts of High-Energy Physics V
edited by Thomas Ferbel

Volume 205—Nuclear Matter and Heavy Ion Collisions
edited by Madeleine Soyeur, Hubert Flocard,
Bernard Tamain, and Madeleine Porneuf

Volume 206—Spectroscopy of Semiconductor Microstructures
edited by Gerhard Fasol, Annalisa Fasolino, and Paolo Lugli

Volume 207—Reduced Thermal Processing for ULSI
edited by Roland A. Levy

Volume 208—Measures of Complexity and Chaos
edited by Neal B. Abraham, Alfonso M. Albano,
Anthony Passamante, and Paul E. Rapp

Volume 209—New Aspects of Nuclear Dynamics
edited by J. H. Koch and P. K. A. de Witt Huberts

Series B: Physics

Measures of Complexity and Chaos

Edited by

Neal B. Abraham and
Alfonso M. Albano

Bryn Mawr College
Bryn Mawr, Pennsylvania

Anthony Passamante

Naval Air Development Center
Warminster, Pennsylvania

and

Paul E. Rapp

Medical College of Pennsylvania
Philadelphia, Pennsylvania

Plenum Press
New York and London
Published in cooperation with NATO Scientific Affairs Division

Proceedings of a NATO Advanced Research Workshop on
Quantitative Measures of Dynamical Complexity in Nonlinear Systems,
sponsored by the Naval Air Development Center, Warminster, Pennsylvania,
and NATO Scientific Affairs Programme, Brussels, Belgium,
held June 22–24, 1989,
at Bryn Mawr College, Bryn Mawr, Pennsylvania

Library of Congress Cataloging-in-Publication Data

NATO Advanced Research Workshop on Quantitative Measures of Dynamical
   Complexity in Nonlinear Systems (1989 : Bryn Mawr College)
     Measures of complexity and chaos / edited by Neal B. Abraham ...
   [et al.].
       p.   cm. -- (NATO ASI series. Series B. Physics : v. 208)
     "Proceedings of a NATO Advanced Research Workshop on Quantitative
   Measures of Dynamical Complexity in Nonlinear Systems, sponsored by
   the Naval Air Development Center, Warminster, Pennsylvania, held
   June 22-24, 1989, at Bryn Mawr College, Bryn mawr, Pennsylvania"-
   -Cip t.p. verso.
     Includes bibliographical references.
     ISBN 978-1-4757-0625-3   ISBN 978-1-4757-0623-9 (eBook)
     DOI 10.1007/978-1-4757-0623-9

     1. Chaotic behavior in systems--Congresses. 2. Nonlinear
   theories--Congresses.   I. Abraham, Neal B. II. Naval Air
   Development Center.   III. NATO Scientific Affairs Programme.
   IV. Title. V. Series.
   Q172.5.C45N378 1989
   003'.75--dc20   89-37104
                                                              CIP

© 1989 Plenum Press, New York
Softcover reprint of the hardcover 1st edition 1989

A Division of Plenum Publishing Corporation
233 Spring Street, New York, N.Y. 10013

All rights reserved

No part of this book may be reproduced, stored in a retrieval system, or transmitted
in any form or by any means, electronic, mechanical, photocopying, microfilming,
recording, or otherwise, without written permission from the Publisher

PREFACE

This volume serves as a general introduction to the state of the art of quantitatively characterizing chaotic and turbulent behavior. It is the outgrowth of an international workshop on "Quantitative Measures of Dynamical Complexity and Chaos" held at Bryn Mawr College, June 22-24, 1989. The workshop was co-sponsored by the Naval Air Development Center in Warminster, PA and by the NATO Scientific Affairs Programme through its special program on Chaos and Complexity.

Meetings on this subject have occurred regularly since the NATO workshop held in June 1983 at Haverford College only two kilometers distant from the site of this latest in the series. At that first meeting, organized by J. Gollub and H. Swinney, quantitative tests for nonlinear dynamics and chaotic behavior were debated and promoted [1]. In the six years since, the methods for dimension, entropy and Lyapunov exponent calculations have been applied in many disciplines and the procedures have been refined. Since then it has been necessary to demonstrate quantitatively that a signal is chaotic rather than it being acceptable to observe that "it looks chaotic". Other related meetings have included the Pecos River Ranch meeting in September 1985 of G. Mayer-Kress [2] and the reflective and forward looking gathering near Jerusalem organized by M. Shapiro and I. Procaccia in December 1986 [3].

This meeting was proof that interest in measuring chaotic and turbulent signals is widespread. Those facing limits of precision or length of data sets are hard at work developing new algorithms and refining the accuracy of old ones. Applications to symbolic dynamics and to spatio-temporal dynamics are also now emerging with "complexity" as the byword for what is even a richer subject than "chaos".

The success of the meeting was in large part guaranteed by the enthusiasm of the participants, but without the tireless efforts of a few key persons, the order of the meeting would have fallen victim to the ever looming chaos. Special thanks go to Ann Daudert, secretary of the physics department at Bryn Mawr College, and her assistant, Linath Lin. We also acknowledge the behind-the-scenes and late-night efforts of the staff of the Bryn Mawr Summer Conference Office under the direction of L. Zernicke. Many others of our colleagues and associates contributed as needed, including M.E. Farrell, G. Alman, H. Lin, and N. Tufillaro. To all of them go our warmest gratitude. With help such as theirs, it will always be more of a pleasure than a burden to organize a meeting.

Finally we should acknowledge special efforts that enlivened the meeting. J. Doran and her staff provided excellent meals and refreshments. L. Caruso-Haviland and a small crew of dedicated performers and technical staff enriched one evening with "Chaotic Metamorphoses", an inspired combination of video, cinematography, choreography, and readings. The program notes for the performance are included as part of these proceedings. Their conference T-shirts, "Complexity and Chaos at Bryn Mawr College", were duly earned.

Perhaps our principal regret (and pleasure) will be the constant task of explaining the scientific meaning of the T-Shirt title in an effort to ride the public relations wave crest.

Neal Abraham
Alfonso Albano
Anthony Passamante
Paul Rapp

August 1989

REFERENCES

[1] N.B. Abraham, J.P. Gollub and H.L. Swinney, "Testing Nonlinear Dynamics", Physica **11D**, 252 (1984).

[2] G. Mayer-Kress, editor, "Dimensions and Entropies in Chaotic Systems - Quantification of Complex Behavior", (Springer-Verlag, Berlin, 1986).

[3] M. Shapiro and I. Procaccia, editors, "Chaos and Related Nonlinear Phenomena" (Physica Scripta, to appear in 1989).

CONTENTS

Complexity and Chaos.. 1
    N.B. Abraham, A.M. Albano, and N.B. Tufillaro

Chaotic Metamorphoses... 29
    L. Caruso-Haviland and P.E. Rapp

I. CHARACTERIZING TEMPORAL COMPLEXITY: CHAOS

A. MEASURING DIMENSIONS, ENTROPIES AND LYAPUNOV EXPONENTS

Measures of Dimensions from Astrophysical Data........................ 33
    H. Atmanspacher, V. Demmel, G. Morfill, H. Scheingraber,
    W. Voges, and G. Wiedenmann

Some Remarks on Nonlinear Data Analysis of Physiological Time Series... 51
    A. Babloyantz

Hierarchies of Relations Between Partial Dimensions and Local
    Expansion Rates in Strange Attractors............................ 63
    R. Badii and G. Broggi

Experimental Study of the Multifractal Structure of the
    Quasiperiodic Set.. 75
    D. Barkley and A. Cummings

Statistical Inference Theory for Measures of Complexity in Chaos
    Theory and Nonlinear Science..................................... 79
    W.A. Brock and W.D. Dechert

Practical Remarks on the Estimation of Dimension and Entropy
    from Experimental Data... 99
    J.G. Caputo

Chaotic Behavior of the Forced Hodgkin-Huxley Axon....................111
    M. Frame

Chaotic Time Series Analysis Using Short and Noisy Dta Sets:
    Applications to a Clinical Epilepsy Seizure......................113
    G.W. Frank, T. Lookman, and M.A.H. Nerenberg

Measuring Complexity in Terms of Mutual Information...................117
    A.M. Fraser

Estimating Lyapunov Exponents From Approximate Return Maps............121
    J.N. Glover

Estimating Local Intrinsic Dimensionality Using Thresholding Techniques.125
    A. Goel, S.S. Rao, and A. Passamante

Seeking Dynamically Connected Chaotic Variables 129
    K. Hartt, and L.M. Kahn

On Problems Encountered with Dimension Calculations....................133
   U. Hübner, W. Klische, N.B. Abraham, and C.O. Weiss

Systematic Errors in Estimating Dimensions from Experimental Data.......137
   W. Lange and M. Möller

Analyzing Periodic Saddles in Experimental Strange Attractors..........147
   D.P. Lathrop and E.J. Kostelich

Time Evolution of Local Complexity Measures and Aperiodic
      Perturbations of Nonlinear Dynamical Systems 155
   G. Mayer-Kress and A. Hübler

Analysis of Local Space/Time Statistics and Dimensions of
      Attractors Using Singular Value Decompositon and
      Information Theoretic Criteria. 173
   A. Passamante, T. Hediger, and M.E. Farrell

Entropy and Correlation Time in a Multimode Dye Laser.................181
   M.G. Raymer

Dimension Calculation Precision with Finite Data Sets..................183
   C.L. Sayers

Chaos in Childhood Epidemics...187
   W.M. Schaffer and L.F. Olsen

Measurement of $f(\alpha)$ for Multifractal Attractors
      in Driven Diode Resonator Systems...............................191
   Z. Su, R.W. Rollins, and E.R. Hunt

Is there a Strange Attractor in a Fluidized Bed? 193
   S.W. Tam and M.K. Devine

Statistical Error in Dimension Estimators.............................199
   J. Theiler

B. OTHER MEASURES

Dynamical Complelxity of Strange Sets 203
   D. Auerbach

Characterization of Complexity by Aperiodic Driving Forces..............209
   D. Bensen, M. Welge, A. Hübler, and N. Packard

Stabilization of Prolific Populations Through Migration and
      Long-lived Propagules...213
   R.E. Byers and R.I.C. Hansell

Complex Behavior of Systems Due to Semi-stable Attractors:
      Attractors That Have Been Destablized but Which Still
      Temporarily Dominate the Dynamics of a System...................219
   R.E. Byers, R.I.C. Hansell and N. Madras

Universal Properties of the Resonance Curve of Complex Systems.........225
   K. Chang, A. Hübler, and N. Packard

The Effects of External Noise on Complexity in Two-dimensional
      Driven Damped Dynamical System..................................229
   J. Fang

Chaos on a Catastrophe Manifold.......................................235
   S.T. Gaito and G.P. King

Topolgical Frequencies in Dynamical Systems...........................241
   R. Gilmore, G. Mindlin, and H.G. Solari

Phase Transitions Induced by Deterministic Delayed Forces...............245
    M. LeBerre, E. Ressayre, and A. Tallet

Mutual Information Functions Versus Correlation Functions
    in Binary Sequences..249
    W. Li

Reduction of Complexity by Optimal Driving Forces.....................253
    T. Meyer, A. Hübler, and N. Packard

Symbolic Dynamical Resolution of Power Spectra.......................257
    M.A. Sepulveda and R. Badii

Relative Rotation Rates for Driven Dynamical Systems.................261
    H.G. Solari and R. Gilmore

Stretching Folding Twisting in the Driven Damped Duffing Device........265
    H.G. Solari, X.-J. Hou, and R. Gilmore

Characterizing Chaotic Attractors Underlying Single Mode Laser Emission
    by Quantitative Laser Field Phase Measurements...................269
    C.O. Weiss and N.B. Abraham

C. CHARACTERIZING HOMOCLINIC CHAOS

Shil'nikov Chaos: How to Characterize Homoclinic and Heteroclinic
    Behavior 281
    F.T. Arecchi, A. Lapucci, and R. Meucci

Time Series Near Codimension Two Global Bifurcations 295
    P. Glendinning

Characterization of Shil'nikov Chaos in a CO_2 Laser Containing a
    Saturable Absorber...299
    D. Hennequin, M. Lefranc, A. Bekkali, D. Dangoisse and P. Glorieux

Symmetry-breaking Homoclinic Chaos 303
    B. Nicolaenko and Z.-S. She

Time Return Maps and Distributions for the Laser with Saturable
    Absorber...309
    F. Papoff, A. Fioretti, E. Arimondo, and N.B. Abraham

D. BUILDING MODELS FROM DATA

Unfolding Complexity in Nonlinear Dynamical Systems...................313
    R. Badii

Inferring the Dynamic; Quantifying Physical Complexity...............327
    J.P. Crutchfield

Symbolic Dynamics from Chaotic Time Series...........................339
    A. Destexhe, G. Nicolis, and C. Nicolis

Modelling Dynamical Systems from Real-world Data.....................345
    A. Mees

Extraction of Models from Complex Data...............................349
    H.G. Schuster

Quantifying Chaos with Predictive Flows and Maps:
    Locating Unstable Periodic Orbits................................359
    L.A. Smith

II. CHARACTERIZING SPATIO-TEMPORAL COMPLEXITY: TURBULENCE

A. THEORETICAL

Defect-induced Spatio-temporal Chaos 367
    P. Coullet

Lyapunov Exponents, Dimension and Entropy in Coupled Lattice Maps.......375
    R.M. Everson

Phase Dynamics, Phase Resettiing, Correlation Functions and
    Coupled Map Lattices...381
    R. Kapral, M.-N. Chee, S.G. Whittington and G.L. Oppo

Characterization of Spatiotemporal Structures in Lasers:
    A Progress Report..395
    G.-L. Oppo, M.A. Pernigo, L.M. Narducci, and L.A. Lugiato

Amplitude Equations for Hexagonal Patterns of Convection
    in Non-Boussinesq Fluids...405
    C. Pérez-Garcia, E. Pampaloni and S. Ciliberto

Fractal Dimensions in Coupled Map Lattices............................409
    A. Politi, G. D'Alessandro, and A. Torcini

Weak Turbulence and the Dynamics of Topological Defects 425
    I. Procaccia

Pattern Cardinality as a Characterization of Dynamical Complexity.......429
    J. Ringland and M. Schell

B. EXPERIMENTAL

Characterizing Spatiotemporal Chaos in Electrodeposition Experiments....433
    F. Argoul, A. Arneodo, J. Elezgaray and G. Grasseau

Characterizing Space-time Chaos in an Experiment of Thermal Convection..445
    S. Ciliberto

Characterizing Dynamical Complexity in Interfacial Waves...............457
    J.P. Gollub

Characterization of Irregular Interfaces: Roughness and Self-affine
    Fractals 461
    M.A. Rubio, A. Dougherty, and J.P. Gollub

The Field Patterns of a Hybrid Mode Laser: Detecting the
    "Hidden" Bistability of the Optical Phase Pattern..................465
    C. Tamm

Contributors..469

Index 475

COMPLEXITY AND CHAOS

N.B. Abraham, A.M. Albano, and N.B. Tufillaro

Department of Physics
Bryn Mawr College
Bryn Mawr, PA 19010-2899, USA

1. INTRODUCTION

Turbulence was one of the key phenomena that motivated the resurgence of interest in nonlinear dynamical systems. It was, after all, investigations into the mechanisms for turbulence that led Ruelle and Takens to invent the term "strange attractor" in 1971. The turbulence that is described by strange attractors is "turbulence in time" (Schuster, 1988) -- deterministic chaos, or *temporal chaos* in current terminology. In the past decade, a vocabulary for the quantitative characterization of temporal chaos has been developed, and has been used to describe and analyze an incredible variety of phenomena in practically all fields of science and engineering. The dimensions of strange attractors, and the entropies and Lyapunov exponents describing motions on them, have been used to analyze heartbeats and brain waves, chemical reactions, lasers, the economy, x-ray emissions of stars, flames, and fluid flow ...

Yet, this vocabulary is not sufficient to describe turbulence, for turbulence is complexity not only in time but also in space. The vocabulary needs to be enlarged to include the quantitative characterization of spatial complexity and its time evolution. So turbulence, once more, is the motivation for efforts to enlarge the scope of nonlinear dynamics to include a description of *spatio-temporal complexity*. These efforts include the use of modal expansions (reminiscent of the old "mode-mode coupling" theories or "dressed-mode analyses"), spatial correlation functions, coupled lattice models, and new approaches based on the analysis of topological defects.

This workshop was an effort to describe the "state of the field" of the quantitative characterization of complexity and chaos. The presentations at the workshop included in these proceedings go a long way towards that description. To enlarge the context and information for the readers, we append to this introduction a bibliography of some of the most recent references and a selection of other key articles. These are organized both by methodology and the subject to which they are applied. Further references can be found in the various chapters of this volume and in the literature citations of other works listed here. (In the following, papers appearing in this volume are followed by an asterisk).

Measures of Complexity and Chaos
Edited by N.B. Abraham *et al.*
Plenum Press, New York

2. CHARACTERIZING TEMPORAL CHAOS

(a) Calculational techniques, precision, error estimates

Dimensions, entropies and Lyapunov exponents have become the standard measures of temporal chaotic behavior. Of these, perhaps because of the possibility of attaching an intuitive geometric significance to them, and because of the commonly made connection between fractal attractors and chaotic behavior, dimensions have been used most frequently. This is also because the algorithms for their computation seem to be the least cumbersome. However, the relative simplicity of these methods can mask a host of possible errors and difficulties. A comprehensive review of the techniques used and the problems encountered in dimension calculations is given in Theiler's review article (Theiler, 1989).

It is now common for a chaotic time series to be described by a spectrum of dimensions, d_q ($-\infty < q < \infty$), rather than by just one. The d_q's are a hierarchy of dimensions introduced by Grassberger (Grassberger, 1983) and by Hentschel and Procaccia (Hentschel and Procaccia, 1983). These are defined by,

$$d_q = \frac{1}{q-1} \lim_{r \to 0} \frac{\log \sum_i p_i^q}{\log r} \quad ,$$

where p_i is the number of points in the ith box of a partition consisting of boxes of size r covering the attractor, typically one that is reconstructed from a time series of a single measured variable by an embedding in an m-dimensional space by the now familiar method of time delays (Packard et al., 1980; Takens, 1980; Mané, 1980). d_0 is the "fractal dimension" or the "capacity" and d_1 is the "information dimension". The "correlation dimension", d_2, is the most widely used member of the hierarchy. Alternatively, the spectrum of scaling indices, $f(\alpha)$, is also used (Halsey et al., 1986), where the variables $(\alpha, f(\alpha))$ are obtained from (q, d_q) by a Legendre transformation:

$$\alpha = \frac{\partial [(q-1)d_q]}{\partial q} \quad , \qquad f(\alpha) = \alpha q - (q-1)d_q \quad .$$

There exist a number of algorithms for calculating dimensions although that due to Grassberger and Procaccia (Grassberger and Procaccia 1983, 1983a, 1983b) for the correlation dimension remains the most widely used because it remains the easiest to implement. The Grassberger-Procacia algorithm is often called a "fixed volume" technique as it involves counting the average number of points in a hypersphere or hypercube as function of the radius, r. "Fixed mass" algorithms (Termonia and Alexandrovicz, 1983; Badii and Politi, 1985), on the other hand, involve the determination of the radius of a hypersphere that contains a given number of points, N, as a function of N. These numbers scale as r^{d_2} and N^{d_1}, respectively. The Grassberger-Procaccia correlation integral, $\log \sum p_i^2$, has also been modified into a statistical test for potential forecastibility or hidden recurrent pattterns in an observed time series (Brock and Dechert*).

Though in some cases it appears that dimensions can be reliably extracted with as few as 500 data points, the minimum sufficient number of data points, and the optimum data sampling rate and embedding delay, all depend critically on the uniformity of the strange attractor and its dimension. Sometimes a total number of points as few as $N = 10^d$ for an attractor of dimension d can be sufficient, but various parameters of the

2

embedding need to be optimized. (Note that Smith (Smith, 1988) has a proof that the lower bound of the number of points to avoid spurious results is $N = 42^d$).

There are problems with the standard methods, however, especially when they are used with noise-corrupted data sets of limited size and limited precision (the only kind obtainable from experiments). Each of these lead to different and often overlapping problems. A number of these problems were considered in this workshop.

In constructing an embedding, one must select the sampling rate for the data acquisition of the time series, the total length of the time series, the delay between successive elements of an embedding vector, the spacing in time between the first elements of successive embedding vectors, and the total number of embedding vectors. Arbitrary selection of these parameters can introduce significant systematic errors in the construction or coverage, of the attractor. One key quantity appears to be the window used in the embedding -- i.e., the time spanned by each embedding vector. It is known that the results of dimension calculations depend rather sensitively on this choice. Current lore suggests "rule of thumb" criteria based on the correlation time determined from the inverse of the bandwidth of the signal's power spectrum or the first zero, minimum, or the decay of the envelope of the autocorrelation function. Fraser* and Schuster* suggest improved criteria for making this choice.

Caputo* and Politi* address the problem of data requirements, Lange* investigates systematic errors due to finite precision and noise and proposes strategies for correcting these errors. Kostelich* presents a procedure for decreasing the effects of noise, while Hunt*, Sayers* and Theiler* address the problem of obtaining realistic error estimates for the dimensions and entropies that are extracted.

Another dimension-seeking strategy is to determine an "intrinsic dimension", a minimal embedding dimension for the time series. One procedure is to use a singular value decomposition (Broomhead and King, 1986) to determine the smallest number of orthogonal directions needed to describe the data. When applied locally to get an average local intrinsic dimension, it is known to be fairly robust relative to noise (Passamante et al., 1989). Passamante, Hediger and Farrell* discuss the use of an information theoretic criterion to determine local dimension. Goel and Rao* present other criteria.

Dimensions and additional topological properties of strange sets are also obtainable from unstable periodic orbits. Some aspects of these procedures are discussed by Schuster*, Auerbach*, Glendinning*, Smith* as well as by Gilmore* and Solari*.

Continuously varying signals can also be characterized by smaller sets of information. For example, a Poincaré section of a periodic signal selects the value of the signal once every clock cycle. Externally driven systems have well defined clock frequencies. For autonomous systems Poincaré sections are defined by when the trajectory in the reconstructed attractor crosses a selected hyperplane. Similar data reduction procedures involve the study of the sequence of peak values of the signals or the sequence of time intervals between peaks. The dimension of such a subset is one less than the dimension of the corresponding continuous trajectory.

The logical and more sophisticated extension of these classification schemes involves the invention of a set of symbols used to represent different types of behavior of the system. The sequences of symbols can be analyzed for the "syntax" and "grammar" of the "language" of the dynamics. Conditional probabilities and the relative probability of unique sequences can be used to define a degree of complexity and an entropy production rate for the system. Questions of selection of a set of symbols and enlargement of that set were addressed by Badii*, Crutchfield* and Fraser*, among others.

3

It has been difficult to calculate Lyapunov exponents from experimental data principally because available procedures are not sufficiently robust against noise in experimental data. Several algorithms have been proposed by a number of investigators (Wolf, 1985; Eckmann et al., 1986; Sato et al. 1987; Stoop and Meier, 1988) but others have had their problems implementing these on experimental results. Most experimental data sets are of insufficient precision, sampling rate, or length to permit the use of these algorithms on them with consistent reliability or success. Use of the algorithms on data generated by numerical simulations are obviously more straightforward and have been more universally successful. Glover* presents a technique that calculates Lyapunov exponents more efficiently using the Poincaré sections.

Although entropies K_q can be calculated by means of the same correlation integral that is used to calculate dimensions d_q, they have not been as widely used as dimensions, either. There are growing indications, though, that entropies may be more robust quantities than dimensions (in that they seem to be invariant under linear filtering of the data, while the dimensions are not similarly invariant [see Lange*]) and may well find more use in the future. Though the relationship between power spectra (and thus correlation functions) and entropies is not rigorously established, evidence has been consistently reported that the bandwidth of power spectra and the decay of the envelope of autocorrelation functions are excellent estimators for the entropies. Applications to measures of the complexity of symbolic sequences will increase as ways are explored to reduce the artificial complexity of continuous variables used to describe low dimensional phenomena.

It is also worth noting that internal consistency of the results can be tested since the entropy is a good estimator for the positive Lyapunov exponent (if there is only one). Furthermore, there is the Kaplan-Yorke conjecture (Kaplan, 1979) on the relationship between the dimension and the Lyapunov exponents,

$$d_1 = j + \frac{\sum\limits_{i}^{j} \lambda_i}{|\lambda_{j+1}|}$$

where j is the largest integer for which $0 \le \lambda_1 + \lambda_2 + \cdots + \lambda_j$. This seems to be particularly accurate for larger systems with only a single positive Lyapunov exponent.

(b) <u>Routes to Chaos and a new standard: Homoclinic and Heteroclinic orbits, Shilnikov chaos</u>

Historically, once some degree of universality was assured, routes to chaos were commonly used as indicators that irregular behavior was indeed dynamical chaos. While these qualitative methods relying on changes in power spectra have fallen in disfavor when more quantitative methods can be applied, the simplicity of a new system following generic routes to chaos is still powerful when one is analyzing experimental data. Nevertheless, countless studies have shown that the universality can break down or become more complex, with truncated period-doubling sequences occuring when there are two controlling parameters rather than one. Quasiperiodicity is equally generic in reaching chaos through locking of the previously incommensurate frequencies and then period doubling of the locked conditions. Only for a few special cases does the incommensurate nature remain until chaos appears with a third frequency.

Another now common route through periodic and chaotic dynamics include sequences of symmetry breaking bifurcations (in systems described by an underlying inversion symmetry) and then glueing bifurcations that restore the symmetry with a higher degree of complexity (Hennequin et al.*).

4

Chaos related to homoclinic or heteroclinic orbits also seems to come in some relatively standard forms. Signatures include the similarity of the topology of successive trajectories around a topologically simple attractor even though the times for successive trajectories may differ widely. Near these homoclinic orbits there are infinite sets of periodic orbits involving different numbers of spirals. Beyond qualitative measures, it now appears that Poincaré plots of return time maps may be the best indicators of the multileaved structures created by the complex homoclinic chaos. Because of the intricate topological structure of these attractors their sensitivity to noise will also be a subject of considerable ongoing investigation. The structure in the return plots will also probably be a major source of data for studies of symbolic dynamics in the future. The new interest in the dynamics of homoclinic and heteroclinic chaos is driven by the experimental observations of such phenomena in chemical reactions (Argoul* and Arneodo*) and lasers (Arecchi* Arimondo*, Glorieux*, Weiss*).

Homoclinic orbits and coherent transients are now also an ordering feature in the study of spatial dynamics. Nikolaenko* demonstrated the importance of local homoclinic behavior in driving spatio-temporal dynamics and Newell (Newell et al., 1988) has also recently focused on the role of local coherent transients and their propagation as a governing feature of turbulence.

(c) Applications to real-world data

The large variety of fields in which dimensions, entropies, and exponents have been used to characterize complex temporal evolution is an indication of the extent to which these quantities have become elements of a scientific vocabulary that is now practically universal. The worksop saw these quantities used to characterize astrophysical data (Atmanspacher*), dendritic growth (Argoul*), electroencephalographic and electrocardiographic data (Babloyantz*), nerve fibers (Frame*), economics (Brock and Dechert*), epidemics (Schaffer*), fluids (Gollub*, Ciliberto*, Nikolaenko*, Sreenivasan), flames (Sreenivasan) and lasers (Arecchi*, Arimondo*, Glorieux*, Tamm*, Raymer*, Weiss*).

As one of the most basic applications of these methods, dimensions have been used to discriminate between chaos and noise. In many situations, it is possible to distinguish those phenomena that result from the combined effects of extremely many independent processes and which therefore may be regarded as stochastic from those that may be described as low-dimensional deterministic processes. Beyond this, dimension calculations have made possible the direct comparison of computational and experimental results.

An example of a comparison between theory and experiment that goes beyond mere matching of the dimensions of experimental data and results of numerical simulations was presented by Weiss* . He described results obtained with a laser working in a region in parameter space where its operation is described by the Lorenz equations, making it possible to make some very meticulous comparisons between theory and experiment.

(d) From data to dynamics: prediction algorithms

It is almost paradoxical that a chaotic time series, characterized by sensitive dependence to initial conditions which renders its distant future values unpredictable in practice, should be the subject of prediction algorithms (Farmer and Sidorowitz, 1987, 1988; Crutchfield and MacNamara, 1987). Yet, since it is deterministic, it is governed by a dynamical law which is discoverable, at least in principle. Badii and Sepulveda*, Crutchfield*, Mees*, Smith*, and Schuster* present various approaches to the short-term prediction problem.

3. CHARACTERIZING SPATIO-TEMPORAL COMPLEXITY

(a) <u>Modal expansions, spatial correlation functions, coupled lattice models</u>

The characterization of spatially complex nonlinear systems is complicated by the fact that sums of solutions are no longer themselves solutions as in the case of linear systems. The time evolution of nonlinear systems can thus no longer be simply described in terms of independently evolving modes. Nevertheless, the system can still be described by expansions over characteristic spatial modes, but one must recognize that the time evolutions of these modes are coupled.

Another alternative is the use of spatial correlation functions. One definition of turbulence, proposed by Heslot et al. (1987), is in terms of the decay of the spatial correlation function in a system with spatio-temporal dynamics which are locally chaotic. However, such a definition will clearly reject as not turbulent systems which are described by a small number of spatial modes with time-dependent amplitudes. Clearly, some intermediate approaches to characterization for systems of moderate size are still needed.

A variety of methods are used by Gollub* to characterize parametrically driven surface waves and by Oppo* to describe lasers with many transverse modes.

Another possibility of describing a spatially inhomogeneous system is to model the system on a lattice in which the dynamics at each lattice point is influenced by interactions with a few near neighbors. Coupled lattice models are discussed by Kapral*, while some problems associated with calculating dimensions for these models are presented by Politi*.

(b) <u>Defect-mediated order-disorder transitions</u>

Defect-mediated turbulence is emerging as a promising paradigm for studying weak turbulence in large aspect ratio systems, i.e., systems in which the size of the basic spatial structure is much less than the size of the system. The inspiration for these ideas comes from analogies with defect-mediated phase transitions in equilibrium systems.

Different theoretical approaches to defect formation in non-equilibrium systems are discussed by Coullet and Procaccia. Coullet* described the usefulness of the Ginzburg-Landau equation in understanding the essential features of defect formation, annihilation, and dynamics. Earlier applications to convective systems were supplemented by illustrations appropriate to large-aperture lasers. In contrast, Procaccia* developed a field theory which described the free dynamics and interdefect forces exhibited at finite range. Aspects of this field theory were illustrated by experimental data from electroconvecting nematics.

Several experimental examples of defect formation in far from equilibrium systems are discussed in Part III of these proceedings. Gollub* discussed the dynamics of parametrically forced surface waves which illustrate temporal chaos at small aspect ratios, and possibly a defect-mediated order-disorder transition at large aspect ratios. Defects might also be found in convective fluid systems such as those described by Ciliberto*, and in lasers with many transverse modes as discussed by Oppo*. The strongest evidence for defect-mediated transitions, however, are to be found in nematics and large aspect-ratio Rayleigh-Bénard convection.

4. PARTING SHOT

Many will continue to doubt the usefulness of quantitative measures of experimental results plagued by noise and data limitations. The work here indicates many examples where such quantitative measures can be used judiciously to confirm or deny the presence of low dimensional chaotic origins of the aperiodic behavior. Where multiple characterizations are possible, types of complexity can be distinguished and definite guidance is provided to those endeavoring to construct adequate models of the behavior.

Given the current state-of-the-art in dimension measurements, a well-defined and *continuous* transition from sharp to broadband spectral structure, when that is experimentally feasible, may still be one of the best practical diagnostics for the onset of chaos in physical systems.

Surely there will be increasing breadth of application of the techniques. New practitioners must be suitably chastened by the systematic errors and difficulties of interpretation, but all should be encouraged by the progress that has been made and by the additional progress which many of these articles portend.

REFERENCES (BY CATEGORY)

Complete bibliographical information is given in the alphabetical listing that follows.

Biology, neuroscience

A. M. Albano, A. I. Mees, G. C. de Guzman and P. E. Rapp (1987).
A. M. Albano, L. Smilowitz, P. E. Rapp, G. C. de Guzman, R. R. Bashore (1988).
A. Babloyantz, (1988).
A. Babloyantz and A. Destexhe (1988).
A. Babloyantz, J.M. Salazar, and C. Nicolis (1985).
R. E. Byers and R.I.C. Hansell, This volume.
R. E. Byers, R.I.C. Hansell, and N. Madras, This volume.
A. Destexhe, G. Nicolis, and C. Nicolis, This volume.
A. Destexhe, J.A. Sepulchre, and A. Babloyantz (1988).
M. Frame, This volume.
G. W. Frank, T. Lookman, M.A.H. Nerenberg, This volume.
W. J. Freeman (1987).
G. J. Mpitsos, H.C. Creech, and S.O. Soinilla (1988).
G. J. Mpitsos, H.C. Creech, C.S. Cohan, and M. Mendelson (1988).
P. E. Rapp, I.D. Zimmerman, A.M. Albano, G.C. de Guzman, and N.N. Greenbaun (1985).
P. E. Rapp, I.D. Zimmerman, T.R. Bashore, A.M. Albano, G.C. de Guzman, and N.N. Greenbaun (1985).
W. M. Schaffer and L.F. Olsen, This volume.
C. A. Skarda and W.J. Freeman (1987).

Books

F. T. Arecchi and R. G. Harrison (1987).
P. Cvitanovic (1986).
J. Gleick (1987).
H. Haken (1988).
Hao Bai-Lin (1984).
J. A. S. Kelso, A. J. Mandell, and M. F. Schlesinger (1988).
G. Mayer-Kress (1986).
H. O. Peitgen and P. H. Richter (1986).
H. G. Schuster (1988).

Chaos
(See also Dimensions and entropies, Fractals)

D. Auerbach, P. Cvitanovic, J-P. Eckmann, G. Gunarante, I.
Procaccia (1987).
D. Auerbach, B. O'Shaughnessy, and I. Procaccia (1988).
A. Destexhe, G. Nicolis, and C. Nicolis, This volume.
J. P. Eckmann, S. Kamphorst, and D. Ruelle (1987).
A. M. Fraser and H. L Swinney (1986).
W. J. Freeman (1987).
R. Gilmore, G. Mindlin, and H. G. Solari, This volume.
Hao Bai-lin (1987).
S. Kim, S. Ostlund, and G. Yu (1989).
D. P. Lathrop and E. J. Kostelich, This volume.
G. Mayer-Kress and A. Hübler, This volume.
T. C. Molteno and N.B. Tufillaro, (1989).
J. Ringland and M. Schell (1989).
R. S. Shaw (1981).
C. A. Skarda and W.J. Freeman (1987).
L. A. Smith, This volume.
H. G. Solari and R. Gilmore, This volume.
H. G. Solari, Xin-Jun Hou, and R. Gilmore, This volume.
N. B. Tufillaro (1989).
H. G. Winful and S. S. Wang (1988).

Complexity
(See also Fractals, Turbulence)

D. Auerbach, This volume.
R. Badii, This volume.
D. Bensen, M. Welge, A. Hübler, N. Packard, This volume.
R. E. Byers and R.I.C. Hansell, This volume.
R. E. Byers, R.I.C. Hansell, and N. Madras, This volume.
S. Ciliberto and J.P. Gollub (1984).
J. M. Gambaude, P. Glendinning, and C. Tresser (1988).
P. Glendinning, This volume.
C. Grebogi, E. Ott, and J.A. Yorke (1987).
C. Grebogi, E. Ott, and J.A. Yorke (1988).
G. H. Gunaratne, M.H. Jensen, and I. Procaccia (1988).
K. Ikeda and K. Matsumoto (1986).
T. Meyer, A. Hübler, N. Packard, This volume.
J. Ringland and M. Schell, This volume.
Y. A. Rzhanov and Y.D. Kalafati (1989).
M. A. Sepulveda and R. Badii, This volume.
D. K. Umberger, C. Grebogi, E. Ott, and D. Afeyan (1989).

Dimensions and entropies
(see also: Chaos, Fractals, Lyapunov exponents).

N. B. Abraham, A.M. Albano, B. Das, G.C. de Guzman, S. Yong, R.S.
Gioggia, G.P. Puccioni, J.R. Tredicce (1986).
A. M. Albano, J. Muench, C. Schwartz, A. I Mees, and P.E. Rapp,
(1988).
A. M. Albano, A.I. Mees, G.C. de Guzman and P.E. Rapp (1987).
A. M. Albano, L. Smilowitz, P.E. Rapp, G.C. de Guzman, R.R.
Bashore (1988).
H. Atmanspacher, V. Demmel, G. Morfill, H. Scheingraber, W. Voges
and G. Wiedenmann, This volume.
P. Atten, J.G. Caputo, B. Malraison and Y. Gagne (1984).
R. Badii and G. Broggi, This volume.
R. Badii, K. Heinzelmann, P.F. Meier, and A. Politi (1988).
R. Badii and G. Broggi (1988).
R. Badii, G. Broggi, B. Derighetti, M. Ravani, S. Ciliberto, A.
Politi, and M.A. Rubio (1988).
R. Badii and A. Politi (1987).

R. Badii and A Politi (1985).

D. Barkley and A. Cummings, This volume.

W. A. Brock and W.D. Dechert, This volume.

G. Broggi (1988).

D. S. Broomhead, R. Jones, and G.P. King (1987).

J. G. Caputo, This volume.

A. Chabra and R.V. Jensen (1989).

A. Destexhe, J.A. Sepulchre, and A. Babloyantz (1988).

M. Ding, C. Grebogi, and E. Ott (1989).

B. Dubuc, J. F. Quiniou, C. Roques-Carmes, C. Tricot, and S. W. Zucker (1989).

S. Ellner (1988).

G. Fahner and P. Grassberger (1987).

J. Fang, This volume.

J. D. Farmer, E. Ott, and J.A. Yorke (1983).

M. Frame, This volume.

G. W. Frank, T. Lookman, M.A.H. Nerenberg, This volume.

A. M. Fraser, This volume.

K. Fukunaga and D.R. Olsen (1971).

D. J. Gauthier, M.S. Malcuit, and R.W. Boyd (1988).

J. A. Glazier, G. Gunaratne, and A. Libchaber (1988).

J. N. Glover, This volume.

A. Goel, S. S. Rao, and A. Passamante, This volume.

P. Grassberger (1988).

P. Grassberger, R. Badii, and A. Politi (1988).

P. Grassberger (1987).

P. Grassberger and I. Procaccia (1983).

P. Grassberger and I. Procaccia (1983a).

T. C. Halsey, M. H. Jensen, L. P. Kadanoff, I. Procaccia, and B. I. Shraiman (1986).

K. Hartt and L. M. Kahn, This volume.

J. W. Havstad and C. L. Ehlers (1989).

H. G. E. Hentschel and I. Procaccia (1983).

H. Herzel (1988)

G-H. Hsu, E. Ott, and C. Grebogi (1988).

U. Hübner, W. Klische, N. B. Abraham, and C. O. Weiss, This volume.

U. Hübner, N. B. Abraham and C. O. Weiss (1989).

F. Hunt, preprint.

K. Ikeda and K. Matsumoto (1987).

M. H. Jensen, L.P. Kadanoff, A. Libchaber, I. Procaccia and J. Stavans (1987).

J. L. Kaplan and J. A. Yorke (1979).

W. Lange and M. Möller, This volume.

M. Le Berre, E. Ressayre, and A. Tallet, This volume.

C.-K. Lee and F. C. Moon (1986).

J. M. Martinerie, A. M. Albano, A. I. Mees, P. E. Rapp, preprint.

A. I. Mees, P. E. Rapp, and L. S. Jennings (1988).

F. Mitschke, M. Möller, and W. Lange (1988).

M. Möller, W. Lange, F. Mitschke, N.B. Abraham, and U. Hübner (1989)

E. Ott, C. Grebogi, and J. Yorke (1989).

N. H. Packard, J. P. Crutchfield, J. D. Farmer, and R. S. Shaw (1980).

A. Passamante, T. Hediger, and M. E. Farrell, This volume.

A. Passamante, T. Hediger, and M. Gollub (1989).

K. Pawelzik and H. G. Schuster (1987).

J. B. Ramsey and H-J. Yuan (1989).

P. E. Rapp, I. D. Zimmerman, A. M. Albano, G. C. de Guzman, and N. N. Greenbaun (1985).

P. E. Rapp, I. D. Zimmerman, T. R. Bashore, A. M. Albano, G. C. de Guzman, and N. N. Greenbaun 1985).

P. E. Rapp, A. M. Albano, and A. I. Mees (1988).

M. G. Raymer, This volume.

A. Renyi (1970).

S. Sato, M. Sano, and Y. Sawada (1987)

C. L. Sayers, This volume.
W. M. Schaffer and L. F. Olsen, This volume.
L. A. Smith (1988).
Z. Su, R. W. Rollins, and E. R. Hunt, This volume.
Z. Su, R. W. Rollins and E. R. Hunt (1987).
F. Takens (1983).
F. Takens (1985).
Y. Termonia and Z. Alexandrowicz (1983).
J. Theiler, This volume.
J. Theiler (1988).
W. Water and P. Schram (1988).
C. O. Weiss and N. B. Abraham, This volume.

Extracting models from data

H. Abarbanel, R. Brown, and J. Kadtke (1989).
D. S. Broomhead and G. P. King (1986a).
J. P. Crutchfield, This volume.
J. P. Crutchfield and B. S. McNamara (1987).
J. P. Crutchfield and K. Young, (1989).
J. D. Farmer and J. J. Sidorovich (1987).
J. D. Farmer and J. J. Sidorovich (1988).
E. Kostelich and J.A. Yorke (1988).
A. Mees, This volume.
H. G. Schuster, This volume.

Fluids
(See also Complexity, Turbulence)

D. Bensimon, B. I. Shraiman, and V. Croquette (1989).
A. Brandstatter and H. Swinney (1987).
S. Ciliberto and J. P. Gollub (1984).
S. Ciliberto and P. Bigazzi (1988).
P. Coullet, This volume.
J. P. Gollub and C. W. Meyer (1983).
J. P. Gollub, This volume.
F. Heslot, B. Castaing, A. Libchaber (1987).
P. Kolodner, D. Bensimon, and C. M. Surko (1988).
B. Nicolaneko and Zhen-Su She, This volume.
C. Pérez-Garcia, E. Pampaloni, and S. Ciliberto, This volume.
I. Procaccia, This volume.
F. Simonelli and J. P. Gollub (1989).
F. Takens (1981)
W. Y. Tam, J. A. Vastano, H. L. Swinney, and W. Horsthemke (1988).

Fractals
(See also Chaos, Complexity, Dimensions and entropies)

M. F. Barnsley (1988).
P. M. Battelino, C. Grebogi, E. Ott, and J. Yorke (1988).
T. Bohr and M. H. Jensen (1987).
A. Hurd (1988).
B. Mandelbrot (1982).
M. A. Rubio, A. D. Dougherty, and J. P. Gollub, This volume.

Homoclinic chaos

V. S. Anischenko and H. Herzel (1988).
J. C. Antoranz and M. A. Rubio (1988).
F. T. Arecchi, This volume.
A. Arneodo, P. Coullet, and C. Tresser (1982).
D. Dangoisse, A. Bekkali, F. Papoff, and P. Glorieux (1988).
S. T. Gaito and G. P. King, This volume.

D. Hennequin, M. Lefranc, A. Bekkali, D. Dangoisse, and P.
    Glorieux, This volume.
D. Hennequin, F. de Tomasi, L. Fronzoni, B. Zambon, and E.
    Arimondo (1989).
F. Papoff, A. Fioretti, and E. Arimondo, and N. B. Abraham, This
    volume.
P. Richetti, J. C. Roux, F. Argoul, and A. Arneodo (1987).

Lyapunov exponents
(See also Chaos, Dimensions and entropies)

J-P. Eckmann, S. O. Kamphorst, D. Ruelle, and S. Ciliberto (1986).
R. M. Everson, This volume.
S. Sato, M. Sano, and Y. Sawada (1987).
R. Stoop and P. F. Meier (1988).
A. Wolf, J. B. Swift, H. L. Swinney, and J. A. Vastano (1985).

Nonlinear optics

N. B. Abraham, F. T. Arecchi, and L. A. Lugiato (1988).
N. B. Abraham, A. M. Albano, T. H. Chyba, L. M. Hoffer, M. F. H.
    Tarroja, S. P. Adams and R. S. Gioggia (1987).
J. C. Antoranz and M. A. Rubio (1988).
F. T. Arecchi (1987).
F. T. Arecchi and R. G. Harrison (1987).
F. T. Arecchi, A. Lapucci, R. Meucci, A. Roversi, and P. Coullet
    (1988).
F. T. Arecchi, W. Gadomski, A. Lapucci, H. Mancini, R. Meucci, and
    J. A. Roversi (1988).
F. T. Arecchi, This volume.
D. K. Bandy, A. N. Oraevsky, and J. R. Tredicce (1988).
P. Coullet, L. Gil, and F. Rocca (1989)
D. Dangoisse, A. Bekkali, F. Papoff, and P. Glorieux (1988).
D. J. Gauthier, M. S. Malcuit, and R. W. Boyd (1988).
D. Hennequin, M. Lefranc, A. Bekkali, D. Dangoisse, and P.
    Glorieux, This volume.
D. Hennequin, F. de Tomasi, L. Fronzoni, B. Zambon, and E.
    Arimondo (1989).
U. Hübner, N. B. Abraham and C. O. Weiss (1989).
G.-L. Oppo, M. A. Pernigo, L. M. Narducci, and L. A. Lugiato, This
    volume.
F. Papoff, A. Fioretti, and E. Arimondo, and N. B. Abraham, This
    volume.
M. G. Raymer, This volume.
Y. A. Rzhanov and Y. D. Kalafati (1989).
C. Tamm, This volume.
J. R. Tredicce, E. J. Quel, A. M. Ghazzawi, C. Green, M. A.
    Pernigo, L. M. Narducci, and L. A. Lugiato (1989).
C. O. Weiss and N. B. Abraham, This volume.
C. O. Weiss, N. B. Abraham and U. Hübner (1988).
H. G. Winful and S. S. Wang, (1988).

Reviews

N. B. Abraham, A. M. Albano, B. Das, T. Mello, M. F. H. Tarroja,
    N. Tufillaro, and R. S. Gioggia (1986).
N. B. Abraham, J. P. Gollub, and H. L. Swinney (1984).
J. P. Crutchfield, J.D. Farmer, N. H. Packard, and R. Shaw (1986).
J-P. Eckmann and D. Ruelle (1985).
J. Guckenheimer and P. Holmes (1983).
E. Ott (1981).
T. S. Parker and L. O. Chua (1987).

Turbulence

(See also Complexity, Fluids)

F. Argoul, A. Arnéodo, J. Elzgaray, and G. Grasseau, This volume.

F. Argoul, A. Arnéodo, G. Grasseau, Y. Gagne, E. J. Hopfinger, and U. Frisch (1989).

D. Bensimon, B. I. Shraiman, and V. Croquette (1989).

A. Brandstatter and H. Swinney (1987).

S. Ciliberto and P. Bigazzi (1988).

S. Ciliberto, This volume.

P. Coullet, L. Gil, and J. Lega (1989).

P. Coullet and J. Lega, (1988).

P. Coullet, This volume.

J. Fang, This volume.

J. P. Gollub, This volume.

J. P. Gollub and C. W. Meyer (1983).

M. Henon (1976).

F. Kasper and H. G. Schuster (1987).

P. Kolodner, D. Bensimon, and C. M. Surko (1988).

W. Li, This volume.

F. Heslot, B. Castaing, A. Libchaber (1987).

A. C. Newell, D. A. Rand, and D. Russell (1988).

B. Nicolaneko and Zhen-Su She, This volume.

G.-L. Oppo, M. A. Pernigo, L. M. Narducci, and L. A. Lugiato, This volume.

C. Pérez-Garcia, E. Pampaloni and S. Ciliberto, This volume.

A. Politi, G. D'Alessandro, A. Torcini, This volume.

I. Procaccia, This volume.

D. Ruelle and F. Takens (1971).

M. Schell, S. Fraser, and R. Kapral, (1982).

F. Simonelli and J. P. Gollub (1989).

F. Takens (1981)

W. Y. Tam, J. A. Vastano, H. L. Swinney, W. Horsthemke (1988).

C. Tamm, This volume.

N. B. Tufillaro, R. Ramshankar, and J. P. Gollub (1989).

REFERENCES (ALPHABETICAL)

H. Abarbanel, R. Brown, and J. Kadtke, "Prediction and system identification in chaotic nonlinear systems: time series with broadband spectra", Phys. Lett A **18**, 401 (1989).

N. B. Abraham, A. M. Albano, B. Das, T. Mello, M. F. H. Tarroja, N. Tufillaro, and R. S. Gioggia, "Definitions of chaos and measuring its characteristics", in: *Optical Chaos*, J. Chrostowski and N. B. Abraham eds. (Proc. SPIE **667**, 1986).

N. B. Abraham, J. P. Gollub, and H. Swinney, "Testing nonlinear dynamics", Physica 11D, **252** (1984).

N. B. Abraham, A. M. Albano, B. Das, G. DeGuzman, S. Young, R. S. Gioggia, G. P. Puccioni, J. R. Tredicce, "Calculating the dimension of attractors from small data sets", Phys. Lett. **114A**, 217 (1986).

N. B. Abraham, *et al.*, "Experimental measurements of transitions to pulsations and chaos in a single mode, unidirectional ring laser", in:*Instabilities and chaos in quantum optics*, F.T. Arecchi and R.G. Harrison, eds. (Springer-Verlag, Berlin, 1987).

N. B. Abraham, F.T. Arecchi, and L.A. Lugiato, eds., *Instabilities and chaos in quantum optics II* (Plenum, New York, 1988).

A. M. Albano, J. Muench, C. Schwartz, A. I Mees, and P. E. Rapp, "Singular-value decomposition and the Grassberger-Procaccia algorithm", Phys. Rev. A **38**, 3017 (1988).

A. M. Albano, A. I. Mees, G. C. de Guzman and P. E. Rapp, "Data requirements for reliable estimation of correlation dimensions", in: *Chaos in Biological Systems*, H. Degn, A.V. Holden, and L.F. Olsen, eds. (Plenum, Oxford, 1987).

A. M. Albano, L. Smilowitz, P. E. Rapp, G. C. de Guzman, R. R. Bashore, "Dimension calculations in a minimal embedding space", in: *Physics of Phase Space*, Y. S. Kim and W. W. Zachary, eds. (Springer, Berlin, 1988).

V. S. Anischenko and H. Herzel, "Noise-induced chaos in a system with homoclinic points", Z. Angew. Math. Mech. **68**, 317 (1988).

J. C. Antoranz and M. A. Rubio, "Hyperchaos in a simple model for a laser with a saturable absorber", J. Opt. Soc. Am. B **5**, 1070 (1988).

F. T. Arecchi, "Instabilities and chaos in single-mode homogenous line lasers", in: *Instabilities and chaos in quantum optics*, F.T. Arecchi and R.G. Harrison, eds. (Springer-Verlag, Berlin, 1987).

F. T. Arecchi, A. Lapucci, R. Meucci, A. Rovers, and P.H. Coullet, "Experimental characterization of Shil'nikov chaos by statistics of return times", Europhys. Lett. **6**, 677 (1988).

F. T. Arecchi, W. Gadomski, A. Lapucci, H. Mancini, R. Meucci, and J.A. Roversi, "Laser with feedback on optical implementation of competing instabilities, Shil'nikov chaos and transient fluctuation enhancement", J. Opt. Soc. Am. B **5**, 1153 (1988).

F. T. Arecchi, "Shil'nikov chaos: How to characterize homoclinic and heteroclinic behaviour", This volume.

F. Argoul, A. Arneodo, J. Elzgaray, and G. Grasseau, "Characterizing spatio-temporal chaos in electrodeposition experiments", This volume.

F. Argoul, A. Arneodo, and P. Richetti, "Experimental evidence of homoclinic chaos in the belusov-Zhabotinskii reaction", Phys. Lett. 120A, 269 (1987).

F. Argoul, A. Arnéodo, G. Grasseau, Y. Gagne, E. J. Hopfinger, and U. Frisch, "Wavelet analysis of turbulence reveals the multifractal nature of the Richardson cascase", Nature 338, 51, (1989).

A. Arneodo, P. Coullet, and C. Tresser, "Oscillators with chaotic behavior: an illustration of a theorem by Shilnikov", J. Stat. Phys. 27, 171 (1982).

A. Arneodo, P. Coullet, C. Tresser, J. Stat. Phys. 27, 171 (1982)

A. Arneodo, P. H. Coullet, E. A. Spiegel, and C. Tresser, Physica 14D, 327 (1985).

H. Atmanspacher, V. Demmel, G. Morfill, H. Scheingraber, W. Voges, and G. Wiedenmann, "Measures of dimensions from astrophysical data", This volume.

P. Atten, J. G. Caputo, B. Malraison and Y. Gagne, "Determination de dimension d'attracteurs pour differents ecoulements", J. Mec. theor. et appl. Numero special, 133 (1984).

D. Auerbach, P. Cvitanovic, J-P. Eckmann, G. Gunarante, I. Procaccia, "Exploring chaotic motion through periodic orbits", Phys. Rev. Lett. 58, 2387 (1987).

D. Auerbach, B. O'Shaughnessy, and I. Procaccia, "Scaling structure of strange attractors", Phys. Rev. A 37, 2234 (1988).

D. Auerbach, "Dynamical complexity of strange sets", This volume.

A. Babloyantz, "Chaotic dynamics in brain activity", in: Dynamics of Sensory and Cognitive Processing by the Brain, E. Basar, ed., (Springer-Verlag, Berlin, 1988).

A. Babloyantz and A. Destexhe, "The creutzfeld-Jakob disease in the hierarchy of chaotic attractors", in: From chemical to biological organization, M. Markus, S. Muller and G. Nicolis, eds., (Springer-Verlag, Berlin, 1988).

A. Babloyantz, J. M. Salazar, and C. Nicolis, "Evidence of chaotic dynamics of brain activity during the sleep cycle", Phys. Lett. 111A, 152 (1985).

R. Badii, "Unfolding complexity in nonlinear dynamical systems", This volume

R. Badii and G. Broggi, "Hierarchies of relations between partial dimensions and local expansion rates in strange attractors", This volume.

R. Badii, K. Heinzelmann, P. F. Meier, and A. Politi, "Correlation functions and generalized Lyapunov exponents", Phys. Rev A. **37**, 1323 (1988).

R. Badii and G. Brocci, "Measurement of the dimension spectrum $f(\alpha)$: fixed mass approach", Phys. Lett. **131A**, 339 (1988).

R. Badii, G. Broggi, B. Derighetti, M. Ravani, S. Ciliberto, A. Politi, and M. A. Rubio, "Dimension increase in filtered chaotic signals", Phys. Rev. Lett. **60**, 979 (1988).

R. Badii and A. Politi, "Renyi dimensions from local expansion rates", Phys. Rev. A **35**, 1288 (1987).

R. Badii and A Politi, "Statistical description of chaotic attractors: the dimension function", J. Stat. Phys. **40**, 725 (1985).

D. Barkley and A. Cummings, "Experimental study of the multifractal structure of the quasiperiodic set", This volume.

M. F. Barnsley, *Fractals Everywhere* (Academic Press, New York, 1988)

P. M. Battelino, C. Grebogi, E. Ott, and J. Yorke, "Multiple coexisting attractors, basin boundaries, and basic sets", Physica D **32**, 296 (1988).

D. Bensen, M. Welge, A. Hübler, N. Packard, "Characterization of complexity in systems by aperiodic driving forces", This volume.

D. Bensimon, B. I. Shraiman, and V. Croquette, "Nonadiabatic effects in convection", Phys. Rev. A **38**, 5461 (1989).

T. Bohr and M. H. Jensen, " Order parameter, symmetry breaking, and phase transitions in the description of multifractal sets", Phys. Rev. A **36**, 4904 (1987).

A. Brandstatter and H. Swinney, " Strange Attractors in Weakly Turbulent Couette-Taylor Flow", Phys. Rev. A **35**, 2207 (1987).

W. A. Brock and W. D. Dechert, "Statistical inference theory for measures of complexity in chaos theory and nonlinear science", This volume.

G. Broggi, "Evaluation of dimensions and entropies of chaotic systems", J. Opt. Soc. Am. B **5**, 1020 (1988).

D. S. Broomhead and G. P. King, "Extracting Qualitative Dynamics from Experimental Data", Physica **20D**, 217 (1986).

D. S. Broomhead, R. Jones, and G. P. King, "Topological dimension and local coordinates from time series data", J. Phys A **20**, L563 (1987).

D. S. Broomhead and G. P. King, "Extracting qualitative dynamics from experimental data", Physica **20D**, 217 (1986a).

R. E. Byers and R. I. C. Hansell, "Stablization of prolific populations through migration and long-lived propagules", This volume.

R. E. Byers, R. I. C. Hansell, and N. Madras, "Complex behavior of systems due to semi-stable attractors: attractors that have been destablized but which still temporaily dominate the dynamics of a system", This volume.

J. G. Caputo, "Practical remarks on the estimation of dimension and entropy from experimental data", This volume.

K. Chang, A. Hübler, N. Packard, "Universal properties of the resonance curve of complex systems", This volume.

A. Chhabra and R. V. Jensen, "Direct determination of $f(\alpha)$ singularity spectrum", Phys. Rev. A **31**, 1872 (1989).

S. Ciliberto and P. Bigazzi, "Spatiotemporal intermittency in Rayleigh-Bernard convection" , Phys. Rev. Lett. **60**, 286 (1988).

S. Ciliberto and J. P. Gollub, "Pattern competition leads to chaos", Phys. Rev. Lett. **52**, 922 (1984).

S. Ciliberto, "Characterizing space-time chaos in an experiment of thermal convection.", This volume

P. Coullet, L. Gil, and F. Rocca, "Optical vortices", Opt. Commun., in press

P. Coullet, L. Gil, and J. Lega, "Defect-mediated trubulence", Phys. Rev. Lett. **62**, 1619 (1989).

P. Coullet and J. Lega, "Defect-mediated turbulence in wave patterns", Europhys. Lett. **7**, 511 (1988).

P. Coullet, "Defect-induced spatio-temporal chaos", This volume.

J. P. Crutchfield, "Inferring the dynamic, quantifying physical complexity", This volume.

J. P. Crutchfield, J. D. Farmer, N. H. Packard, and R. Shaw, "Chaos", Sci. Am. **255**, 46 (1986).

J. P. Crutchfield and B. S. McNamara, "Equations of motion from a data series", Complex Systems **1**, 417 (1987).

J. P. Crutchfield and K. Young, "Inferring statistical complexity", Phys. Rev. Lett. **63**, 10 (1989).

P. Cvitanovic, *Universality in Chaos*, (Adam Hilger, Bristol, 1986).

D. Dangoisse, A. Bekkali, F. Papoff, and P. Glorieux, "Shilnikov dynamics in a passive Q-switching laser", Europhys. Lett. **6**, 335 (1988).

A. Destexhe, G. Nicolis, and C. Nicolis, "Symbolic dynamics from chaotic time series", This volume.

A. Desteche, J. A. Sepulchre, and A. Babloyantz, "A comparative study of experimental quantification of deterministic chaos", Phys. Lett. **132A**, 101 (1988).

M. Ding, C. Grebogi, and E. Ott, "Dimensions of strange nonchaotic attractors", Phys. Lett. **137A**, 167 (1989).

M. Dorfle, "Spectrum and eigenfunctions of the Frobenius-Perron operator of the tent map", J. Stat. Phys. **40**, 93 (1985).

B. Dubuc, J. F. Quiniou, C. Roques-Carmes, C. Tricot, and S. W. Zucker, Evaluating the fractal dimension of profiles", Phys. Rev. A **39**, 1500 (1989).

J. P. Eckmann, S. Oliffson Kamphorst, and D. Ruelle, "Recurrence plots of dynamical systems", Europhysics Lett. **4**, 973 (1987).

J-P. Eckmann, S. O. Kamphorst, D. Ruelle, and S. Ciliberto, "Lyapunov exponents from a time series", Phys. Rev. A **34**, 4971 (1986).

J-P. Eckmann and D. Ruelle, "Erogdic theory of chaos and strange attractors", Rev. Mod. Phys. **57**, 617 (1985).

S. Ellner, "Estimating attractor dimensions from limited data: a new method with error estimates", Phys. Lett. **133A**, 128 (1988).

R. M. Everson, "Lyapunov exponents, dimension and entropy in coupled lattice maps", This volume.

G. Fahner and P. Grassberger, "Entropy estimates for dynamical systems", Complex Systems **1**, 1093 (1987).

J. Fang, "Evolution of the irreversible beam dynamical variable and applications", This volume.

J. Fang, "The effects of external noise on complexity in two dimensional driven damped dynamical system", This volume.

J. D. Farmer and J. J. Sidorovich, "Predicting chaotic time series", Phys. Rev. Lett. **59**, 845 (1987).

J. D. Farmer and J. J. Sidorovich, "Exploiting chaos to predict the future and reduce noise", in: *Evolution, Learning and Cognition* (World Scientific, Singapore, 1988).

J. D. Farmer, E. Ott, and J. A. Yorke, "The dimension of chaotic attractors", Physica **7D**, 153 (1983).

M. Frame, "Chaotic behavior of the forced Hodgkin-Huxley axon", This volume.

G. W. Frank, T. Lookman, M. A. H. Nerenberg, "Chaotic time series analysis using short and noisy data sets: application to a clinical epilepsy seizure", This volume.

A. M. Fraser, "Measuring complexity in terms of mutual information", This volume.

A. M. Fraser and H. L Swinney, "Independent Coordinates for Strange Attractors from Mutual Information", Phys. Rev. A **33**, 1134 (1986).

W. J. Freeman, "Simulation of chaotic EEG patterns with a dynamic model of the olfactory system", Biol. Cybern. **56**, 139 (1987).

K. Fukunaga and D. R. Olsen, "An algorithm for finding intrinsic dimensionality of data", IEEE Trans. Comput. **C-20**, 176 (1971).

S. T. Gaito and G. P. King, "Chaos on a catastrophe manifold", This volume.

J. M. Gambaude, P. Glendinning, and C. Tresser, "The gluing bifurcation: I. Symbolic dynamics of closed curves", Nonlinearity 1, 203 (1988).

P. Gaspard, R. Kapral, and G. Nicolis, J. Stat. Phys. 35, 697 (1984).

D. J. Gauthier, M. S. Malcuit, and R. W. Boyd, "Polarization instabilities of counterpropagating laser beams in sodium vapor", Phys. Rev. Lett. 61, 1827 (1988).

R. Gilmore, G. Mindlin, H. G. Solari, "Topological frequencies in dynamical systems", This volume.

J. A. Glazier, G. Gunaratne, and A. Libchaber, "f(α) curves: experimental results", Phys. Rev. A 37, 523 (1988).

J. Gleick, *Chaos, Making a New Science* (Viking, New York, 1987).

P. Glendinning and C. Sparro, J. Stat. Phys. 35, 645 (1984)

P. Glendinning , "Time series near codimension two global bifurcations", This volume.

J. N. Glover, "Estimating lyapunov exponents from approximate return maps", This volume.

A. Goel, S. S. Rao, and A. Passamante, "Estimating local intrinsic dimensionality using thresholding techniques", This volume.

J. P. Gollub, "Characterizing dynamical complexity in interfacial waves", This volume.

J. P. Gollub and C. W. Meyer, "Symmetry-breaking instabilities on a fluid surface", Physica 6D, 337 (1983).

P. Grassberger, "Finite sample corrections to entropy and dimension estimates", Phys. Lett. 128A, 369 (1988).

P. Grassberger, R. Badii, and A. Politi, "Scaling laws for invariant measures on hyperbolic and non-hyperbolic attractors", J. Stat. Phys. 51, 135 (1988).

P. Grassberger, "Are there really climatic attractors?", Nature 323, 609 (1987).

P. Grassberger and I. Procaccia, "Characterization of Strange Attractors", Phys. Rev. Lett. 50, 346 (1983).

P. Grassberger and I. Procaccia, "Estimation of the Kolmogorov Entropy from a Chaotic Signal", Phys. Rev. A 28, 2591 (1983a).

P. Grassberger and I. Procaccia, "Measuring the Strangeness of Strange Attractors", Physica 9D, 189 (1983).

C. Grebogi, E. Ott, and J. A. Yorke, "Unstable periodic orbits and the dimensions of multifractal chaotic attractors", Phys. Rev. A 37, 1711 (1988).

C. Grebogi, E. Ott, and J. A. Yorke, "Unstable periodic orbits and the dimension of chaotic attractors", Phys. Rev. A **36**, 3522 (1987).

J. Guckenheimer and P. Holmes, "Nonlinear Oscillations, Dynamical Systems, and Bifurcation of Vector Fields" in: *Applied Mathematical Sciences* **42** (Springer-Verlag, New York, 1983).

G. H. Gunaratne, M. H. Jensen, and I. Procaccia, "Universal strange attractors on wrinkled tori", Nonlinearity **1**, 157 (1988).

H. Haken, *Information and self-organization: a marcroscopic approach to complex systems*, Springer Series in Synergetics **40** (Springer-Verlag, Berlin, 1988).

T. C. Halsey, M.H. Jensen, L.P. Kadanoff, I. Procaccia, and B.I. Shraiman, "Fractal Measures and their Singularities: The characterization of strange sets", Phys. Rev. A **33**, 1141 (1986).

Hao Bai-lin, "Bifurcation and chaos in the periodically forced Brusselator" (Collected Papers Dedicated to Professor Kazuhisa Tomita, Kyoto University, 1987).

Hao Bai-Lin, *Chaos*, (World Scientific, Singapore, 1984).

K. Hartt and L. M. Kahn, "Seeking dynamically connected chaotic variables." This volume.

J. W. Havstad and C. L. Ehlers, "Attractor dimension of nonstationary dynamical systems from small data sets", Phys. Rev. A **39**, 845 (1989).

D. Hennequin, M. Lefranc, A. Bekkali, D. Dangoisse, and P. Glorieux, "Characterization of Shilnikov chaos in a CO2 laser containing a saturable absorber", This volume.

D. Hennequin, F. de Tomasi, L. Fronzoni, B. Zambon, and E. Arimondo, "Shilnikov chaos and noise in a laser with saturable absorber", Opt. Commun., in press (1989).

M. Henon, "A two-dimensional mapping with a strange attractor", Comm. Math. Phys. **50**, 69 (1976).

H. G. E. Hentschel and I. Procaccia, "The infinite number of generalized dimensions of fractals and strange attractors", Physica **8D**, 435 (1983).

H.-P. Herzel and B. Pompe, "Effects of noise on a nonuniform chaotic map", Phys. Lett. **122A**, 121 (1987).

H. Herzel, "Stabilization of chaotic orbits by random noise", Z. Angew. Math. Mech. **68**, 582 (1988).

F. Heslot, B. Castaing, A. Libchaber, " Transition to turbulence in helium gas", Phys. Rev. A **36**, 5870 (1987).

A. V. Holden, ed., *Chaos*, (Princeton University Press, Princeton, 1986).

G-H. Hsu, E. Ott, C. Grebogi, "Strange saddles and the dimensions of their invariant manifolds", Phys. Lett. **127A** , 199 (1988).

U. Hübner, W. Klische, N. B. Abraham, and C. O. Weiss, "On problems encountered with dimension calculations", This volume.

U. Hübner, N.B. Abraham and C.O. Weiss, "Dimensions and Entropies of Chaotic Intensity Pulsations in a Single-Mode FIR NH3 Laser", Physical Review A, to be published 1989

F. Hunt, "Error analysis and convergence of capacity dimension algorithms", preprint.

A. Hurd, "Resource letter FR-1: Fractals", Am. J. Phys. **56,** 969 (1988).

K. Ikeda and K. Matsumoto, "Study of a high-dimensional chaotic attractor", J. Stat. Phys. **44** , 955 (1986).

K. Ikeda and K. Matsumoto, "High-dimensional chaotic behavior in systems with time-delayed feedback", Physica **29D,** 223 (1987).

M. H. Jensen, L. P. Kadanoff, A. Libchaber, I. Procaccia and J. Stavans, "Global universality at the onset of chaos: Results on a forced Rayleigh-Benard experiment", Phys. Rev. Lett. **55,** 22798 (1987).

J. L. Kaplan and J. A. Yorke, "Functional differential equations and approximations of fixed points", in *Lecture Notes in Mathematics*, **13,** H.O. Peitgen and H.O. Walther, eds. (Springer, Berlin, 1979).

F. Kasper and H. G. Schuster, "Easily calculable measure for the complexity of spatiotemporal patterns", Phys. Rev. A. **36,** 842 (1987).

S. Kim, S. Ostlund, G. Yu, "Fourier analysis of multi-frequency dynamical systems", Physica **31D,** 117 (1989).

P. Kolodner, D. Bensimon, and C. M. Surko, "Traveling-wave convection in an annulus", Phys. Rev. Lett. **60,** 1723 (1988).

E. Kostelich and J. A. Yorke, "Noise reduction in dynamical systems", Phys. Rev. A **38,** 1649 (1988).

W. Lange and M. Moller, "Systematic errors in estimating dimensions from experimental data", This volume.

D. P. Lathrop and E. J. Kostelich, "Analyzing periodic saddles in experimental strange attractors", This volume.

M. Le Berre, E. Ressayre, and A. Tallet, "Phase transitions induced by deterministic delayed forces", This volume.

C.-K. Lee and F. C. Moon, "An optical technique for measuring fractal dimensions of planar Poincare maps", Phys. Lett. **114A,** 222 (1986).

W. Li, "Mutual information functions versus correlation functions in binary sequences", This volume.

B. Mandelbrot, *The Fractal Geometry of Nature*, (Freeman, San Francisco, 1982)

J. M. Martinerie, A. M. Albano, A. I. Mees, P. E. Rapp, "Mutual information, strange attrators and optimal estimation of dimension", preprint, submitted to Phys. Rev A.

G. Mayer-Kress and A. Hübler, "Time evolution of local complexity measures and aperiodic perturbations of nonlinear dynamical systems", This volume.

G. Mayer-Kress, ed., *Dimensions and Entropies in Chaotic Systems - - Quantification of Complex Behavior*, Springer Series in Synergetics **32** (Springer-Verlag, Berlin, 1986).

A. Mees, "Modeling dynamical systems from real-world data", This volume.

A. I. Mees, P.E. Rapp, and L.S. Jennings, "Singular-value decomposition and embedding dimension", Phys. Rev. A **37**, 4518 (1988).

T. Meyer, A. Hübler, N. Packard, "Reduction of complexity by optimal driving forces", This volume.

F. Mitschke, M. Moller, and W. Lange, "Measuring filtered chaotic signals", Phys. Rev. A. **37**, 4518-(1988).

M. Möller, W. Lange, F. Mitschke, N.B. Abraham, and U. Hübner, "Errors from Digitizing and Noise in Estimating Attractor Dimensions", Phys. Lett. A **138**, 176-182 (1989)

T. C. Molteno and N. B. Tufillaro, "Torus doubling and chaotic string vibrations: experimental results", J. Sound Vib., in press.

G. J. Mpitsos, H. C. Creech, and S. O. Soinilla, "Evidence for chaos in spike trains of neurons that generate rhythmic motor patterns", Brain Res. Bull. **21**, 529 (1988).

G. J. Mpitsos, H. C. Creech, C. S. Cohan, and M. Mendelson, "Variability and chaos: Neurointegrative principles in self-organization of motor patterns", in: *Dynamic Patterns in Complex Systems*, J.A.S. Kelso, *et al.*, eds. (World Scientific, Singapore, 1988).

A. C. Newell, D. A. Rand, and D. Russell, "Turbulent dissipation rates and the random occurrence of coherent events", Phys. Lett. **132A**, 112 (1988).

B. Nicolaneko and Zhen-Su She, "Symmetry breaking homoclinic chaos", This volume.

G.-L. Oppo, M. A. Pernigo, L. M. Narducci, and L. A. Lugiato, "Characterization of spatiotemporal structures in lasers: a progress report", This volume.

E. Ott, C. Grebogi, and J. A. Yorke, "Theory of first order phase transitions for chaotic attractors of nonlinear systems", Phys. Lett. **135A**, 342 (1989).

E. Ott, "Strange attractors and chaotic motion of dynamical systems", Rev. Mod. Phys. **53**, 655 (1981).

N. H. Packard, J. P. Crutchfield, J. D. Farmer, and R. S. Shaw, "Geometry from a time series", Phys. Rev. Lett. **45**, 712 (1980).

F. Papoff, A. Fioretti, and E. Arimondo, and N. B. Abraham, "Time return maps and distributions for laser with saturable absorber", This volume.

T. S. Parker and L. O. Chua, "Chaos: A tutorial for engineers", Proc. IEEE **75**, 982 (1987).

A. Passamante, T. Hediger, and M. E. Farrell, "Analysis of local space/time statistics and dimensions of attractors using singular value decompostion and information theoretic criteria", This volume.

A. Passamante, T. Hediger, and M. Gollub, "Fractal dimension and local intrinsic dimension", Phys. Rev. A **39**, 3640 (1989).

K. Pawelzik and H. G. Schuster, "Generalized dimensions and entropies from a measured time series", Phys. Rev. A **35**, 481 (1987); Errata, Phys. Rev. A **36**, 4529 (1987).

H. O. Peitgen and D. Saupe, eds., *The Science of Fractal Images*, (Sprigner Verlag, New York, 1988)

H. O. Peitgen and P. H. Richter, *The beauty of fractals*" (Springer-Verlag, Berlin, 1986)

C. Pérez-Garcia, E. Pampaloni and S. Ciliberto, "Amplitude equations for hexagonal patterns of convection in non-Boussinesq fluids", This volume.

A. Politi, G. D'Alessandro, A. Torcini, "Fractal dimensions in coupled map lattices", This volume.

I. Procaccia, "Weak turbulence and the dynamics of topological defects", This volume.

J. B. Ramsey and H-J. Yuan, "Bias and error bars in dimension calculations and their evaluation in some simple models", Phys. Lett. **134A**, 287 (1989).

P. E. Rapp, I. D. Zimmerman, A. M. Albano, G. C. de Guzman, and N. N. Greenbaun, "Dynamics of spontaneous neural activity in the simian motor cortex: the dimension of chaotic neurons", Phys. Lett. **110A**, 335 (1985).

P. E. Rapp, I. D. Zimmerman, T. R. Bashore, A.M. Albano, G.C. de Guzman, and N. N. Bashore, "Experimental studies of chaotic neural behavior: cellular activity and electroencephalographic signals", in: *Nonlinear oscillations in biology and chemistry*. H. G. Othmer, ed. (Springer-Verlag, Berlin, 1985).

P. E. Rapp, A. M. Albano, and A. I. Mees, "Calculation of correlation dimensions from experimental data: Progress and problems", in: *Dynamic Patterns in Complex Systems*. J. A. S. Kelso, et al, eds. (World Scientific, Singapore, 1988).

M. G. Raymer, "Entropy and correlation time in a multimode dye laser", This volume.

A. Renyi, *Probability Theory*, (North-Holland, Amsterdam, 1970).

P. Richetti, J. C. Roux, F. Argoul, and A. Arneodo, "From quasiperiodicity to chaos in the Belousov-Zhabotinskii reaction. II Modeling and theory", J. Chem. Phys. 86, 3339 (1987)

J. Ringland and M. Schell, "Pattern cardinality as a characterization of dynamical complexity", This volume.

J. Ringland and M. Schell, "The Farey tree embodied -- in bimodal maps of the interval", Phys. Lett. **136A**, 379 (1989).

O. E. Rössler, Phys. Lett. **71A**, 155 (1979).

M. A. Rubio, A. D. Dougherty, and J. P. Gollub, "Characterization of irregular interfaces: Roughness and self-affine fractals. This volume.

D. Ruelle and F. Takens, "On the nature of turbulence", Commun, Math. Phys. **20**, 167 (1971).

Y. A. Rzhanov and Y. D. Kalafati, "Spatial structure multistability in an array of optically bistable elements", Opt. Commun. **70**, 161 (1989).

S. Sato, M. Sano, and Y. Sawada.,"Practical methods of measuring the generalized dimension and the largest lyapunov exponent in high dimensional chaotic systems", Prog. Theor. Phys. **77**, 1 (1987)

C. L. Sayers, "Dimension calculation precison with finite data sets", This volume.

W. M. Schaffer and L. F. Olsen, "Chaos in childhood epidemics", This volume.

M. Schell, S. Fraser, and R. Kapral, "Diffusive dynamics in systems with translational symmetry: a one-dimensional-map model", Phys. Rev. A **26**, 504 (1982).

H. G. Schuster, *Deterministic Chaos*, (VCH Verlagsgesselschaft, Weinheim, 1988).

H. G. Schuster, "Extraction of models from complex data", This volume.

M. A. Sepulveda and R. Badii, "Symbolic dynamical resolution of power spectra", This volume.

R. . Shaw, "Strange attractors, chaotic behavior, and information flow", Z. Naturforsch. **36A**, 80 (1981)

L. P. Shil'nikov, Dokl. Akad. Nauk SSSR **160**, 558 (1965).

L. P. Shil'nikov, Mat. Sbornik 77, (119), **461** (1968).

L. P. Shil'nikov, Mat. Sbornik 81, (123), **92** (1970).

L. P. Shil'nikov, (Eng. Trans.: Sov. Math. Doklady **3**, 394 (1962).

L. P. Shil'nikov, (Eng. Trans.: Sov. Math. Doklady **6**, 163 (1965).

L. P. Shil'nikov, (Eng. Trans.: Sov. Math. Doklady **8**, 54 (1967).

L. P. Shil'nikov, Math USSR Sbornik **6**, 427 (1968).

L. P. Shil'nikov, Math USSR Sbornik **10**, 91 (1970).

F. Simonelli and J. P. Gollub, "Surface wave mode interactions: effects of symmetry and degeneracy", J. Fluid Mech. **199**, 471 (1989).

C. A. Skarda and W. J. Freeman, "How brains make chaos in order to make sense of the world", Behavioral and Brain Sciences **10**, 161 (1987).

L. A. Smith, "Quantifying chaos with predictive flows and maps: locating unstable periodic orbits", This volume.

L. A. Smith, "Intrinsic limits on dimension calculations". Phys. Lett. **133A**, 283 (1988).

H. G. Solari and R. Gilmore, "Relative rotation rates from driven dynamical systems", This volume.

H. G. Solari, Xin-Jun Hou, and R. Gilmore, "Stretching, folding and twisting in driven damped Duffing device", This volume.

R. Stoop and P. F. Meier, "Evaluation of Lyapunov exponents and scaling functions from time series", J. Opt. Soc. Am. B **5**, 1037 (1988).

Z. Su, R. W. Rollins, and E. R. Hunt, "Measurement of $f(\alpha)$ for multifractal attractors in driven diode resonator systems", This volume.

Z. Su, R. W. Rollins and E. R. Hunt, "Measurement of $f(\alpha)$ spectra of attractors at transitions to chaos in driven diode resonator systems", Phys. Rev. A **36**, 3515 (1987).

F. Takens, "Invariants related to dimensions and entropy", in: *Atas do 13 Coloqkio Brasiliero do Matematica* (Rio de Janerio, 1983).

F. Takens, "On the numerical determination of the dimension of an attractor", in: *Dynamical Systems and Bifurcations, Groningen, 1984*. Lecture Notes in Mathematics **1125** (Springer-Verlag, Berlin, 1985).

F. Takens, "Detecting strange attractors in turbulence", in: *Dynamical Systems and Turbulence, Warwick, 1980*. Lecture Notes in Mathematics **898**, (Springer-Verlag, Berlin, 1981)

W. Y. Tam, J. A. Vastano, H. L. Swinney, W. Horsthemke, "Regular and chaotic chemical spatiotemporal patterns", Phys. Rev. Lett. **61**, 2163 (1988).

C. Tamm, "The field patterns of a hybrid mode laser: detecting the "hidden" bistability of the optical phase pattern", This volume.

Y. Termonia and Z. Alexandrowicz, "Fractal dimensions of strange attractors from radius versus size of arbitrary clusters", Phys. Rev. Lett. **51**, 1265 (1983).

J. Theiler , "Statistical error in dimension estimators", This volume.

J. Theiler, "Lacunarity in a best estimator of fractal dimension", Phys. Lett. **133A**, 195 (1988).

J. R. Tredicce, E. J. Quel, A. M. Ghazzawi, C. Green, M. A. Pernigo, L. M. Narducci, and L. A. Lugiato, "Spatial and temporal instabilities in a CO_2 laser", Phys. Rev. Lett. **62**, 1274 (1989).

C. Tresser, Ann. Inst. Henri Poincaré **40**, 441 (1984).

C. Tresser, J. Physique **45**, 837 (1984)

N. B. Tufillaro, R. Ramshankar, and J. P. Gollub, "Order-disorder transition in capillary ripples", Phys. Rev. Lett. **62**, 422 (1989).

N. B. Tufillaro, "Nonlinear and chaotic string vibrations", Am. J. Phys. **57**, 408 (1989).

D. K. Umberger, C. Grebogi, E. Ott, and D. Afeyan, "Spatiotemporal dynamics in a dispersively coupled chain of nonlinear oscillators", Phys. Rev. A. **39**, 4835 (1989).

W. Water and P. Schram, "Generalized dimensions from near-neighbor information", Phys. Rev. Lett. **47**, 1400 (1988).

C. O. Weiss and N. B. Abraham, "Characterizing chaotic attractors underlying single mode laser emission by quantitative laser field phase measurement", This volume.

C. O. Weiss, N. B. Abraham and U. Hübner, "Homoclinic and Heteroclinic Chaos in a Single-Mode Laser", Phys. Rev. Lett. **61**, 1587-1590 (1988).

H. G. Winful and S. S. Wang, "Stability of phase locking in coupled semiconductor laser arrays", Appl. Phys. Lett. **53**, 1894 (1988).

A. Wolf, J.B. Swift, H.L. Swinney, and J.A. Vastano, "Determining Lyapunov exponents from a time series", Physica **16D**, 285 (1985).

ADDENDUM (February, 1990)

S. Adachi, M. Toda, and K. Ikeda, "Potential for mixing in quantum chaos", Phys. Rev. Lett. **61**, 655 (1988).

S. Adachi, M. Toda, and K. Ikeda, "Quantum-classical correspondence in many-dimensional quantum chaos", Phys. Rev. Lett. **61**, 659 (1988).

P. M. Battelino, C. Grebogi, E. Ott and J. A. Yorke, "Chaotic Attractors on a 3-Torus, and Torus Break-up", Physica D **39**, 299 (1989).

P. Cvitanovic, "Invariant measurement of strange sets in terms of cycles", Phys. Rev. Lett. **61**, 2729 (1988).

P. Cvitanovic, G.H. Gunaratne, and I. Procaccia, "Topological and metric properties of Hénon-type strange attractors", Phys. Rev. A **38**, 1503 (1988).

W. L. Ditto, S. Rauseo, R. Cawley, C. Grebogi, G.-H. Hsu, E. Kostelich, E. Ott, H.T. Savage, R. Segnan, M.L. Spano, and J .A. Yorke, "Experimental observation of crisis-induced intermittency and its critical exponent", Phys. Rev. Lett. **63**, 923 (1989).

A. M. Fraser, "Reconstructing attractors from scalar time series: a comparison of singular system and redundancy criteria", Physica **34D**, 391 (1989).

G. H. Gunaratne and I. Procaccia, "Organization of Chaos", Phys. Rev. Lett. **59**, 1377 (1987).

G. H. Gunaratne, P. S. Linsay, and M. J. Vinson, "Chaos beyond onset: a comparison of theory and experiment", Phys. Rev. Lett. **63**, 1 (1989).

K. Ikeda, S. Adachi, and M. Toda, "Absorption of light by quantum chaos system: can quantum chaos be an origin of dissipation?", Kyoto preprint.

K. Ikeda, and K. Matsumoto, "Information theoretical characterization of turbulence", Phys. Rev. Lett. **62**, 2265 (1989).

K. Ikeda, K. Otsuka, and K. Matsumoto, "Maxwell-Bloch turbulence", to appear in Suppl. Prog. Theor. Phys. (1989).

E. J. Kostelich and H. L. Swinney, "Practical considerations in estimating dimension from time series data", Physica Scripta **40**, 436 (1989).

D. P. Lathrop and E. J. Kostelich, "Characterization of an experimental strange attractor by periodic orbits", Phys. Rev. A **40**, 4028 (1989).

S. D. Meyers, J. Sommeria and H. L. Swinney, "Laboratory study of the dynamics of jovian-type vortices", Physica **37D**, 515 (1989).

M. Mizuno and K. Ikeda, "An unstable mode selection rule: frustrated optical instability due to competing boundary conditions", Physica **36D**, 327 (1989).

L. M. Narducci, E. J. Quel, and J. R. Tredicce, eds., *Lasers and Quantum Optics* (World Scientific, Singapore, 1990).

K. Otsuka and K. Ikeda, "Cooperative dynamics and functions in a collective nonlinear optical element system", Phys. Rev. A **39**, 5209 (1989).

B.-S. Park, C. Grebogi, E. Ott, and J. A. Yorke, "Scaling of fractal basin boundaries near intermittency transitions to chaos", Phys. Rev. A **40**, 1576 (1989).

F. J. Romeiras, C. Grebogi and E. Ott, "Multifractal properties of snapshot attractors of random maps", Phys. Rev. A **41**, 784 (1990).

M. Toda, S. Adachi, and K. Ikeda, "Dynamical aspects of quantum-classical correspondence in quantum chaos", to appear in Suppl. Prog. Theor. Phys, No. 98 (1989).

J. A. Vastano, J. E. Pearson, W. Horsthemke, and H. L. Swinney, "Turing patterns in an open reactor", J. Chem. Phys. **88**, 6175 (1988).

Linda Caruso-Haviland

Director, Dance Program
Bryn Mawr College
Bryn Mawr, PA 19010

Paul E. Rapp

Department of Physiology and Biochemistry
The Medical College of Pennsylvania
Philadelphia, PA 19129

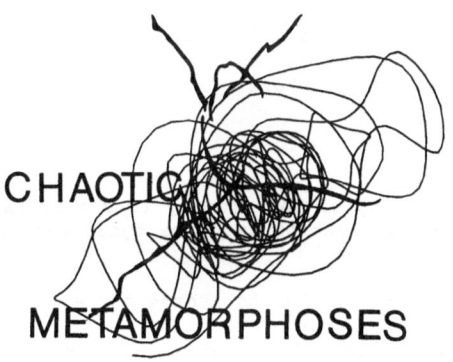

CHAOTIC METAMORPHOSES

A WORK IN PROGRESS

A fusion of poetry, dance and music
incorporating chaotic dynamical
structures and concepts.

Measures of Complexity and Chaos
Edited by N.B. Abraham *et al.*
Plenum Press, New York

29

Chaotic Metamorphoses is a thirty-five minute performance formed by the fusion of poetry, dance and music and incorporates chaotic dynamical structures and concepts. It includes three dancers, music, poetry, film and video displays. Its first performance was for the participants in this workshop on June 22, 1989. Though the composition continues to evolve, it currently contains the following elements.

The Second Coming (A composition for three dancers)

This is an allegorical performance by three dancers of a nonlinear compositional form. It is accompanied by a recorded reading of the poem of the same title by W. B. Yeats.

Chaotic Vortex (A composition for solo dancer)

The accompanying music is a sound collage. The primary tonal sequence was generated from an electroencephalographic signal recorded from a subject performing mental arithmetic. Additional sounds include Gondwanaland didgeridoo and synthesizer duets, vocalizations and fragments of poetry that appear elsewhere in the composition. The dance is patterned after the dynamical structure of fluid vortices such as the Jovian Red Spot.

The Brain is Wider than the Sky

A recorded reading of the poem of the same title by Emily Dickinson is presented without visual accompaniment to a darkened auditorium.

Music of the Mind (A composition for three dancers)

The visual image projected on the dancers is a filmed video sculpture created by constructing a strange attractor from electroencephalographic data on an Evans & Sutherland graphics computer. The EEG signal was obtained from a subject performing mental arithmetic. The music is Baroque Variations by Lukas Foss. The dance was evoked by the rhythms and kinetic forms of the electroencephalographic video sculpture.

The Wall Between

This is a recorded reading of the poem by Trisha F. Harper played to an empty stage illuminated by a lighting sequence synchronized with the poem.

Logistic Bifurcations (A composition for three dancers)

The tonal sequence and video-displayed images were generated by a sequence of period-doubling bifurcations of the logistic equation. The dance is generated by sucessive bifurcations: steady state, period-2, period-4, period-8, chaos, and reverse bifurcations returning the systems to a steady state. As the tonal sequence proceeds through successive period-doubling bifurcations, the dancers move synchronously but in increasingly complex geometric patterns. After the bifurcation to chaos, their movement displays a progressive divergence characteristic of the exponentially diverging phase space trajectories of chaotic systems. The composition is palindromic. The dancers reunite and converge to their inital configuration.

Acknowledgments: The performance was conceived by Paul Rapp and Linda Caruso-Haviland and choreographed by Linda Caruso-Haviland. The dancers were Karen Anderson, Renee Banson Shapiro and Linda Caruso - Haviland. Film was by Michael Lawrence Productions and video by R & CH Productions. Technical direction: Toni Vahlsing, Technical Consultant: Carmen Slider, Stage Manager:

Jean-Luc Hannink, Crew: Tania Schmah, Audio-Visual Director: Ralph del Giudice, Audio-Visual Assistant: Gwen Bonebrake, Sound Crew: Rob Wozniak, Slide Projector: Michele Francl, Magister Cybernetica: Joseph S. Waldron. Mr. Rapp appeared courtesy of Canticle Productions.

MEASURES OF DIMENSIONS FROM ASTROPHYSICAL DATA

H. ATMANSPACHER, V. DEMMEL, G. MORFILL,
H. SCHEINGRABER, W. VOGES, AND G. WIEDENMANN

Max-Planck-Institut für extraterrestrische Physik
D-8046 Garching, FRG

1. Introduction

The complexity of a system may have numerous aspects, and the problems to define complexity in a generally relevant manner seem to increase self–similarly with the intensity of corresponding efforts. In this sense it is certainly a complex task to provide a compulsory concept of the notion of complexity. In the present contribution we deal with dimensions as measures of complexity. Mathematically the concept of dimensions reflects the scaling properties of point distributions on a given support. Speaking in terms of physical systems, this support is usually a vector space. Studying the structural properties of a system refers simply to structures in position space, whereas functional properties of a system are related to the structure of its dynamics in phase space.

These two kinds of structure determine the organization of the article. The first part (Sec.2) intends to demonstrate the usefulness of dimensional measures for the functional complexity of a specific class of astrophysical systems. In detail, deterministic chaos in matter–accreting systems will be discussed from an observational and from a theoretical viewpoint. In the first respect it has to be emphasized that the established procedures of an attractor reconstruction based on a single variable time series are of particular importance in an astrophysical context, since the only information source (variable of the system) presently available is the electromagnetic radiation from an object. Concerning the theoretical aspect of our work, a novel accretion scenario will be described which is complex enough to provide chaotic luminosities, and also realistic enough to explain observed effects not understood previously.

The second part (Sec.3) deals with structure in position space, namely with the large scale structure of the universe. In recent years this field has witnessed an increasing interest in fractal and multifractal concepts. These concepts present definite relationships to the standard correlation approach. Results as well as regimes of relevance and possible pitfalls will be compared. Particular emphasis will be put on the problem of estimating a proper homogeneization distance.

This article is designed to be a progress report. No attempt has been made either to present the topics in their full width, or to provide a complete selection of references. Therefore, it should be mentioned that a lot of interdisciplinary work combining nonlinear dynamics and astrophysics has been documented elsewhere.[1]

Measures of Complexity and Chaos
Edited by N.B. Abraham *et al.*
Plenum Press, New York

2. Chaos in matter–accreting systems

The process of matter accretion plays an important role in astrophysics. It is believed to have considerable impact on a wide range of phenomena, e.g. the formation of our planetary system as well as on the formation of galactic and extragalactic structures. A further situation which is firmly related to matter accretion is found in compact objects like white dwarfs, neutron stars, and black holes. Here matter accretion is certainly of dominating influence due to the enormous gravitational action taking place.

In a simple picture, matter falling onto a compact star is decelerated at its surface, where the kinetic energy is converted into electromagnetic radiation. Corresponding to temperatures in the order of 10^7 K, its maximum is in the X-ray regime. Since the earth atmosphere is optically thick in this wavelength range, the resulting X-ray luminosity of the compact object can only be detected outside the atmosphere, e.g. by satellite. Much information in this respect has been gaind from the EXOSAT mission between 1983 and 1986. A recent conference proceedings volume (with an emphasis on neutron stars)[2] contains numerous results and reviews.

The X-ray luminosity as a function of time is of high value for an evaluation of rotational and orbital properties of the compact object. It has additionally been used to classify X-ray emitting objects according to their power spectrum. Many of these sources show a continuous power spectrum with distinct $1/f$ components and quasi-periodic contributions.[3] It is therefore very appealing to look for dynamical invariants of the accretion process which would allow to distinguish between random (stochastic) behavior and deterministic chaos. This is particularly important in view of the high degree of uncertainty about dynamical details of the matter accretion process.

In the following Sec.2.1 we shall describe observational indications for chaos, obtained for the example of the accreting neutron star Her X-1. Section 2.2 discusses a route toward an accretion scenario capable of producing deterministic chaos. Although further progress has to be made to compare model and observations in detail, the results are encouraging since they explain additional observed effects for which no theoretical basis existed so far.

2.1 OBSERVATIONAL EVIDENCE FOR CHAOS IN HER X-1

The neutron star Her X-1 and its visible companion HZ Her form a low mass X-ray binary system showing several specific features. Her X-1 is surrounded by an accretion disk consisting of gaseous matter flowing in from HZ Her. The strong surface magnetic field of $\approx 4 \cdot 10^{12}$ Gauss channels the accreted material onto the polar regions of Her X-1, where its energy is converted into X-ray radiation.

The emitted X-ray radiation as observed by the EXOSAT satellite shows three different periods of regular temporal variability: (i) the 1.24 sec rotational period of Her X-1, (ii) the 1.7 day orbital period of the system Her X-1/HZ Her, and (iii) the 35 day period ascribed to a warping accretion disk. In addition to the regular periodic behavior, irregular temporal variability has been found on each of the mentioned time scales. For different geometrical configurations, this irregular behavior has to be distinguished according to four source modes (see Fig.1):
(A) Unobscured phase: the X-ray radiation is observed without any obscuration effects due to the accretion disk or due to HZ Her and its atmosphere.
(B) Totally obscured phase (eclipse): the X-ray source is shaded by HZ Her.
(C) Partially obscured phase (absorption dips): strong intensity variations, probably caused by absorption of radiation in the inhomogeneous accretion disk.
(D) Background as obtained in off – source observations with an average count rate comparable to (B).

The main emphasis of the present contribution is on source mode (A) since it reflects the primary process of radiative transport in the atmosphere of Her X-1, a

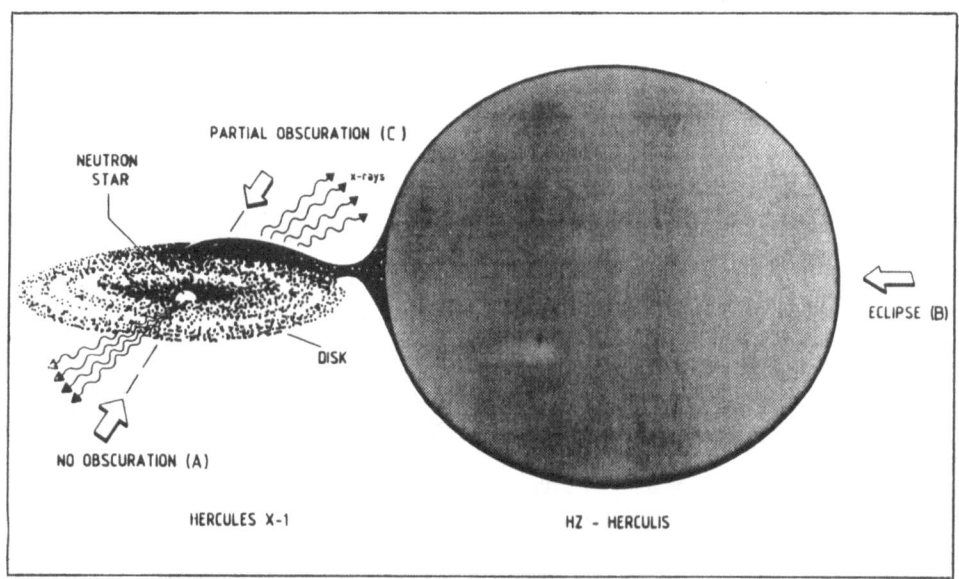

Figure 1. The geometrical configuration of the binary system Her X-1/HZ Her. The source modes (A) - (C) are indicated as described in the text.

model of which we are going to describe in Sec.2.2. In Figs.2a and 2b the X-ray luminosity as observed in mode (A) is compared with a time series for mode (B). In order to check the existence and strangeness of an attractor behind these time series we applied the procedure due to Grassberger and Procaccia.[4] Since the analysis and its methodical details have been reported elsewhere,[5,6,7,8,9,10] let us just introduce our notations.

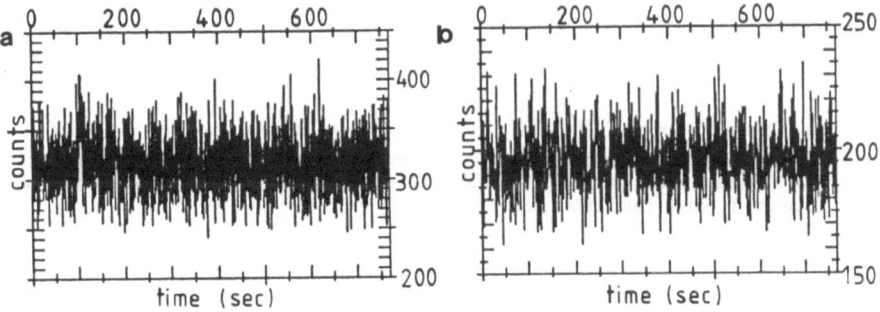

Figure 2. Temporal evolution of the X − ray count rates from Her X-1 (a: mode A) and background radiation (b: mode B). Both time series show 1000 points with a resolution $\tau = 770$ msec.

The relevant invariants are determined for an attractor reconstructed[11] from the observed time series with temporal resolution τ. The time delay used to generate a d – dimensional vector is denoted as Δt. According to a correlation period of ≈ 10 sec for the mode (A) signal we took $\tau = 770$ msec and $\Delta t = \tau$. (The results of the analysis turned out to be quite insensitive to different time delays $\tau < \Delta t < 30\tau$.) The time series contained up to 2000 data points, usually we took $N = 1000$ (for the sufficiency of this relatively small number see below). The correlation integral $C_d^{(2)}(r)$

is given by:[4]

$$C_d^{(2)}(r) = \lim_{r \to 0} \frac{1}{N^2} \sum_{i,j=1}^{N} H(r - |x_i - x_j|_d) \tag{1}$$

where H is the Heaviside step function and x_i is the position of a point belonging to the reconstructed attractor in the d – dimensional phase space. The superscript (2) indicates the relation of $C^{(2)}$ to a two – point (pair) correlation. From $C_d^{(2)}(r)$ the correlation dimension $D^{(2)}$ of the attractor is obtained according to:

$$D^{(2)} = \lim_{r \to 0} \lim_{d \to \infty} \frac{\log_2 C_d^{(2)}(r)}{\log_2 r} \tag{2}$$

A phase space dimension $d \approx 2D^{(2)} + 1$ has been shown[12] to be adequate for a sufficient embedding of the reconstructed attractor.

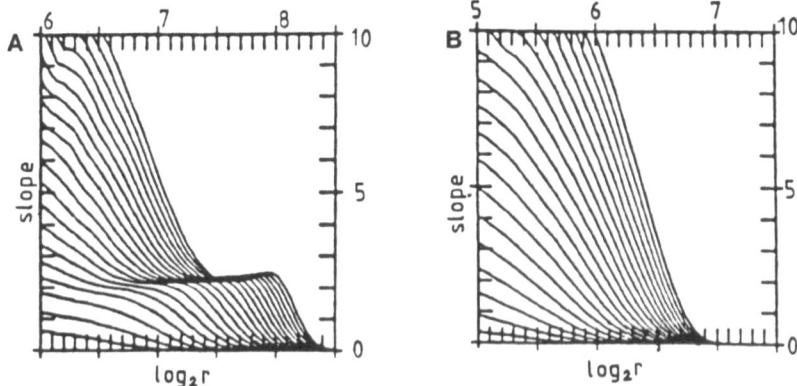

Figure 3. Plot of the local slopes $\frac{d \log_2 C_d^{(2)}(r)}{d \log_2 r}$ vs $\log_2 r$ as obtained from an attractor analysis of both time series shown in Fig.2. Dimensions of the reconstructed phase space are shown from $d = 1$ to 20, $N = 1000$, $\tau = 770$ msec. Mode (A) provides a distinctly converging slope above a background noise of $\approx 40\%$. Mode (B) is purely random as expected.

Figure 3 shows the obtained slopes $\frac{d \log_2 C_d^{(2)}(r)}{d \log_2 r}$ vs $\log_2 r$ for the signals plotted in Fig.2. There is a clear indication for a plateau at a slope 2.3 in mode (A) equivalent to convergence in the limit $d \to \infty$.[1] The mode (B) signal (as well as mode (D)) provides no convergence at all as it should be expected for a truely random background. The physical relevance of mode (C) refers to the dynamics of the accretion disk. During this mode we found indications for a tendency of convergence at slopes between 8 and 10. However, such high values are not reliable for data sets as used here.

---

[1] We have checked whether this plateau could be caused by an influence of the periodic 1.24 sec signal due to the rotation of Her X-1. It turned out that a simulation of the periodic signal plus the proper amount of white noise did not provide the same feature as in Fig.3a. However, replacing white noise by coloured noise[13] provides correlation integrals which do not contradict those obtained here. Coloured noise signalizes the relevance of chaotic components in the process investigated.

Since the low – dimensional attractor in mode (A) is most interesting with respect to a basically novel accretion scenario (Sec.2.2) we estimated the dynamical entropy $K^{(2)}$ which is also available from the correlation integrals.[14] It turned out that $K^{(2)} > 0$ for all mode (A) time series investigated, thus providing further confirmation for low – dimensional deterministic chaos.

Moreover, the corresponding chaotic attractor has been found to be characterized by a multifractal structure.[6] This has been shown by an analysis using the generalized correlation integral:[15]

$$C_d^{(q)}(r) = \lim_{r \to 0} \left[\frac{1}{N} \sum_{i=1}^{N} \left(\frac{1}{N} \sum_{j=1}^{N} H(\Theta - |\mathbf{x_i} - \mathbf{x_j}|d) \right)^{q-1} \right]^{1/q-1} \tag{3}$$

reproducing Eq.(1) for $q = 2$. The spectrum of dimensions $D^{(q)}$ is then obtained by:

$$D^{(q)} = \lim_{r \to 0} \lim_{d \to \infty} \frac{\log_2 C_d^{(q)}(r)}{\log_2 r} \tag{2}$$

The spectrum $D^{(q)}$ as calculated for $-20 < q < 20$ is plotted in Fig.4a. The difference between dimensions of different order q measures the degree of inhomogeneity of the attractor in the sense of clustered regions ($q > 0$) and rarefied regions ($q < 0$). Hence, $D^{(q)}$ provides a criterion for a sufficient length of the analyzed time series. If a time series does not contain enough data points to reflect the inherent scaling properties of the attractor sufficiently, then it can be expected that the degree of inhomogeneity is higher than for a sufficient number of points. For the Her X-1 attractor it has been found that no significant difference arises for $N = 1000$ and $N = 2000$ points. This fact provides further evidence for the relevance of a chaotic attractor in mode (A).

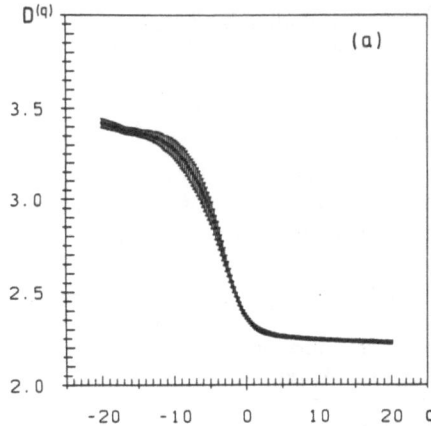 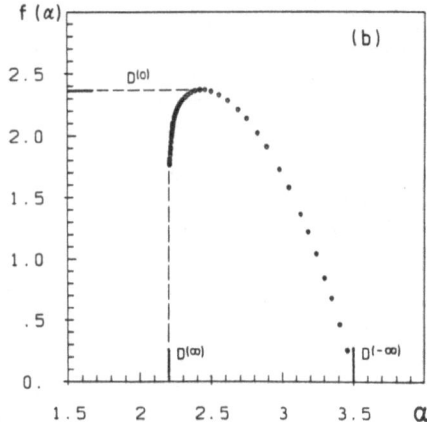

Figure 4. a) Spectrum of dimensions $D^{(q)}$ and b) $f(\alpha)$ spectrum of the Her X-1 attractor. (b) represents $f(\alpha)$ as calculated for $-4.5 < q$. $D^{(0)}$, $D^{(-\infty)}$, and $D^{(\infty)}$ are indicated. The most characteristic features of $f(\alpha)$ are its strong asymmetry and the positive value of $f(\alpha_{min}) \approx 1.75$.

The spectrum of dimensions may be transformed into a $f(\alpha)$ spectrum, charac-terizing the scaling indices $f(\alpha)$ of subsets with scaling index α.[16,17] For the Her X-1 attractor the $f(\alpha)$ spectrum as derived by Legendre transform of $(q-1)D^{(q)}$ is displayed in Fig.4b.[6] (A different method to obtain $f(\alpha)$ will be sketched in Sec.3.2.) The $f(\alpha)$ spectrum shows a strong asymmetry suggesting that the attractor is dominated by its concentrated subsets of almost equal density (α_{min}). Hence the dominant properties of the attractor can be extracted using a low number of data points: most of them are situated in the dense subsets.

2.2 A DISCRETIZED ANISOTROPIC ACCRETION SCENARIO

It is self–evident that the determination of quantitative estimates of attractor invariants enfolds its physical meaning in being related to physical quantities governing the dynamics of the system. In this context, the low dimensionality of an attractor is helpful in reducing the number of dynamical degrees of freedom relevant to the system. Corresponding procedures of mode reduction by truncation (as in case of the Lorenz system[18]) or by mode coupling (as in case of multimode laser dynamics[19,20,21]) enable a simplified treatment of systems conventionally described by many degrees of freedom. The dimension of the attractor sets a lower limit to the number of modes to be kept.

The dynamics of accretion onto compact objects is a problem dealing with radiation (magneto)hydrodynamics. The principal transport equations involved are the conservation laws and the radiative transfer equation. These are supplemented by another set of equations governing details of the physical processes taking place. The resulting system of partial differential equations is even more complicated if the atmosphere of the star is optically thick (which is the case for high accretion rates). In this case, a two–fluid description is appropriate: infalling matter and outgoing radiation are dynamically interacting.

With an increasing mass accretion rate the radiation pressure slowing down the infalling material also increases. In a hydrostatic approximation the energy gain from the conversion of kinetic energy into radiation has an upper limit. This limit is the so–called Eddington luminosity given by $L_E = 1.37 \cdot 10^{38}$ erg/sec, providing an Eddington mass accretion rate of $\dot{M}_E = 9.46 \cdot 10^{17}$ g/sec for typical neutron star parameters.

Observations of luminosities significantly above L_E in various objects suggest modifications of the general assumptions leading to the estimate of L_E as an upper limit. Based on a number of recent publications,[10,22,23,24,25,] we shall subsequently discuss how a two–fluid description naturally provides anisotropic accretion (although spherically symmetric on a temporal average) as such a modification. The capability of the resulting dynamics for super – Eddington luminosities will be addressed, and prospects for a chaotic luminosity as a function of time will be described.[2]

The scenario is basically one of two vertically adjacent fluids in the gravitational field of the neutron star. The radiation pressure decelerates the infalling plasma at a certain distance above the star surface, which depends on the mass accretion rate. Consequently the decelerated accreted plasma forms a dense boundary layer between both fluids. Since the density of this layer exceeds the density of the radiating photon gas below, the situation is Rayleigh Taylor unstable. For this reason one can assume[27] that optically thick clumps of a diameter corresponding to the most instable wavelength will develop and fall down to the surface. The kinetic energy of a clump is then converted into radiation ascending as a radiation bubble of high energy density. This bubble will be able to penetrate the dense layer, producing a channel in the accreting material, or it may escape through already existing channels. A schematic picture of this scenario is sketched in Fig.5, visualizing the high degree of anisotropy

---

[2] As mentioned above, this model does not exactly apply to the situation in Her X-1, since it considers spherical accretion instead of disk accretion. However, preliminary indications for low–dimensional chaos in the spherically accreting system 4U 1700-37 have also been reported.[26]

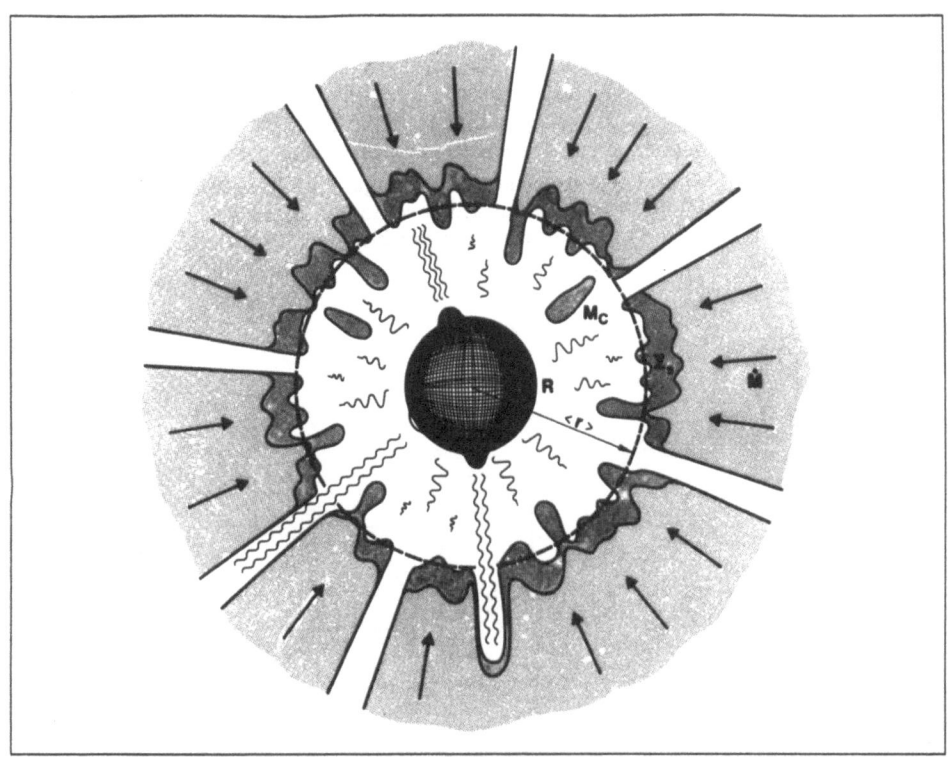

Figure 5. Illustration of the anisotropic accretion scenario described in the text.

of the scenario.

For each of the individual processes mentioned above it is possible to define a relevant time scale: the generation time of a clump, the free fall time of a clump down to the star surface, the rise time of a bubble, the penetration time of a bubble channeling into the accretion flow, the escape time of a bubble through existing channels, and finally the relaxation time of a channel. Using reasonable estimates for these time scales, a temporal discretization of the scenario can be achieved. It enormously reduces the computational effort for a quantitative treatment of the scenario. Details about the discretization and other precise properties and physical parameters of the model are reported separately.[25]

In addition to the temporal discretization it turns out necessary to introduce a partition of the spherical surface of the Rayleigh Taylor unstable shell around the neutron star in order to define probabilities for the individual processes defined above. The required cell size corresponds to the size of clumps and bubbles which are assumed to be approximately equal.

In a first step the case of randomly distributed Rayleigh Taylor instabilities over the instable shell will be considered. This case of spontaneous clump formation leads to an iterative map for the number C of channels at a given time step i:

$$C_{i+1} = \frac{1}{x}(1 - xC_i)(1 - ax)^{C_i} \equiv f(C_i) \tag{5}$$

where $1/x$ is the number of cells (depending on the distance between star surface and instable shell and on the cell size), and a is an estimator for the escape efficiency of bubbles arriving at cells adjacent to channels. Both quantities can be uniquely determined by the mass accretion rate $\dot{M} = \dot{M}/\dot{M}_E$. The map according to Eq.(5), parametrized by $\dot{M}$, is shown in Fig.6a. The form of $f(C_i)$ allows only for asymptotically periodic solutions. For a channel relaxation time which is much larger than the iteration intervall (thus introducing a higher order Markov chain) one obtains stationary (fixed point) solutions in addition to the limit cycle. Most important, the **average luminosity** in this scenario exceeds the Eddington limit already at $\dot{M} = 2$, a feature which is possible only because of the anisotropic dynamics taken into account. However, no chaotic evolution of the luminosity can be expected from Eq.(5).

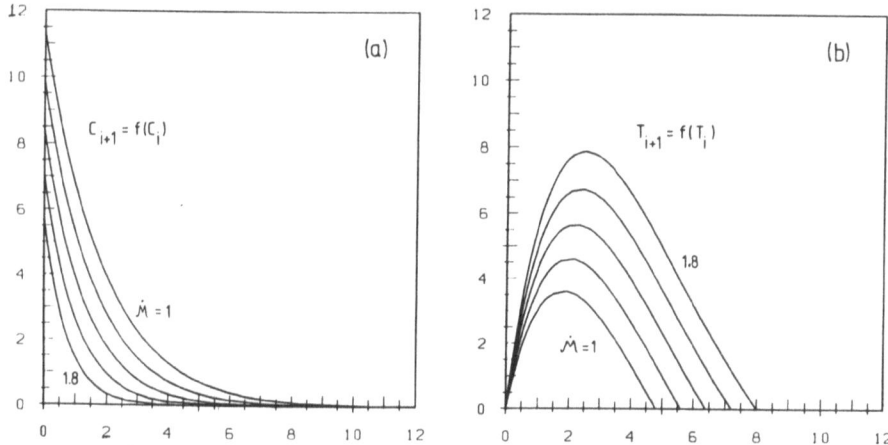

Figure 6. a) Iterative map for spontaneous clump formation according to Eq.(5) for $\dot{M} = 1.0, 1.2, 1.4,$ 1.6, 1.8; $a = 4$. b) Iterative map for stimulated clump formation according to Eq.(6) for $\dot{M} = 1.0, 1.2,$ 1.4, 1.6, 1.8; $ax = 0.2$. From the curves it is clear that only Eq.(6) can give rise to chaotic luminosities. However, both iteration schemes produce super–Eddington luminosities.

In the latter respect, it is interesting to consider a further effect originating from bubbles penetrating into the dense layer and thus producing channels. Obviously these bubbles do compress the plasma in the vicinity of the newly formed channels. In this manner, they stimulate the process of clump formation just in the neighbouring cells. This leads to a feedback mechaniam induced by stimulated clump formation. The number T of channels produced by stimulated clumps is governed by an iterative map of the form:

$$T_{i+1} = \dot{M}T_i(1 - xT_i)(1 - ax)^{T_i} \equiv f(T_i) \tag{6}$$

which differs from (5) by an additional factor $\dot{M}T_i$. This factor is crucial in producing a quadratic maximum of $f(T_i)$ as shown in Fig.6b. Hence, $f(T_i)$ has a Ljapunov spectrum equivalent to that of the standard quadratic map, so that Eq.(6) is indeed capable of producing a chaotic evolution of channels for sufficiently high mass accretion rates.

Although Eq.(6) does not include spontaneously generated clumps, it seems plausible that T_i dominates the process consisting of both spontaneous and stimulated contributions in the large $\dot{M}$ limit. It can therefore be expected that the evolution of T_i might give rise to a chaotic luminosity as shown in Fig.2a.

Finally, we should like to mention that spontaneous as well as stimulated clump formation via Rayleigh Taylor instability can also be treated as a cellular automaton with specific rules simulating the degree of spatial correlations around a channeling bubble on the instable shell. Such types of models are presently investigated, and we suspect that the relevant CA produces chaotic spatial patterns on the instable shell. If this can be confirmed, then the presented scenario shows a nice example how the existence of temporal chaos may intimately be related to chaotic spatial patterns.

3. Galaxy statistics

Since the discovery of the microwave background radiation[28] back in 1965 the universe is known to be almost homogeneous with respect to its radiative density on distance scales $\gg$ 100 Mpc (1 Mpc = $3.4 \cdot 10^{24}$ cm). The mw background originates from the time of radiation– matter decoupling at $t \approx 10^{13}$ sec on a cosmological time scale, so that the homogeneity of the background radiation provides indirect evidence for a similar degree of homogeneity in the matter distribution at that early age of the universe. On the other hand, the distribution of galaxies and clusters of galaxies representing the distribution of matter on comparably small scales (in the order of tens of Mpc) is extremely inhomogeneous. Obviously there is a transition from inhomogeneity to homogeneity at some intermediate distance scale. The investigation of this transition as well as the type of structures representing the inhomogeneities are of utmost importance for various cosmological questions. For instance, such fundamental quantities as the average matter density (deciding whether the universe is closed or open) may at some time be extracted from detailed knowledge about the matter distribution. Based on standard methods of statistical physics, the homogeneity and related properties of the galaxy distribution are commonly exploited using the concept of correlations.

3.1. PROBING HOMOGENEITY BY CORRELATION MEASURES

A great deal of pioneering work concerning correlation analyses of the galaxy distribution is due to P.J.E. Peebles and coworkers who started extensive studies of catalogs containing angular galaxy positions (as projected onto a celestial sphere) in the early 70s. These angular 2-D positions (Zwicky catalog,[29] Lick catalog,[30] and Jagellonian catalog[31]) represent projections of positions in the actual 3-D distribution which has recently become available to some extent by the CfA red shift surveys.[32,33]

Essential features of the two – point correlation function (i) a power–law scaling regime of 2-D correlations according to

$$\xi(r) \propto r/r_0^{-0.77(4)} \tag{7}$$

on a distance scale smaller than 1 h^{-1} Mpc (1° in the Zwicky sample), and (ii) a characteristic correlation length of $r_0 \approx 5~h^{-1}$ Mpc (5° in the Zwicky sample), often interpreted as an indication for a homogeneization of the distribution on distance scales larger than this value.[3]

With particular respect to the second point, the homogeneization of the universe concerning its matter density, Pietronero[34] has recently introduced a formal argument questioning the relevance of the standard correlation analysis in certain situations. It has indeed been shown[34,35] that the abrupt decrease of the correlation function could under certain circumstances (to be discussed below) be an effect of the limited size of the galaxy sample considered, whereas the actual distribution would still follow a power law as it does for small distances. Such a behavior would imply that any "average" density depends on the specific sample investigated and cannot be regarded to be of general relevance.

---

[3] h is the present value of the Hubble parameter in units of 100 km sec^{-1} Mpc^{-1}. Observational data strongly suggest $0.4 < h < 1$.

For instance, a corresponding distribution could be hierarchical. In order to match the almost homogeneous situation at very large distances, a decreasing degree of power–law exponents for increasing distance scales would be required in this case. For an overview of early and recent work on a hierarchically structured universe we recommend the text of Mandelbrot.[36] If the matter distribution in the universe were hierarchical on particular distance scales, then it would have to satisfy the formal properties of a fractal or multifractal distribution on these scales. This also means that the concept of a homogeneous matter distribution would have to undergo a serious modification.

The sample size dependence in the standard correlation analysis results from the definition

$$\xi^{(2)}(r) = \frac{< n(\mathbf{x})n(\mathbf{x} + r) >}{< n >^2} - 1 \tag{8}$$

for two–point correlations in an isotropic distribution where r is the distance between two objects situated at $\mathbf{x}$ and $\mathbf{x} + r$. Since averages are defined as

$$< ... >= \frac{1}{V_s} \int_{V_s} d\mathbf{x} \ ... \tag{9}$$

the correlation function $\xi(r)$ contains an implicit dependence on the sample size, here characterized by the 3-D sample volume $V_s \propto R_s^3$.

For a constant density (equivalent to a distribution homogeneous with respect to its density for $r_h < r \ll R_s$, r_h being some homogeneization distance) this implicit dependence does not matter at all. However, if the density is still changing at distances in the order of R_s (as it does in a fractal or multifractal distribution) then the sample size dependence causes a more or less abrupt decrease of $\xi(r)$, thus apparently simulating a homogeneization. This case is particularly unfortunate, since the combined effect of inhomogeneity and sample size dependence of $\xi(r)$ appears misleadingly as if it would confirm the assumption of homogeneity and sample size independence which is in turn required to justify the relevance of $\xi(r)$ at large distances.[4] Obviously this situation represents a vicious circle of "self – consistency" and might be partly responsible for recent debates about the relevance of a new look at galaxy statistics.

The sample size dependence can easily be removed using the two point correlation measure:[34]

$$\Gamma^{(2)}(r) = \frac{< n(\mathbf{x})n(\mathbf{x} + r) >}{< n >} \tag{10}$$

Both $\xi(r)$ and $\Gamma(r)$ are differential measures, somehow approaching zero or a constant value, resp., for $r = r_h$. In contrast to $\xi(r)$, $\Gamma(r)$ is not dimensionless, it is a so–called "average conditional density".[39,34] Because of statistical advantages of smaller errors at larger distances one introduces the integral correlation measure

$$I^{(2)}(r) \propto \int_0^r d^d r' \Gamma^{(2)}(r') \tag{11}$$

yielding a dimensionless quantity again. Now, any homogeneization in the distribution expresses itself in the behavior of the slope of $I(r)$. At $r = r_h$ the slope approaches

---

[4] In the terminology of the literature on galaxy statistics these assumptions are covered by the so–called "fair sample hypothesis". In recent articles[37,38] this hypothesis has been seriously questioned for various galaxy samples. It is an essential advantage of the approach described here that it is independent from such a hypothesis since it requires no a priori knowledge about large scale averages, in particular about an average density.

the topological dimension of the support of the point set investigated, i.e. 2 in case of 2-D positions.

The integral measure $I(r)$ (sometimes quoted as $J_3(r)$ in the literature on galaxy statistics) as given in Eq.(11) is equivalent to the correlation integral $C_3^{(2)}(r)$ for $d = 3$ in Eq.(2) the slope of which determines the correlation dimension (Eq.(2)) of a structure, be it in phase space or in position space. It is self–suggesting to evaluate q–point correlations in the galaxy distribution by a generalized $I^{(q)}(r)$, completely analogous to Eq.(3), and eventually to obtain an $f(\alpha)$ spectrum providing information about possible multifractal properties of the galaxy distribution.

In the following, new results concerning $I^{(2)}(r)$ as well as the $f(\alpha)$ spectrum will be described for the 2-D Zwicky catalog. In this case distances are angular distances Θ, and the correlation measures are $I(\Theta)$, $\Gamma(\Theta)$, and (following the notation in the relevant literature) $w(\Theta)$.

3.2. Correlation dimension and $f(\alpha)$ for the Zwicky sample

The Zwicky catalog contains the angular positions of galaxies up to a photographic magnitude $m = 15.5$. In order to avoid selection effects, investigators usually restrict themselves to the marked area in Fig.7 (for more details see figure caption). The remaining number of galaxies is still $N_{tot} = 3087$.

In addition to implicit selection effects it is necessary to exclude boundary effects. This point is of particular significance since considerable part of our interest focuses on the large scale behavior for angular distances $\Theta > 5°$ where boundary effects gain increasing influence. For this reason, $\Gamma(\Theta)$ and $I(\Theta)$ have been determined using only those galaxies with a distance smaller than Θ from the catalog boundary as reference galaxies. The number N_{ref} of reference galaxies decreases with increasing Θ. Since $I(\Theta)$ is given as an average over the pairs of galaxies within a ball around each reference galaxy, its error will be dominated by the small number N_{ref} for large angular distances Θ. For small Θ the Poisson error of the counted number of pairs dominates.

Applying the described boundary correction, the correlation measures are to be reformulated as:

$$\Gamma^{(2)}(\Theta) = \frac{1}{N_{ref}(N_{tot} - 1)} \frac{1}{2\pi \sin \Theta} \frac{dN(\Theta)}{d\Theta} \tag{12}$$

$$I^{(2)}(\Theta) = 2\pi \int_0^\Theta \Gamma(\Theta') d\cos \Theta' \tag{13}$$

and, in extension of (13):

$$I^{(q)}(\Theta) = \left[\frac{1}{(N_{tot} - 1)} \sum_{i=1}^{N_{tot}-1} \left(\frac{1}{N_{ref}} \sum_{j=1}^{N_{ref}} H(\Theta - |\theta_i - \theta_j|) \right)^{q-1} \right]^{1/q-1} \tag{14}$$

where θ_i characterizes the angular position of the ith galaxy.

Another way of boundary corrections would be to consider all galaxies as reference galaxies and to normalize the resulting counts with respect to the total area around each particular reference galaxy. We did not use this procedure instead of its applicability for larger Θ, since it contains assumptions (like local isotropy) about the actual galaxy distribution outside the catalog, thus again introducing a "fair sample" like argument.[35]

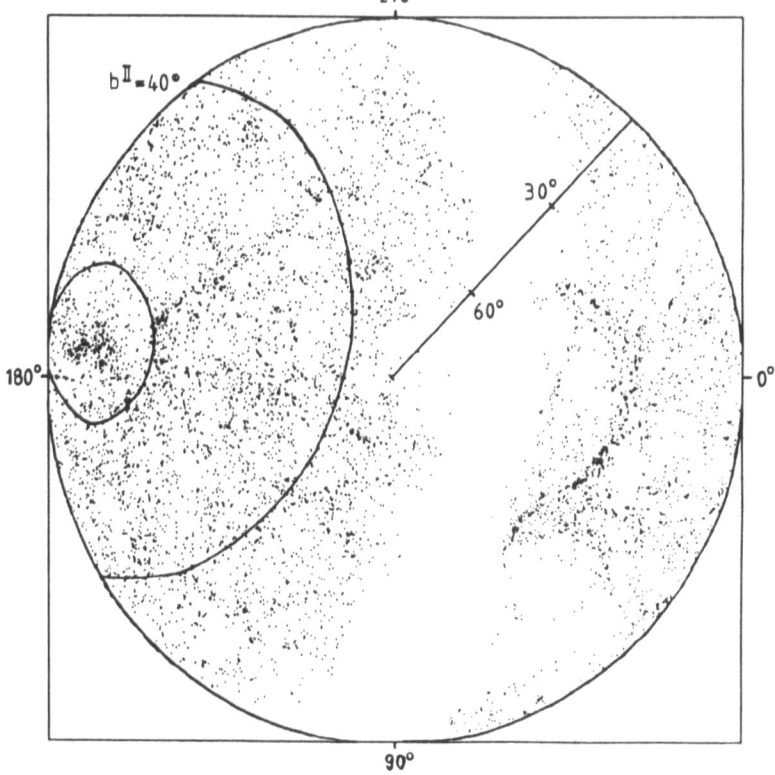

Figure 7. Distribution of galaxies according to the Zwicky catalog containing the angular positions of galaxies with apparent magnitude $m \leq 15.5$ as projected onto a celestial sphere. The figure shows a planar projection of galaxies with $m \leq 15.0$ in the northern hemisphere (declination $\delta \geq 0^\circ$, right ascension $0^\circ < \alpha < 360^\circ$). For statistical analyses, the Virgo cluster around $\delta = 13.5^\circ$, $\alpha = 187^\circ$ as well as all objects with galactic latitude $b^{II} < 40^\circ$ are excluded in order to avoid selection effects.

A further point concerns the non–Euclidean geometry of the support of the sample. Since a homogeneous distribution on a spherical surface does not scale like Θ^2 but like $(1 - \cos \Theta)$, it has been checked that the deviations in the relevant angular range remain marginal.

Figure 8a shows $I^{(2)}(\Theta)$ with (dots) and without boundary corrections (dashes) in a doubly logarithmic plot. The corresponding slopes $\frac{d \log I}{d \log \Theta}$ are given in Fig.8b. It is clear that missing boundary corrections cause a significant decrease of the slope for large Θ. For the corrected (dotted) curve two scaling ranges can be identified using χ^2 test procedures.[35] For $\Theta < 1^\circ$, a slope corresponding to a correlation dimension $D^{(2)} = 1.34(5)$ turns out. Since $D^{(2)}$ and the scaling exponent δ of $w(\Theta)$ are related by $\delta = 2 - D^{(2)}$ on a two–dimensional support, there is good agreement with the established value $\delta = 0.77(4)$.[40]

Figure 8b also shows a strong indication for a second scaling range which has not been observed using the standard approach $w(\Theta)$. It provides a slope $D^{(2)} = 1.86(1)$ for $5^\circ < \Theta$. This value of $D^{(2)}$ is significantly smaller than a value of 2 characterizing a homogeneous distribution. There is an obvious discrepancy between this result and the commonly quoted homogeneization distance of $\approx 5^\circ$. A second scaling range

44

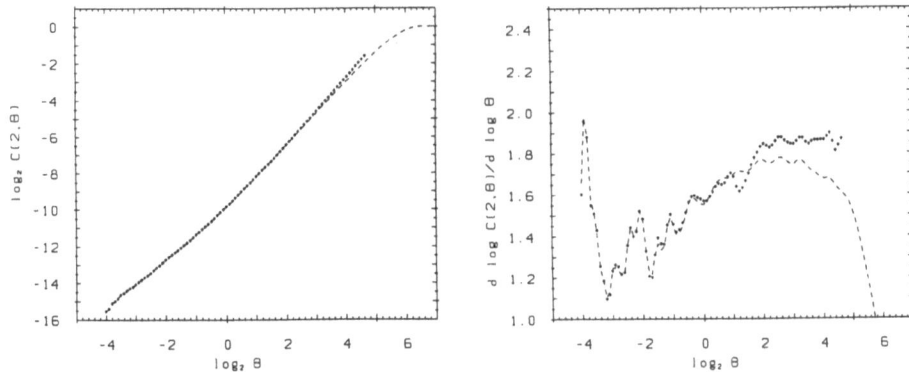

Figure 8. a) Log – log plot of the correlation integral $I(\Theta)$ for $q = 2$ with (dotted line) and without (dashed line) boundary corrections as described in the text. b) Local slopes $\frac{d\log I(\Theta)}{d\log\Theta}$ as a function of $\log\Theta$ with (dotted line) and without (dashed line) boundary corrections. A linear scaling range extending from $\Theta \approx 5^o$ up to $\Theta \approx 20^o$ is clearly visible in the boundary corrected curve. The value of the slope (corresponding to the dimension $D(2)$) is 1.86.

would also question the uniqueness of any correlation length r_o.

It is our suggestion to interpret the discrepancy along the lines described in Sec.3.1. In this sense the drop of $w(\Theta)$ would be caused by the sample size dependence of the standard analysis, thus simply obscuring the scaling regime $\Theta > 5^o$. This explanation is further supported by the observation that $w(\Theta)$ tends to flatten out around 1° before the abrupt decrease sets in.[41] For readers interested in these and more details we recommend an extensive comparison of $w(\Theta)$ and $I(\Theta)$ given elsewhere.[35] A recent analysis[42] of the CfA red shift catalog[32] also shows no indication of any homogeneization within the available distance scales.

The existence of two different power law scaling ranges with increasing dimension for increasing angular regime would be consistent with a stepwise approach toward a homogeneous matter distribution as the microwave background indicates for very large distances. This implies that the universal matter distribution cannot be described as a simple fractal characterized by one single dimension measure. Instead, the analysis yields scaling regimes with different correlation dimensions $D(2)$. It is now interesting to explore whether the distributions within these particular regimes are single fractals or multifractals, i.e. whether their spectrum of dimensions is constant or monotonically decreasing. In the first case, the corresponding $f(\alpha)$ spectrum would give rise to a single value $\alpha = f(\alpha) = D(q) = $ const. Otherwise, a $f(\alpha)$ spectrum over an extended range of α would be obtained.

Among different methods to extract $f(\alpha)$ from a point distribution, the fixed mass approach in combination with the histogram method[43] has been shown to be preferable.[44] In contrast to the Legendre transform of $(q-1)D(q)$, this procedure starts with an evaluation of the probability distribution $P_N(\alpha)$ (in a histogram representation) of scaling (crowding) indices:

$$\alpha_i = \log\frac{N}{N_{tot}}\bigg/\log\frac{\Theta_i(N)}{\Theta_{tot}} \qquad (15)$$

where $\Theta_{tot} = 42.4^o$ corresponds to a circular sample covering the same area as the restricted Zwicky sample (1.646 sterad), and $\Theta_i(N)$ is the angular diameter of a circle around a reference galaxy containing exactly N galaxies (the fixed mass). Dealing with the scaling regime $5^\circ < \Theta < 20^\circ$, we choose $N = 180$ corresponding to $\Theta = 14^\circ$ for the largest values of α. The resulting histogram is shown in Fig.9a (drawn line: with boundary corrections; dashed line: without boundary corrections). The $f(\alpha)$ spectrum is then achievable by vertically shifting and rescaling $P_N(\alpha)$:

$$f(\alpha) = \alpha \left(const - \log \frac{P_N(\alpha)}{N_{ref}} \bigg/ \log \frac{N}{N_{tot}} \right) \tag{16}$$

According to the generic properties of $f(\alpha)$,[16] the constant is uniquely determined by the condition:

$$f(\alpha(q = 1)) = \alpha \tag{17}$$

The resulting values $f(\alpha)$ are plotted in Fig.9b, where the drawn line represents a parabolic fit to the values. The dashed straight line shows $f(\alpha) = \alpha$ A proper error analysis is described in detail elsewhere.[44]

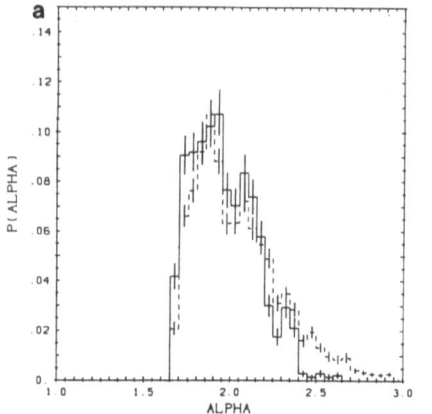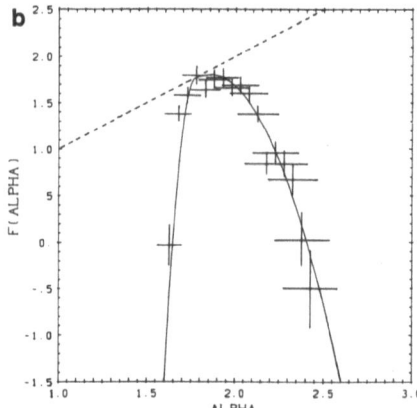

Figure 9. a) Histograms representing the probability distribution $P_N(\alpha)$. The drawn (dashed) line shows the number of occurences of a value α at fixed $N = 180$ with (without) boundary corrections. b) $f(\alpha)$ as obtained from the histogram plotted in a). The constant in Eq.(16) is determined for $D(1) = 1.76$.

The $f(\alpha)$ spectrum reflects the total spectrum of correlation exponents ($q > 0$, left branch) and anticorrelation exponents ($q < 0$, right branch). The asymmetry between both branches gives a visual impression (and quantitative measure as well) of the scaling behavior of dense subsets (left) and rarefied subsets (right). Of course, these subsets can be assigned to clustered regions and voids in the context of galaxy statistics. The $f(\alpha)$ spectrum also provides evidence that, in addition to the different scaling regimes, the distribution within one single regime corresponds to a multifractal. Further work on the multifractal nature of the galaxy distribution according to 2-D and 3-D catalogs and its astrophysical interpretation is in progress.

4. Conclusions

The usefulness of dimensional measures (as they emerged from statistical physics and nonlinear dynamics) for astrophysical applications has been demonstrated in the preceding sections.

As a measure of the functional complexity of the process of matter accretion onto compact objects, attractor dimensions have been determined from the X-ray luminosity of the neutron star Her X-1. The low value of this dimension encourages theoretical efforts to model appropriate accretion scenarios. First attempts along this line have been performed using a two–fluid (radiation – matter) description of the star atmosphere. Under the assumption of a spherically symmetric accretion on a temporal average, an anisotropic scenario has been derived giving rise to temporal chaos if the distribution of the anisotropies is generated by a nonlinear feedback mechanism. (This issue is probably interesting for hydrodynamic problems in less exotic environments as well.) Independent observational confirmation for the model has been established, since it predicts super–Eddington luminosities which have not been understood so far.

In the context of galaxy statistics, dimensions have been used as measures for the structural complexity of the galaxy distribution. A reinvestigation of the Zwicky sample of galaxies has shown that the standard correlation approach is of restricted relevance for large distance scales if the distribution is "piecewise" fractal or multi-fractal on the corresponding scales. Using a modified correlation formalism it has been shown that there is definitely no homogeneization at scales as small as $5h^{-1}$ Mpc. Moreover, a second scaling range in addition to the well – known small scale power law has been discovered. In this regime the distribution is a multifractal with a dimension $D^{(2)}$ larger than that obtained for the small scale regime. This result might be an indication for a stepwise transition toward the homogeneity of the microwave background on very large scales $\gg 100$ Mpc. A proper value of the distance for which the matter distribution becomes homogeneous would of course have important consequences for the value of the density parameter determining the future of our universe.

REFERENCES

1. J. R. Buchler, J. M. Perdang, and E. A. Spiegel, eds., *Chaos in Astrophysics*, (Reidel, Dordrecht, 1985)

2. H. Ögelman and E. P. J. van den Heuvel, eds., *Timing Neutron Stars* (Kluwer, Dordrecht, 1989)

3. G. Hasinger and M. van der Klis, *Astron. Astrophys.* **225**, 79 (1989)

4. P. Grassberger and I. Procaccia, *Phys. Rev. Lett.* **50**, 346 (1983)

5. W. Voges, H. Atmanspacher, and H. Scheingraber, *Ap. J.* **320**, 794 (1987)

6. H. Atmanspacher, H. Scheingraber, and W. Voges, *Phys. Rev.* **A 37**, 1314 (1988)

7. H. Atmanspacher, H. Scheingraber, and W. Voges, in: *Timing Neutron Stars*, eds. H. Ögelman and E. P. J. van den Heuvel (Kluwer, Dordrecht, 1989) p.219

8. H. Atmanspacher, H. Scheingraber, and W. Voges, in: *Data Analysis in Astronomy III*, eds. V. di Gesu and L. Scarsi (Plenum Press, New York, 1989)

9. W. Voges, H. Atmanspacher, and H. Scheingraber, *Adv. Space Res.* **8**, (2)497 (1988)

10. G. E. Morfill, H. Atmanspacher, V. Demmel, H. Scheingraber, and W. Voges, in: *Timing Neutron Stars*, eds. H. Ögelman and E. P. J. van den Heuvel (Kluwer, Dordrecht, 1989) p.71

11. N. H. Packard, J. P. Crutchfield, J. D. Farmer, and R. S. Shaw, *Phys. Rev. Lett.* **45**, 712 (1980)

12. F. Takens, in *Dynamical Systems and Turbulence, Lecture Notes in Mathematics* **898**, eds. D. A. Rand and L. S. Young (Springer, Berlin, 1981), p.366

13. J. P. Norris and T. A. Matilsky, *Ap. J.* **346**, 912 (1989)

14. P. Grassberger and I. Procaccia, *Phys. Rev.* **A 28**, 2591 (1983)

15. K. Pawelzik and H. G. Schuster, *Phys. Rev.* **A 35**, 481 (1987)

16. T. C. Halsey, M. H. Jensen, L. P. Kadanoff, I. Procaccia, and B. I. Shraiman, *Phys. Rev.* **A 33**, 1141 (1986)

17. J. Feder, *Fractals* (Plenum, New York, 1988) Chap.6.4

18. E. N. Lorenz, *J. Atmos. Sci.* **20**, 130 (1963)

19. H. Atmanspacher and H. Scheingraber, *Phys. Rev.* **A 34**, 253 (1986)

20. H. Atmanspacher, H. Scheingraber, and V. M. Baev, *Phys. Rev.* **A 35**, 142 (1987)

21. Yu. M. Ajvasjan, V. V. Ivanov, S. A. Kovalenko, V. M. Baev, E. A. Sviridenkov, H. Atmanspacher, and H. Scheingraber, *Appl. Phys.* **B 46**, 175 (1988)

22. G. E. Morfill, V. Demmel, and H. Atmanspacher, *Mitt. Astron. Ges.* **68**, 251 (1987)

23. V. Demmel, G.E. Morfill, and H. Atmanspacher, in: *Timing Neutron Stars*, eds. H. Ögelman and E. P. J. van den Heuvel (Kluwer, Dordrecht, 1989) p.749

24. V. Demmel, H. Atmanspacher, and G. Morfill, *Adv. Space Res.* **8**, (2)583 (1988)

25. V. Demmel, diploma thesis, 1987

26. H. Doll and W. Brinkmann, *Astron. Astrophys.* **173**, 86 (1986)

27. Y.-M. Wang, M. Nepveu, and J. A. Robertson, *Ap. J.* **135**, 66 (1984)

28. A. A. Penzias and R. W. Wilson, *Ap. J.* **142**, 419 (1965)

29. F.Zwicky, E. Herzog, P. Wild, M. Karpowicz, and C. T. Kowal, *Catalogue of Galaxies and Clusters of Galaxies*, Vols.1-6 (California Institute of Technology, Pasadena, 1962-68)

30. C. D. Shane and C. A. Wirtanen, *Proc. Amer. Phil. Soc.* **94**, 13 (1950)

31. K. Rudnicki, T. Z. Dworak, P. Flin, B. Baranowski, and A. Sendranowski, *Acta Cosmologica* **1**, 7 (1973)

32. J. P. Huchra, M. Davis, D. Latham, and J. Tonry, *Ap. J. Suppl.* **52**, 89 (1983)

33. J. P. Huchra, V. de Lapparent, M.J. Geller, M.J. Kurtz, E. Horine, J. Peters, and S. Tokarz, 1989, to be published

34. L. Pietronero, *Physica* **144 A**, 257 (1987)

35. G. Wiedenmann and H. Atmanspacher, *Astron. Astrophys.* **229**, 283 (1990)

36. B. B. Mandelbrot, *Fractals and Multifractals: Noise, Turbulence, and Galaxies* (Springer, New York, 1989)

37. V. de Lapparent, M. J. Geller, and J. P. Huchra, *Ap. J.* **322**, 44 (1988)

38. M. Davis, A. Meiksin, M. A. Strauss, L. N. da Costa, and A. Yahil, *Ap. J. (Letters)* **333**, L9 (1988)

39. B. B. Mandelbrot, *The Fractal Geometry of Nature* (Freeman, San Francisco, 1982)

40. P. J. E Peebles, *The Large-Scale Structure of the Universe* (Princeton University Press, 1980)

41. P. J. E. Peebles and M. G. Hauser, *Ap. J. Suppl.* **28**, 19 (1974)

42. P. H. Coleman, L. Pietronero, and R. H. Sanders, *Astron. Astrophys.* **200**, L32 (1988)

43. P. Grassberger, R. Badii, and A. Politi, *J. Stat. Phys.* **51**, 135 (1988)

44. H. Atmanspacher, H. Scheingraber, and G. Wiedenmann, *Phys. Rev. A* **40**, 3954 (1989)

SOME REMARKS ON NONLINEAR DATA ANALYSIS

OF PHYSIOLOGICAL TIME SERIES

A. Babloyantz

Service de Chimie Physique
Université Libre de Bruxelles
CP231 Campus Plaine, Blvd. Triomphe
B-1050 Bruxelles, Belgium

1. INTRODUCTION

The analysis of model as well as experimental time series has become very popular in recent years. Various procedures have been developed to evaluate such quantities as dimensions [1-3], entropies [1-2,4] or Lyapunov exponents [1-2,5]. These quantities furnish important information about the dynamics of some experimental systems where typically very few variables can be measured and not much is known about the complex dynamics underlying the time evolution of the system.

A good example of such a system is the mamalian brain which is made of some 10^{10} neurons connected together in an intricate network. The electrical activity of this network can be measured from the scalp and constitutes the electroencephalogram or EEG. Although we have a good understanding of individual neuronal function and various regions of the brain are chartered in functional units, not much is known about the interrelationship and coherent activity of the mamalian brain. Whatever information about the activity of the cortex, local or global, could be found is highly desirable.

The researchers working in the area of nonlinear analysis of time series can be divided roughly into three groups. The first group has produced concepts, algorithms and theorems which have grounded the time series analysis on sound mathematical basis. Unfortunately these studies always involve simple maps or differential equations of only few degrees of freedom. Such systems are noise free and are well defined. The second group has tried to apply the proposed algorithms to real experimental data. Such an approach necessitates a good understanding of the concepts involved, and the way they can be used. In experimental situation we are not in the presence of infinite data sets and some parameters must be adjusted for which no satisfactory procedures have been proposed. There is a third group of researchers who have applied these methods to problems for which the methods were not suitable, or not enough data was available, or the time series represented a high dimensional dynamics which could not be assessed under the given conditions. Unfortunately it is the work of this third group which has shed a discredit on the power, applicability and usefulness of dynamical methods to the study of experimental time series.

The aim of this paper is to show that with a good understanding of the methods and a great care in their application, the time series analysis can become a powerful tool for assessing dynamics of complex systems such as the EEG.

The key to success is to ask the right questions about appropriate systems for which dynamical methods can shed some light on the complexity of the system. For example the dynamical analysis could be used for the study of electrical brain waves provided that the major aim is not to provide exact values for dynamical parameters. In such systems the important

Measures of Complexity and Chaos
Edited by N.B. Abraham *et al.*
Plenum Press, New York

question is to assess the nature of the system's dynamics nad its differential change when brain waves make transitions between several states. Such an approach is followed in this paper where we analyze and compare different stages of normal and pathological brain activity.

In the second section, we show how, with the help of a simple graphical tool, the recurrence plot [6], we can assess the stationarity of time series, an absolute requirement for the dynamical analysis. In the third section we introduce another simple graphical tool called differential interval plot which distinguishes unambiguously between a noise prone periodic signal and a low dimensional chaotic dynamics. In the fourth section, we report a comparative study of the EEG for several key stages of the human brain.

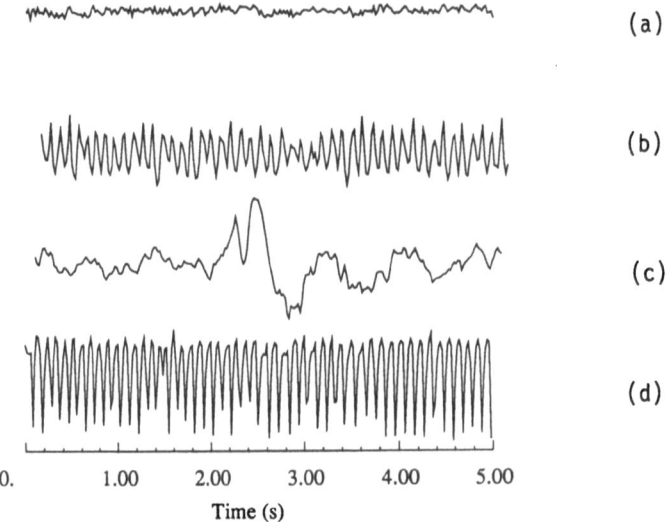

Figure 1. 5 seconds EEG recordings of a normal human brain. (a) eyes open - resting (1200 Hz sampling), (b) alpha rhythm: eyes closed - resting (1200 Hz sampling), (c) deep sleep (100 Hz sampling). (d) 5 seconds of intracortical EEG recorded from a rabbit during an experimentally induced epileptic seizure (1024 Hz sampling).

2. RECURRENCE PLOTS

The dynamical analysis of experimental time series requires stationary data sets. Usually in all applications one assumes the stationarity of the phenomenon under consideration. However this is not always the case. Therefore the first step in every analysis is to check for the stationarity of time series.

The full value of such an investigation appears in the study of electroencephalogram (EEG). The EEG measured from the scalp reflects the global activity of a portion of the cerebral cortex. Although global, the EEG shows characteristic features for well defined stages of the cerebral activity. It is customary to distinguish between beta waves (eyes open), alpha waves (eyes closed and relaxing state) and various stages of the sleep such as for example sleep stage four (deep sleep). Severe pathologies such as seizures or coma's exhibit also characteristic wave forms. Figure 1 shows the EEG corresponding to alpha waves, beta waves, sleep stage four and an epileptic seizure recorded from the surface of a rabbit brain.

In normal conditions, regularly the brain dynamics switches between its various states of activity. Therefore if a dynamical analysis is performed, it is of crucial importance to determine how long a given state of activity, say alpha waves, remain stationary.

Recently Eckmann and Ruelle [6] proposed a novel, powerful and elegant graphical tool, called a "recurrence plot" for the diagnoses of the presence of drift and hidden periodicities in the time evolution of dynamical systems which are unnoticeable otherwise.

The first step in the construction of the recurrence plot is the digitization of the EEG at regular time intervals. From this time series m vectors are constructed from a single lead V(t) by introducing a time lag τ leading to the V(t), V(t+τ), V(t+2τ) ... V(t+(m-1)τ) variables. It has been shown that the topological properties of phase portraits constructed by the above procedure are equivalent to the original phase portraits constructed from known variables of the system [7,8].

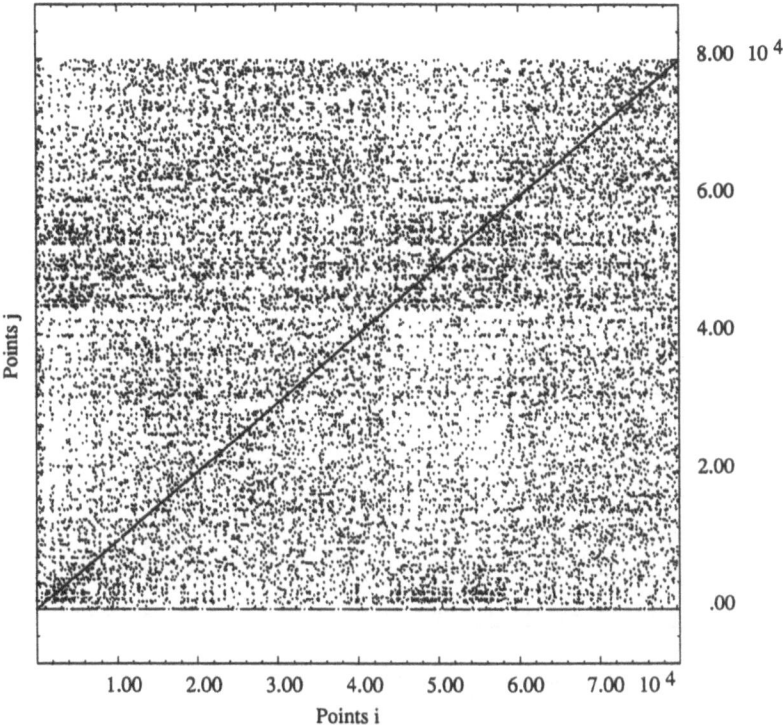

Figure 2. Recurrence plot obtained from 1280 seconds of deep sleep EEG (62.5 Hz sampling).

Once the phase portrait is constructed, one chooses a point x(i) on a trajectory and considers a ball of radius r(i) centered on this point. r(i) is chosen such as it contains a reasonable number of other points x(j) on the orbit. The recurrence plot is constructed as an array of dots in a NxN square, where a dot is drawn at (i,j) whenever x(j) is sufficiently close to x(i) (see Fig.1). Recurrence plots tend to be fairly symmetric with respect to the diagonal i-j. If x(i) is close to x(j) then x(j) is also close to x(i). However there is no complete symmetry as one does not require r(i)=r(j) but rather a fixed number of points in the ball r(i). Points i and j represent time, therefore the recurrence plot embodies natural and subtle time correlation information which locally shows a texture and in the same time exhibits a global topology.

If all characteristic times (the usual periodicities of EEG) are short compared with the total recording time, then the global aspect of the recurrence plot is homogeneous.

Recurrence plots were introduced recently for EEG analysis by Babloyantz [9]. It was shown that a long stretch of deep sleep was stationary. Figure 2 shows such a recurrence plot obtained from a 1280 second recording of an episode of deep sleep. It is seen that the data set is stationary and does not contain a slowly varying drift. For a time series with drift the recurrence plot is much paler as the distance from the diagonal increases (see Fig.2 from ref. [6]).

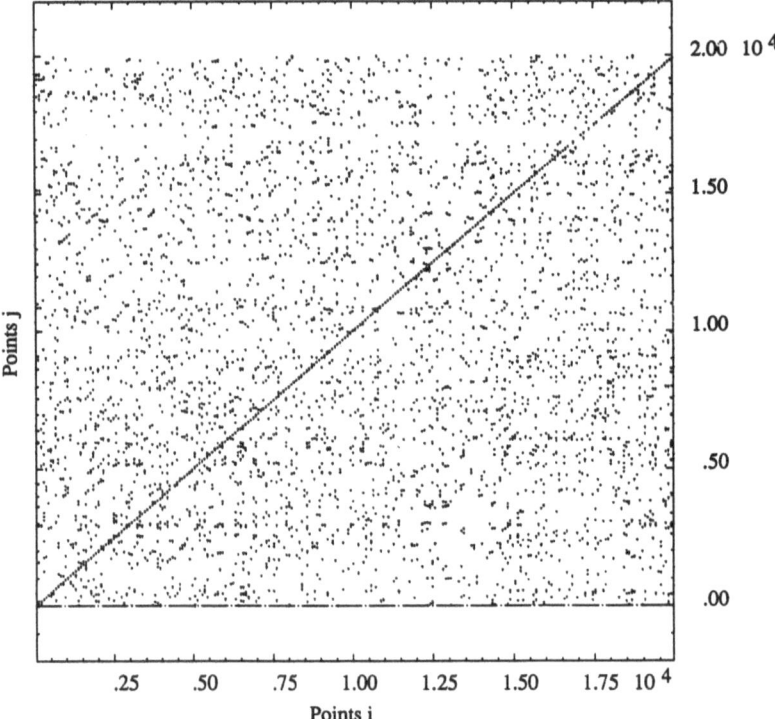

Figure 3. Recurrence plot obtained from 80 seconds of alpha rhythm EEG (250 Hz sampling).

Figure 3 shows the recurrence plot obtained from alpha waves. Although the plot is not as homogeneous as for the case of deep sleep, nevertheless this portion of the record may be considered as stationary. A non stationary recurrence plot was obtained for the case of the rabbit epileptic seizure (Fig. 4). The remarkable fact is that such a non stationary behavior is not apparent when the EEG of fig. 1 is examined for at least 10 seconds. Moreover, as we shall see later, the phase portrait is more like a limit cycle than that of alpha waves. Therefore, we may conclude that the recurrence plot is a simple and powerful tool for assessing stationarity of time series.

We turn now to a second simple graphical tool for the analysis of experimental data.

3. DIFFERENTIAL TIME INTERVAL PLOT

In some physiological cases, one encounters experimental data that at first sight seem to show periodic activity contaminated with random noise. Such data may also result from low dimensional chaotic dynamics. Therefore it is important to have tools which distinguishes between these two radically different dynamics in an unambiguous fashion.

In principle the Poincaré sections could provide a discriminating factor. However unfortunately sometimes these seemingly periodic signals may have a fractal dimension greater than three. In such cases, the Poincaré sections do not provide any concluding results. Here we describe a simple graphical tool introduced previously [10], which can discriminate between a noisy periodic signal and a deterministic chaotic dynamics.

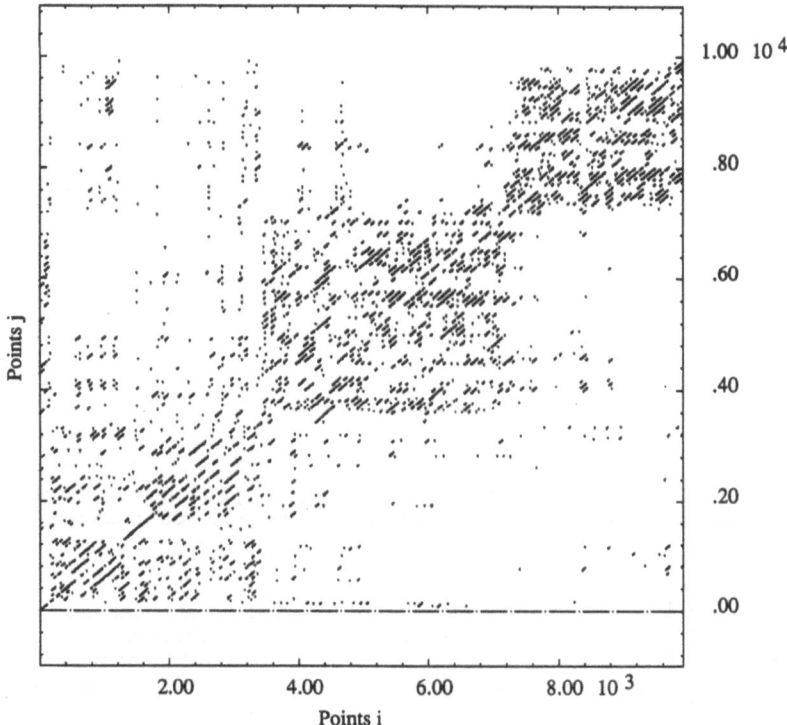

Figure 4. Recurrence plot obtained from 10 seconds of a rabbit epileptic seizure (described in Fig.1d).

Let us illustrate the procedure with an example. Figure 5 shows the EEG recorded from a patient suffering from Creutzfeldt-Jakob coma. In this terminal disease under the action of slow viruses, the brain gradually degenerates and the EEG shows an extremely coherent activity. The EEG of fig.5 was recorded from such a patient a few days before brain death.

A trained eye of a physicist will find an obvious periodicity in the data and might conclude that the record shows a noise prone limit cycle.

Let R_i represent the time interval between two successive minima in the record of fig.5. We consider the consecutive intervals R_i, R_{i+1}, R_{i+2}, $\cdots$ A two dimensional representation may be constructed by plotting $\Delta R_{i+1} = R_{i+2} - R_{i+1}$ vs. $\Delta R_i = R_{i+1} - R_i$ (in three dimensions the ΔR_{i+2} axis is necessary) using the totality of the data length.

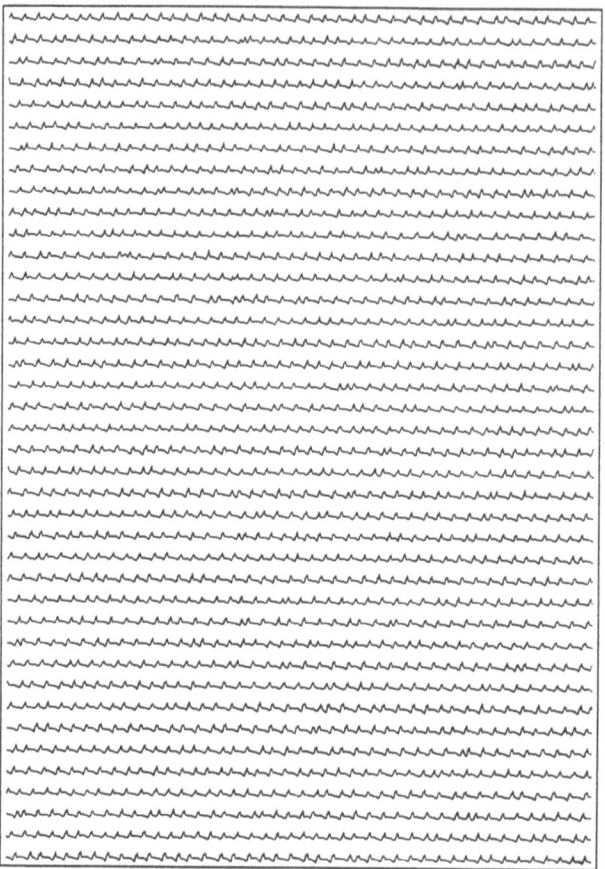

Figure 5. 20 minutes EEG recoding of a patient suffering from Creutzfeldt-Jakob disease. Each line is of 30 seconds duration (250 Hz sampling).

In this graph a limit cycle would be represented as a point at the origin as in this case all intervals are equal. On the other hand, a periodic motion with additive noise will show a cloud of points scattered homogeneously around the center. The differential time interval plot (DTI) of such a system is shown in fig.6 (from ref. [10]). When such plots are constructed from the data recorded in a Creutzfeldt-Jakob patient, an obvious structure is seen (fig.7). This structure indicates that the dynamics underlying the brain activity is not periodic but is much more complex. This fact has been confirmed by the dimension analysis which showed the presence of deterministic chaos characterized by a correlation dimension of 3.8 [11]. Strong correlations have also been found between successive intervals by an independent method [12] which relies on symbolic dynamics.

A similar DTI plot may be constructed using the time series obtained from the measurement of the electrical activity of the heart [10], the so-called electrocardiogram (ECG). Even in healthy subjects, the cardiac interbeat interval is not constant and shows an obvious random distribution. However again the DTI plot exhibits a remarkable structure (see Fig. 8) which shows that not every succession of time intervals have equal probability. Moreover it also indicates that the signal is not a noise prone periodic motion.

With these two examples, we see that the DTI plots are very simple and efficient graphical tools that can discriminate between periodic and chaotic motion without the help of complex algorithms that most often are subject to criticism.

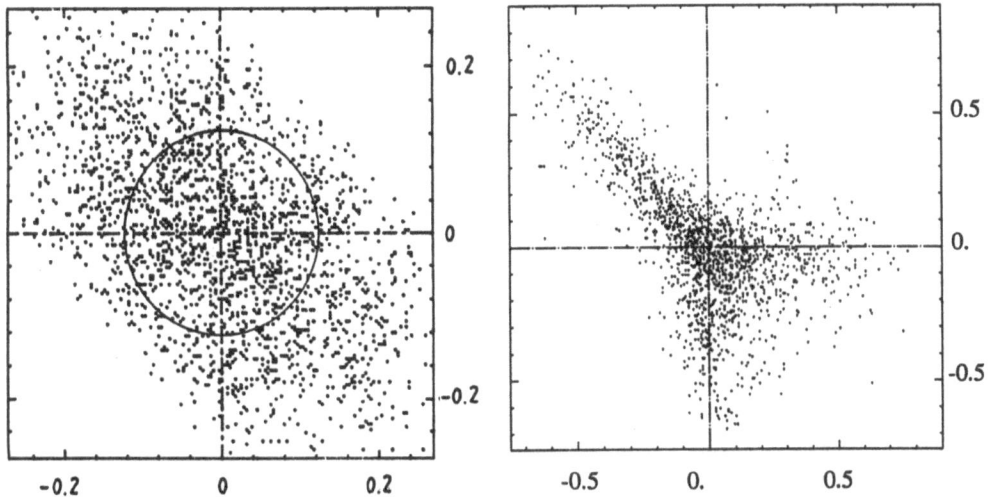

Figure 6. Differential time interval (DTI) plot obtained from the successive minima of a randomly perturbed limit cycle [10].

Figure 7. DTI plot obtained from the minima of the EEG of the Creutzfeldt-Jakob disease.

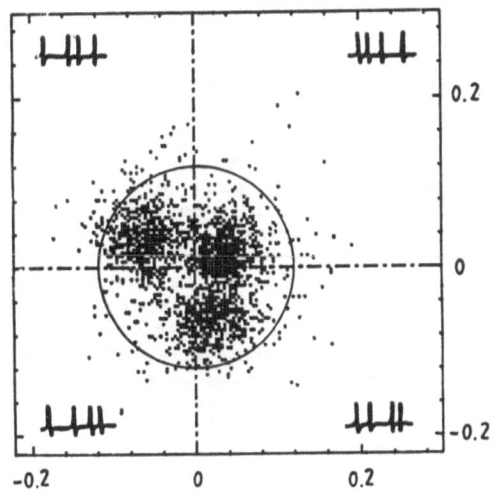

Figure 8. DTI plot obtained from a normal human heart during rest [10]

4. CORRELATION DIMENSIONS

The two simple graphical tools of preceding sections are welcome additions to several other techniques which we describe below for assessing the system's dynamics.

Once with the help of recurrence plots, it is decided that a sufficiently long stretch of data set is stationary and that it does not represent a simple periodic motion, it is worthwhile to proceed further.

Although the phase portrait in some cases may indicate the presence of chaotic dynamics, this is not always the case. Figure 9 shows the phase portrait of alpha waves, deep sleep and rabbit epilepsy. The alpha waves unambiguously exhibit a chaotic attractor. However the existence of a chaotic attractor cannot be guessed from the phase portrait of the deep sleep. Let us note that the attractor of fig. 9c corresponding to the non-stationary EEG from rabbit epileptic seizure is the most coherent looking attractor. In this case, one is tempted to infer the presence of a periodic motion contaminated with noise. These examples show the limitations and usefulness of phase portraits.

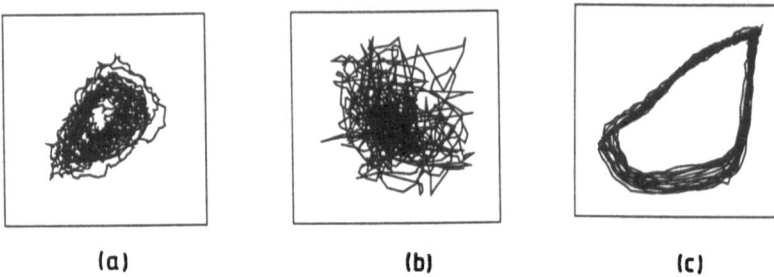

(a)　　　　　　　　　　　(b)　　　　　　　　　　　(c)

Figure 9. Projection from a 3-dim phase space reconstruction of the EEG episods shown in Fig.1. (a) eyes closed, (b) deep sleep, (c) Rabbit epileptic seizure.

Fortunately in the study of nonlinear time series we can use other techniques such as power spectra, autocorrelation functions, Lyapunov exponents, Kolmogorov entropies and correlation dimensions.

In the presence of chaotic dynamics, the broad band power spectra are not of great usefulness. Autocorrelation functions furnish qualitative information. Lyapunov exponents, Kolmogorov entropies and dimensions are much more promising. These quantities provide us with quantitative measures which characterize a given attractor. Several algorithms are developed which enable us to assess these quantities from experimental time series [3,5].

As we stated already, the algorithms are tested on simple maps or low dimensional systems described by few coupled nonlinear differential equations. However with experimental time series, which are usually finite, various parameters such as data length, delay time and sampling frequency may influence the numerical values of dimensions or Lyapunov exponents.

Due to these difficulties, after an early enthusiasm for these methods, presently we are in the presence of equally enthusiastic criticism for their use. In an attempt at calculating the quantitative dynamical parameters, the first questions which arise are: what kind of system and for which purposes?

It is obvious that if one is in the presence of a well defined physical system which is expected to behave always in the same fashion in every latitude, it is legitimate to attempt to measure these quantities with a high degree of accuracy. In these cases, data sets of order of million points are probably necessary [13,14].

However in many problems, useful information about the system's dynamics could be obtained with limited data sets. Let us take the case of the EEG which results from the average activity of some 10^{10} neurons. Until recently, the only way to assess the dynamics of such a complex system was to construct models comprising a very modest number of neurons. The collective behavior of such systems were assumed to account for the EEG activity. The comparison was visual and did not shed much light on the dynamics of the brain. Therefore any kind of dynamical information, qualitative as well as quantitative, from the recorded EEG is highly desirable even if not very accurate, and prone to some degree of error.

By the nature of the system under consideration, a high accuracy in the measurement of any parameter characteristic of brain dynamics is not necessary and even probably meaningless.

The brain is not a uniquely defined physical system. Every brain has probably its own way of operating, although they are all made of similar neurons. If we continue to believe that learning, therefore our daily experience, influences the connections between the neurons, then in the study of the cortex we are faced with differently connected networks. Thus we cannot expect identical collective behavior in all brains. Therefore it is truly remarkable that we still could find stages of the human brain activity that presents enough similar characteristics that with naked eye, an easy classification into beta waves, alpha waves, deep sleep,..., is possible. However this does not mean that alpha waves of every individual should obey exactly the same dynamics and that the corresponding attractors are identical.

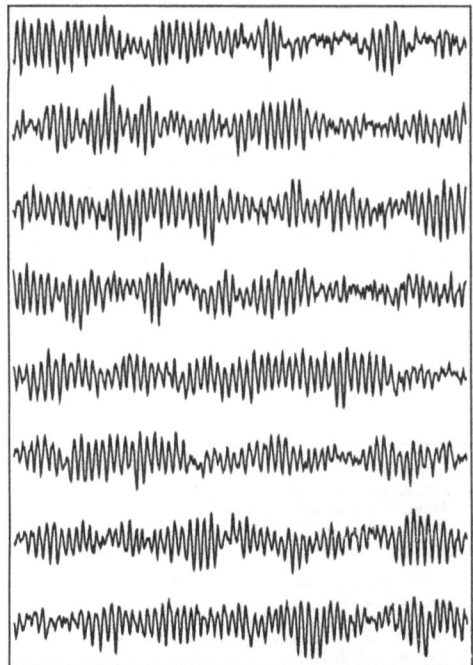 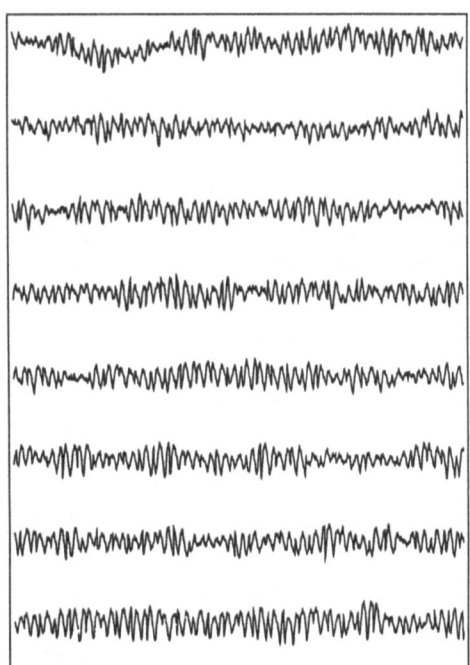

Figure 10. (left) Alpha rhythm EEG of a normal human subject. (right) the same eight months later (250 Hz sampling).

One may for example wonder whether a given individual's alpha waves remain exactly the same over several months. In fig.10 we report two alpha waves recorded from the same individual, under similar experimental conditions, at eight months interval. The difference between the two recordings is obvious. Therefore, we should not expect, even if we could evaluate the dynamical parameters with 100% accuracy, to find the same value for every alpha wave. Thus the absolute value of these quantities are not of much importance.

However if the dynamical methods could convince us that for several key stages of brain activity, we are in the presence of chaotic dynamics of relatively low dimension, we have accomplished an important step toward the understanding of brain dynamics. Chaotic dynamics would imply that 10^{10} nonlinear interconnected units obey dynamics that can be described with only few degrees of freedom.

The existence of chaotic dynamics in alpha waves, deep sleep and two pathological states was observed by several groups using independent techniques (see Table 1).

In the study of human EEG it is also important to quantify the change in dynamics when activity state changes. In this comparative study the absolute value of parameters are not of much importance. We are only interested in relative changes. The important fact is to choose the activity states so different that the change for example in correlation dimension is much larger than errors that can be made at various stages of analysis.

Such a comparative approach has been adopted by our group. We have chosen well defined stages of EEG activity and investigated the change in dynamical properties of beta waves (eyes open) [15], alpha waves (eyes closed) [15,16], deep sleep [17,18] and pathologies such as petit-mal type epilepsy [19] and Creutzfeldt-Jakob coma [11].

The quantitative measurements consisted of the evaluation of the correlation dimension with the help of the Grassberger-Procaccia algorithm [3]. Prior to this analysis, several embedding techniques for phase space reconstruction were used [16]: laging method, singular value decomposition [20] and the use of multichannel recordings. Topological dimensions [21] were also evaluated in these various phase space constructions. The results are exhibited in Table 1 which shows that various embedding techniques as well as different dimensions point to the existence of chaotic attractors in the brain. It also shows how these attractors change as the brain switches from one state to the next.

Gallez & Babloyantz [22] evaluated the Lyapunov exponents from alpha waves and deep sleep. For both cases, positive Lyapunov exponents were found showing again the presence of chaotic dynamics. With the help of these exponents, we also evaluated the correlation dimensions using the Kaplan-Yorke and Mori conjectures [23]. It was reassuring to find comparable values to the ones obtained previously with a quite different approach (see Table 1).

Table 1. Correlation dimensions from scalp EEG.

Description	Correlation dimension	Authors
Awake, eyes open	(*) 6.5 8.6 - 8.7 9.7 ± 0.7 (*) 8.1 (*)	Babloyantz et al. (1985) Dvorak & Siska (1986) Mayer-Kress & Layne (1987) Babloyantz & Destexhe (1987a) Cerf et al. (1988)
Awake, eyes closed (occipital leads)	5.7 5.1 ± 5.4 4 - 7 6.1 ± 0.5 3.7 - 6.3 6.3-7.4 5.71 - 7.63 3.4 - 4.3	Dvorak & Siska (1986) Layne et al. (1986) Mayer-Kress & Layne (1987) Babloyantz & Destexhe (1987a) Cerf et al. (1988) Babloyantz & Destexhe (1988b) Gallez & Babloyantz (1989) Rapp et al. (1989)
Sleep stage 2	5 - 5.03 6.8 ± 6.1	Babloyantz et al. (1985) Mayer-Kress & Holzfuss (1987)
Sleep stage 4	4.05 - 4.4 5.9 ± 4.4 4.8 ± 0.3 4.44 - 6.12	Babloyantz & al. (1985) Mayer-Kress & Holzfuss (1987) Babloyantz & Destexhe (1987b) Gallez & Babloyantz (1989)
REM sleep	(*) 6.4 ± 5.1 8.2 (*)	Babloyantz et al. (1985) Mayer-Kress & Holzfuss (1987) Babloyantz & Destexhe (1987b)
Petit-Mal seizure	2.05 ± 0.09	Babloyantz & Destexhe (1986)
C.J. disease	3.7 - 5.4 4.1 - 5.3	Babloyantz & Destexhe (1988a) Gallez & Babloyantz (1989)
Fluroxene anesthesia	6.8 - 8.6	Layne et al. (1986)
Awake - mental tasks	5.5 - 6.4 4.8 ± 0.9	Mayer-Kress & Layne (1987) Rapp et al. (1989)

(*) the attractor is high dimensional and it is probable that the algorithms are not applicable

5. CONCLUSIONS

As the content of table 1 shows, our own work and the analysis performed by several other groups, indicates the existence of chaotic dynamics in several well defined stages of the activity of the human brain. The existence of positive Lyapunov exponents for the corresponding attractors confirms these findings.

The key criticism to the dimension analysis is two fold:

1. It is commonly believed that the EEG is not stationary, therefore all the existing algorithms cannot be used with confidence. It is correct to say that the EEG changes constantly but it is also true that for a length of time the signals are well defined. Here we have shown how the stationarity of the time series can be assessed and that sufficiently long stretches of EEG can be shown to be stationary. This type of analysis shows also the presence of superimposed activity on a data set that cannot be seen otherwise [9].

2. The evaluation of the dimensions from time series is prone to errors at several stages of the analysis. We believe such errors can be reduced substantially by careful selection of sampling frequencies, delay times, and data length. In the absence of absolutely reliable methods for the choice of these parameters, we suggest that the correlation dimension must be evaluated by varying these parameters once at a time systematically, till a region is found where no variation in the correlation dimension are seen with parameter change.

In any dynamical analyses, we must keep in mind that most characteristic stages of the EEG are probably multifractals. Therefore the correlation dimensions are average values over superimposed dynamics. However this does not seem to be the case for deep sleep which shows a homogeneous data set.

In this paper we also showed that one must be cautious in classifying a seemingly periodic signal as a limit cycle. Such periodic signals may in fact represent chaotic dynamics of low dimension. DTI is a simple tool for discriminating between periodic and chaotic time series.

ACKNOWLEDGMENTS

We are indebted to A. Destexhe for his help in the preparation of the manuscript. Profs. Pockberger and Petsche are acknowledged for giving us access to their data on rabbit epilepsy.

REFERENCES

1.   J.P Eckmann and D. Ruelle, Rev. Mod. Phys. 57, 617 (1985)
2.   H. Schuster: in "Deterministic Chaos" (Physik-verlag, Weinheim, 1984)
3.   P. Grassberger and I. Procaccia: Phys. Rev. Lett. 50, 346 (1983)
4.   P. Grassberger and I. Procaccia: Phys. Rev. A 28, 2591 (1983)
5.   J.P Eckmann , S.O. Kamphorst, D. Ruelle and S. Ciliberto: Phys. Rev. A 34 , 4971 (1986)
6.   J.P Eckmann and D. Ruelle: Europhys. Lett. 4, 973 (1987).
7.   N.H. Packard, J.P. Crutchfield, J.D. Farmer, and R.S. Shaw: Phys. Rev. Lett. 45, 712 (1980).
8.   F.Takens: in "Dynamical systems and turbulence", Eds Rand, D.A. and Young, L.S., Lect. Notes in Math. 898, 366, Springer, Berlin (1981)
9.   A. Babloyantz: submitted to Neuroscience (1989)
10. A.Babloyantz and A.Destexhe, Biological Cybernetics 58, 203 (1988)
11. A.Babloyantz and A.Destexhe: in "From Chemical to Biological Organization", Ed by M. Markus, S. Müller and G. Nicolis, Springer-Verlag (1988)

12. A. Destexhe, G. Nicolis and C. Nicolis, this volume; A. Destexhe, Phys Lett A, in press, (1989)
13. L.A. Smith: Phys. Lett. A 133, 283 (1988)
14. J.P Eckmann and D. Ruelle: Preprint (1989)
15. A. Babloyantz & A. Destexhe: in "Temporal disorder in human oscillatory systems" Eds. L. Rensing, U. an der Heiden and M.C. Mackey, Springer Series in Synergetics 36,48 (1987a)
16. A. Destexhe, J.A. Sepulchre & A. Babloyantz, Phys. Lett. A 132, 101 (1988)
17. A.Babloyantz, C.Nicolis & M.Salazar: Phys. Lett. 111A, 152 (1985)
18. A. Babloyantz & A. Destexhe: in Proceedings of the first IEEE International Conference on Neural Networks, M. Caudill and C. Butler (eds), Vol IV, 31 (1987b)
19. A. Babloyantz & A. Destexhe: Proc. Natl. Acad. Sci. USA 83, 3513 (1986)
20. D.S. Broomhead and G.P. King, Physica 20D, 217 (1986)
21. D.S. Broomhead and G.P. King: J. Phys. A: Math. Gen. 20, L563 (1987)
22. D. Gallez and A. Babloyantz, submitted to J. Neurophysiol. (1989)
23. H. Mori: Prog. Theor. Phys 63, 1044 (1980)
24. I. Dvorak & J. Siska: Phys. lett. 118A, 63 (1986)
25. G. Mayer-Kress & S.P. Layne: in "Perspectives in Biological Dynamics and Theoretical Medicine", proceedings, Ann N.Y. Acad. Sci. 504, 64 (1987)
26. R. Cerf, Z. Farssi, B. Burgun, J.M. Trio & D. Kurtz: C. R. Acad. Sci. Paris 307, 715 (1988)
27. S.P. Layne, G. Mayer-Kress & J. Holzfuss: in "Dimensions and Entropies in Chaotic Systems" Ed G.Mayer-Kress (Springer Berlin 1986)
28. P.E. Rapp, T.R. Bashore, J.M. Martinerie, A.M. Albano & A.I. Mees: Preprint (1989)
29. G. Mayer-Kress & J. Holzfuss: in "Temporal disorder in human oscillatory systems" Eds. L. Rensing, U. an der Heiden and M.C. Mackey, Springer Series in Synergetics 36, 57 (1987)

HIERARCHIES OF RELATIONS BETWEEN PARTIAL DIMENSIONS

AND LOCAL EXPANSION RATES IN STRANGE ATTRACTORS

R. Badii and G. Broggi

Fakultät für Physik, Universität Konstanz, D-7750 Constance, W.Germany
Physik-Institut der Universität, Winterthurerstr. 190, CH-8057 Zurich, Switzerland

ABSTRACT

Connections between local partial dimensions and Lyapunov exponents in nonlinear dynamical systems are studied by using symbolic dynamics. Equations for the probabilities of symbol sequences are derived, based on the structure of the logic tree. These show that the dimension spectrum $f(\alpha)$ cannot be obtained in closed form from the sole knowledge of the local Lyapunov exponents.

1 INTRODUCTION

In recent years, relations among generalised dimensions, entropies and Lyapunov exponents have been introduced [1,2,3,4], expressed as averages over suitable coverings of the invariant measure on strange attractors, providing a global description of the scaling behaviour of nonlinear dynamical systems. A more complete characterisation of chaotic attractors requires the investigation of their microscopic properties. Local, coarse-grained, definitions of dimension and metric entropy can be given in terms of the natural invariant measure on the attractor, and the associated probability distributions can be evaluated, either analytically or numerically [4]. The observables of interest are then measured by suitably regrouping local quantities and by performing the infinite-resolution limit (arbitrarily fine-grained partitions). This is best obtained by labelling each partition element with its symbolic dynamical encoding and by determining parental hierarchies on a logic tree [5,6].

In the present contribution, we show that the relation between static (dimensions) and dynamic (Lyapunov) exponents is nonlinear and involves an open hierarchy of equations. Therefore, the determination of, e.g., the dimension spectrum [7] from the knowledge of the Lyapunov exponents requires additional input for one or more local dimensions. depending on the system. However, our procedure yields a set of equations which is valid for maps of any dimensionality (including one-dimensional noninvertible transformations) and for strange repellers as well. Finally, we discuss a recently proposed method [8] for a closed-form evaluation of the dimension spectrum of two- or higher-dimensional maps in terms of the Lyapunov exponents of the periodic orbits. We show that this procedure implicitly assumes the factorisation of probabilities and the constancy of local dimensions along a periodic symbolic orbit.

Measures of Complexity and Chaos
Edited by N.B. Abraham *et al.*
Plenum Press, New York

2 SYMBOLIC DYNAMICAL ENCODING OF LOCAL LYAPUNOV EXPO-
NENTS AND PARTIAL DIMENSIONS

In order to recall the definitions of local (partial) dimensions and Lyapunov exponents[4], we consider the exactly solvable (piecewise linear) generalised baker transformation[9], defined by

$$(x_{i+1}, y_{i+1}) = \begin{cases} (r_1 \cdot x_i, \ y_i/p_1) & \text{if } y_i \leq p_1 , \\[2mm] (r_2 \cdot x_i + 1 - r_2, \ (y_i - p_1)/p_2) & \text{if } y_i > p_1 , \end{cases} \tag{1}$$

where $r_1 + r_2 < 1$ and $p_1 + p_2 = 1$ $(0 \leq x, y \leq 1)$. The equation for the y-variable is the asymmetric Bernoulli shift and is not affected by the equation for the x-variable. The generating partition is obtained by cutting the unit square with a horizontal line at $y = p_1$, so that each point $\vec{x} = (x, y)$ in phase-space can be univocally represented by an infinite sequence of symbols

$$S_\infty \equiv \dots s_{i-1} s_i s_{i+1} \cdots , \tag{2}$$

with

$$s_i \equiv \frac{1 + \text{sgn}(y_i - p_1)}{2} \in [0, 1] . \tag{3}$$

Therefore, any finite sequence S_n corresponds to a subset of the unit square which shrinks to a point (or to a line) for $n \to \infty$. In a generic system, however, it is necessary to choose a covering and assign a symbolic sequence to every element of it. In particular, we take ellipsoids (rectangles, for the baker map) $B(\vec{\varepsilon}, \vec{x})$ of axes $\vec{\varepsilon} = (\varepsilon_1, \varepsilon_2)$, centred at $\vec{x}$. The size ε_2 along the contracting direction (x) must be small enough to guarantee that the first n_2 preimages of all the points in the ellipsoid generate the same backward symbolic sequence H_{n_2}. This requirement yields $\varepsilon_2 \sim e^{n_2 \lambda_2(\vec{x}, -n_2)}$, where $\lambda_2(\vec{x}, -n_2)$ is the second effective Lyapunov exponent[4], computed over the first n_2 preimages of $\vec{x}$. Analogously, along the expanding direction, the size will be such that all points yield the same forward symbol sequence T_{n_1}, of length n_1: i.e., $\varepsilon_1 \sim e^{-n_1 \lambda_1(\vec{x}, n_1)}$. In the following, H_n (T_m) will denote a backward (forward) sequence of length n (m). The procedure leads, in general, to a covering in which both the mass P (integral of the invariant natural measure[13]) and the size $\vec{\varepsilon}$ of the balls fluctuate with the position[10]. In this way, a finite sequence $H_{n_2} s_0 T_{n_1} = s_{-n_2} \dots s_0 \dots s_{n_1}$ labels the ball $B(\vec{\varepsilon}, \vec{x})$ (the symbol s_0 corresponds to the point $\vec{x}$ itself and can be included in either H or T).

Taking the n-th image of the generating partition $(y = p_1)$, the attractor of the baker map can be covered with 2^n vertical strips of height 1 and widths

$$\varepsilon_2(T_n) \equiv e^{n \lambda_2(T_n)} = r_1^k r_2^{n-k} , \tag{4}$$

where k is the number of 0s in the symbol-sequence T_n $(k \in [0, n])$. Each strip can then be split horizontally into 2^n rectangles (each one being the n-th preimage of an element of the generating partition) of height

$$\varepsilon_1(H_n) \equiv e^{-n \lambda_1(H_n)} = p_1^l p_2^{n-l} , \tag{5}$$

where l is the number of 0s in H_n $(l \in [0, n])$. Accordingly, we can define the partial local (coarse-grained) dimensions (α_1, α_2) at a generic point $\vec{x}' \in B(\vec{\varepsilon}, \vec{x})$ through the scaling relation[11,12]

$$P(\varepsilon_1, \varepsilon_2) \sim \varepsilon_1^{\alpha_1} \varepsilon_2^{\alpha_2} , \tag{6}$$

where $P(\varepsilon_1, \varepsilon_2)$ is the mass in rectangle $B(\vec{\varepsilon}, \vec{x})$, and both the sides ε_j and the dimensions α_j $(j = 1, 2)$ carry the proper sequence label: H_n for the unstable direction $(j = 1)$ and T_n for the stable one $(j = 2)$. Since the masses in these strips are given by $P(\varepsilon_1(H_n), 1) = p_1^l p_2^{n-l}$ and $P(1, \varepsilon_2(T_n)) = p_1^k p_2^{n-k}$, we obtain

$$\alpha_1(H_n) = \ln P(\varepsilon_1(H_n), 1) / \ln \varepsilon_1(H_n) = 1 \tag{7}$$

64

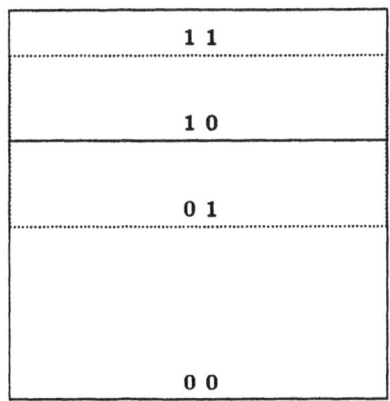
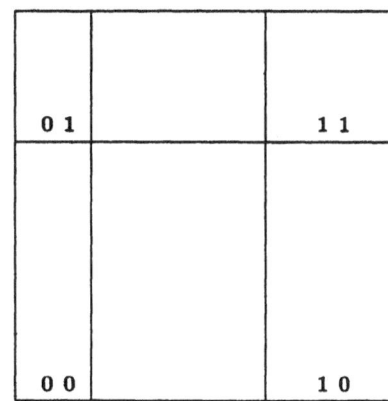

Figure 1. Phase-space of the baker map: the generating partition (solid line) and its first refinement (dashed lines) are displayed together with the symbol sequences of their elements, both at time $n = 0$ (left, a) and at time $n = 1$ (right, b). Each rectangle in the left picture maps onto the corresponding rectangle (i.e., with the same symbolic label) in the right one.

and

$$\alpha_2(T_n) = \ln P(1, \varepsilon_2(T_n)) / \ln \varepsilon_2(T_n) \in [\alpha_2^-, \alpha_2^+], \qquad (8)$$

where $\alpha_2^{+(-)} = \max(\min)\{\ln p_1 / \ln r_1, \ln p_2 / \ln r_2\}$ ($\alpha_1 = 1$ for any sequence S, since the invariant measure $\rho(\vec{x})$ is uniform along y). The above discussion, in which a constant-n procedure was adopted, applies to all situations in which the grammar is described by complete trees [10]: the general case will be treated in the next section.

3 LOGIC TREES AND RELATIONS BETWEEN LOCAL EXPONENTS

In this section we relate the local dimensions α_j to the local Lyapunov exponents λ_j which have been defined above as the logarithms of the corresponding box-sizes divided by the length n of the orbit. We first recall the expression of the conservation of mass during the time evolution along a trajectory labelled by S_n, for a hyperbolic system [13]. Consider a rectangle with sides of lengths $(\varepsilon_1, \varepsilon_2)$, centred round the initial point $\vec{x}$ of the trajectory, which evolves, in n iterations, into a rectangle of side-lengths $\varepsilon_1' \equiv \varepsilon_1 e^{n\lambda_1(S_n)}$ and $\varepsilon_2' \equiv \varepsilon_2 e^{n\lambda_2(S_n)}$. Since the total mass in the rectangle is conserved by the map (invertibility condition) and scales as $\varepsilon_1^{\alpha_1(S_n)} \cdot \varepsilon_2^{\alpha_2(S_n)}$ (transversality condition between stable and unstable manifold of the orbit), it is also equal to $\varepsilon_1'^{\alpha_1(S_n)} \cdot \varepsilon_2'^{\alpha_2(S_n)}$. Substituting the expressions for the primed lengths, we obtain the local Kaplan-Yorke relation [14,4]

$$\alpha_1(S_n)\lambda_1(S_n) + \alpha_2(S_n)\lambda_2(S_n) = 0. \qquad (9)$$

This equation, of course, does not yield both values $\alpha_1(S_n)$ and $\alpha_2(S_n)$ but only fixes, e.g., the ratio α_1/α_2. Due to the hyperbolicity of the system, however, the invariant measure along the expanding direction is non-singular (the distribution is even uniform, for the baker map), and $\alpha_1(S_n) = 1$, for any S_n, as seen above. Instead of using this "external information", it would be desirable to obtain the values of all partial dimensions (possibly in closed form) from the Lyapunov exponents.

The baker transformation is the prototype of a hyperbolic dynamical system with the specially useful characteristic that the dynamics of the y-variable does not depend on that of x. Therefore, the partial local dimensions $\alpha_1(\vec{x})$ along the expanding direction (corresponding to y) must depend on the dynamical law for y alone. This, of course, reduces the problem to the study of a $1-d$ (noninvertible) map: the Bernoulli shift. The addition of an equation for x renders the map two-dimensional and invertible: this introduces the partial dimensions α_2, which can be readily evaluated from the knowledge of the α_1s, through the Kaplan-Yorke relation. However, the α_1s must be first computed independently, from the $1-d$ Bernoulli map. This discussion applies to all hyperbolic systems, due to the transversality between stable and unstable manifolds. It is therefore clear that any relation yielding α_1 cannot contain quantities depending on the properties of the stable manifold (e.g., α_2 or λ_2), at variance with the results of ref. [8]. In the nonhyperbolic case, instead, α_1 and α_2 loose their individuality near the homoclinic tangencies (stable and unstable directions exchange their role)[15], and a "global" notion of dimension has to be used[16], or the interpretation of α_1 and α_2 must be changed.

Let us now discuss whether it is possible to derive, as proposed in ref. [8], a closed set of relations for α_1, which takes into account the hierarchical structure of the map (also for parameter values yielding a strange repeller). To this end, we recall the relation linking the probability of a symbol sequence to the corresponding local dimension and Lyapunov exponent[4]:

$$P(S_n) \sim e^{-n\alpha_1(S_n)\lambda_1(S_n)} \quad \text{for } n \to \infty. \tag{10}$$

In fact, considering again the case of the baker map, a thin horizontal strip (see fig.1a) of height $e^{-n\lambda_1(S_n)}$, around some point $\bar{x}$ of the S_n-orbit, contains all and only the points that share the same symbol sequence S_n for the next n iterates. The probability $P(S_n)$ is, hence, given by the mass in such a strip, i.e., by the (infinitesimal) size along the expanding direction raised to the corresponding partial dimension (eq.10). The contracting direction is, of course, irrelevant: the same result, indeed, can be derived for $1-d$ maps. In higher-dimensional cases, the expression $\alpha_1\lambda_1$ must be substituted by $\sum_{j=1}^{j^+} \alpha_j\lambda_j$, where j^+ is the number of positive (local) Lyapunov exponents[10]. For hyperbolic systems, since the bounded variation principle holds[17], the Lyapunov exponent in eq.(10) can be computed by iterating any point in the initial strip for n steps: the result will be independent of the initial condition, in the limit $n \to \infty$. In generic systems, however, the local Lyapunov exponent in eq.(10) should be defined carefully. Given a finite trajectory of length n, a local exponent λ might be identified with the effective Lyapunov exponent[4], or with the exponent of the n-periodic orbit with the same symbolic sequence as the orbit under investigation[8,18] or, finally, with the Lyapunov exponent of any other trajectory which shares the same symbolic sequence in that time interval. For the correct application of eq.(10), λ must be defined in terms of the logarithm of the box-length in the expanding direction[4]. Alternatively, it has also been suggested to substitute the effective Lyapunov exponent of eq.(10) with that of the periodic orbit with the same symbolic sequence[8].

Although eq.(10) might be employed to determine the value of α once the probability $P(S_n)$ and the Lyapunov exponent $\lambda(S_n)$ have been evaluated, the achievable precision is not high, due to the influence (at low n) of unknown prefactors and to the statistical fluctuations. Therefore, it is desirable to introduce additional relations among the probabilities $P(S_n)$, organised in a hierarchical way, in order to improve the convergence and estimate the errors as a function of n.

To this end, following the approach of ref. [5], we consider an input symbol sequence (2) and we organise hierarchically all its subsequences S_n by creating a "logic" tree which contains all relevant parental relations. In general, the levels of the hierarchy are allowed to have infinite population: as the subsequence length n increases, the levels are gradually filled. The procedure is such that the degree of "incompleteness" is quantitatively measurable at each step and is a relevant intrinsic property of the system under investigation. The approach is authentically hierarchical because the degree of understanding about the system which one has gained at a certain step of the procedure is ideally measured by making predictions on the composition of the next hierarchical level of the tree. These predictions are then compared with the "actual" structure of the tree (i.e., with what is obtained by letting n go to infinity). The information gain in this organisation process is taken as a measure of the system's complexity[5]; if there exists an $n_0 < +\infty$ such that for $n > n_0$ all predictions match, the system is simple; if new surprises are encountered at each step, it is complex.

In the following, the generation mechanism of the symbolic string is assumed to be sequential (i.e., the sequence is extended by adding symbols to the end of it). This is the case, for instance, of the output of a chaotic dynamical system. For the construction of the tree, we first compute the probabilities $P(S_n)$ of all subsequences S_n (also called orbits) with length $n \in [1, n_{max}]$, as one usually does in evaluating the metric entropy $K_1 \equiv \lim_{n \to \infty} - \sum P(S_n) \ln P(S_n)/n$. We then classify the sequences in the as follows. A string S_n is called **allowed** if its probability is non-zero. A string S_n is called **periodic** if the probability of its longest available iteration (with length $< n_{max}$) is non-zero (for instance, if we are analysing the system with "resolution" $n_{max} = 7$, and we have to decide whether the sequence 011 is periodic or not, we have to measure P(0110110)). In a dynamical system, the latter sequences correspond to unstable periodic orbits of order $n \in [1, n_{max}]$. If a sequence is not periodic, some **rule** exists which prevents it from repeating itself indefinitely many times. In the hierarchical approach, both periodic and non-periodic orbits will play a role, so that the analysis can also be performed when no periodic sequences exist as, e.g., at the accumulation point of the period-doubling cascade[5]. Periodic sequences (i.e., allowed without any restrictions) will be actually allocated on the tree, while non-periodic ones (called **phantoms**) will be only "virtually" placed on it. Symbolic strings that cannot be decomposed in substrings which are allowed and periodic are called **primitives**. The method is based on the reconstruction of the original symbolic signal as a combination of primitive elements.

Any application of the procedure is obviously carried out by considering a finite sequence $S_N = s_1 s_2 \cdots s_N$ (with, typically $N \approx 10^8$); the tree is then constructed by considering all subsequences S_n in order of increasing length $n = 1, \ldots, n_{max}$ (with, e.g., $n_{max} \leq 22$), together with their probabilities (frequencies). Beginning with $n = 1$, periodic sequences (if any) are allocated at the level $l = 1$ of the tree, whereas phantoms (if any) are virtually placed at the same level (see figure 1 in ref.[5]). Phantoms will reappear indeed, but only as a part of longer strings containing one or more primitives. Once all sequences of length n have been considered (for $n = 1$, all elementary symbolic items), one is in a position to make a prediction about the orbits of length $n+1$, which are constructed by joining a **head** H, consisting of any allocated sequence or phantom, with a **tail** T, chosen from the primitives only. The signal is reconstructed as a combination of primitive blocks of symbols, which appear randomly in time, according to the measured transition probabilities[19]. The primitives are the elementary components of all sequences generated by the system. Phantom-phantom combinations are not considered, since they are by construction iterations of non-periodic sequences which may not exist at higher n. For a generic n, one constructs in this way sequences S_n composed of a head of length m ($m < n$) and a tail of length $n - m$. Predicted sequences may exist or not; the latter is actually the interesting case, called a **surprise**. By considering surprises, new understanding about the system is obtained, since additional rules governing it are learned. Notice that a newly formed sequence S_n can be discarded a priori if it contains a forbidden subsequence (which is known, because all strings of length up to $n - 1$ have already been examined). Predicted orbits which actually exist are classified and allocated on the tree. They may be (i) primitives (i.e. indecomposable in periodic substrings) and therefore placed at level $l = 1$ as a part of the system's **dictionary**; or (ii) periodic HT-combinations or (iii) phantoms; in all cases they will be allocated on the tree at a level l corresponding to the level of the head increased by 1. In general, as $n \to \infty$, new elements are added to the dictionary, while the lower levels are gradually filled, though remaining incomplete for any finite n. Notice that a finite number of primitives is found, in general, also when no Markov partition exists. The definitions of metric and topological complexity based on the hitherto described hierarchical approach are given in ref.[5], together with some examples.

Once a level is complete (which happens, in the general case, only asymptotically) the sum of the probabilities of the sequences belonging to it always equals 1[5]:

$$\sum_{j=1}^{\nu(l)} P(S_{n_j}^{(j)}) = 1 \ , \ \ (l = 1, 2, \ldots) \ , \tag{11}$$

where $\nu(l)$ is the asymptotic number of sequences allocated at level l. The finite-n deviation from completeness of level l is measured by

$$\Delta(l, n) \equiv 1 - \sum_{j=1}^{\nu(l,n)} P(S_{n_j}^{(j)}) \ , \tag{12}$$

where $\nu(l, n)$ is the number of sequences allocated at level l, after consideration of all orbits up to length n. The asymptotic behaviour of the deviation $\Delta(l, n)$ with n is described by

$$\Delta(l, n) \sim e^{-n/\sigma_l} \ , \tag{13}$$

where the non-negative parameter σ_l measures the **stiffness** of level l. If the mass becomes exactly 1 at some finite value $n = n_0$, the stiffness is 0. Systems for which the levels are filled very slowly have a large stiffness.

Relations (11), which are valid separately for each level l, can be rewritten as

$$\sum_{j=1}^{\nu(l)} \exp\left[-n_j \alpha_1(S_{n_j}^{(j)}) \lambda_1(S_{n_j}^{(j)})\right] = 1 \ , \ \ (l = 1, 2, \ldots) \ , \tag{14}$$

where eq.(10) has been used. Each of these l independent equations (which approximate the exact relations (11)) is not sufficient to determine the dimensions, since the ($\nu(l)$) unknowns outnumber the equations. However, it can be used to improve the estimate of the α_1s at a given level obtained directly from eq.(10), because it imposes an additional bound to the fluctuations of the local dimensions. Notice the resemblance of eq.(14) with the self-similarity relation

$$r_1^{d_2(0)} + r_2^{d_2(0)} = 1 \ ,$$

which yields the box dimension $d_2(q = 0)$ [13] along the contracting direction of the baker map (1). We can now write the required hierarchy of equations which relate local dimensions α_1 to Lyapunov exponents λ_1 by considering the parental relations among symbolic orbits allocated on the tree. Given the probabilities $P(H)$ and $P(T)$, the probability of the daughter sequence HT is predicted to be approximately equal to $P_0(HT) = P(H)P(T)$ [5]. Since this assumption is, in general, very well verified also for non-hyperbolic systems, we can exploit it, together with (10), to introduce additional constraints on the local exponents α_1 and λ_1:

$$n_{HT}\alpha_1(HT)\lambda_1(HT) = n_H \alpha_1(H)\lambda_1(H) + n_T \alpha_1(T)\lambda_1(T) \ , \tag{15}$$

where n_S is the length of sequence S. Notice that the system consisting of eqs (14) and (15) represents an open hierarchy, since the determination of $\alpha_1(H)$, for instance, requires consideration of the sequence HT, which belongs to the next level. However, it is possible to start an iterative method by assigning, as an input, $\nu(1) - 1$ α_1-values (generally equal to 1) at the top of the tree and by generating all others through a cascade process, using eq.(15). This procedure is exact for the baker map and for symmetric one-hump maps at the boundary-crisis point [20] (with arbitrary order of the maximum), and yields accurate results for asymmetric maps [21]. In generic cases, errors introduced together with the input values propagate undisturbed downwards along the tree, under the effect of eq.(15). This undesired effect can be counteracted by using eq.(14) at every intermediate step, in conjunction with relation (10) which, as mentioned above, yields α_1 from λ_1, apart from the effect of the prefactor. Moreover, the estimate of the value $P_0(HT)$ can be improved when necessary: this introduces additional terms in eq.(15).

The discussion of the previous paragraphs also applies to the more general case of a strange repeller ($\alpha_1(S) < 1$). We consider the tent map

$$y_{n+1} = \begin{cases} y_n/p_1 & \text{if } y_n \le p_1/(p_1 + p_2) , \\ \\ (1 - y_n)/p_2 & \text{if } y_n > p_1/(p_1 + p_2) , \end{cases} \tag{16}$$

with $p_1 + p_2 < 1$. The generating partition is defined by $y = p_1/(p_1 + p_2)$. All points belonging to the two intervals $[0, p_1]$ and $[1 - p_2, 1]$ map onto the unit interval, while the central part escapes from it in one step. At the n-th iteration, a number 2^n of segments of width $p_1^k p_2^{n-k}$ is left (k is the number of iterations spent in the interval $[0, p_1]$). Asymptotically, the only non-escaping points belong to a non-uniform Cantor set which can also be obtained by backward iteration of the map (every point has two pre-images). At variance with strange attractors, in this case there is a certain arbitrariness in the choice of the invariant measure. The natural measure $\rho(y)$ is obtained by weighing the preimages proportionally to their own width (p_1 or p_2), at each step. More precisely, $\rho(y)$ is defined as the fixed-point solution of the modified Frobenius-Perron equation [22,23]

$$\rho_{n+1}(y') = e^\beta \sum_{y \in F^{-1}(y')} \frac{\rho_n(y)}{|F'(y)|} , \tag{17}$$

where $F'(y)$ is the derivative of the map. A finite limit $\rho(y) > 0$ only exists for $\beta = -\ln(p_1 + p_2) \ge 0$. The measure $\rho(y)$ can be seen as the limit, for $n \to \infty$, of piecewise constant densities [24]

$$\rho_n(y) = \begin{cases} e^{n\beta} & \text{if } F^n(y) \in [0, 1] , \\ \\ 0 & \text{otherwise.} \end{cases} \tag{18}$$

The exponent β is called escape rate and is related to the fraction f^n of mass which has not yet escaped a neighbourhood of the repeller after n steps: $f^n = e^{-n\beta}$. In eq.(17), the term e^β is a global compensation factor for the outflow of probability from the repeller ($\beta = 0$, for attractors). An interval characterised by a symbol sequence S_n (with k symbols equal to 0) contains the mass

$$P(S_n) = p_1^k p_2^{n-k} e^{n\beta} \equiv e^{-n[\lambda(S_n) - \beta]} \tag{19}$$

and has the length

$$\varepsilon(S_n) = p_1^k p_2^{n-k} \equiv e^{-n\lambda(S_n)} , \tag{20}$$

where $\lambda(S_n)$ is the Lyapunov exponent of the orbit S_n. The local dimension $\alpha(S_n)$ is, then, given by

$$\alpha(S_n) \equiv \ln P(S_n)/\ln \varepsilon(S_n) = 1 - \beta/\lambda(S_n) . \tag{21}$$

For $\beta = 0$ (strange attractor), we recover the value $\alpha = 1$. It is now easy to verify that eq.(14) is also valid for natural measures on repellers, and that it is just the modified Frobenius-Perron equation (17). In fact, due to the relation among $\alpha(S_n)$, β and $\lambda(S_n)$ (eq.21), eq.(17) can be rewritten as

$$P(H) \sim e^{-n_H \alpha(H)\lambda(H)} = \sum_T P(HT) \sim \sum_T e^{-n_{HT}\alpha(HT)\lambda(HT)} , \tag{22}$$

where the sum is over all allowed tails T of the initial sequence H. In the case of map (16), eq.(22) yields the correct result $p_1 + p_2 = e^{-\beta}$. Notice that probabilities factorise also in this case, if the natural measure (18) is chosen (see eq.(19) giving the probability of any orbit).

4   CLOSED-FORM RELATIONS BETWEEN DYNAMICAL AND GEOMETRICAL EXPONENTS

As discussed in section 2, it is not clear in general how to relate static to dynamic quantities in an unambiguous way. Although it is known that the local metric entropy κ can be written as a product of the local dimension α times a local Lyapunov exponent λ, the definition of the latter quantity needs a careful consideration in the case of nonhyperbolic systems.

In order to determine the local dimensions α_i ($i = 1, 2$), it has been recently proposed to consider three different periodic orbits [8] and not just one (as in the local Kaplan-Yorke relation (9)). In particular, let us consider some refinement of the generating partition of the baker map (see fig.1) and label a given periodic orbit with the symbol sequence $S_n \equiv HT$, where H stands for a left subsequence of S_n and T represents the remaining part of S_n. For example, a period-two orbit exists, the points of which belong to the elements labelled by $S_2 \equiv HT \equiv 01$ (see fig.1b) and by $S_2' \equiv (HT)' \equiv 10$: then, $H \equiv 0$ and $T \equiv 1$ (and vice versa for H' and T').

In order to find an additional equation to combine with the Kaplan-Yorke relation (9), the following procedure has been adopted in ref. [8] : the probability $P(HT)$ to observe a given symbol-sequence HT is computed in two independent ways, which are then compared. The first expression for $P(HT)$ is given in terms of the first Lyapunov exponent $\lambda_1(HT)$ of the periodic orbit labelled by the same sequence HT [13]:

$$P(HT) \sim e^{-n_{HT}\lambda_1(HT)}, \tag{23}$$

where the sequence HT is supposed to be of length $n_{HT} \gg 1$. The second expression is obtained by noticing that the sides of the (rectangular) element HT (of length ε_1 and ε_2, along the unstable and stable direction, respectively) are given by the Lyapunov exponents of the orbits labelled by H and T, in the following way (see fig.1b, for orbits of period 1 and 2):

$$\varepsilon_1(HT) \sim e^{-n_H\lambda_1(H)},$$

$$\varepsilon_2(HT) \sim e^{n_T\lambda_2(T)}. \tag{24}$$

This relation is exact, in the case of the baker map, even for short sequences HT ($n_{HT} = 2$, in fig.1b). Then, the probability $P(HT)$ is estimated as [8]

$$P(HT) \sim \varepsilon_1(HT)^{\alpha_1(HT)} \cdot \varepsilon_2(HT)^{\alpha_2(HT)}, \tag{25}$$

for $\varepsilon_1(HT), \varepsilon_2(HT) \to 0$. Notice that the lengths of the sides of the rectangle (24) are raised to the partial dimensions corresponding to the orbit HT, even though they are estimated from the two orbits H and T. By comparing eqs (23) and (25), and using eq.(24), we obtain the desired additional relation for the partial dimensions [8]:

$$- n_H\lambda_1(H)\alpha_1(HT) + n_T\lambda_2(T)\alpha_2(HT) = -n_{HT}\lambda_1(HT). \tag{26}$$

However, it is easy to verify that this relation is not satisfied by the baker map. In fact, in that case, $n_{HT}\lambda_1(HT) = n_H\lambda_1(H) + n_T\lambda_1(T)$, for any H and T (the same holds for the second Lyapunov exponent), and $\alpha_1(HT) = 1$. By substituting these relations in eq.(26), we find

$$\alpha_2(HT) = -\lambda_1(T)/\lambda_2(T), \tag{27}$$

in contradiction with the Kaplan-Yorke relation (9) (with $S_n = HT$). Similarly, for the cyclically permuted sequence TH, repeating the whole procedure we have:

$$- n_T\lambda_1(T)\alpha_1(TH) + n_H\lambda_2(H)\alpha_2(TH) = -n_{HT}\lambda_1(TH), \tag{28}$$

where, of course, partial dimensions and Lyapunov exponents with argument TH are the same as those with argument HT, since the periodic orbit is the same. Equation (28) is equivalent to

$$\alpha_2(HT) = -\lambda_1(H)/\lambda_2(H), \tag{29}$$

again in contradiction with eq.(9). Notice that eqs (27) and (29) would be correct if the arguments of α_2 were T and H, respectively, and not HT.

This inconsistency derives from assumption (25) (also the labelling of ε_1, ε_2 in eq.(24) is incorrect). In fact, the HT-orbit (period two in figure 1b) has one point in a box of size $e^{-\lambda_1(H)} \cdot e^{\lambda_2(T)}$ and one in a box of size $e^{-\lambda_1(T)} \cdot e^{\lambda_2(H)}$; therefore, a cyclic permutation of the symbols ($HT \to TH$) should leave the probability invariant ($P(HT) = P(TH)$). However, this is not true for eq.(25), after substitution of eq.(24); as a consequence, the l.h. sides of eqs (26) and (28) are different, in general. Notice that complete agreement would have been found if, in the l.h.s. of eqs (26) and (28), each α had the same argument as the corresponding Lyapunov exponent (i.e., $\lambda_1(H)\alpha_1(H)$, $\lambda_2(T)\alpha_2(T)$, etc.). In that case, because of the Kaplan-Yorke relation for orbits H and T, eqs (26) and (28) would be completely equivalent. Furthermore, as pointed out in the introduction, any equation for α_1 should only involve the first Lyapunov exponent λ_1 (possibly of various periodic orbits) and not λ_2, unless the latter appears in the product $\lambda_2 \alpha_2$ (which can be substituted by $-\alpha_1 \lambda_1$, making use of the Kaplan-Yorke relation). Finally, eq.(23) is only true for hyperbolic attracting maps and should be substituted by eq.(10) in the general case ($\alpha_1 \neq 1$).

5 APPLICATION TO EXPERIMENTAL SIGNALS

In this section we discuss possible applications of symbolic-dynamical techniques to the characterisation of experimental signals, having as a goal obtaining dimension and Lyapunov spectra more accurately than by employing standard statistical methods based on either constant-size or constant-mass partitions[25]. In fact, optimal convergence is achieved, in general, by using partitions whose elements have variable size and mass, and that cannot be obtained from a constant-n preimage of the generating partition. With the procedure exposed in section 3, the partition is automatically refined in correspondence to the homoclinic tangencies (or to the images of the maxima in one-dimensional maps)[5]. The method, moreover, allows one to define new indicators (complexity; scaling functions for probabilities, for characteristic lengths, for multipliers etc.) that describe memory effects over several (possibly infinitely many) levels of the logic tree[21].

The application of symbolic-dynamical techniques to experimental systems is extremely problematic because it requires the preliminary identification of a generating partition in a suitable Poincaré map. The determination of a generating partition, however, is a very difficult task even when the equations of motion are given, since no general procedure is known[26]. Moreover, the phase-space of experimental signals must be reconstructed in E-dimensional embedding spaces ($E > 2D + 1$, where D is the attractor's dimension), thus giving rise to problems of statistics and precision which add themselves to unavoidable noise and discretisation effects. Since it is not known how to localise homoclinic tangencies[26] in such spaces, it has been proposed to extract all periodic orbits up to a given length n_{max} and to construct a partition which assigns to them different symbolic itineraries[8]. This project, however, has not been realised but in the trivial case of data generated by the Hénon map, for which time is discrete and a two-dimensional reconstruction is exact[8]. The lengths of periodic orbits in a continuous-time system are, in general, different from one another and unknown; any fixed-point search method (e.g., a Newton's method) must be integrated by a procedure which yields the proper periodicity by determining accurately the intersection of the trajectories with a suitable local Poincaré surface[27]. A certain degree of arbitrariness is also introduced by the criterion itself which is used to identify periodic orbits from returns in finite neighbourhoods of the initial point on the local Poincaré surface. Precision problems are amplified by the repulsiveness of the neighbourhood of the trajectories and by the presence of noise. Furthermore, all orbits must be distinguished from one another (by introducing additional adjustable distance-parameters) and regrouped according to their length. A global Poincaré surface must then be chosen in such a way that it intersects all "period-1" orbits, so that they yield the same topological return time (one step). Successively, all period-1 points must then be numbered with integers and their neighbourhoods taken as elements of the partition. All points belonging to longer cycles are then labelled according to the neighbourhood to which they belong. Coinciding sequences may be found since the above defined elements do not yet constitute a generating partition. The generating partition, in fact, has elements which do not contain simple fixed points but only points belonging to longer cycles as, for instance, the segment [0,1/2] (labelled by 0) in the above discussed example of the roof map. At this point, the symbolic dynamics of the system can be easily extracted. The dimension spectrum $f(\alpha)$ and the spectrum of Lyapunov exponents can be finally obtained from the knowledge of the local quantities associated with the periodic orbits. For the computation of the dimensions one applies the iterative approach which was described in the preceeding section with reference to two-dimensional maps. Notice that the estimation of the Lyapunov exponents is not straightforward because of the existence of spurious values, corresponding to non-physical directions [13]. The major improvements introduced by this approach are expected in the estimation of generalised entropies.

ACKNOWLEDGEMENTS

We especially thank I. Procaccia for many illuminating discussions. We benefitted from discussions with D. Auerbach, P. Cvitanović, J.-P. Eckmann and A. Politi. Exchange of electronic mail with G.H. Gunaratne has been particularly helpful. This work has been partly supported by a grant of the Von Humboldt Foundation and by the Swiss National Science Foundation.

REFERENCES

[1] P. Grassberger and I. Procaccia, Physica **13D**, 34 (1984).

[2] P. Grassberger, in *Chaos in Astrophysics*, J. Perdang et al. Eds. (Reidl, Dortrecht, 1985).

[3] R. Badii and A. Politi, Phys.Rev. **A35**, 1288 (1987).

[4] P. Grassberger, R. Badii and A. Politi, J.Stat.Phys. **51**, 135 (1988).

[5] R. Badii, *Unfolding Complexity in Nonlinear Dynamical Systems*, this issue; *Quantitative Characterization of Complexity and Predictability*, submitted for publication.

[6] M.J. Feigenbaum, M.H. Jensen and I. Procaccia, Phys.Rev.Lett. **57**, 1503 (1986).

[7] T.C. Halsey, M.H. Jensen, L.P. Kadanoff, I. Procaccia and B. Shraiman, Phys.Rev. **A33**, 1141 (1986).

[8] G.H. Gunaratne and I. Procaccia, Phys.Rev.Lett. **59**, 1377 (1987); G.H. Gunaratne, M.H. Jensen and I. Procaccia, Nonlinearity 1, 157 (1988); P. Cvitanović, G.H. Gunaratne and I. Procaccia, Phys.Rev. **A38**, 1503 (1988); D. Auerbach, P. Cvitanović, J.P. Eckmann, G.H. Gunaratne and I. Procaccia, Phys.Rev.Lett. **58**, 2387 (1987).

[9] J.D. Farmer, E. Ott and J.A. Yorke, Physica **7D**, 153 (1983).

[10] R. Badii, Riv. Nuovo Cim. **12**, N° 3, (1989).

[11] V.N. Shtern, Dokl.Akad.Nauk. SSSR **270**, 582 (1983).

[12] P. Grassberger, Phys.Lett. **107A**, 101 (1985).

[13] J.P. Eckmann and D. Ruelle, Rev.Mod.Phys. **57**, 617 (1985).

[14] J.L. Kaplan and J.A. Yorke, Lect.Not.Math. **13**, 730 (1979); P. Fredrickson, J.L. Kaplan, E.D. Yorke and J.A. Yorke, J.Diff.Eqs. **49**, 185 (1983).

[15] S. Newhouse, in *Chaotic Behaviour of Deterministic Systems*, G. Ioos, R.H.G. Helleman and R. Stora Eds., North-Holland (1983).

[16] A. Politi, R. Badii and P. Grassberger, J.Phys. **A21**, L763 (1988).

[17] D. Rand, *The Singularity Spectrum for Hyperbolic Cantor Sets and Attractors*, University of Arizona, preprint (1986);
T. Bohr and D. Rand, Physica **25D**, 387 (1987).

[18] C. Grebogi, E. Ott and J.A. Yorke, Phys.Rev. **A37**, 1711 (1988).

[19] M.A. Sepúlveda, R. Badii and E. Pollak, unpublished.

[20] C. Grebogi, E. Ott and J.A. Yorke, Physica **7D**, 181 (1983).

[21] R. Badii, unpublished.

[22] G. Pianigiani and J.A. Yorke, Trans.Am.Math.Soc. **252**, 351 (1979).

[23] T. Tél, Phys.Lett. **119A**, 65 (1986).

[24] H. Kantz and P. Grassberger, Physica **17D**, 75 (1985).

[25] *Dimensions and Entropies in Chaotic Systems*, G. Mayer-Kress Ed., *Springer Series in Synergetics*, **32** (Springer, Berlin, 1986).

[26] P. Grassberger and H. Kantz, Phys.Lett. **113A**, 235 (1985).

[27] G. Broggi and R. Badii, unpublished.

EXPERIMENTAL STUDY OF THE MULTIFRACTAL STRUCTURE

OF THE QUASIPERIODIC SET

Dwight Barkley

Applied Mathematics
California Institute of Technology
Pasadena, California 91125

Andrew Cumming

AT&T Bell Laboratories
Murray Hill, New Jersey 07974

A periodically driven relaxation-oscillator circuit is used to experimentally study
the multifractal structure of the quasiperiodic set at the transition to chaos. Using
the thermodynamic formalism to quantify the complex scaling of this set, we are
able to efficiently compare experimental results with numerical results obtained
from the sine circle map.

Fractal sets arise in a variety of physical settings such as fluid turbulence,[1,2] chaotic
dynamics,[3,4] and fractal growth processes.[5] Unlike the simple triadic cantor set commonly
used to illustrate fractal structure, the sets which arise in nature are, in general, not self-
similar and cannot by characterized by a single number such as the fractal dimension,
D_0. Such non-self-similar sets are called multifractals.[1-7] To quantify the complexity of
multifractals one typically needs to specify an infinity of numbers, such as the generalized
dimensions,[3,8] D_q, labeled by the continuous subscript q, or the spectrum of scaling indices,[4]
$f(\alpha)$, where α takes on a range of values.

The thermodynamic formalism[4,9,10] provides a convenient way of quantifying the complexity of multifractals and allows for the effective comparison of theory and experiment.[11]
The analogy to thermodynamics arises, in part, because different descriptions are possible
for multifractals, and these descriptions are related via Legendre transformations in the same
way as are variables in thermodynamics. Moreover, it has recently been shown[9,10,12] that
there is a microscopic theory, i.e. a statistical mechanics, for certain multifractals, and this
has put the thermodynamic analogy on firm foundation.

The multifractal set of interest here is the quasiperiodic set at the transition to chaos,
that is, the complement to the set of mode lockings on the critical line.[13,14] This set is dominated by an infinity of scalings which are presumably universal for generic systems undergoing
the transition from quasiperiodicity to chaos.[15] To experimentally test this universality we
have measured the quasiperiodic set for a periodically driven operational-amplifier oscillator.
Using the thermodynamic formalism, we compare experiment with results obtained from the
sine circle map.

Measures of Complexity and Chaos
Edited by N.B. Abraham *et al.*
Plenum Press, New York

Before discussing the experimental details, we recall the thermodynamic treatment.[4,10,14] Consider the covering of a fractal measure by intervals of length l_i. Each interval contains a weight or probability p_i. Then define the partition function by

$$\Gamma(q,\tau) = \sum_{i=1}^{N} \frac{p_i^q}{l_i^\tau}. \tag{1}$$

In principle one takes the limit $N \to \infty$; in practice, we consider only finite N. Setting $\Gamma(q,\tau) = 1$ we have an implicit equation for either $\tau(q)$ or $q(\tau)$. If we consider τ as a function of q, then we can obtain the generalized dimensions $D_q = \tau(q)/(q-1)$, or by Legendre transformation of $\tau(q)$, the spectrum of singularities $f(\alpha) = q\alpha - \tau(q)$, where $\alpha = d\tau/dq$. Alternatively, if we think of q as a function of τ, we can consider $s(\mu) = q(\tau) - \tau\mu$, where $\mu(\tau) = dq/d\tau$. We shall follow this latter choice here. In the thermodynamic analogy, $q(\tau)$ plays the role of free energy, $\mu(\tau)$ the role of entropy, and $s(\mu)$ the role of internal energy.

The starting point for the thermodynamics of the (critical) quasiperiodic set is a collection of $N+1$ mode-locked intervals $\{P_i/Q_i\}$, labeled by their winding number. This set of intervals is thought of as a set of holes in the critical line on which the quasiperiodic set does not live. The covering, $\{l_i\}$, for the quasiperiodic set is the complement to the set of holes, i.e. interval l_i is the gap between neighboring mode lockings P_i/Q_i and P_{i+1}/Q_{i+1}.

Several different measures have been put forth for the quasiperiodic set,[4,14] as represented by different definitions of the probabilities, p_i. Here we follow Halsey et al.[4] and take the p_i to be the change in (dressed) winding number across covering interval l_i (normalized such that $\sum p_i = 1$). This is the most natural measure from an experimental point of view.

The experimental apparatus that we have used to investigate the quasiperiodic set is an operational-amplifier relaxation oscillator which is driven by a sine wave whose frequency and amplitude can be varied. Complete details can be found in Refs. 16 and 17. We first experimentally determined the critical line (drive voltage as a function of drive frequency) for the transition to chaos (see Fig. 1 of Ref. 16). We then scanned the critical line by making 5 Hz steps in the drive frequency and adjusting the drive amplitude so as to remain on the critical line. At each step the period and winding number of the driven oscillator were obtained by a computer. From these data we were able to identify 237 mode-locked intervals on the critical line.

From the experimentally measured mode lockings we can compute the l_i's and p_i's. Then from definition (1) we can, in principle, find $q(\tau)$. However, $q(\tau)$ computed directly from (1) converges slowly as $N \to \infty$. We improve the convergence by considering the ratio of partition functions:[4,14]

$$\Gamma(q,\tau)/\Gamma'(q,\tau) = 1, \tag{2}$$

where primed and unprimed functions refer to a coarse and fine covering of the quasiperiodic set, respectively. The coarse and fine coverings are determined from mode lockings whose periods Q_i do not exceed maximum periods Q' and Q. Numerical studies of circle maps[14] show that $Q' \simeq [Q/\sqrt{2}]$ gives good convergence. (Square brackets denote integer part.) We have found that this also works well for the experimental data. Given Q and Q' we can find the function $q(\tau)$ for the experimental data directly from (1) and (2).

The largest uncertainty in the experimental results is due to the variation in $q(\tau)$ with the choice of Q and Q'. We have found that the uncertainty due to finite-precision measurements is negligible when compared with the variation due to the choice of Q and Q'. To quantify the experimental uncertainty we have computed mean values of $q(\tau)$ and $\mu(\tau)$ for $42 \le Q \le 48$ with Q' in the range $[Q/\sqrt{2}] \pm 2$, and have based the error bars on the variation

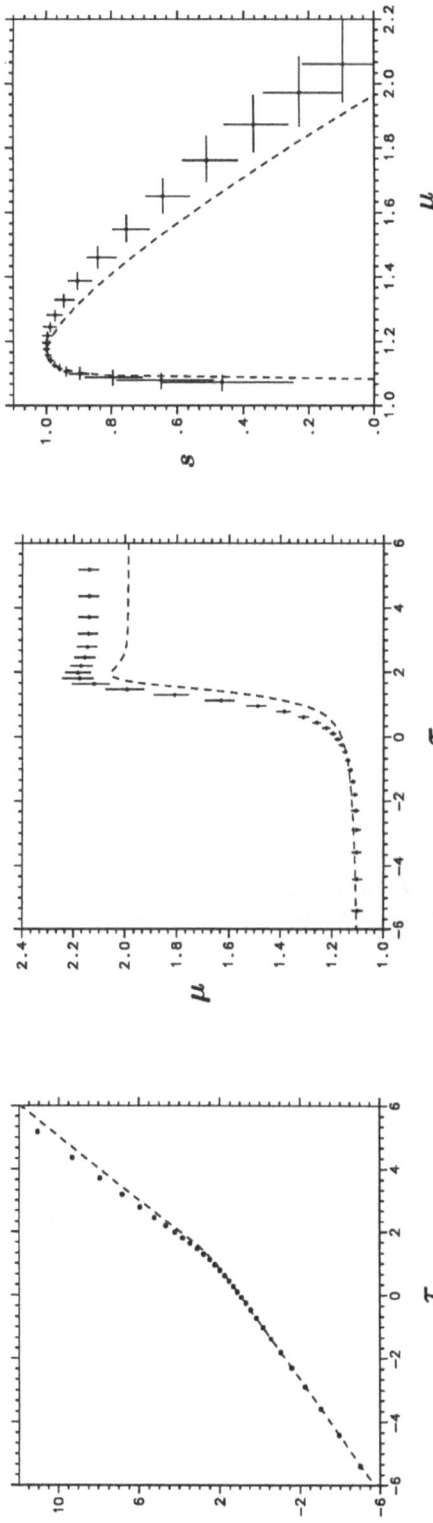

Figure 1. The free energy, $q(\tau)$, entropy, $\mu(\tau) = dq/d\tau$, and internal energy, $s(\mu) = q(\tau) - \tau\mu$, from experiment and the sine circle map (dashed). The results for the sine circle map were obtained by numerically computing all mode lockings with $Q_i \leq 180$ and using (1) and (2) with $Q = 180$ and $Q' = 127$. The error bars for $q(\tau)$ are not shown as they are barely distinguishable from the points. The agreement between experiment and circle map is good throughout the region of negative τ and strongly supports universal scaling of the critical quasiperiodic set. The disagreement between experiment and circle map at large τ is the result of finite-N effects (see text).

of these quantities over this range of Q and Q'. The uncertainties in $s(\mu)$ have been obtained by error propagation through the Legendre transformation.

The thermodynamic functions are shown in Fig. 1. The agreement between experiment and theory (sine circle map) is good throughout the region of negative τ and strongly supports universal scaling of the critical quasiperiodic set. Experimentally the entropy, μ, at large τ is greater than for the circle map. This disagreement is undoubtedly the results of finite-N effects and is consistent with the extremely slow convergence of harmonic series within the Farey partition.[14,15] Note that even with $Q = 180$ (corresponding to $N = 4940$), $s(\mu)$ for the circle map does not cross the μ-axis at $\mu = 2$, as it does in the $N \to \infty$ limit.[15]

The rapid change in the slope of the free energy $q(\tau)$ at $\tau \simeq 1.3$ is indicative of a phase transition.[12,14] We do not pursue this here, but note that by examining the "specific heat," $c(\tau) = d^2q/d\tau^2$, we have been able to provide experimental evidence of a first-order phase transition.[18]

We have presented here an experimental study of the multifractal structure of the quasiperiodic set. Using the thermodynamic formalism as a tool to quantify the complex scaling of this set, we have shown that there is good agreement between our experimental results and numerical results from the sine circle map. This agreement provides evidence that the scaling of the quasiperiodic set is universal at the transition from quasiperiodicity to chaos.

ACKNOWLEDGEMENTS

D.B. acknowledges support from ONR grant number N00014-85-K-0205 SRO.

REFERENCES

1. B. B. Mandelbrot, J. Fluid Mech. **62**, 331 (1974).
2. U. Frisch and G. Parisi, in *Turbulence and Predictability in Geophysical Fluid Dynamics and Climate Dynamics*, edited by M. Ghil, R. Benzi, and G. Parisi (North-Holland, Amsterdam, 1985), p. 84.
3. H. G. E. Hentschel and I. Procaccia, Physica **8D**, 435 (1983).
4. T. C. Halsey, M. H. Jensen, L. P. Kadanoff, I. Procaccia, and B. I. Shraiman, Phys. Rev. A **33**, 1141 (1986).
5. T. C. Halsey, P. Meakin, and I. Procaccia, Phys. Rev. Lett. **56**, 854 (1986).
6. D. J. Farmer, Physica **4D**, 366 (1982).
7. R. Benzi, G. Paladin, G. Parisi, and A. Vulpiani, J. Phys. A **17**, 3521 (1984).
8. A. Renyi, *Probability Theory*, (North-Holland, Amsterdam, 1970).
9. M. J. Feigenbaum, M. H. Jensen, and I. Procaccia, Phys. Rev. Lett. **57**, 1503 (1986).
10. M. H. Jensen, L. P. Kadanoff, and I. Procaccia, Phys. Rev. A **36**, 1409 (1987).
11. M. H. Jensen, L. P. Kadanoff, A. Libchaber, I. Procaccia, and J. Stavans, Phys. Rev. Lett. **55**, 2798 (1985).
12. D. Katzen, and I. Procaccia, Phys. Rev. Lett. **58**, 1169 (1987).
13. M. H. Jensen, P. Bak, and T. Bohr, Phys. Rev. Lett. **50**, 1637 (1983).
14. R. Artuso, P. Cvitanović, and B. G. Kenny, Phys. Rev. A **39**, 268 (1989).
15. P. Cvitanović, B. Shraiman, and B. Söderberg, Phys. Scr. **32**, 263 (1985).
16. A. Cumming and P. S. Linsay, Phys. Rev. Lett. **59**, 1633 (1987).
17. A. Cumming, Ph.D. Thesis, Massachusetts Institute of Technology (1988).
18. D. Barkley and A. Cumming (to be published).

STATISTICAL INFERENCE THEORY FOR MEASURES OF

COMPLEXITY IN CHAOS THEORY AND NONLINEAR SCIENCE

W. A. Brock and W. D. Dechert

Dept. of Economics, University of Wisconsin and University of Houston

Madison, WI 53706 and Houston, TX 77004

1 Introduction

Concepts such as the Grassberger and Procaccia (1983a,b) correlation dimension, Kolmogorov entropy, and measures of sensitive dependence on initial conditions are making their way into economics. But they have been rather drastically modified to cope with the special issues that economic applications raise. In this paper we will survey some literature in economics that deals with these issues. In order to save space and concentrate on what we know best we will focus on our own work. Citations will be given to help the reader locate the work of others. First we will take up the question of testing for low dimensional deterministic chaos in economics and finance where the vast bulk of professional belief is centered on nonstationary and stochastic data. Second, in Section 2 we will deal with some of the technical underpinnings of the methodology we discuss in Section 1. Third, in Section 3, we briefly mention applications and take up unresolved issues not covered in Sections 1 and 2.

1.1 On Proving the Existence of Chaos in Economics

To set the stage consider the problem of testing a time series of observations such as stock returns for the presence of deterministic chaos of low dimension. The first issue that we face is that in economic time series such as stock returns there are strong theoretical arguments against the presence of low dimensional deterministic chaos. One of the most famous is the random walk hypothesis. This says that returns on assets that are traded on well organized markets like stock markets must be unpredictable in the short term as well as the long term. Since a deterministic chaos (especially a low dimensional deterministic chaos) is predictable in the short term it can't describe asset returns. Hence the burden of proof is on those economists who believe that stock returns are low dimensional deterministic chaotic. See Brock (1988) for discussion and references.

How does the scientist evaluate claims to have found chaos in economics when reconstruction (H. Swinney (1985)), i.e., "you know it when you see it," is impossible due to data limits? That is the subject we take up here. First, let us agree that the issue of

Measures of Complexity and Chaos
Edited by N.B. Abraham *et al.*
Plenum Press, New York

whether economic data is high dimensional chaos or genuinely stochastic is not possible to resolve with short data sets. If there are a large number of active modes a short time series just can not give enough information to uncover their dynamics. In view of this fundamental undecidability between a large number of active modes and an infinite number of active modes for a finite data set we are going to assume that the real issue is evaluating evidence of low dimensional chaos. We will take "low" to be less than 7 or 8, just to focus the discussion. Turn to the issue of burden of proof.

In any evidentiary proceeding, for example, in the Law, there is the issue of who bears the burden of proof. Due to the beliefs of almost all economists that economic data are nonstationary and stochastic, we are placing the burden of proof on the claim to have found evidence of low dimensional deterministic chaos in economic data. Due to the nonstationarity, shortness, and quality of economic data there are some basic issues that must be faced when claiming to have found evidence of low dimensional deterministic chaos in economic data.

We should say something about the length and quality of economic time series before we proceed. In macroeconomics many series such as Gross National Product are constructed by government statisticians and are available at limited frequencies such as quarterly. So it is obvious that there is a data limitation problem in macroeconomics. What is not obvious is that there is a data limitation problem in finance.

Many people look at the mass of financial data that is available (for example, stock returns are available at tic by tic frequencies) and assert that there is no data limitation problem in economics. What they overlook is a host of subtle issues: (i) at high frequency stock returns dynamics are infected by institutional detail such as movements in the bid/ask spreads of the specialists. Phenomena such as bid/ask bounce create artifacts of temporal dependence in returns that have little to do with the dynamics of the stock price itself. Hence analysts try to wash out some of this artificial dependence by looking at returns over lower frequencies such as weekly or interpreting the high frequency dynamics in the context of some microstructural model. (ii) Over longer periods of time stock returns do not seem to be generated by the same laws of motion. Outside shocks, technical change, and institutional innovations have such an impact on stock returns that some financial analysts believe that you cannot write down any time stationary model of stock returns (Black (1976)). Others such as Fama (1976) are willing to postulate stationarity for periods as long as five years but no longer. Hence, even in finance, we are seriously data limited. No one doubts that the laws of fluid dynamics or chemical reactions will remain stationary across centuries. Almost all economists believe that any sector of the economy is buffeted enough by omitted variables that the law of motion remains stationary only for a fairly short period of time.

A stylized test for low dimensional deterministic chaos might go like this: (i) first estimate a usable notion of attractor dimension such as the correlation dimension (Grassberger and Procaccia (1983a,b), Swinney (1985)) and see if it is low. (ii) estimate a measure of sensitive dependence on initial conditions such as the largest Lyapunov exponent and see if it is positive (Swinney (1985)). (iii) Estimate the Kolmogorov entropy and see if the estimate is consistent with chaos. (iv) Adduce indirect evidence of a deterministic map generating the series under scrutiny by devices such as phase portraits. (v) Reconstruct the underlying map using, for example, Poincaré sections (Swinney (1985)). We take up problems in applying this procedure to economic time series seriatim.

First, if the real data (detrended log linearly or by first differences, with other nonstationarities removed by dummy variables or whatever) is generated by a low order autoregression with near unit roots the estimated dimension of this autoregression may be low over a large range of scaling even though you have an infinite dimensional stochastic process rather than low dimensional deterministic chaos. More precisely suppose that the time series $\{x_t\}_{t=1}^n$ of real numbers is a sample from the stochastic process

$$X_t = \alpha_0 + \alpha_1 X_{t-1} + \cdots + \alpha_n X_{t-k} + \epsilon_t \tag{1.1}$$

where the sequence $\{\epsilon_t\}$ is independently and identically distributed (IID) with mean zero and finite variance. If the roots of the associated polynomial of equation (1.1) are near the boundary of the unit circle, $\{X_t\}$ will be a very persistent process and it will look low dimensional. This problem was pointed out by Brock (1986) and a diagnostic was presented there. The diagnostic was to estimate the correlation dimension of the estimated residuals of a model like equation (1.1) and compare it to the estimated dimension of the original data. As long as the number of lags, k, is small this procedure is a useful diagnostic. Unfortunately this procedure tends to reject chaos too many times even when the data are truly chaotic. This problem caused us to invent a better method which is explained in detail in Section 2 below.

In any event, in economics and finance, there are plausible stochastic alternatives to low dimensional deterministic chaos and these plausible alternatives must be discarded in order to bear the burden of proof necessary to make a convincing argument that low dimensional deterministic chaos is present. After all, it is a commonplace in macro economics that many, if not most, macroeconomic time series have "Granger's typical spectral shape" of a downward sloping spectrum at low frequencies (cf. Sargent (1987) for the received views in macroeconomics). This is consistent with a highly persistent stochastic process, which, in turn, is consistent with low dimension estimates over a wide range of scaling. Somewhat similar problems plague inference using estimates of the largest Lyapunov exponent (Brock and Sayers (1988)). Turn now to a better method than dimension estimation and estimation of largest Lyapunov exponents for detecting noisy chaos in short time series.

The basic idea is to fit models consistent with professional practice to the time series under scrutiny. In order to be specific consider the class of time series models below.

Consider the data generator

$$y_t = f(Y_{t-1}, I_{t-1}) + \sigma(Y_{t-1}, I_{t-1})u_t \tag{1.2}$$

where I_t is an information set available to the statistician at time t, Y_{t-1} denotes past values of y, and $\{u_t\}$ is IID with mean zero and finite variance. The popular GARCH-M (e.g. Bollerslev (1985), and Engle (1982)) class fits into the framework of equation (1.2).

The basic idea is to assume that f and σ can be estimated $\sqrt{n}$ consistently from a sample of length n, (this means that the estimated functions $\hat{f}_n, \hat{\sigma}_n$ converge to the true functions f, σ at a speed of $\sqrt{n}$ as $n \to \infty$) and test

$$u_{t,n} = \frac{y_t - \hat{f}_n}{\hat{\sigma}_n} \tag{1.3}$$

for IID. Under the null model $\{u_{t,n}\}$ converges to $\{u_t\}$. Note that

$$u_{t,n} = \frac{f + \sigma u_t - \hat{f}_n}{\hat{\sigma}_n}$$

$$= \frac{f - \hat{f}_n}{\hat{\sigma}_n} + \frac{\sigma - \hat{\sigma}_n}{\hat{\sigma}_n} u_t + u_t$$

$$\equiv d_{t,n} + u_t \tag{1.4}$$

We are now in a position to use the BDS statistical test for chaos which will be explained in Section 2 below. The test is a test of the null hypothesis of IID against dependent alternatives that has high power against the alternative of chaos. In practice it is applied to the estimated residuals (1.4) of fitted models (1.2). It has a first order asymptotic distribution that is normal with mean zero and variance one under the null of IID. The use of the BDS test of models of the form (1.2) by applying it to the estimated standardized residuals (1.4) is asymptotically justified by

Theorem 1.1 *The first order asymptotic distribution of the BDS statistic is the same if you calculate the statistic at the estimated standardized residuals $\{u_{t,n}\}$ and the true ones $\{u_t\}$.*

Proof: See Brock and Dechert (1988).

In practice many of the models (1.2) are linear in information for conditional means and conditional variances. We then examine the estimated standardized residuals of these models for evidence of "extra short term predictable structure." Looking at standardized residuals is called "looking at prewhitened series" in the trade. Linear prewhitening is the process of passing your series through a linear filter in an attempt to get linear dependence out. Most economists practice variations on linear prewhitening. Our approach is to prewhiten the series with filters like (1.2) that the world's best econometricians believe adequately model the series. Then we test the residuals for the presence of extra short term predictable structure with a statistical test that has very high power against low dimensional deterministic chaos. Blake LeBaron has done many experiments where he passes chaotic series such as 7–dimensional Mackey–Glass through filters like (1.2) and uses tests such as BDS and nonlinear one-step-ahead forecastability on the standardized residuals (1.4). The conclusion is that it is very difficult to "destroy" a true chaos by passing it through models of the form (1.2) that economists like to fit. Based on this observation and fitting models of the form (1.2) to stock returns he concludes that the evidence for low dimensional chaos in stock returns is weak, but, he does conclude that there is evidence for "extra structure," which may be due to left out nonstationarity, in stock returns. See LeBaron (1988a,b,c). This is an important point. Let us put it another way.

If low dimensional chaos is present, even improper prewhitening that adds estimation noise, generates residuals with so much temporal dependence that general non parametric tests for dependence such as Brock, Dechert, Scheinkman, and LeBaron (1988) can easily detect this dependence. Exhaustive Monte Carlo studies by LeBaron (1988a,b,c) and Hsieh and LeBaron (1988a,b,c) have demonstrated that one has to work fairly hard to transform a low dimensional chaos enough by standard time series prewhitening to destroy enough of the dependence in the residuals to keep the BDS test from detecting the dependence. But it can be done.

For example, a series consisting of every m th iterate of the tent map will appear purely random (IID) to the BDS test, for m large enough but it is hard to imagine any econometrician doing this kind of transformation in an attempt to linearly (or even nonlinearly) prewhiten. Other tests for predictable left out dependence are available.

For example, if low dimensional deterministic chaos is present one should be able to forecast it to some extent by using techniques such as nearest neighbor above and beyond linear techniques. As we pointed out above LeBaron's work (1988a,b,c) has shown that linearly prewhitened 7-dimensional Mackey–Glass time series can still be forecasted using nearest neighbor technique. We have not seen anyone successfully forecast a linearly prewhitened aggregated macro economic or financial time series out of sample. A review of nonlinear models in macro economics such as threshold and regime change type models is in Potter (1989). For data limited economic contexts one could look upon the requirement that the investigator adduce evidence of nonlinear forecastibility as an imperfect substitute for the requirement that one reconstruct the chaotic map when claiming a discovery of low dimensional chaos. Phase portraits are not enough.

One must be careful in using phase portraits and Lyapunov exponent estimation to infer low dimensional deterministic chaos since low order autoregressions can look like chaos. (Brock and Sayers (1988)). Under specific null models one can estimate standard errors of statistics for nonlinear science (such as an estimator of the largest Lyapunov exponent, Kolmogorov entropy, etc) using the bootstrap work surveyed by Efron and Tsibirani (1986). More will be said about this strategy later. Brock and Sayers discuss difficulties in using tools such as phase portraits and Lyapunov exponent estimates to distinguish low dimensional deterministic chaos from other alternatives like near unit root low order autoregressions. This difficulty brings us to the work of Ramsey, Sayers, and Rothman.

Ramsey, Sayers, and Rothman (1988) have evaluated claims of low dimensional deterministic chaos in economic data including monetary aggregates. They conclude that the burden of proof that low dimensional deterministic chaos is present in any economic data set has not been met. We agree.

Turn now to a discussion of technical methods that economists have used in doing statistical inference theory for complexity concepts in nonlinear science.

2 U–Statistics and Error Bars for Measures of Dynamic Complexity

When one estimates an object such as, for example, a correlation integral or a correlation dimension the error bar (the standard error for statisticians) depends on the nature of the underlying process. For example a noisy chaos will give different standard errors for a correlation integral estimate depending on the nature of the noise. With a little bit of work, a recently well developed theory in mathematical statistics, viz. U–statistic theory for weakly dependent processes, can be applied to develop a precise theory of error bars as well as tests for noisy low dimensional chaos in limited data sets. We do this below.

2.1 U-Statistics

A U-statistic is just a generalized average. The theory of U-statistics for IID processes is well known and is found in Serfling (1980), Chapter 5. Denker and Keller (1983) extend the theory to treat a broad class of weakly dependent processes that includes noisy chaos.

We first explain the projection method of U-statistic theory. This is a device to reduce a complicated looking U-statistic (which is just a generalized average) to a simple average so that standard central limit theorems may be applied. Let us explain.

Let $h(x, y)$ be a symmetric real valued function, called a "kernel", where x, y are in R^m. Put $h_1(x) = Eh(x, Y)$, where E is the mathematical expectation with respect to the distribution function of random variable Y. The function h_1 is called the projection of h. In the special case of the correlation integral for a time series embedded in m-space $h(x, y) \equiv H(\epsilon - ||x - y||)$, where $H(r)$ is the Heavyside function which is 1 when $r > 0$ and 0 otherwise. Here $|| \cdot ||$ denotes a norm in R^m. We will use the max norm for this article. Consider the statistic (called a U-statistic)

$$U_n = \frac{1}{\binom{n}{2}} \sum\sum_{1 \leq s < t \; len} h(u_s, u_t) \tag{2.1}$$

where $\binom{n}{2}$ is the binomial coefficient, and $\{u_t\}$ are IID. Any statistic which can be written in the form (2.1) where the "kernel" h is symmetric is called a U-statistic. Put $\theta = Eh$. Then if $E[h^2] < \infty$, and, if the variance of h_1 is positive, (cf. Denker and Keller (1983)), we have,

$$U_n - \theta = \frac{2}{n} \sum_{t=1}^{n} [h_1(u_t) - \theta] + R_n(u_1, \ldots, u_n) \tag{2.2}$$

where $\sqrt{n} R_n$ converges to 0 in distribution. Therefore, the double sum statistic can be written as a simple average of IID random variables up to a term R_n that goes to zero in distribution when magnified by $\sqrt{n}$. Hence $\sqrt{n}[U_n - \theta] \xrightarrow{D} N(0, 2\text{Var}(h_1))$. We will assume that $\text{Var}(h_1) > 0$ throughout this paper. More delicate CLT's are available when $\text{Var}(h_1) = 0$ in Serfling (1980, Chapter 5).

2.2 U-Statistic Central Limit Theorems for Dependent Processes

U-statistic central limit theory is available for general dependent processes, i.e. the case when $\{u_t\}$ are not IID. We will follow Serfling (1980), and Denker and Keller (1983) in developing the extended U-statistic theory that we need. Recall that a measurable function $h : R^m \to R$ is called a kernel for $\theta = Eh$ if it is symmetric in its m arguments. A U-statistic for estimating θ is given by

$$U_n = \frac{1}{\binom{n}{m}} \sum_{1 \leq t_1 < t_2 <} \sum \cdots \sum_{< t_m \leq n} h(x_{t_1}, \ldots, x_{t_m}) \tag{2.3}$$

U-statistics are interesting because (i) they have many of the desirable properties that averages of IID variables possess, including central limit theorems and laws of large numbers, (ii) they are minimum variance estimators of θ in the class of all unbiased estimators of θ (Serfling (1980, p176)), (iii) they converge rapidly to normality (Serfling (1980, p193, Theorem B). We will only use the case $m = 2$. So for now on m is fixed at 2.

The projection method of Hoeffding is applied by Denker and Keller (1983) to obtain the decomposition

$$U_n = \theta + \frac{2}{n} \sum_{t=1}^{n} h_1(x_t) + R_n \tag{2.4}$$

$$h_1(x) = E[h(X_1, X_2) \mid X_1 = x] \tag{2.5}$$

Put $g_1(x) = h_1(x) - \theta$. We state part of one of Denker and Keller's theorems below.

Suppose that $\{X_t\}$ is strictly stationary with distribution function F and let $\mathcal{F}_i^j$ be the σ-algebra generated by $\{X_t \mid i \leq t < j\}$ for $1 \leq i \leq j \leq \infty$. The sequence $\{X_t\}$ is called absolutely regular if

$$\beta_n = \sup_{i \in \mathbb{N}} E[sup\{|P(A \mid \mathcal{F}_1^i) - P(A)| \mid A \in \mathcal{F}_{i+n}^\infty\}]$$

converges to zero. The variance of the leading term in equation (2.4) is

$$\sigma_n^2 = E\left[\sum_{t=1}^{n} g_1(X_t)\right]^2 \tag{2.6}$$

and the asymptotic variance is

$$\sigma^2 = E[g_1(X_1)^2] + 2 \sum_{t>1} E[g_1(X_1)g_1(X_t)] \tag{2.7}$$

provided this sum converges absolutely. In this case, $\sigma = \lim_{n \to \infty} \sigma_n/\sqrt{n}$.

Theorem 2.1 (Denker and Keller) *If $\sigma^2 > 0$ and for some $\delta > 0$*

i) $\sum \beta_n^{\delta/(2+\delta)} < \infty$,

ii) $\sup_{i<j} E[h(X_i^m, X_j^m)^{2+\delta}] < \infty$

then

$$\sqrt{n} \frac{U_n - \theta}{2\sigma_n/\sqrt{n}} \xrightarrow{D} N(0,1)$$

Other mixing conditions besides (1) including two by Denker and Keller yield similar results. The point is that we need some type of condition on the rate of decay of dependence over time, i.e. a mixing condition, in order to get the central limit theorem for dependent processes. Condition (1) seems as useful as any. This will be satisfied for a noisy chaos because decay of dependence is exponential for chaotic systems.

2.3 The Distribution of Functions of U–Statistics

It turns out that some interesting quantities from dynamical systems theory can be written as functions of U-statistics. Asymptotic standard errors are easy to calculate for any function of given statistics by the "delta method" (Serfling (1980, Chapter 3)).

Let us show how this method works by looking at the BDS (1986) test for chaos, which is built up from ingredients supplied by the Grassberger–Procaccia (1983a,b) correlation integral for a time series embedded in m dimensional space. The test is typically applied to the residuals of fitted models. Recall that the Grassberger–Procaccia correlation integral for a scalar time series $\{x_t\}$ embedded in m-space is defined by

$$C_{m,n}(\epsilon) = \frac{1}{\binom{n}{2}} \sum\sum_{1 \le s < t \le n} H(\epsilon - \|x_s^m - x_t^m\|) \tag{2.8}$$

where $x_t^m = (x_t, \dots, x_{t+m-1})$, $\|x_t^m - x_s^m\| = \max(|x_t - x_s|, \dots, |x_{t+m-1} - x_{s+m-1}|)$, and $H(r) = 1$ for $r > 0$ and 0 otherwise is the Heavyside function. Brock and Dechert (1988) showed (i) $C_{m,n}(\epsilon) \to C_m(\epsilon)$ as $n \to \infty$ a.s. for ergodic processes, and (ii) if $\{x_t\}$ is IID, $C_m(\epsilon) = C_1(\epsilon)^m$. This result motivates the following general development.

Look at the statistics:

$$\theta_m = \frac{1}{\binom{n}{2}} E\left[\sum\sum_{1 \le s < t \le n} h(x_s^m, x_t^m, m)\right]$$

$$A_{m,n} = \frac{1}{\binom{n}{2}} \sum\sum_{1 \le s < t \le n} [h(x_s^m, x_t^m, m) - \theta_m] \tag{2.9}$$

$$B_n = \frac{1}{\binom{n}{2}} \sum\sum_{1 \le s < t \le n} [h(u_s, u_t) - \theta] \tag{2.10}$$

$$W_{m,n} = A_{m,n} - B_n^m \tag{2.11}$$

here $u_t^m = (u_t, \dots, u_{t+m-1})$, $h(u_s^m, u_t^m, m) = h(u_s, u_t)h(u_{s+1}, u_{t+1}) \cdots h(u_{s+m-1}, u_{t+m-1})$, and $h(x, y)$ is a symmetric kernel which in Brock, Dechert, Scheinkman and LeBaron (1988) was taken to be the indicator of the event $\{|x - y| < \epsilon\}$, i.e., $h(x,y) = H(\epsilon - |x - y|)$. Note that both A and B are U–statistics and the composite statistic W is the function $F(A, B) = A - B^m$.

At the risk of repeating let us explain the intuition of what is going on here. Under the hypothesis of IID, $\theta_m = \theta^m$, and $W_{m,n} \to 0$, almost surely as $n \to \infty$. This suggests that $\sqrt{n}W_{m,n}$ should converge to a normal distribution with mean zero and finite variance. This is what the Brock, Dechert, Scheinkman and LeBaron paper (1988) showed. Astute choices of the kernel function $h(x, y)$ should improve the ability of this test to detect temporal dependence. Hsieh and LeBaron (1988) have shown by extensive Monte Carlo work that the choice $H(\epsilon - |x - y|)$ works well when ϵ is chosen between one half to one and one half times the standard deviation of your data.

For any kernel h such that $E|h| < \infty$, the projection method (Denker and Keller (1983), Serfling (1980)) yields

$$A_{m,n} = \frac{2}{n}\sum_{t=1}^{n}[h_m(u_t^m) - \theta_m] + R_{m,n} \tag{2.12}$$

$$B_n = \frac{2}{n}\sum_{t=1}^{n}[h_1(u_t) - \theta] + R_n. \tag{2.13}$$

Here, $h_1(u_t) = E[h(u_s, u_t)|u_s]$, and $h_m(u_t) = E[h_m(u_t^m, u_s^m)|u_s^m]$ for $|s - t| > m$. These functions take a simple form when h is the indicator of the event, $\{|x - y| < \epsilon\}$, and the max norm is used to measure the distance between m-dimensional vectors. In order to see the form of these projection functions let $\{u_t\}$ be IID with distribution function F and h set equal to the indicator function of the event $\{|x - y| < \epsilon\}$. Then

$$h_1(u) = F(u + \epsilon) - F(u - \epsilon) \tag{2.14}$$

$$h_m(u_1^m) = \prod_{i=1}^{m}[F(u_i + \epsilon) - F(u_i - \epsilon)]. \tag{2.15}$$

Return to the problem of putting the BDS statistic $W_{m,n}$ into a simple form to work with. Assume h is C^2 with $\|h''\|_4 < \infty$. Then symmetry of h may be used to show that the above remainder terms go to zero in probability when multiplied by $\sqrt{n}$ even when they are evaluated at $u_{t,n} = u_t + d_{t,n}$ provided that $\sqrt{n}d_{t,n} \to ax_t$, as $n \to \infty$ where $\{x_t\}$ is a stationary stochastic process satisfying modest regularity conditions, and is measurable with respect to the sigma algebra generated by past information which is denoted by $\mathcal{F}_{t-1}$. The argument will be given in Theorem 2.2 below. This fact plus use of central limit theory for transformations of asymptotically normal statistics (Serfling (1980, Chapter 3)), called the delta method suggests the statistic

$$\sqrt{n}W_n^* = \frac{2}{\sqrt{n}}\sum_{t=1}^{n}[h_m(u_t^m) - \theta_m - m\theta^{m-1}(h_1(u_t) - \theta)] \tag{2.16}$$

should converge to $N(0, \sigma_m)$ where σ_m is the same even if evaluated at $\{u_t + d_{t,n}\}$ instead of $\{u_t\}$.

2.4 The Main Theorem

Since the theorem to follow shows that use of the BDS statistic is asymptotically valid even when evaluated at estimated residuals of a null models like those treated in Section 1 and since we do not care about the parameters of the null model but, yet, they must be estimated, we call the theorem the "nuisance parameter theorem."

Theorem 2.2 (Nuisance parameter theorem): *Assume that the symmetric kernel h is C^2 with $\|h''\|_4 < \infty$. Assume $Eh_1' = 0$. Then*

$$\sqrt{n}W_{m,n}^* \xrightarrow{d} N(0, \sigma_m) \tag{2.17}$$

$$\sigma_m^2 = E\left[g_m(u_1^m)g_m(u_1^m) + 2\sum_{t=2}^{m}g_m(u_1^m)g_m(u_t^m)\right] \tag{2.18}$$

where W^* is evaluated at $\{u_t + d_{t,n}\}$, $\sqrt{n}d_{t,n} \to ax_t$ and x_t is $\mathcal{F}_{t-1}$-measurable.

Proof: See Brock and Dechert (1988b). The idea is to show that all remainder terms go to zero in probability when multiplied by $\sqrt{n}$, when evaluated at the estimated residuals. Hence we only have to expand the expression (2.18) in a Taylor series about u_t to obtain the limit distribution. Doing this we obtain

$$\sqrt{n}W_{m,n}^* \xrightarrow{d} N(0, \sigma_m^2) + aE[g_m'(u_t^m) \cdot x_t^m] \tag{2.19}$$

Simple algebra and use of iterated expectations shows that the condition $E[h_1'] = 0$ implies the last term is zero. $\qquad\square$

Corollary 2.3 *Assume the parameters of the null models like those treated in Section 1 are estimated $\sqrt{n}$ consistently. We have: (i) $\sqrt{n}[C_{m,n}(\epsilon) - C_{1,n}(\epsilon)]^m]$ converges to a normal distribution with mean 0 and variance V, where V is independent of whether you evaluate at the true IID residuals are the estimated ones. (ii) Statistics of the form $\sqrt{n}[F(C_{1,n}(\epsilon), \ldots, C_{k,n}(\epsilon)) - F(C_1(\epsilon), \ldots, C_k(\epsilon))]$ have the same convergence properties.*

Proof: See Brock and Baek (1988).

Remark: Note that the Grassberger and Procaccia (1983a,b) lower bound for the Kolmogorov entropy,

$$K_m^{GP}(\epsilon) = \ln\left(\frac{C_m(\epsilon)}{C_{m+1}(\epsilon)}\right)$$

is covered by the corollary. Hence an asymptotic standard error for K^{GP} under the null of IID can be worked out using methods of Theorem 2.2. Similar methods are useful for working out asymptotic standard errors under weakly dependent null hypotheses as well as IID. See Brock and Baek (1988) for details.

Note that for the case $h(x,y) = H(\epsilon - |x - y|)$ the condition $E[h'_1] = 0$ is satisfied. To see this just compute using (2.14).

The formula (2.18) for the variance σ_m simplifies for the case $h(x,y) = H(\epsilon - |x - y|)$:

$$\sigma_m^2 = 4\left[K^m + 2\sum_{j=1}^{m-1} k^{m-j}C^{2j} + (m-1)^2 C^{2m} - m^2 KC^{2(m-1)}\right] \qquad (2.20)$$

where

$$C = E[H(\epsilon - |X - Y|)] = \int[F(\xi + \epsilon) - F(\xi - \epsilon)]dF(\xi) \qquad (2.21)$$

and

$$K = E[H(\epsilon - |X - Y|)H(\epsilon - |Y - Z|)] = \int[F(\xi + \epsilon) - F(\xi - \epsilon)]dF(\xi) \qquad (2.22)$$

Consistent estimators are given for C and K by Brock, Dechert and Scheinkman (1986) and Hsieh and LeBaron (1988a,b). They are

$$\hat{C}_n(\epsilon) = \frac{1}{n^2}\sum_{s=1}^{n}\sum_{t=1}^{n} H(\epsilon - |x_s - x_t|)$$

$$\hat{K}_n(\epsilon) = \frac{1}{n^3}\sum_{r=1}^{n}\sum_{s=1}^{n}\sum_{t=1}^{n} H(\epsilon - |x_r - x_s|)H(\epsilon - |x_s - x_t|)$$

Improved versions of these estimators are used in the Dechert and LeBaron software packages cited in the references.

This theorem is simple but very useful. For example consider a very special case of the null models treated in Section one above:

$$y_t = \beta y_{t-1} + u_t, \qquad \{u_t\} \text{ IID} \qquad (2.23)$$

then $d_{t,n} = (\beta - \beta_n)y_{t-1}$. The same method of proof shows that the limiting distribution of the W^* statistic can be derived by replacing the second term of the RHS of (2.19) by $N(0, V(\beta_n))E[g'_m(u_t^m) \cdot y_{t-1}^m]$, where $V(\beta_n)$ is the variance of the estimate of β. But this last term is zero by the same argument as used in the nuisance parameter theorem. The same argument works even if (2.23) is nonlinear, provided that the parameters are estimated by a $\sqrt{n}$ consistent procedure such as maximum likelihood.

Remark: In Brock, Dechert, Scheinkman and LeBaron (1986) each series was normalized so that the mean was zero and the variance unity before the BDS test was applied at $\epsilon = 0.5$ and 0.75. For kernels that are functions of $|x - y|$ the mean estimate cancels but the nuisance parameter σ_n remains. In order to take care of this put $u_{t,n} = u_t/\sigma_n$ and replace u_t in Theorem 2.2 by u_t/σ, expand $1/\sigma_n$ in a Taylor series around σ, assume $\sqrt{n}(\sigma_n - \sigma) \overset{d}{\to} N(0, V(\sigma))$, as $n \to \infty$, carry on as in the proof of Theorem 2.2 to obtain

$$\sqrt{n}C^* \to N(0, \sigma_m^2) - \frac{1}{\sigma^2}N(0, V(\sigma))E[g'_m \cdot u_t^m]. \tag{2.24}$$

Now use the definition of g_m, and the null hypothesis of $\{u_t\}$ IID to show that $E[g'_m \cdot u_t^m] = 0$. Hence the "nuisance" parameter σ does not change the limit distribution of the BDS statistic. A similar argument applies for an estimated mean except that $E[g'_m \cdot u_t^m]$ is replaced by $E[g'_m \cdot 1^m]$ where 1^m denotes an m-vector of 1's. It is easy to show that $E[g'_m \cdot 1^m] = 0$. Note that the condition $E[h'_1] = 0$ is not needed.

In practice the BDS statistic is normalized by the square root of a consistent estimator of the variance W. If the null model is correct then this normalized statistic converges to the normal distribution with mean zero and variance one. From now on we will use the normalized version.

2.5 Summary

This section of the paper has shown that it is valid to use the BDS statistic as a specification test for econometric models that can be estimated root T consistently and where the innovations are IID. This includes all linear models with IID innovations. Thus it is asymptotically valid to use the test in the following way. Fit a linear time series model to your data and use BDS to test whether the estimated residuals are IID. If you fail to reject IID then this is evidence consistent with the hypothesis that the data is linear. If you reject the IID hypothesis then this is evidence consistent with the hypothesis that the data are nonlinear.

It is also consistent with nonstationarity and temporal dependence of higher order conditional moments of the data such as conditional variance and conditional skewness. If conditional variance can be estimated $\sqrt{n}$ consistently then Brock and Dechert (1988) showed that the nuisance parameter theorem can be extended to this case also. Simply divide the estimated residuals by the square root of the $\sqrt{n}$ consistent estimator and test the resulting "standardized" residuals for IID. If you have very strong priors as to the possible direction of departure from IID use a test that maximizes power against that alternative. If, as is usually the case, you do not have such strong priors then use a test with power against general alternatives such as the BDS test. Turn now to applications and issues that should be treated in the future.

3 Applications, Summary and Open Questions

Some sample applications are given here. Much of what follows is taken directly from

Brock and Dechert (1988). Brock and Sayers (1988) have conducted some of the tests discussed below on macroeconomic data such as U.S. industrial production and have found some evidence of nonlinearity or left out nonstationarity. We concentrate on financial applications here.

First, Hsieh (1987) has used the BDS test to test for the adequacy of fit of GARCH (Generalized Auto Regressive Conditionally Heteroscedastic) models to exchange rates of five currencies against the U.S. dollar. He did this by using the fact that under the null hypothesis that the GARCH model was correct then the one step ahead forecast errors divided by the conditional standard deviations, call these the standardized forecast errors, must form an IID process. Theorem 2.2 justifies Hsieh's procedure asymptotically. The theorem applies provided you have estimated the parameters of the GARCH model using a root n consistent procedure and a member of the GARCH class describes the data. This was exactly the procedure that Hsieh carried out. Second, we illustrate this procedure by examining Bollerslev's (1985) fit of a GARCH-M model to monthly returns on the S&P index.

3.1 Auto Regressive Conditional Heteroscedasticity

Bollerslev fits the model

$$y_t = y_{t|t-1} + \epsilon_t \tag{3.1}$$
$$h_{t|t-1} = a_0 + a_1\epsilon_{t-1}^2 + b_1 h_{t-1|t-2} \tag{3.2}$$
$$y_{t|t-1} = b_0 + \delta\sqrt{h_{t|t-1}} \tag{3.3}$$

where ϵ_t is $N(0, h_{t|t-1})$, and $\{y_t\}$ is the monthly returns on the S&P. This is an example of a Generalized Auto Regressive Conditional Heteroscedastic model with a conditional mean, $y_{t|t-1}$, which is dependent on current conditional volatility (GARCH-M).

The idea is intuitive: looking at Capital Asset Pricing Models (CAPM's) or Arbitrage Pricing Models (APT's) suggests that aggregate returns should be high on average when volatility is high in order to compensate investors for bearing higher risk. Furthermore, stylized facts dating back to Mandelbrot and others are these: (i) unconditional distributions of stock returns are fat tailed; (ii) volatilities are persistent; (iii) large linear forecast errors tend to come together but appear unpredictable in sign; and (iv) lower frequency returns "look more normal" than high frequency returns. An intertemporal version of the central limit theorem appears to be at work. The GARCH-M class of models is an attempt to parameterize this kind of data in a way to do statistical inference.

Bollerslev's results are:

$$
\begin{aligned}
b_0 &= -0.0340 \quad (0.0220) \\
\delta &= 1.2565 \quad (0.6892) \\
a_0 &= 0.0002 \quad (0.0001) \\
a_1 &= 0.0561 \quad (0.0340) \\
b_1 &= 0.8038 \quad (0.1066)
\end{aligned}
$$

Asymptotic standard errors are in parentheses. While the results are not overwhelmingly strong, the persistence parameter b_1 for volatility is surely statistically significant.

Table 1. BDS Statistics

Data	ϵ	C_m	BDS	m	n
ew	.02815	.081	7.28	2	696
vw	.02815	.059	5.39	2	696
ew:1	.02815	.056	4.77	2	336
ew:2	.02815	.11	3.77	2	360
vw:1	.02815	.047	6.32	2	336
vw:2	.02815	.074	1.95	2	360
1/5	.02815	.116	8.86	2	696
5/5	.02815	.08	5.94	2	696
1/5:1	.02815	.071	6.33	2	336
1/5:2	.02815	.17	4.13	2	360
5/5:1	.02815	.061	6.32	2	336
5/5:2	.02815	.10	1.16	2	360

vw (ew) is the monthly returns on value weighted (equal weighted) CRSP index, 1926–1985. 1/5 (5/5) is the monthly returns on the smallest quintile (largest quintile) of CRSP firms. ew:1 is based on the years 1926-1953 and ew:2 is based on the years 1954-1985, etc.

Note that the ratio of the coefficient to the standard error is much larger for b_1 than it is for δ. This indicates that it is easier to predict the conditional variance than the conditional mean in the GARCH-M context. Of course all this is dependent on the GARCH-M being the correct model for S&P returns, which might be false. We do a diagnosis of the quality of the fit in equation 3.4 below. This will be an example of the application of some of the tools advanced in this paper.

Bollerslev set $r_t = ln(P_t/P_{t-1})$ where P_t is the monthly S&P 500 common stock composite price index. He pre whitened r_t by putting $y_t = (1 + .268B)^{-1}r_t$ where B is the backshift operator, $Bx_t = x_{t-1}$. The GARCH-M model (3.1)–(3.3) was fit to y_t.

We calculated BDS statistics for the GARCH-M residuals of Bollerslev's model and failed to reject the null hypothesis that they were IID. In Brock (1987) a more powerful diagnostic than the BDS test was applied and the GARCH-M model was rejected. Hence the evidence for the model is mixed for this application.

The BDS test has been applied in the form of Theorem 2.2 in LeBaron (1988a) where he finds that GARCH-M models fit weekly returns on the value weighted CRSP index poorly for 1962–1974, but, they fit adequately for the period 1975–1985. We conduct an application to returns on several portfolios below.

Some results of the BDS test for temporal dependence are recorded in Table 1. Recall that under the null hypothesis of IID, the BDS statistic is normally distributed with mean zero and variance one. Therefore values of the BDS statistic greater than two may be viewed as significantly greater than zero.

Monte Carlo studies in Hseih and LeBaron (1988a,b,c) that, for sample sizes 500 or less one should insist on a value of BDS greater than the usual 5% value in order to be significant at the 5% level. Values greater than 3.5 for $m = 2$ are significant at the 5% level, for the sample sizes in Table 1.

Five conclusions can be drawn from Table 1. First, evidence for non randomness as measured by the size of the BDS statistic is strongest for small firms in the first period which extends from 1926 to the mid 1950's. Second, evidence for non randomness is weakest for large firms in the second period. Third, evidence for non randomness is uniformly weakest for large firms. Fourth, the evidence for non randomness may be due to predictability of variance, e.g. auto regressive conditional heteroscedasticity. Fifth, the evidence for non randomness may be due to uncovered predictability of higher order conditional moments such as skewness and kurtosis. More generally it may be due to predictability of the conditional characteristic and moment generating functions. Joint evidence from the Lo and MacKinlay (1987) tests and the BDS tests suggest that the evidence of predictable structure uncovered here in stock returns data is not due to the predictability of conditional variance of the type captured by GARCH and ARCH models.

This is so because the Lo and MacKinlay test cannot detect ARCH and GARCH processes because the auto-covariances of such processes are zero. To put it another way the Lo and MacKinlay test has zero power against ARCH and GARCH type dependence. Yet the Lo and MacKinlay test rejects with much the same pattern as does the BDS test. That is to say their strongest rejections of the random walk hypothesis (first differences of the logarithm of stock prices being uncorrelated) occur for small firms over the 1926–1985 period. Weakest rejections occur for large firms over the post World War II (approximately 1948–1985) period. Turn now to a general use of the BDS statistic as a specification test and to testing the adequacy of ARCH models to capture the non randomness that appears to exist in financial data.

The null hypothesis of IID is a "straw man" hypothesis that is not very interesting to time series econometricians. One might think that it would have some interest in finance. After all the classical random walk hypothesis asserts that the first differences of the logarithm of stock prices satisfies the random walk process,

$$ln(P_t) - ln(P_{t-1}) = \mu + \epsilon_t \tag{3.4}$$

where $\{\epsilon_t\}$ is purely random, i.e. IID. Versions of this hypothesis are tested with variance ratio statistics in Lo and MacKinlay (1987). Their tests of the random walk hypothesis are based upon the principle that $Var(ln(P_t) - ln(P_{t-k})) = kVar(\epsilon)$ where $Var(\epsilon)$ is the common variance of ϵ_t. Various versions of the random walk hypothesis are tested by Lo and MacKinlay (1987) where $\{\epsilon_t\}$ is only required to be uncorrelated with some regularity conditions on how fast the dependence in $\{\epsilon_t\}$ dies out as $t \to \infty$. The pattern of strength of rejection of the random walk hypothesis they found is consistent with the pattern reported in Table 1 above. But notice that Lo and MacKinlay do not require the maintained hypothesis that $\{\epsilon_t\}$ is IID. We have shown above how the BDS test can be adapted to non IID cases.

3.2 Non-Forecastible Non-Stationarity v. Predictable Extra Structure

A major problem with the BDS test (and many tests for temporal dependence in statistics) is that it is triggered by non-forecastible nonstationarity as well as potentially forecastible "left out structure." Recall when we discussed the procedure used by natural scientists (cf. Swinney (1985)) to test for chaos, we said that reconstruction was difficult in economics. To our knowledge no one in economics has even come close to reconstructing a chaos in economic data like what was done for weak turbulence and the Belousov-Zhabotinskii chemical reaction as in Swinney (1985). Nevertheless we can examine an imperfect substitute for reconstructibility: evidence of one step ahead nonlinear forecastibility.

Consider the following thought experiment. Extract all forecastible structure known to economists by fitting models like those in Section 1 or GARCH-M models as above. Let $\{u_{t,n}\}$ denote the estimated standardized residuals which converge to an IID sequence $\{u_t\}$ as $n \to \infty$ under the null model. Consider any nonlinear forecasting method such as nearest neighbors (LeBaron (1988a,b,c)) and any statistical measure, $M(n, \{u_{t,n}\})$, of how well you are doing at forecasting one-step-ahead, say. Now create an IID series $\{u_{t,sf}\}$ of length n from $\{u_{t,n}\}$ by shuffling $\{u_{t,n}\}$ as in Scheinkman and LeBaron (1988a,b). This IID series will have the same unconditional moments as the original series. Forecast the shuffled series and evaluate M. Do this R times. Get a distribution of the M measures for these R experiments. Tick off a 5distribution. See if the M measure for your actual data $\{u_{t,n}\}$ lies in this rejection region. If it does reject the null hypothesis of no forecastible structure in these residuals in favor of the alternative that there is forecastible left out structure in these residuals. Note how this procedure is related to the bootstrapping procedures of Efron and Tsibirani (1986).

While the above procedure has not been explored enough for us to make any predictions on its usefulness it should have the potential (unlike the BDS test) to discriminate between one-step-ahead forecastible structure and non-forecastible nonstationarity. Parenthetically we emphasize one-step-ahead forecastibility because it is impossible to forecast a chaos with a finite resolution state measurement apparatus H steps-ahead if H is large.

We feel that bootstrapping type methods should be useful for estimating standard errors and setting up confidence intervals for other measures of complexity such as estimated largest Lyapunov exponents. We have studies underway that are exploring this possibility.

3.3 Summary

This paper has explored methods used by economists to test the estimated residuals of received models used in economics and finance for the presence of left out short term predictable structure and left out nonstationarity. The basic theory of the methods was outlined in Section 2. Applications to financial markets were outlined in Section 3.

References

Ashley, R., and Patterson, D. (1986)[1] "Linear Versus Nonlinear Macroeconomics: A Statistical Test," Department of Economics, VPI&SU, Blacksburg, VA 24061, *International Economic Review* (forthcoming).

Ashley, R., Patterson, D., and Hinich, M. (1986), " A Diagnostic Test for Nonlinearity

---

[1]This list of references contains not only those cited in the paper but also related references.

and Serial Dependence in Time Series Fitting Errors," *Journal of Time Series Analysis*, 7, #3, 165-178.

Baek, E. (1988), PhD Thesis, Department of Economics, University of Wisconsin, Madison.

Baek, E. (1987), "Contemporaneous Independence Test of Two IID Series," Department of Economics, University of Wisconsin, Madison.

Barnett, W,. and Chen, P. (1988), "The Aggregation Theoretic Monetary Aggregates are Chaotic and Have Strange Attractors: An Econometric Application of Mathematical Chaos," in Barnett, W., Berndt, E., and White, H. (Eds.), *Dynamic Econometric Modelling*, Cambridge University Press, Cambridge, 199-245.

Baumol, W., and Benhabib, J. (1989), "Chaos: Significance, Mechanism, and Economic Applications," *Journal of Economic Perspectives* Vol 3, No. 1, Winter.

Black, F. (1976), "Studies of Stock Price Volatility changes," *Proceedings of the 1976 Meetings of the Business and Economics Statistics Sections*, American Statistical Association, 177-181.

Bollerslev, T. (1986), "Generalized Autoregressive Conditional Heteroskedasticity," *Journal of Econometrics*, 31, 307-327.

Bollerslev, T. (1985), "A Conditionally Heteroskedastic Time Series Model for Security Prices and Rates of Return Data," forthcoming *Review of Economics and Statistics*

Bollerslev, T. (1988), "On the Correlation Structure for the Generalized Autoregressive Conditional Heteroskedastic Process," *Journal of Time Series Analysis*, Vol. 9, No. 2, 121-132.

Brock, W. A. (1988), "Nonlinearity and Complex Dynamics in Economics and Finance," in Anderson, P.W., Arrow, K.J., Pines, D. (Eds.), *The Economy as an Evolving Complex System*, Vol V, *Santa Fe Institute Studies in the Sciences of Complexity*, Redwood City, Ca.: Addison Wesley, 77-97.

Brock, W. A., and Dechert, W. D. (1988), "Theorems on Distinguishing Deterministic from Random Systems," in Barnett et al op. cit., 247-265.

Brock, W. A., and Dechert, W. D. (1988b), "A General Class of Specification Tests: The Scalar Case," *Business and Economics Statistics Section the Proceedings of the American Statistical Association* (forthcoming).

Brock, W., and Sayers, C. (1988), "Is The Business Cycle Characterized by Deterministic Chaos?" *Journal of Monetary Economics* July, 1988, 71-90.

Brock, W. A., Dechert, W. D., Scheinkman, J., and LeBaron, B. (1988), "A Test for Independence Based Upon the Correlation Dimension," Working Paper, The Univ. of Wisconsin, Madison, The Univ. of Houston, and The Univ. of Chicago, Department of Economics.

Brock, W. A., Dechert, W. D., and Scheinkman, J. (1986), "A Test for Independence

Based On the Correlation Dimension," Department of Economics, University of Wisconsin, Madison, University of Houston, and University of Chicago.

Brock, W. A., and Baek, E. (1988), "The Theory of Statistical Inference for Nonlinear Science: Gauge Functions, Complexity Measures, and Instability Measures," Department of Economics, The University of Wisconsin, Madison.

Dechert, W. D. (1987), "A Program to Calculate BDS Statistics for the IBM PC," Department of Economics, University of Houston.

Denker, M., and Keller, G. (1983), "On U-Statistics and Von Mises Statistics for Weakly Dependent Processes," *Z. Wahrscheinlichkeitstheorie verw. Gebiete*, 64, 505-522.

Eckmann, J., and Ruelle, D. (1985), "Ergodic Theory of Chaos and Strange Attractors," *Review of Modern Physics*, 57, No. 3, 617.

Efron, B., and Tibshirani, R. (1986), "Bootstrap Methods for Standard Errors, Confidence Intervals, and Other Measures of Statistical Accuracy," *Statistical Science*, Vol 1, No.1, 54-77.

Engle, R. (1982), "Autoregressive Conditional Heteroscedasticity With Estimates of The Variance of U. K. Inflations," *Econometrica*, 50, 987-1007.

Engle, R. (1987), "Multivariate ARCH with Factor Structures-Cointegration in Variance," Department of Economics, University of Calif. at San Diego.

Engle, R., Hendry, D., Trumble, D. (1985), "Small-sample properties of ARCH estimators and tests," *Canadian Journal of Economics*, 18, #1, February.

Fama, E. (1976), *Foundations of Finance*, New York: Basic Books.

Fama, E., and French, K., "Permanent and Temporary Components of Stock Prices," Center for Research on Security Prices Working Paper # 178.

Gallant, A. R., and Tauchen, G. (1986), "Seminonparametric Estimation of Conditionally Constrained Heterogeneous Processes: Asset Pricing Applications," Department of Economics, North Carolina State University and Duke University.

Gallant, A., and White, H. (1988), *A Unified Theory of Estimation and Inference For Nonlinear Dynamic Models*, Basil Blackwell, Oxford.

Granger, C., and Andersen, A. (1978), *An Introduction to Bilinear Time Series Models*, Vandenhoeck & Ruprecht in Gottingen, Germany.

Granger, C. (1987), "Stochastic or Deterministic Non-linear Models? A Discussion of the Recent Literature in Economics," University of Calif. at San Diego.

Grassberger, P., and Procaccia, I. (1983a), "Measuring the Strangeness of Strange Attractors," *Physica D*, 9, 189.

Grassberger, P., and Procaccia, I. (1983b), "Estimating the Kolmogorov entropy from a Chaotic Signal," *Physical Review A*, 28, 2591.

Hinich, M., and Patterson, D. (1985), "Evidence of Nonlinearity in Stock Returns," *Journal of Business and Economic Statistics*, January, 69-77.

Hinich, M., and Patterson, D. (1986), "A Bispectrum Based Test of The Stationary Martingale Model," University of Texas at Austin and Virginia Polytechnic Institute.

Hinich, M. (1982), "Testing for Gaussianity and Linearity of a Stationary Time Series," *Journal of Time Series Analysis*, 3, No. 3, 169-176.

Hsieh, D. (1987), "Testing for Nonlinear Dependence in Foreign Exchange Rates," Graduate School of Business, The University of Chicago. *Journal of Business*, forthcoming.

Hsieh, D., and LeBaron, B. (1988a), "Small Sample Properties of the BDS Statistic, I" Graduate School of Business, The University of Chicago.

Hsieh, D., and LeBaron, B. (1988b), "Small Sample Properties of the BDS Statistic, II" Graduate School of Business, The University of Chicago.

Hsieh, D., and LeBaron, B. (1988c), "Small Sample Properties of the BDS Statistic, III" Graduate School of Business, The University of Chicago.

LeBaron, B. (1987b), "A Program to Calculate BDS statistics for the MacIntosh PC," Department of Economics, The University of Chicago and the University of Wisconsin, Madison.

LeBaron, B. (1988), PhD Thesis, The University of Chicago, Department of Economics.

LeBaron, B. (1988a), "Nonlinear Puzzles in Stock Returns," Department of Economics, The University of Wisconsin at Madison.

LeBaron, B. (1988b), "Stock Returns Nonlinearities: Comparing Tests and Finding Structure," Department of Economics, The University of Wisconsin at Madison.

LeBaron, B. (1988c), "The Changing Structure of Stock Returns," Department of Economics, The University of Chicago.

Loève, M. (1963), *Probability Theory*, Van Nostrand, Princeton.

Lo, A., and MacKinlay, C. (1987), "A Simple Specification Test of The Random Walk Hypothesis," Wharton School, University of Penn.

Mayer-Kress, G. (Ed.) (1986), *Dimensions and Entropies in Chaotic Systems*, New York, Springer-Verlag.

Potter, S. (1989), "Nonlinear Time Series and Economic Fluctuations," Department of Economics, The University of Wisconsin, Madison.

Priestley, M. (1981), *Spectral Analysis And Time Series, Vols. I, II*, Academic Press, New York.

Ramsey, J. (1988), "Economic and Financial Data as Nonlinear Processes," C. V. Starr Center for Applied Economics, Department of Economics, New York University.

Ramsey, J., Sayers, C., and Rothman, P. (1988), "The Statistical Properties of Dimension Calculations Using Small Data Sets: Some Economic Applications," Department of Economics, New York University and the University of Houston, *International Economic Review*, forthcoming.

Sakai, H. and H. Tokumaru (1980), "Autocorrelations of a Certain Chaos," *IEEE Transactions on Acoustics, Speech and Signal Processing*,V.I. ASSP- 28, No. 5, 588-590.

Sargent, T. (1987), *Macroeconomic Theory*, Second Edition, Academic Press, New York.

Sayers, C. (1986), PhD Thesis, Department of Economics, University of Wisconsin, Madison.

Scheinkman, J., and LeBaron, B. (1986), "Nonlinear Dynamics and Stock Returns," University of Chicago, forthcoming, *Journal of Business*.

Scheinkman, J., and LeBaron, B. (1987), "Nonlinear Dynamics and GNP Data," University of Chicago, Barnett, W., cf. LeBaron for exact cite??

Serfling, R. (1980), *Approximation Theorems of Mathematical Statistics*, Wiley, New York.

Tong, H. (1983),*Threshold Models in Non-linear Time Series Analysis*, Lecture Notes in Statistics, #21, Springer Verlag, New York.

White, H. (1987), "Specification Testing in Dynamic Models," in T. Bewley (Ed.), *Advances in Econometrics*, New York: Cambridge University Press.

PRACTICAL REMARKS ON THE ESTIMATION OF

DIMENSION AND ENTROPY FROM EXPERIMENTAL DATA

Jean Guy Caputo

LESP, INSA de Rouen
BP 8, 76131 Mont-Saint-Aignan cedex, France

INTRODUCTION

For non-linear dissipative dynamical system, invariant sets have volume zero in phase-space, and trajectories possess the property of sensitivity to initial conditions. The geometry and dynamics can be characterised by quantities, dimensions and entropies which will be invariant under a large class of coordinate changes. To calculate these quantities from experimental data one needs to imbed the data in R^n. These projections generically are diffeomorphisms for n sufficiently large [1,2]. In practice, they are done using time-delays see [3] for example or building vectors from measurements in different locations. However the geometrical structure of these projections sets will determine the values obtained in practice for the dimensions and entropies. For simple model systems these sets can be studied in a semi- quantitative way using the Grassberger-Procaccia correlation integral [4]. This yields some criterions on the choice of the distance in R^n, the different parameters : the number of vectors N and averages m ,the sampling period, the embedding dimension and the delay. Deterministic chaotic data are projected onto manifolds which are locally the product of cantor sets and smooth manifolds, it will be shown that the curvature of these sets caused by the divergence of trajectories becomes predominant for the determination.

I. THE METHOD

In most of the following we will assume we have a 1-dimensional time-series $X(t)$, $t\epsilon[\Delta t, N\Delta t]$, in practice X is an integer because we are dealing with experimental data. In the following it is shown that the influence of the discretisation becomes neglegible as soon as the dimension is over 3. Using the time-delay technique we build a vector time-series in R^n , $\mathbf{X}(t) = (X(t), X(t - \tau), ...X(t - (n - 1)\tau))$ $\tau = p\Delta t$ time-delay, p delay. Adapting the algorithm proposed by Grassberger and Procaccia [4] we calculate

$$C(r) = \frac{1}{m} \sum_{i=1}^{m} C_i(r)$$

Measures of Complexity and Chaos
Edited by N.B. Abraham *et al.*
Plenum Press, New York

where

$$C_i(r) = \frac{1}{N-1}\{number \ j, \|\mathbf{X_i} - \mathbf{X_j}\| \le r\}$$

(proportion of points contained in a sphere of radius r) In [4] m=N , but it is possible to approximate it even with m much smaller than N. In fact choosing m=N can induce spurious correlations for oversampled data [5]. For the cases presented below N and m are such that $1000 \le N \le 100000$ and $100 \le m \le 10000$. It is expected that $C(r) = r^\nu$ for small r, ν being an approximation of the order 2 Renyi dimension [6] which is a lower bound for the information dimension of the set of points. ν is estimated by a least square fit on a certain region of the C(r) plot or by the average local slope. Because of this power-law behavior a given value of N and m will fix the accessible region in r; it will shrink exponentially with the dimension of the attractor. N is also the parameter which fixes the length of the orbit so that it needs to be sufficiently large so that all the accessible phase-space is explored. Figure 1 shows a C(r) plot for increasing values of N for the Henon [7] map embedded in R^3. The C(r) plots keep evolving for r small up to $N = 1.5 \ 10^5$. However it is possible [8] to have a rough estimate of the dimension with as few points as 500. This is probably due to the smallness of the dimension and to the divergence of nearby trajectories which allows a typical orbit to cover the set in a few hundred iterations.

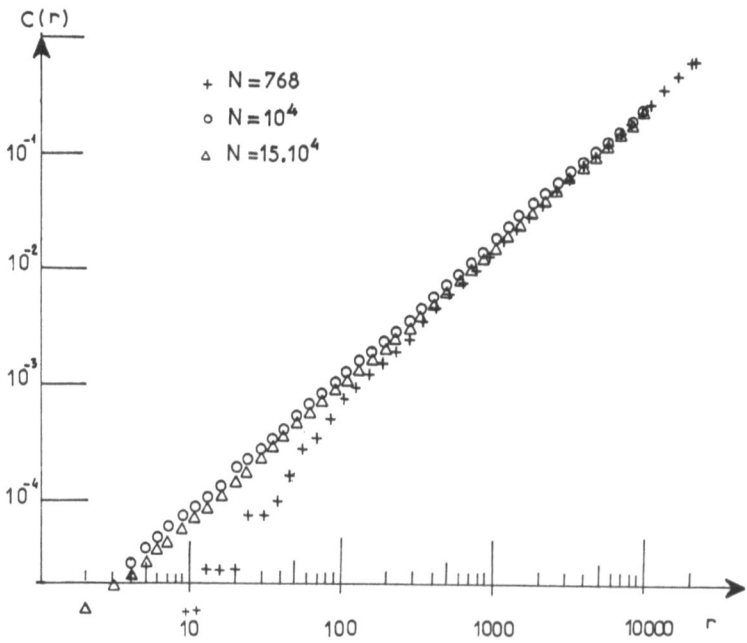

Figure 1. C(r) curves in log-log coordinates for the Henon system
a=1.4, b=0.3, $10^2 \le m \le 1.5 \ 10^2$, $\Delta t = 1$, p=1, n=3

Keeping N fixed and increasing m yields a much faster saturation in the C(r) curve as is shown in figure 2. It is clear from this example that N needs to be as large as possible, m is an averaging parameter which can be kept much smaller. This can also be seen in a study done on the Mackey-Glass system [9] with lag =100 by Havstadt and Ehlers [10]. They compared estimates of the dimension obtained from a sample of 12800 vectors and from averages obtained from two different sets of samples with N=200. The dimension is clearly underestimated for the small data size. For this system the dimension is large (above 7) and the entropy small (10^{-3}) [11] so that a large value of N is necesssary to cover the set.

The algorithm uses the maximum norm and dicards all distances greater than a fixed value r_0. This allows great speed, a typical run with N=100000, m=2000 and embedding dimension n=25 takes about 2 hours cpu on a vax 780. Theiler [12] suggested improving the Grassberger-Procaccia algorithm by boxing the data points prior to the calculation of the C(r) function. This technique is quite efficient for small embedding dimensions where the cost of boxing and searching is minimal. For large embedding dimensions, Theiler suggests to use prisms instead of boxes, however the computational costs at high embedding dimension becomes prohibitive so that the dimension of the prism must be decreased.

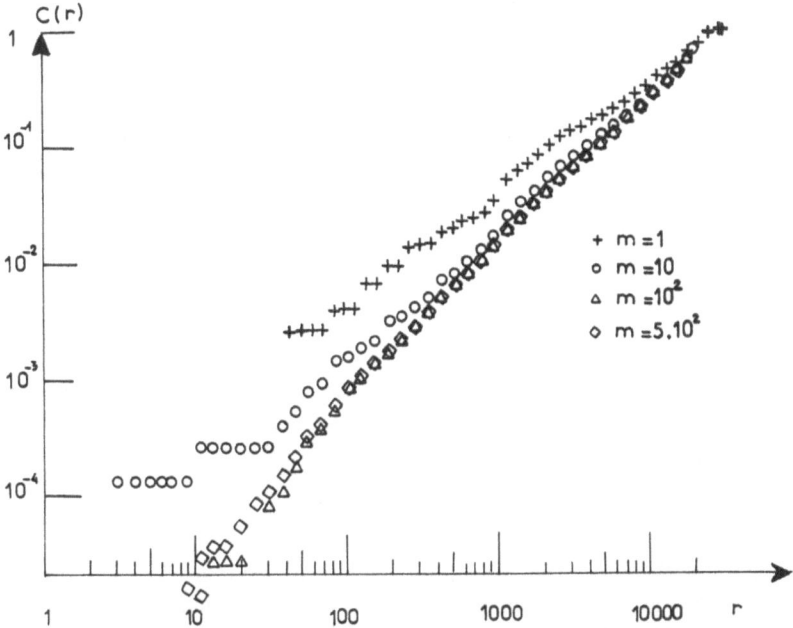

Figure 2. C(r) curves in log-log coordinates for the Henon system
(a=1.4, b=0.3) $N = 768$, $\Delta t = 1$, p=1, n=3

Using the maximum norm is equivalent to using 1 dimensional prisms. The choice of the maximum norm can also be justified by convergence rate for the entropy as we will see further on.

Mayer-Kress [13] showed that calculating the C(r) correlation function could lead to a reducing of the scaling region. He advocates instead averaging on the exponent associated to individual $C_i(r)$. However as the data is embedded in R^n with n large the inhomogeneities are greatly reduced; in [13] it is clear that the error bars get smaller as n increases. We have not tested this method because the effects pointed out in the following are linked to the reconstruction and not to the particular way of estimating the dimension.

II. NON CHAOTIC MODEL SYSTEMS

II.1 periodic data

The simplest such data which can be studied in some detail is a sine data. In this case, the data points lie on a plane in R^n. Despite of this side-effect, the study showed some elementary precautions that must be taken in order to avoid commensurability problems at small r scales for all periodic data [14]. These include the choice of the ratio of the period of the data to the sampling period and the calculation of the argument of the data in double precision.

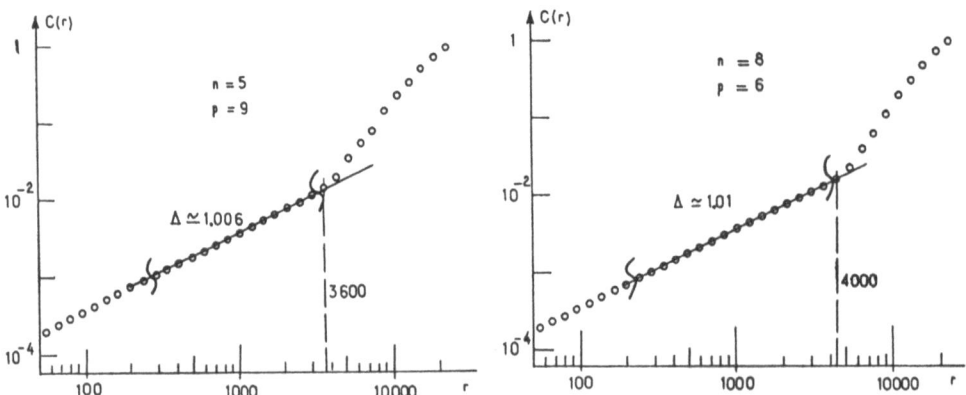

Figure 3. C(r) curves in log-log coordinates for two sine waves of frequencies f and f/10
$N = 1.5 \ 10^5$, $\Delta t = 1(\approx \frac{T_0}{40})$, m=12

For general periodic data, the presence of subharmonics can lead to errors in the estimation of the dimension. In Figure 3, the sum of a sine of period 1 and of a sine of period 10 and of same amplitude is analysed. The C(r) curve presents two distinct regions, in the first the slope is 1 and in the second there is a curvature and the maximum slope is 2.4. Geometrically, the data lies on a closed orbit on a torus. For small values of r one resolves the one-dimensional structure and for larger values the envelope. In this example, the value corresponding to the cross- over can be calculated [14]. For more complicated situations where a great number of subharmonics are present such as a flow undergoing a sequence of period doubling bifurcations the topology of the orbits will be so complex [15] that no information will be obtained from the C(r) curve.

One can also estimate the entropy for these systems. In [16] it is shown that the order 2 Renyi entropy, a lower bound for the metric entropy can be estimated from the correlation function: $C_b(r) \approx r^\nu e^{-b\Delta t K_2}$ $r \to 0$ $b \to \infty$. The authors of [16] use the euclidian norm, we have used the maximum norm [17] for two main reasons. First because the vectors $X(t)$ are reconstructed from time-delays on a single variable, and the norm is the maximum norm calculating r-coincidences on sets of vectors will be equivalent to doing it on a big vector spanning the previous set of vectors. Second with the maximum norm, the C(r) curve for a periodic data set will practically not evolve for embedding dimensions n and delays τ such that $n\tau \geq 10\ T_0$ where T_0 is the period of the orbit [14]. Instead for the euclidian norm which is unbounded, the distances between vectors (sets of vectors) will continue to grow as the embedding dimension is increased, so that the C(r) curves will always be shifted. From these considerations it is expected that the convergence of the estimators of the entropy will be slower with the euclidian norm.

II.2 multi-periodic data

Bi-periodic data can be generated by adding two sine waves of different frequencies. Geometrically ,this is the same situation as for periodic data except that now the orbit is dense on the torus. When the two components have same amplitude, the scaling region in the C(r) curve will be well defined.

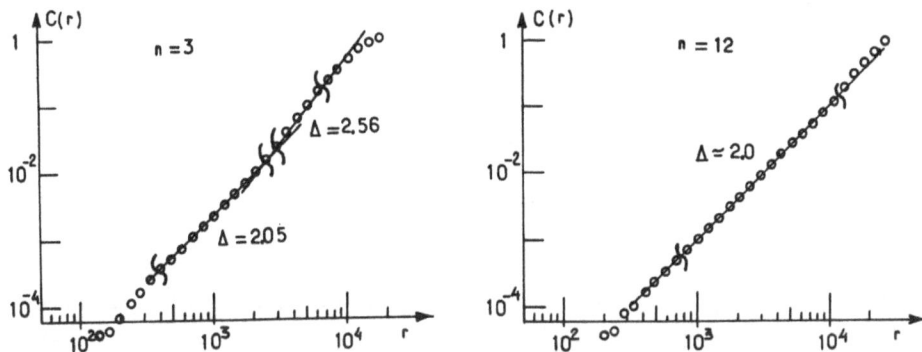

Figure 4. C(r) curves in log-log coordinates for two sine waves of irrationally related frequencies and same amplitude
$N = 1.5\ 10^5$, $\Delta t = 1 (\approx \frac{T_0}{40})$, p=9, m=1

Figure 4 shows the C(r) curves for such a data set for two different embedding dimensions $n = 3, 12$. For the first, the C(r) curve does not behave linearly in log-log scales. When the embedding dimension is increased and we get passed Takens's criterion [2], the scaling behavior becomes apparent and the value of the exponent in agreement with what is expected.

When the amplitudes of the periodic components are different, the curves present a knee shape. In figure 5 such a curve is shown for the sum of 2 sine waves of amplitude 1 and 3 %. The r value corresponding to the cross-over is 6 % of the diameter of the set of points. The value 2.11 for the slope in the small r region which overestimates clearly the dimension indicates an curvature effect (see section III). The value for r large can be explained using ideas introduced by Takens [18] for estimating the ν exponent and its error bar. He assumes that the density of probability of finding a distance $r_{ij} = \|X_i - X_j\|$ between r and $r + dr$ is $ar^{\alpha-1}dr$ so that C(r) varies as r^α. An estimator of the exponent α can be obtained through the average of $log(r)$ because

$$\int_0^1 ar^{\alpha-1} log r\, dr = -\frac{1}{\alpha}$$

Let us the assume that for data analysed $C(r) = C_1 r^{\alpha_1}$ for $r \leq r_0$ and $C(r) = C_2 r^{\alpha_2}$ for $r_0 \leq r \leq 1$. Because $C(1)=1$ and $C(r)$ is continuous $C_2 = 1$ and $C_1 = r_0^{\alpha_2 - \alpha_1}$. This will give us an estimator of the exponent at $r=1$ and for any $r > r_0$ by replacing r_0 by $\frac{r_0}{r}$ [14].

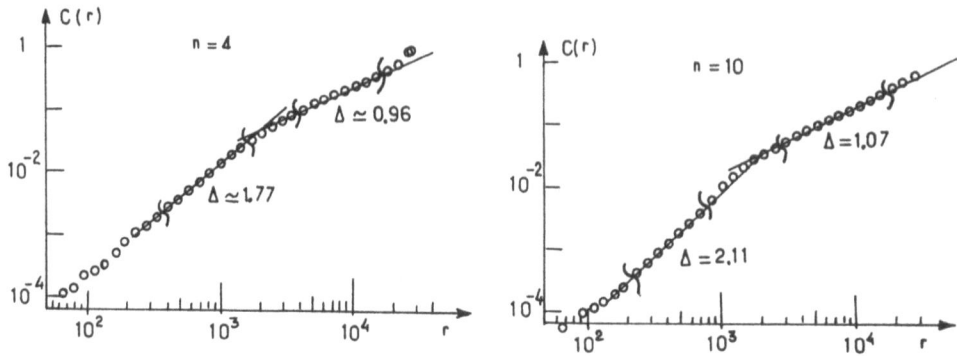

Figure 5. $C(r)$ curves in log-log coordinates for two sine waves of irrationally related frequencies and very different amplitudes
$N = 1.5 \ 10^5$, $\Delta t = 1 (\approx \frac{T_0}{40})$, p=9, m=12

Table 1 presents the exponent computed from the analytic formula with r_0= 3 % diameter of the set of points and the local slope computed from the experimental data for two different values of r. There is a good agreement because the geometry is relatively simple: for $r \leq r_0$ we are on a torus and the distribution of distances is such that $\alpha_1 = 2$ (neglecting any curvature effects), for $r \geq r_0$ $\alpha_2 = 1$. We have tested multi-periodic data sets with amplitudes that are relatively close; in that case, the method above fails except for the large values of r [14]. At this point we do not know exactly why, but we suspect that the variations assumed for $C(r)$ may be too simplistic and more complex geometric effects need to be taken into account because the amplitudes of the different components of the data may be too close.

Table 1. comparison between ν estimated experimentally and from a scaling argument derived from [18]

	ν experimental				ν analytic
	p = 9			p = 5	
r	n = 5	8	10	n = 3	
r = 1768	1.40	1.35	1.38	1.34	1.39
r = 2546	1.24	1.21	1.19	1.16	1.24

As regards the entropy for multi-periodic data sets, the C(r) curves will not saturate as is the case for periodic data because the signal does not reproduce itself. The quantity $F_q(b,r) = \frac{1}{q}\log\frac{C_b(r)}{C_{b+q}(r)}$ when $b \to \infty$ is an estimator for the entropy. In [17], the authors calculated $F_q(b,r)$ for up to 18 typical time-scales for an experimental bi-periodic data set and it decreased from .5 to .002. When the entropy is zero, there is no downward shift of the C(r) curves so it is possible to characterize very high dimensional multi-periodic sets (see section III). Then to estimate the possibilities of these embedding methods for the determination of dimension and entropy it is necessary to study systems with infinite entropy: stochastic systems.

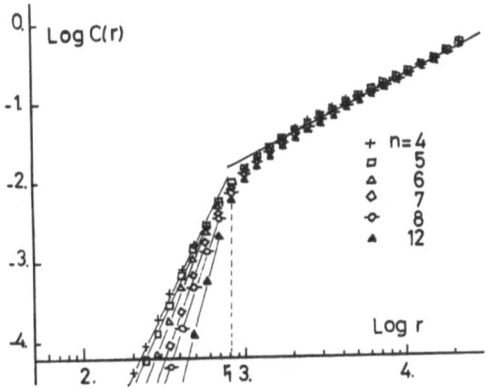

Figure 6. logC(r) vs. logr curves for a noisy limit-cycle $a = 1$. $b = 5$. 10^{-2} for different embedding dimensions
$N = 1.5 \ 10^5$, $\Delta t = 1(\approx \frac{T_0}{40})$, p=9, m=12

II.3 stochastic data

The first data analysed is white noise. In theory it has no correlation time so there is no influence of the delay or the sampling period; this is well verified in practice for white noise generated by standard random number generators. In [19], the authors applied the embedding technique to such data and found that the exponent was systematically below the embedding dimension for $n \geq 5$ due to edge effects. Choosing the centers on which the averaging was done far from the edge of the hypercube containing the data and increasing N to $1.5 \ 10^5$ delayed this effect till n=10. This effect can also be found in figure 3 of reference [20] where a fixed mass method of determination of the dimension is used. The author plots an estimate of the dimension as a function of the number of points for different neighbors. The dimension steadily decreases as the order of the neighbor increases because there are more and more edges between the points so that the boxes contain fewer and fewer points.

This is an edge effect and it shows that these are geometric effects linked to the distribution of points in phase-space. For such data, it is possible to predict the minimum distance from the center of the hypercube. The points have a radial distribution in r^{n-1} so that $\frac{1}{N} = \int_0^{r_{min}} \frac{r^{n-1}}{r_{max}^n} dr$. The C(r) curve can be calculated by using the isotropy of the data in the hypercube of R^n. A weighted average of the C(r) curves corresponding to each edge of dimension i $0 \leq i \leq n-1$ with the probability of being in that edge gives the global C(r) curve [14].

When the data is a correlated random noise the edge effects are not so drastic [14], the vectors are not distributed with the previous density because consecutive coordinates are not independant. It is then expected that the exponent will be larger in the correlated case than for the uncorrelated one. This is again what is observed in figure 4 in [20] for uniform and gaussian noise.

In [19], the influence of a random component on deterministic data was analysed in order to model the influence of experimental noise on the determination of the dimension. The data analysed was a noisy limit-cycle : $X(t) = \frac{a\sin(\omega t)+b\xi(t)}{a+b}$. Figure 6 shows a series of C(r) curves for such a data for different embedding dimensions. There are two regions in the curves: one below r_1 with an exponent which increases with the embedding dimension though slower due to the edge effects discussed above [19] and another above r_1 with an exponent $\nu = 1.05$. The fact that the exponent is larger than 1 can be modeled again using Takens's scaling arguments [18]. α_2 is set to be infinity, there is however a problem on the choice of r_0. For low embedding dimensions there is a good agreement with values obtained with $r_0 = r_{max}\frac{b}{a+b}$ and for high embedding dimensions with values obtained with $2r_0$ [14], this is due to the fact that for large embedding dimensions, the vectors are bunched against the sides of a hypercube in R^{n-1} suspended along the limit-cycle. The typical distance for the randomness to be felt is then $2r_0$. For a given r scale, the correction to the exponent due to a random component is proportional to $(\frac{r}{r_0})^{D_2}$ [18,14] so that for high dimensional data, where the values of r remain large it becomes neglegible.

The study of stochastic data will give some conclusions towards the understanding of high dimensional attractors. However the entropy for the latter is finite so that the manifold support of the data will be curved in R^n. For stochastic data which can be seen from a practical point of view as a deterministic data sampled at a very large sampling period the manifold will be folded so much that there will be no structure apparent on the scale observed.

III. CHAOTIC MODEL SYSTEMS

III.1 embedding dimension, time-delay and the entropy

In [21], it was shown by comparing data obtained from the Lorenz model [22] with data from a convection experiment that a careful estimation of the dimension and the entropy was possible at least on low-dimensional systems. When the embedding dimension and the delay were small the scaling region was to be found for very small values of r because the vectors were aligned along the diagonal in R^n. When for a fixed sampling period the embedding dimension and (or) the delay are increased the curves C(r) develop a systematic curvature in log-log coordinates. The local slope is in a first approximation a function only of the total length of time $T = (n-1)p\Delta t$ representing a vector in phase-space as can be seen on figure 7. This behavior as a function of T is to be linked with the dependance of C(r) on the entropy $C_n(r) \approx r^\nu e^{-(n-1)p\Delta t K_2}$ $r \to 0$ $b \to \infty$ [17]. A given value of T will correspond to an upper bound r_0 of the scaling region. For a system with a given dimension and entropy the number of data points N and centers m will select the accessible scaling range in the (r,C(r)) diagram depending on n, p and Δt. The noise range gives a lower bound in r and so does the sampling period.

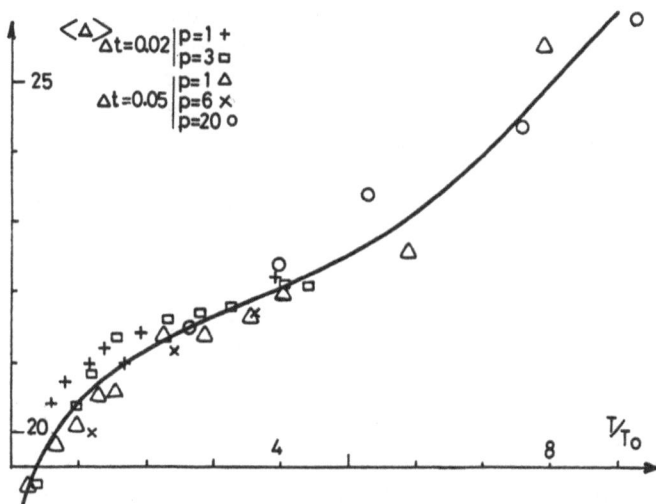

Figure 7. average local slope for $10^2 \leq r \leq 10^3$ as a function of T normalised by a typical time-scale T_0 for the lorenz system for different values of the embedding dimension, sampling period Δt and delay p.

III.2 sampling period

For high dimensional attractors, increasing n and p without changing the sampling period can lead to the generation of 'tails of slope 1 in the C(r) curve [19]. For those values of n,p and N there will be a value of r below which only consecutive points will be neighbors. To get more coincidences one could increase N but one should increase it by enormous amounts because of the exponential dependance of C(r). The remedy is when increasing T to increase the sampling period or to discard consecutive vectors [23]. Because of the behavior of C(r) there is a relation between the critical values of n, p, Δt, K_2 and the absolute value of the derivative of the data [14].

III.3 examples

Figure 8 shows C(r) curves for the Franceschini system [24] obtained after a systematic variation of the different parameters [14]. Only the curves for n=30 to 60 give an exponent in agreement with the Kaplan-Yorke dimension derived from the lyapunov exponents calculated in [25]. For n=15 and 20 there is a good scaling behavior but the dimension is underestimated. This effect may be the cause for the disagreement between the correlation dimension obtained for the Mackey-Glass system in [16,10] and the one in [11]. From the C(r) curves in figure 8 it is possible to extract an estimate of K_2 in agreement with the sum of the positive lyapunov exponents.

Because of the dependance of C(r) on the entropy, the main parameter is T. Fraser and Swinney advocate fixing the time-delay to an ideal value given by the minimum of mutual information [26]; this could lead to an underestimation of the dimension. It would be preferable to vary all parameters, except for maybe Δt for which there is a loose criterion, and fix their interesting ranges. The dependance of C(r) on the entropy is a big draw-back of the time-delay method, there is no such problem for vectors built from measurements in different locations. High dimensional non-chaotic systems such as a torus T_{20} can be characterized with a relatively small number of points but much smaller dimensional chaotic data would require 10 times as many points [14].

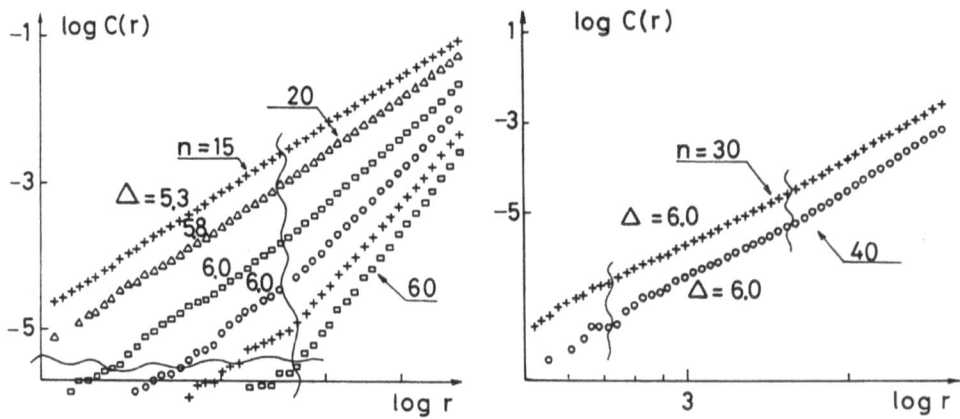

Figure 8. logC(r) vs. logr curves for the Franceschini model, parameters as in [25] $N = 10^4$, $\Delta t \approx \frac{T_0}{5}$, p=1, m=600
the second set of curves is for $N = 1.5\ 10^5$ and $m = 1.2\ 10^3$

REFERENCES

[1] R. Mane
in "dynamical systems and turbulence 1980"
Lecture Notes in Math. Springer-Verlag, New-York **898**, 230 (1981)

[2] F. Takens
in Ref. 366

[3] H.Froehling ,J. P. Crutchfield, D. Farmer, N.Packard and R. shaw
Physica **3D** 605 (1981)

[4] P. Grassberger and I. Procaccia
Phys. Rev. Lett. **50** 346

[5] C. Lausberg
PhD. doctoral dissertation, University of Grenoble (1987)

[6] P.Grassberger
Phys. Lett. **97A** 227 (1983)
G. Hentchell and I. Procaccia
Physica **8D** 435 (1983)

[7] M. Henon
Comm. Math. Phys. **50** 69 (1976)

[8] N.Abraham, A Albano, R. Das ,G. DeGuzman, S.Yong, R. Gioggia, G. Puccioni
and G. Tredicce
Phys. Lett. **114A** 217 (1986)

[9] M. Mackey and L. Glass
Science **197** 287 (1977)

[10] J. Havstadt and C. Ehlers
Phys. Rev. A **39**,nb. 2 845 (1989)

[11] D. Farmer
Physica **4D** 366 (1982)

[12] J. Theiler
Phys. Rev. A **36**,nb. 9 4456 (1987)

[13] G. Mayer-Kress
in "directions in Chaos, Vol. 3 of World Scientific Series on Directions in Condensed
Matter Physics" edited by Hao Bai Lin World Scientific, Singapore (1987)

[14] J.G. Caputo, B. Malraison and P. Atten
to be published

[15] J. D. Crawford and S. Omohundro
Physica **13D** 161 (1984)

[16] P. Grassberger and I. Procaccia
Phys. Rev. A **28**, 2591 (1983)

[17] J.G. Caputo and P. Atten
Phys. Rev. A **35**,nb. 3 1311 (1987)

[18] F. Takens
pre-print

[19] P. Atten, J.G. Caputo, B. Malraison and Y. Gagne
Journal de Mecanique Theorique et Appliquee Special Issue: Bifurcations and Chaotic
behavior
133 (1984)

[20] G. Broggi
J. Opt. Soc. Am. B , Vol **5** ,nb. 5, 1020 (1988)

[21] J.G. Caputo, B. Malraison and P. Atten
in "dimensions and entropies in chaotic systems"
Springer Series in Synergetics , Springer-Verlag, **32**, 180 (1986)

[22] E. Lorenz
J. Atm. Sci., Vol. **20**, 130 (1963)

[23] J. Theiler
pre-print

[24] V. Franceschini
Physica **6D** 285 (1983)

[25] R. Tavakol and A. S. Tworkovski
Physics Letters **102A** 273 (1984)

[26] A. Fraser and H. Swinney
Phys. Rev. A **33** 1134 (1986)

CHAOTIC BEHAVIOR OF THE FORCED HODGKIN-HUXLEY AXON

Michael Frame

Department of Mathematics
Union College
Schenectady, New York 12308

Using the method of Benetin, et. al., [3] and of Shimada and Nagashima [9], the Liapunov exponents of the stimulated Hodgkin-Huxley equations were estimated for different forcing amplitudes and frequencies. From the Kaplan-Yorke conjecture [7], the dimension of the underlying attractor in (V, n, m, h)-space was approximated. The Hodgkin-Huxley equations were taken in the form (see [6]) in which the unstimulated axon is not in a state of self-sustained oscillation. In this sense, the work of ([1],[2], [5]) is complemented.

A four point Runge-Kutta procedure with step size of 0.001 was used to generate a solution trajectory. The method of [3] and [9] consists of using the Jacobian of the differential equation to transport the tangent vectors along the trajectory. To keep track of the growth rate of n-dimensional volumes along the trajectory, and to avoid the problem of all basis vectors collapsing to that vector with the largest growth rate, the Gram-Schmidt method is applied to orthogonalize the transported vectors. The logarithms of the ratios of the lengths of the orthogonalized vectors to their original lengths, averaged along the trajectory, approximate the Liapunov exponents. The orthogonalized vectors are normalized and used as the new tangent basis. To calculate the dimension, the Liapunov exponents are sorted by size and the Kaplan-Yorke conjecture [7] applied.

Some regions in the stimulating amplitude-stimulating frequency plane were found in which chaotic behavior occurs. The "stability" of this chaotic behavior was investigated by tracking the largest Liapunov exponent and observing what appears to be a positive asymptotic value. For example, in the units of [6], for a stimulating frequency of 1.30 and amplitude of 111, the largest Liapunov exponent approaches a value around .016. At these parameter values, the dimension of the underlying attractor is about 2.15. Several graphical explorations in various state space planes (for example, m-h and m-n) were made and an intricate pattern of nested spirals was obtained.

For a fixed amplitude, increasing the frequency generally reduced the dimension (though not always: some interesting local variations were observed), eventually to one. This corresponds to the attractor's collapsing to a limit cycle, following the periodic stimulation. The boundary between periodic and chaotic behavior appears to have a complicated nature. Though we have not yet assembled enough information to test this quantita-

Measures of Complexity and Chaos
Edited by N.B. Abraham *et al.*
Plenum Press, New York

111

tively, nevertheless, the fractal boundary Moon observed [8] between periodic and chaotic motion in a forced mechanical oscillator makes plausible the existence of a fractal boundary between these regimes for the forced Hodgkin-Huxley axon. More details will appear elsewhere [4]. We shall continue this project, with the eventual aim of estimating the dimension of the boundary between the chaotic and periodic regimes. What physiological significance, if any, there is to these observations remains to be seen.

REFERENCES

1. K. Aihara, G. Matsumoto, Chaotic oscillations and bifurcations in squid giant axon, in: "Chaos," A. Holden, ed. Princeton University Press, Princeton (1986).

2. K. Aihara, G. Matsumoto, Y. Ikegaya, Periodic and non-periodic responses of a periodically forced Hodgkin-Huxley oscillator, J. Theoret. Biol. 109: 249 (1984).

3. G. Bennetin, L. Galgani, J. Strelcyn, Lyapunov characteristic exponents for smooth dynamical systems and for Hamiltonian systems; a method for computing all of them, Meccanica 15:9 (1980).

4. M. Frame, K. Bodden, Strange attractors in the forced Hodgkin-Huxley axon.

5. H. Hayashi, M. Nakao, K. Hirakawa, Chaos in the self-sustained oscillation of an excitable biological membrane under sinusoidal stimulation, Phys. Lett. A 88:265 (1982).

6. A. Hodgkin, A. Huxley, A quantitative description of membrane current and its applications to conduction and excitation in nerve, J. Physiol. 117:500 (1952).

7. J. Kaplan, J. Yorke, Chaotic behavior of multidimensional difference equations, in: "Functional differential equations and approximation of fixed points" H. Peitgen, H. Walther, eds. Springer-Verlag, Berlin (1979).

8. F. Moon, Fractal boundary for chaos in a two state mechanical oscillator, Phys. Rev. Lett. 53:962 (1984).

9. I. Shimada, T. Nagashima, A numerical approach to ergodic problem of dissipative dynamical systems, Progr. Theoret, Phys. 61:1605 (1979).

CHAOTIC TIME SERIES ANALYSIS USING SHORT AND NOISY DATA SETS:

APPLICATION TO A CLINICAL EPILEPSY SEIZURE

G.W. Frank, T. Lookman, M.A.H. Nerenberg
Department of Applied Mathematics
University of Western Ontario
London Canada N6A 5B9

In recent years the methods of chaotic time series analysis have been applied to data sets from a number of experimental systems. In this note we report on work in progress on the usefulness of chaotic time series analysis as a potential diagnostic tool in the classification of epileptic seizure activity [1]. Seizure episodes are classified by correlation dimension and estimated largest Lyapunov exponent. The exponent is found using a modified version of Wolf's method [2]. We begin with a brief discussion of some of these modifications before proceeding to a discussion of their application to an epilepsy data set.

Lyapunov exponent

The method of delays [3] can be used to reconstruct the strange attractor of a nonlinear dynamical system from a univariate time series. The largest Lyapunov exponent associated with the recovered flow may be found using Wolf's method [2], which monitors the evolution of infinitesimal displacement vectors within the recovered attractor. Such vectors are constructed from a benchmark point on the trajectory and a carefully selected neighbour point with a large temporal separation. The two points are then allowed to evolve under the action of the reconstructed phase flow and the magnitudes of the initial and evolved displacement vectors examined for evidence of the exponential growth characterizing chaos. In order for the method to provide a useful estimate it is important to examine *only* those expansion rates in the direction of maximum instability, that is tangent to the unstable manifold. Displacement vectors are pulled in this direction by the unstable dynamics, and those vectors which are allowed to evolve a sufficient time are assumed to lie along the unstable tangent subspace.

A replacement process is needed to prevent evolving vectors from becoming large enough to sample global nonlinear attractor structures such as folds. Ideally, replacement points should be chosen to lie directly along the direction of greatest instability, but data sparseness owing to finite time series makes such replacements extremely unlikely. Replacement points must instead be selected from a pool of candidates which populate a narrow cone centered on the previously evolved displacement vector. In order to ensure only local dynamical behaviour contributes to the exponent estimate, we must discard points which lie outside the region over which dynamics are assumed approximately linear, and closer than anticipated average noise levels. Both these quantities may be estimated from the graph of the correlation function [5,6].

The key to good exponent estimation is the careful selection of a candidate from the replacement cone which best estimates the direction of greatest expansion. We offer a few brief observations on replacement selection, which will be more fully developed in a forthcoming paper [6].

Measures of Complexity and Chaos
Edited by N.B. Abraham *et al.*
Plenum Press, New York

(1) Introduce a priority function to assign a goodness of fit' weighting to each candidate within the replacement cone, based on both its proximity to the evolved benchmark and angular deviation from the evolved displacement vector. Wolf's original implementation simply chose the nearest neighbour within the cone, regardless of its alignment. This choice often leads to slow convergence of the estimated Lyapunov exponent owing to an accumulation of random orientation errors. We have found that weighting candidates with respect to the angular deviation of the displacement vector yields greatly improved convergence properties. It is often better to use a more distant but better aligned point than one which is nearby but poorly aligned. In practice a weighting function which uses both radial and angular separation is often used [6].

(2) The subsequent evolution of the selected replacement point must be considered. Rapid divergence can occur when neighbouring trajectories diverge for topological reasons other than chaos. An example of such a catastrophic divergence occurs when two neighbouring trajectories evolve through opposite lobes of the Lorenz attractor. The resulting trajectory divergence can provide an enormous (and erroneous) contribution to the Lyapunov exponent estimate. This can also result from geometric properties of the reconstruction, such as degenerate structure resulting from a mistakenly low-dimensional reconstruction or a poorly chosen delay time. Such misleading replacements can largely be avoided by evolving both the candidate replacement point and the benchmark site forward in time, and monitoring the angular separation of vectors tangent to the original and perturbed trajectory segments. Rapidly increasing angular separation signals the occurrence of catastrophic replacement, and the candidate is rejected. Screening candidate replacement points in this way was found to improve both the convergence and stability properties of exponent estimates.

(3) The determination of the largest Lyapunov exponent must be regarded as an asymptotic process. Finding an exact exponent value is possible only in the ideal limit of an infinite amount of noiseless data. Exponent estimates from experimental systems must be interpreted in the context of the constraints imposed by short, noisy data sets. Intermediate results based on limited time series length and quality must be extrapolated to this ideal limit before a meaningful exponent estimate is possible [1,6].

Epilepsy data set

We turn now to an analysis of an epilepsy data set. Previous work by a number of authors has indicated the possibility of dynamical chaos within various states of brain function [7,8]. Chaotic time series analysis of data sets obtained from electroencephalogram (EEG) recordings suggests the existence of very low-dimensional ($d_c < 3$) behaviour associated with intense absence (*petit mal*) seizure events [7,8]. Such events are usually short-lived, characterized by highly coherent spike-wave complexes lasting only a few seconds, with a characteristic frequency of approximately 3 Hz. The brevity of the seizure event places a severe restriction on the analysis of such seizures. As the total sampling time is usually less than 20 orbits of the recovered 'attractor,' it is unclear whether computations based on such sparse data can provide accurate estimates of dynamical quantities such as correlation dimension or Lyapunov exponents.

Data sparseness does not, however, preclude the usefulness of chaotic time series analysis as a potential diagnostic tool. To demonstrate this we applied these methods to data taken from markedly different seizure events obtained within a clinical environment. The seizure episodes studied were of much longer duration, consisting of approximately 15,000 measurements at 200 Hz, and were no longer simple absence seizures, but generalized epileptic seizures of both petit and grand mal states. While steps were taken to minimize noise contamination from extraneous sources, noise levels within the data set reflect those encountered in common practice. These events were more complex states extending over a longer time interval than previous analyses, and are thus more representative of events encountered within a clinical setting. An unambiguous indication of chaos within such seizures would suggest the feasibility of chaotic time series analysis as a diagnostic device.

The seizures were tested for evidence of chaos by computing the correlation dimension and the largest Lyapunov exponent. We present details of only one of several episodes studied. EEG data sets were taken using specifics of [1]. Our primary data set was from a 35 year old male with primary generalized seizures involving the cortex bilaterally, symmetrically and diffusely. Fourier transforms of the signal revealed continuous activity at all observable frequencies. The autocorrelation function suggested an average rate of information dissipation on the order of $k \sim 1 bit/sec$. This figure estimates the total rate of information dissipation from all sources, both random and chaotic [6].

The method of delays was used to generate multidimensional embeddings of the univariate time series. The delay time was chosen as the first zero of the autocorrelation function. Visual inspection showed this delay time provided a good unfolding of the reconstruction. A transformation to Karhunen-Loeve coordinates was performed, followed by a rescaling of data to a unit hypercube. This provided a uniform distance measure for all directions within the reconstruction.

The Grassberger-Procaccia algorithm [4] was used to obtain estimates of the correlation dimension, found to be $d_c \sim 6.5$ across a range of delay times. The graph of the correlation function also provided estimates of average noise levels and scaling lengths over the reconstruction [4,5]. Both quantities are relevant to a meaningful estimation of the largest Lyapunov exponent.

Our implementation of Wolf's method was applied to the time series data. A weighting function using both the angular and radial separations of the candidate replacement point from the evolved displacement vector was used to allow good estimation of the direction of greatest instability. The data density within the reconstruction was sufficiently high to allow replacement points to be selected from within a cone with a 20-30 degree apex with a 7-dimesional embedding. Narrow cone angles must be used when working with higher-dimensional embeddings as the method can break down when data diffuses to the point where extremely large cone search angles must be used. Care was taken to avoid selecting replacements which diverged catastrophically over short time intervals.

The estimated Lyapunov exponent was found for evolution times between 1-10 seconds and embedding dimensions from 3 to 12. Behaviour for each embedding dimension suggested a noisy signal containing a deterministic component. Extrapolation of results to infinite evolution times yielded small but consistent positive largest Lyapunov exponents on the order of $\lambda \sim 0.1 bit/sec$ [1]. This figure is consistent with the total average rate of information dissipation within the system, as found by the autocorrelation function. Exponent estimates remained approximately constant for embedding dimensions in excess of 6, consistent with dynamics being (locally) confined to an (approximately) 6-dimensional subspace. The possiblity of chaos can also be assessed by using Wolf's method with a time-reversed time series. One would expect from the irreversibility of chaotic dynamics that Wolf's method would give different results with time reversed ordering when applied to a chaotic time series. We find this is indeed the case for the seizure data, with the estimated Lyapunov exponent of the time-reversed ordering substantially different from that of a normal time ordering. A further comparison with a time series of 'normal' resting states from the same individual showed this state to be characterized by a largest Lyapunov number statistically indistinguishable from zero and correlation dimension in excess of 12.

Conclusion

The primary restriction on the analysis of high-dimensional data sets is the enormous amount of data necessary to provide sufficiently dense attractor coverings. Current estimates of data requirements suggest the number of data points grows rapidly with the number of dimensions. In the face of such constraints, the interpretation of our estimates of the correlation dimension and Lyapunov exponent for such a high-dimensional data set is not clear. Nevertheless, we have demonstrated that the methods used here can distinguish between normal and seizure states using noisy data from short time series. The data displays an inherent sense of time ordering consistent with chaos. We have also found a consistent indication of chaotic dynamics of similar complexity found using two completely different methods.

REFERENCES

1. Frank, G.W., T. Lookman, M.A.H. Nerenberg, C. Essex, submitted to *Nature*.

2. Wolf, A. J.B. Swift, H.L. Swinney, J.A. Vastano, Physica **16**D (1985), 285.

3. Packard, N.H., J.P. Crutchfield, J.P. Farmer, R.S. Shaw, Phys. Rev. Lett. **45** (1980), 712.

4.   Grassberger, P., I. Procaccia, Phys. Rev. Letts. **50**, 5 (1983), 346.

5.   Ben-Mizrachi, A., I. Procaccia, P. Grassberger, Phys. Rev. A **29**, 2 (1984), 975.

6.   Frank, G.W., T. Lookman, M.A.H. Nerenberg, C. Essex, manuscript in preparation.

7.   Babloyantz, A., A. Destexhe, Proc. Natl. Acad. Sci. USA, **83** (1986), 3515.

8.   Rapp, P.E., I.D. Zimmerman, A.M. Albano, G.C. Guzman, N.N. Greenbaum, T.R. Bashore, *Nonlinear Oscillations in Biology and Chemistry* (H.G. Othmer, ed.) Springer-Verlag, Berlin.

MEASURING COMPLEXITY IN TERMS OF MUTUAL INFORMATION

Andrew M. Fraser

Center for Nonlinear Dynamics
Department of Physics
University of Texas
Austin, Texas 78712

An alternative definition of the KS entropy h_μ based on mutual information is proposed. The new definition is designed to handle experimental noise more gracefully than the standard definition. An example is used to illustrate the difference.

The entropy of the distribution of a random variable b (discrete, continuous, vector, or scalar) is defined to be

$$H(B) = \langle -\log P(b) \rangle .$$

For the joint distribution of two random variables a and b one can define the joint entropy, conditional entropy, and mutual information by

$$
\begin{aligned}
H(A,B) &= \langle -P(a,b) \rangle \\
H(A|B) &= H(A,B) - H(B) \\
I(A;B) &= H(A) + H(B) - H(A,B)
\end{aligned}
$$

respectively[1]. Boltzmann introduced these kinds of quantities for use in the kinetic theory of gasses and Shannon built the mathematical theory of communication or information theory on them. A primary interest of Shannon's was characterizing an information source X. As an example one can think of a teletype being used to send a sequence of characters from a finite alphabet. If $x(1)$ and $x(2)$ are iid (identically independently distributed) then $H_n(X) = nH_1(X)$ where $H_n(X) = H(X(1), X(2), \ldots, X(n)) = \langle -\log\left[P\left(x(1), x(2), \ldots, x(n)\right)\right] \rangle$. In general there are statistical relationships between characters that are close to each other and the entropy of the source is defined as

$$
\begin{aligned}
h(X) &= \lim_{n\to\infty} \frac{1}{n} H_n(X) \\
&= \lim_{n\to\infty} H'_n(X)
\end{aligned}
$$

where $H'_n(X) = H_{n+1}(X) - H_n(X)$.

By applying Shannon's idea of source entropy to dynamical systems, Kolmogorov[2] and Sinai[3, 4] were able to solve a long standing classification problem in ergodic theory. A dynamical system operates on a continuous variable, i.e., $x(t+1) = \phi(x(t))$. By dividing the possible values of x into a *partition* consisting of a finite number of disjoint subsets, $\alpha = \{a_1, a_2, \ldots, a_k\}$ one can convert a sequence of x's into a sequence of a's with $x(t) \in a(t)$. The entropy of a map ϕ and measure μ with respect to a partition α is defined

$$h_\mu(\phi, \alpha) = \lim_{n\to\infty} \frac{1}{n} H_n(A)$$

Measures of Complexity and Chaos
Edited by N.B. Abraham *et al.*
Plenum Press, New York

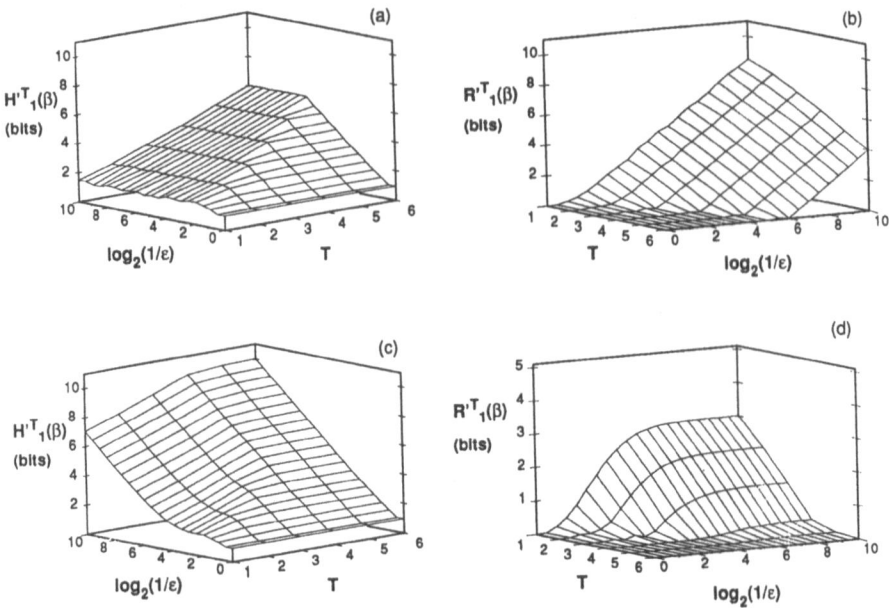

Figure 1. An intuitive illustration of conditional entropy $H'_n(\beta)$. $H'_n(\beta)$ is roughly the log of the number of different boxes into which a trajectory might fall after evolving one step from a state with a particular specification. Plots (a) and (b) illustrate the regime in which the partition resolution ϵ is small but much larger than the noise scale η, (c) and (d) illustrate the regime in which η is larger than ϵ, and (e) and (f) illustrate the regime in which the partition is so coarse or incomplete that almost nothing can be predicted about where a trajectory will go.

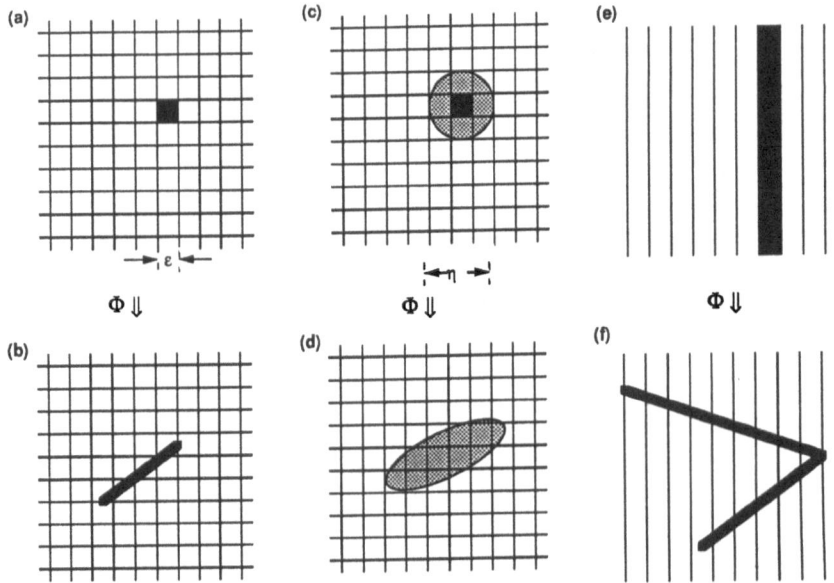

Figure 2. The dependence of $R_1'^T(\beta)$ and $H_1'^T(\beta)$ on partition resolution ϵ, delay time T, and noise is shown. The dynamical system is the 1-d map $s_{n+1} = \phi(s_n) = 2s_n$ mod 1. (a) and (b) are results for noise-free measurements, while (c) and (d) show the effect of 5% measurement noise. Without noise $R_1'^T(\beta)$ increases linearly with $\log(1/\epsilon)$ and $H_1'^T(\beta)$ is well behaved. When noise is added $R_1'^T(\beta)$ behaves well and $H_1'^T(\beta)$ diverges in the limit of fine partitions (large $\log 1/\epsilon$).

Table 1. Approximate Dependence of $H_1^{IT}(\beta)$ and $R_1^{IT}(\beta)$ on Partition Resolution ϵ.

Resolution	$H_n^{IT}(\beta)$	$R_n^{IT}(\beta)$
small ϵ	$Th_\mu + \log(\eta/\epsilon)$	$\log(1/\eta) - Th_\mu$
mid ϵ	Th_μ	$\log(1/\epsilon) - Th_\mu$
large ϵ	$\log(1/\epsilon)$	0

and the KS entropy $h_\mu(\phi)$ is obtained by finding an α that gives the largest possible value

$$h_\mu(\phi) = \sup_\alpha h_\mu(\phi, \alpha).$$

Note that $h_\mu(\phi^T) = Th_\mu(\phi)$. In numerical studies, when searching for an α to attain the supremum, one often uses a rectangular grid with a spacing ϵ and then makes ϵ small. Such a procedure yields the operational definition

$$h_\mu(\phi) = \lim_{\epsilon \to 0} h_\mu(\phi, \alpha_\epsilon).$$

Following suggestions of Rob Shaw[5], I developed an alternative approach to estimating h_μ, described in [6]. Let

$$R_n' = I(x(n+1); x(t+1), x(t+2), \dots, x(t+n)) = H_n(X) + H_1(X) - H_{n+1}(X)$$

and let R_n^{IT} indicate that the map ϕ^T is being considered. This notation yields the operational definition

$$h_\mu(\phi) = \lim_{n \to \infty} \lim_{\epsilon \to 0} \frac{R_n^{IT} - R_n^{It}}{t - T}.$$

The standard definition of h_μ is the rate at which information is required to specify the time series. For a noisy time series, unless an ad hoc minimum length scale is defined, that rate is infinite because it requires an infinite amount of information to select a real number out of a continuum, and such a selection is required for each element of the time series. The mutual information based definition of h_μ measures how predictability fades as separations in time increase. It does not diverge due to measurement noise, but in the limit of zero noise, infinities of mutual information appear which preclude its use for defining complexity. Thus the KS entropy and a mutual information based measure of complexity are complementary.

The idea of defining complexity as the rate at which predictability decays can be transferred from time series to spatial patterns yielding a quantity that we would call the spatial entropy. Spatial entropy, like its temporal counterpart, is coordinate invariant. Notice that the spatial entropy is zero for either a completely random field or a uniform field. This property is desirable because experimentally the only difference between the two fields is one of scale (the contrast knob on a camera for instance). If the coherence of a spatial field is the degree to which one part of the field can be used to predict what another part of the field will look like, then spatial entropy is the decay rate in space of the coherence.

Acknowledgment. This work was supported by the Department of Energy Office of Basic Energy Sciences.

References

[1] R. G. Gallager. *Information Theory and Reliable Communication.* John Wiley and Sons, New York, 1968.

[2] A. N. Kolmogorov. A new metric invariant of transient dynamical systems and automorphisms in lebesgue spaces. *Dokl. Akad. Nauk. SSSR*, 119:861–864, 1958. English summary in Mathematical Reviews, vol. 21, pp. 386, 1960.

[3] Y. Sinai. On the concept of entropy for a dynamic system. *Dokl. Akad. Nauk. SSSR*, 124:768–771, 1959. English summary in Mathematical Reviews, vol. 21, pp. 386-387, 1960.

[4] Y. Sinai. *Introduction to Ergodic Theory.* Princeton University Press, Princeton, 1976.

[5] R. Shaw. Strange attractors, chaotic behavior, and information flow. *Z. Naturforsch*, 36a(1):80–112, Jan. 1981.

[6] A. M. Fraser. Information and entropy in strange attractors. *I.E.E.E. Transactions on Information Theory*, 35(2):245–262, March 1989.

ESTIMATING LYAPUNOV EXPONENTS FROM APPROXIMATE RETURN MAPS

J.N. Glover

Department of Mathematics, University of Western Australia
Nedlands 6009 Australia

Introduction

Phase portrait reconstructions from experimental time series are widely used for the calculation of dimensions, Lyapunov exponents and entropy, all of which can be used to characterise a dynamical system and investigate the effect of varying external parameters (Schuster 1984). Often it is desirable to reduce the dimensionality of the system by the method of Poincaré sections (Hirsch and Smale 1974). The reduction in the amount of data available for calculating quantities such as correlation dimension is compensated for by the improvement in quantitative and qualitative deductions that can be made (Brandstater and Swinney 1987). An alternative approach, sometimes possible in low dimensions, is to consider an approximate one-dimensional return map of the flow, for example by plotting successive minima of some component (Sparrow 1982, Chapman et.al. 1989). If the trajectories end up near an attractor with Hausdorff dimension slightly greater than two, then the resulting return map is close to one dimensional and the comprehensive theory of one-dimensional maps can be used to characterise bifurcations in the system as a parameter is varied. In some cases this approach can be made rigorous and when it works it is simpler than using the reconstructed flow. In this report the two approaches are related: the first produces a 'Poincaré section of a Reconstruction' while the second produces a 'Reconstruction of a Poincaré section'. The one-dimensional approximate return map can then be used to estimate the largest Lyapunov exponent of a Poincaré map of the flow. Only the case in $\mathbb{R}^3$ is considered and when the measured signal is the first component of the state vector $x(t)$.

Return Maps and Poincaré Sections

Consider the dynamical system defined by the first order autonomous differential equation

$$\dot{x} = f(x(t)) \tag{1}$$

where $f : \mathbb{R}^3 \to \mathbb{R}^3$ is a C^1 vector field.

An experiment is a set of realisations of a map $\Phi : \mathbb{R}^3 \to \mathbb{R}$ which generates a measured output signal $\phi(t)$. Here Φ is the projection of $x(t)$ onto the first component, namely,

$$\phi(t) = \Phi(x(t)) = x_1(t) \tag{2}$$

Suppose that there is a one dimensional return map G which approximates succesive minima of $\phi(t)$ which we denote by the sequence $\{\phi_k\}_{k=1}^{\infty}$, that is

$$\phi_{k+1} = G(\phi_k) + \epsilon_k. \tag{3}$$

where $\epsilon_k \ll \phi_{k+1}$ is a small error term.

The pairs of points (ϕ_k, ϕ_{k+1}) will lie within $\epsilon = \max \{|\epsilon_k|\}$ of the graph of G given by $(w, G(w))$ where w and ϕ_k lie in the same range. If ϵ is sufficiently small then the sequences $\{G^{k-1}(\phi_1)\}$ and $\{\phi_k\}$ should remain close for reasonably long times.

Measures of Complexity and Chaos
Edited by N.B. Abraham *et al.*
Plenum Press, New York

If ϕ_k is a minima of $x_1(t)$ occuring at $t = t_k$, then $\dot{x}_1(t_k) = 0$ and $\ddot{x}_1(t_k) > 0$. At $t = t_k$ the trajectory $x(t)$ intersects a set,

$$S = \{x : f_1(x) = 0, h(x) > 0\} \tag{4}$$

where

$$h(x) = \sum_{i=1}^{3} f_i(x) \frac{\partial f_1(x)}{\partial x_i}. \tag{5}$$

Direct substitution shows $f_1(x(t_k)) = 0$ and $h(x(t_k)) > 0$. This set S does not contain any fixed points of the flow. In addition assume that there are no critical points of $df_1(x)$ in S and that the flow intersects S transversally. This ensures that S is a Poincaré section of the flow and there is a well defined C^1 map P relating successive intersections of $x(t)$ with S,

$$x_{k+1} = P(x_k), \tag{6}$$

where the subscript k denotes the kth intersection of $x(t)$ with S and not its kth component.

If we measure only the first component of the discrete two-dimensional time series $\{x_k\}_{k=1}^{\infty}$ then this is the same as the sequence of minima $\{\phi_k\}_{k=1}^{\infty}$. It follows from Takens embedding theorem (Takens 1980) that a delay reconstruction of $\{\phi_k\}_{k=1}^{\infty}$ is an embedding of the orbit $\{x_k\}_{k=1}^{\infty}$ on the Poincaré section S. Such an reconstruction is given by the sequence of vectors $\{v_k\}_{k=1}^{\infty}$ where $v_k = (\phi_k, \phi_{k+1})$. This is precisely the sequence which we used to find the approximate return map G; this establishes a simple relation between a Poincaré section of the flow and this return map G.

The Rössler Attractor

To illustrate the above ideas consider the Rössler equations

$$\begin{aligned} \dot{x}_1 &= -x_2 - x_3 \\ \dot{x}_2 &= x_1 + ax_2 \\ \dot{x}_3 &= b + x_3(x_1 - c) \end{aligned} \tag{7}$$

for $a = 0.15$, $b = 0.2$ and $c = 10.0$. For these parameter values trajectories are believed to be chaotic, and hence the largest Lyapunov exponent is greater than zero. Figure 1 shows a plot of successive minima of $x_1(t)$ and these appear to be well approximated by a one-dimensional graph. From the discussion above figure 1 can also be viewed as a reconstruction of a Poincaré section of the flow. From the equations (7) the equation of the Poincaré section, S, can be obtained explicitly,

$$S = \{x : x_2 + x_3 = 0; h(x) > 0\} \tag{8}$$

where

$$h(x) = -x_1 - ax_2 - b + x_3(x_1 - c) \tag{9}$$

Figure 2 shows an orbit on the Poincaré section S. While this contains the same dynamical information as the reconstruction this information cannot be obtained from a static representation of the data. The dynamics of the reconstruction however can be obtained from a static representation by using the one-dimensional approximation.

Estimating Lyapunov Exponents

The embedded orbit shown in figure 1 is generated by a discrete dynamical system of the form (6) and hence we can estimate dynamical quantities such as Lyapunov exponents and entropy. We now show that the largest Lyapunov exponent λ_1, can be estimated from the one-dimensional map G. Since λ_1 measures the average amount of stretching of a region of the attractor over long periods we expect to be able to estimate it from G. This is because the difference between $(\phi_k, G(\phi_k))$ and (ϕ_k, ϕ_{k+1}) is determined by the contraction of the phase space onto the attractor and this is measured by the second Lyapunov exponent λ_2, which is independant of λ_1. In approximating the two dimensional dynamical system by G we are implicitly assuming that $\lambda_2 \ll 0$. It is reasonable to assume that for such systems the single Lyapunov exponent λ_G, of the attractor of G is a good estimate of λ_1. This is defined as

$$\lambda_G = \lim_{n \to \infty} \frac{1}{n} \log(|G'(w_n)G'(w_{n-1}) \cdots G'(w_1)|) \tag{10}$$

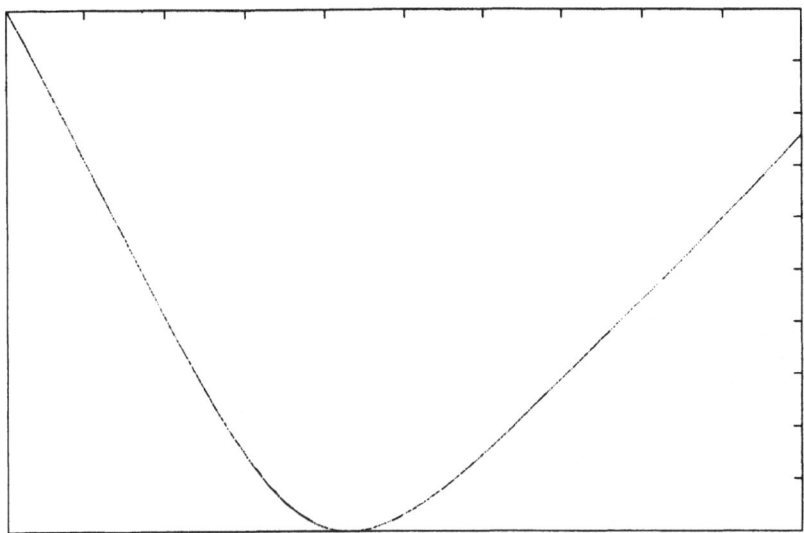

Figure 1. A plot of successive minima of the first component of a trajectory of equation(7). Alternatively, an embedding of the Poincaré section shown if figure 2.

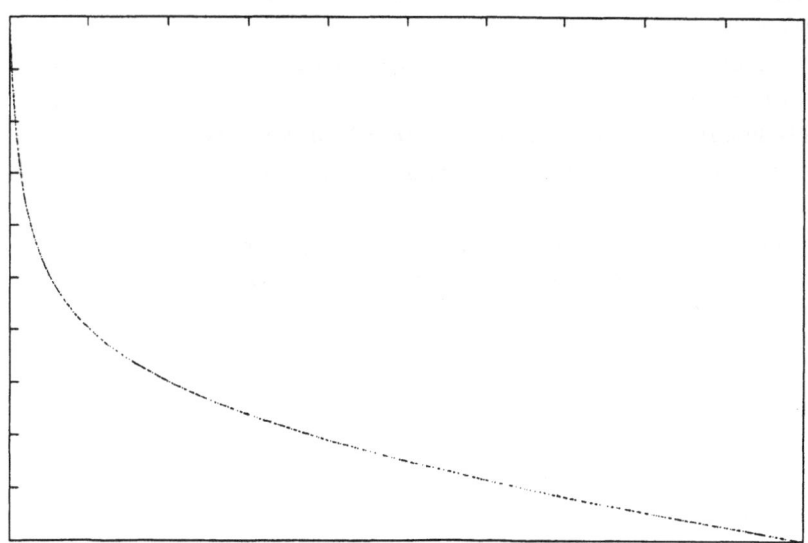

Figure 2. A Poincaré section of the Rössler attractor.

for almost all initial conditions w_1. An equivalent defintion is

$$\lambda_G = \int \log(|G'(w)|)d\mu(w) \tag{11}$$

where $\mu(w)$ is the invariant measure on the attractor. To estimate this integral a histogram is made of the relative number of times that the sequence $\{\phi_k\}_{k=1}^\infty$ is in a subinterval B_i of size h. For a large number of points this converges to an estimate of $\mu(B_i)$. If the attractor has diameter L then h should be chosen so that $\epsilon \ll h \ll L$. This ensures that the fine structure of the attractor does not influence our estimate of λ_G. To estimate the contribution of $G'(w)$ to the integral we take the values of ϕ_k closest to the upper and lower endpoints of B_i and use a linear approximation $G'_i \equiv (\phi_{k,i}^u - \phi_{k,i}^l)/h$. The Lyapunov exponent is estimated by

$$\lambda_G = \sum_i \mu(B_i)G'_i. \tag{12}$$

For the Rössler attractor example λ_G was calculated for an embedding of 4000 successive minima of $x_1(t)$. The estimate was found to be stable over a large range of values of h. For example: for 5 bins ($h \approx 2.0$) the estimate was $\lambda_G = 0.588$; for 20 bins ($h \approx 0.5$) the estimate was $\lambda_G = 0.532$; for 100 bins ($h \approx 0.1$) the estimate was $\lambda_G = 0.531$. These values suggest that $\lambda_1 \approx 0.55$.

Conclusion

We have demonstrated that a relationship between Poincaré sections and approximate return maps of some attractors enables us to not only deduce qualitative properties of a chaotic system but also to estimate dynamical quantities which characterise the chaotic behaviour. This approach, where it works, is considerably simpler than methods which attempt to estimate these quantities directly from a Poincaré section of the flow.

References

Brandstater A., Swinney H.L., 1987 *Strange Attractors in Weakly Turbulent Cuoette-Taylor Flow,* Phys. Rev.A, 2207 (**35**).

Chapman P.B., Glover J.N., Mees A.I. 1989 *The Dynamics of the Rikitake Equations from the Stiff Limit* submitted for publication.

Hirsch M.W., Smale S.,1974 *Differential Equations, Dynamical Systems , and Linear Algebra*, Academic Press, New York.

Schuster H.G.,1984 *Deterministic Chaos: An Introduction*, VCH, New York, 117-139.

Sparrow C.,1982 *The Lorenz Equations: Bifurcations,Chaos and Strange Attractors*, Apppl. Math. Sci. **41**. Springer-Verlag, New York.

Takens F.,1980 *Detecting Strange Attractor in Turbulence.*, Dynamical Systems and Turbulence. Lecture Notes in Mathematics. Vol. 898, Rand D.A. and Young L.S., eds, 365-381. Springer-Verlag, New York.

ESTIMATING LOCAL INTRINSIC DIMENSIONALITY USING THRESHOLDING TECHNIQUES

A. Goel, S.S. Rao, and A. Passamante *

EE Department   *Naval Air Development Center
Villanova University   Warminster, PA 18974
Villanova, PA 19085

The problem of determining the number of dominant (signal related) singular values in estimating Local Intrinsic Dimensionality (LID) using Singular Value Decomposition (SVD) is considered. Earlier a method for estimating the LID using the SVD was proposed when the observed data is corrupted by noise. Problems are encountered when the Signal to Noise Ratio (SNR) gets very high or very low. For noisy data the algorithm will produce higher dimensionality even when the observed system has low dimension. A signal/noise separation criterion is proposed based on the analysis of the perturbation matrix to identify the number of dominant singular values. Results are presented for some standard chaotic signals and compared to the previously used approach, showing the superiority of the criterion used at high SNR's.

I. INTRODUCTION

In recent years, identification of chaotic systems has become a very important task due to the inherent low dimensionality of the chaotic systems. Modelling a system that is chaotic needs fewer parameters than the corresponding stochastic system. Earlier a method was proposed using the SVD to estimate the LID which is an upper bound on the fractal dimensionality of the reconstructed attractor.[1,2] In the earlier proposed method, identification of the dominant singular values turned out to be the key step in estimating the LID when the data is corrupted by noise.

SVD has become a very powerful tool in the analysis of signals corrupted by noise because of it's robustness.[3] In this paper we examine a thresholding technique to estimate the true rank of the data matrix by finding the number of dominant singular values, which in turn specifies the local intrinsic dimensionality.

II. BACKGROUND AND ANALYSIS OF THE PERTURBATION MATRIX

In this paper we follow the work of ref. 2 and begin our analysis after the SVD of the local data matrix A which is composed of the embedded data vectors.

The key step in the procedure mentioned above is to find a thresholding criterion to differentiate between the significant (signal) singular values and insignificant singular values of the data matrix, thus identifying the rank of it. We developed a criterion using perturbation analysis results used in the field of matrix computations.

Let the data matrix under consideration be A and the perturbation matrix be δA. The perturbed matrix B can be represented as the summation of the data matrix A and the perturbation matrix δA:

Measures of Complexity and Chaos
Edited by N.B. Abraham *et al.*
Plenum Press, New York

$$B = A + \delta A \tag{2.1}$$

Elements of matrix δA are the errors due to estimation and the noise effect in the data. Let's assume that the data matrix A is of dimension m*n and of rank r, where r is less than or equal to the lesser of m and n. The perturbation matrix is usually of full rank and the effect of adding it to the data matrix A can be seen as increasing the rank of the data matrix. The singular value decomposition is a very good tool for finding the effective rank of the perturbed matrix B.

If the noise under consideration is white and gaussian in nature and the perturbation is acute, that is, if the signal plus noise subspace makes an acute angle with the signal subspace, then P. A. Wedin's criterion (ref. 4) says that

$$\| B - A \|_2 < \frac{1}{\| A^+ \|_2} \tag{2.2}$$

where superscript + denotes the pseudo-inverse. After manipulaing this equation, it can be shown that

$$\| A^1 - A \|_2 \leq 2 * \| \delta A \|_2 < \frac{1}{\| A^{1+} \|_2} \tag{2.3}$$

where A^1 denotes an estimate of A. We can reformulate the above inequality in the following form :

$$\frac{\| A^1 \|_2 \| A^{1+} \|_2}{\| A \|_2} 2 * \| \delta A \|_2 < 1 \tag{2.4}$$

Simplifying the above equation leads us to the following terms:

$$\frac{\dfrac{\sigma_1^2}{\sigma_r^2} * \| \delta A \|_2}{\sigma_1^2} < 1 \quad or \quad \| \delta A \|_2 < \frac{\sigma_r^2}{2} \tag{2.5}$$

In the field of spectral estimation (ref. 5), the same concept was used to obtain the frequency estimates when a signal consisting of one or more sinusoids is corrupted by the additive white gaussian noise. The approximation made after this step is that the two norm of the perturbation matrix can be approximated as the first insignificant singular value of the data matrix, thus giving us the following criterion to identify the rank of the data matrix:

$$\| \delta A \|_2 = \sigma_{p+1}^2 \quad or \quad \sigma_p^2 > 2 * \sigma_{p+1}^2 \tag{2.6}$$

This inequality shows that the rank of the matrix is determined by comparing the successive singular values. The first singular value is compared with the second one and, if the first one is twice or more than the second, then the comparision is stopped and the matrix under consideration is identified to be of rank 1. Otherwise, the second and third are compared and, if the inequality is true, then the rank is identified to be 2. In this way the process goes on until the inequality is found to be true.

III. SIMULATION RESULTS

Simulations were performed to test the quality of the low rank approximant using three standard chaotic signals : Lorenz, Henon and Two-torus. The equations used for generating Lorenz, Henon and Two-torus data are given as follows:
Lorenz:

Table 1. Dimensionality Results

Type	SNR (dB)	MDL Criterion [q=30] d	<D>/R	[q=40] d	<D>/R	[q=50] d	<D>/R	New Criterion [q=30] d	<D>/R	[q=40] d	<D>/R	[q=50] d	<D>/R
Lorenz	5	2.56	0.20	1.84	0.20	2.40	0.21	2.40	0.1558	2.28	0.1609	1.68	0.1650
	10	2.20	0.14	1.72	0.15	1.68	0.15	1.92	0.1083	1.84	0.1119	1.24	0.1146
	15	1.92	0.10	1.88	0.10	2.20	0.10	2.72	0.0617	2.44	0.0638	1.96	0.0657
	30	3.68	0.03	3.24	0.03	3.08	0.04	2.12	0.0156	2.00	0.0164	1.76	0.0170
Henon	5	2.00	0.31	1.44	0.33	1.44	0.34	2.48	0.3053	1.52	0.3168	1.36	0.3262
	10	1.84	0.23	1.88	0.24	1.68	0.25	2.16	0.2430	2.04	0.2541	2.04	0.2625
	15	2.12	0.16	2.24	0.17	2.28	0.18	1.56	0.1618	1.20	0.1704	1.20	0.1779
	30	2.76	0.07	2.80	0.08	3.00	0.09	1.48	0.0935	1.48	0.1112	1.52	0.1236
2-Torus	5	2.44	0.24	1.84	0.25	1.84	0.26	2.2	0.1980	1.92	0.2047	1.56	0.21
	10	2.04	0.18	2.28	0.19	2.64	0.20	2.88	0.1379	1.80	0.1441	1.60	0.1485
	15	3.24	0.13	3.08	0.14	3.04	0.14	2.16	0.1021	2.36	0.1069	1.92	0.1110
	30	3.20	0.08	3.12	0.09	3.16	0.10	2.04	0.0555	2.04	0.0640	2.20	0.0698

$$\dot{x} = a(y - x), \quad \dot{y} = cx - y - xz, \quad \dot{z} = -bz + xy \qquad (3.1)$$

where $a = 10$, $b = 8/3$, and $c = 28$.

Henon:

$$x_{i+1} = bx_{i-1} + 1 - ax_i^2 \qquad (3.2)$$

where $a = 1.4$ and $b = 0.3$.

Two-torus:

$$x(i) = \sin(2 * \pi * f_1 * i) + \sin(2 * \pi * f_2 * i) \qquad (3.3)$$

where f_1 and f_2 are independent and incommensurate frequencies with values as $\sqrt{2}$ and $\sqrt{7}$.

The thresholding technique described above was used to determine the rank (d) of the data matrix, which in turn determined the local intrinsic dimensionality of the reconstructed attractor. Simulation results are shown in Table 1 using three different numbers of points in the local region, namely $q = 30$, 40, and 50 and $<r>/R$ is the local radius as a percentage of the maximum attractor radius. SNR's considered are 5, 10, 15, and 30 dB. The results are compared with the previously used MDL approach.[6,7]

IV. CONCLUSIONS

Simulations showing the comparision between the MDL approach and the new criterion clearly indicate the superiority of the new criterion over the MDL, particularly at high SNR's. Even below 10 dB the results are comparable with that of MDL.

Work is continuing in the area of determining an exact two-norm solution of the perturbation matrix so as to get the exact result as opposed to an approximation. Simulations performed till now clearly indicate that a better thresholding technique can be achieved to get the SNR down to 0 dB and give more accurate and reliable results at all of the presented SNR's.

REFERENCES

1) A. Goel, S.S. Rao, and A. Passamante, "Identification of Chaotic Systems using Fractal Dimension of the Reconstructed Attractor", 32nd Midwest Symposium on Circuits and Systems, Univ. of Illinois, 14-15 Aug 1989.

2) A. Passamante, T. Hediger, and M. Gollub, "Fractal Dimension and Local Intrinsic Dimension", Physical Review A, Vol. 39, No. 7, p. 3640, April 1989.

3) A.M. Albano, J. Muench, C. Schwartz, A.I. Mees, and P.E. Rapp, "Singular Value Decomposition and the Grassberger-Procaccia Algorithm", Physical Review A., Vol. 36, No. 1, p. 340, Jan 1988.

4) P.A. Wedin, "Perturbation Bounds in Connection with Singular Value Decomposition", BIT 12, pp. 99-111, 1972.

5) S.S. Rao and D.C. Gnanaprakasam, "A Criterion for Identifying Dominant Singular Values in the SVD based Methods of Harmonic Retrieval", Vol. 4, Paper # E6.5, ICASSP 1988.

6) A. Passamante, T. Hediger, and Mary Eileen Farrell, "Analysis of Local Space/Time Statistics and Dimensions of Attractors Using Singular Value Decomposition and Information Theoretic Criteria," in this proceedings.

7) M. Wax and T. Kailath, "Detection of Signals by Information Theoretic Criteria," IEEE Transactions on Acoustics, Speech, and Signal Processing, Vol. 33, No. 2, p. 387, April 1985.

SEEKING DYNAMICALLY CONNECTED CHAOTIC VARIABLES[†]

K. Hartt and L. M. Kahn

Department of Physics
The University of Rhode Island

Kingston, RI 02881-0817

It is of interest to determine whether two scalar time series can
belong to the same chaotic dynamical system. In geophysical and biological
systems, where the observables relevant to modeling may need to be uncovered, and also in extended physical systems such as turbulent ones, the
question of dynamical connectedness is important. Different parts of an
extended turbulent system can possess different values of the correlation
dimension D_2 [1], and one easily finds different systems with nearly equal
D_2. Therefore, equality of D_2 is neither a necessary nor sufficient condition for dynamical connectedness. Also, the cross-correlation function can
be misleading. In Fig. 1 we give the plots of the cross-correlation functions for (x_1,x_2) of the Roessler and (x_1,x_3) of the Lorenz attractors [2].
The Lorenz case shows a cross-correlation amplitude of only a few percent.
Surprisingly, the amplitude is slightly increased when a multiplicative
noise signal of up to 0.1 is injected. The Roessler case, with little correlation at nearly equal times, is more typical of strongly correlated systems. Clearly, a small cross-correlation function is an indecisive test of
whether two different observables are dynamically connected. Consequently,
for example, the interpretation of the disappearance of the cross-correlation function at two different points in a Rayleigh-Bénard cell at the
onset of soft turbulence [3] is ambiguous.

We have developed a test for connectedness of two time series, which
up to now we have only tested on subsystems where there are no spatial degrees of freedom. It is simple and easily described in terms of mutual information [4]. By way of introducing our test, we briefly review the development of our algorithm, which is of fixed-size rather than fixed-mass type
[5]. The delay coordinate embedding techniques we use [6] were developed
rapidly [4, 7] to see a variety of applications [8]. With noise or small
data sets, some optimal procedures have been found. For small data sets,
the maximum likelihood method can point to low dimensional chaos [9]. Mutual information provides an optimized value of the delay time T [5] and
can be calculated in terms of $\log C_m^1(\ell)$ [10]. Perhaps more crucial than
delay time is the window length for small data sets [11]. It is important
to seek convergence at a low embedding dimension M. One way is to do a
singular value decomposition and to discard the smallest components [11].
Another is to compute a topological stability index for nearest-neighbor
distances as a function of T and M and to choose those values that minimize this index [12]. The best choice of optimization technique is probably sensitive to the amount and quality of the data.

[†]Research supported by Federal Sea Grant 2-11066 at URI

Measures of Complexity and Chaos
Edited by N.B. Abraham *et al.*
Plenum Press, New York

129

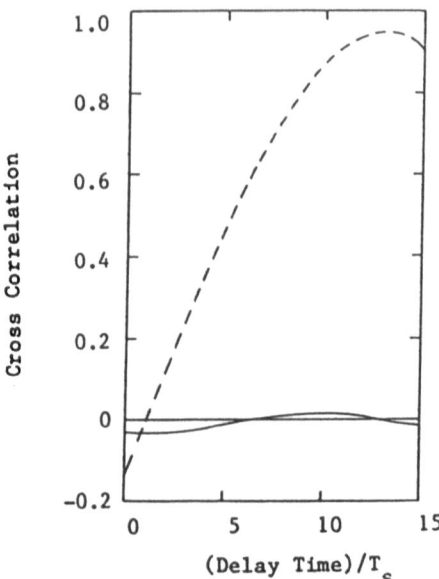

Fig. 1. Cross-correlation function for (x_1, x_2) of the Roessler attractor
(- - -) and (x_1, x_3) of the Lorenz attractor (———). The abscissa
is the delay time in units of sampling time for the two series.

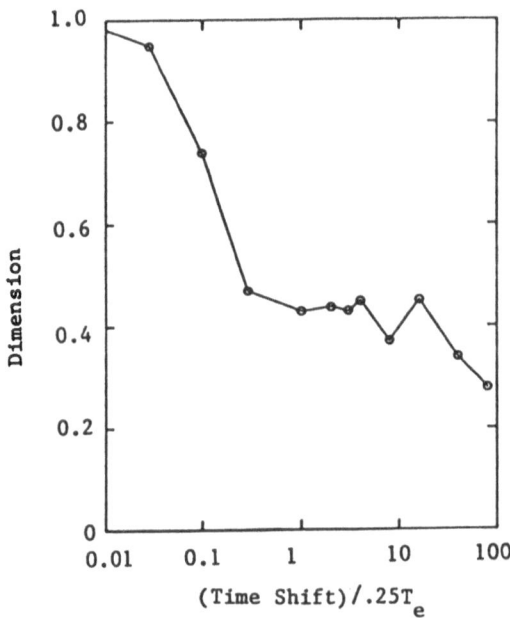

Fig. 2. Mutual information dimension associated with two scalar Roessler
time series, as a function of the time shift τ between the starts
of the series. τ is in units of $T_e/4$. Values were normalized to
unity by dividing D_{mi} by our computed information dimension of
the Roessler attractor.

130

We follow the notation of Albano et al., and we use as an example
the Roessler attractor as optimized in Ref. 12. For simplicity, we employ
the maximum norm. A trajectory in an M-dimensional space is reconstructed
from a scalar time series $\{y(t_j)\}$, in which equal time intervals $t_j = j\,T_s$
are employed:

$$\vec{y}(p) = [y(1+(p-1)J),\; y(1+(p-1)J+L),\ldots,\; y(1+(p-1)J+(M-1)L)]$$

T_s is called the sampling interval, M the embedding dimension, and L the
lag. The window length is $(M-1)L$ and the epoch length is $T_e = T_s(N_t-1)$
where N_t is the number of data collected at the sample interval T_s. We
shall refer to J as the optimized sampling parameter. In our example, T_s
is chosen such that $J = 1$ is optimum if one uses the topological stability index criterion [12]. We use log-log plots to find dimensions [7].

We construct a composite system from the two scalar time series $\{y(i)\}$
and $\{z(i)\}$ to be compared, by creating two M-dimensional vector series,
$\{\vec{y}(i)\}$ and $\{\vec{z}(i)\}$ as before. We concatenate them to form one 2M-dimensional
vector series with elements

$$\vec{X}(p) = [y(1+(p-1)J),\; \ldots,\; y(1+(p-1)J+(M-1)L),\; z(1+(p-1)J),\; \ldots,$$

$$z(1+(p-1)J+(M-1)L]$$

or just

$$\vec{X}(p) = (\vec{X}_a(p),\; \vec{X}_b(p)).$$

In an expanded version of this paper [13] we show:

I. This composite system yields the mutual information between the two
series, which is [4]

$$I_{2m} = I_{2m}(\{\vec{X}_a\},\{\vec{X}_b\},\tau) = H_m(\{\vec{X}_a\}) + H_m(\{\vec{X}_b\}) - H_{2m}(\{\vec{X}_a\},\{\vec{X}_b\},\tau)$$

where we have introduced a time shift τ between the starts of the two series. The H_m and H_{2m} become in the limits $\ell \to 0$ and M, $N_t \to \infty$ just the
Kolmogoroff entropies. The function I_{2m} answers the question, "How many
bits of information on the average does the time series $\{\vec{X}_a\}$ contain about
the elements of the time series $\{\vec{X}_b\}$?"

II. Associated with $\exp[-I_{2m}(\{\vec{X}_a\},\{\vec{X}_b\},\tau)]$ one can define a mutual information dimension $D_{mi} \geq 0$ which equals $D_{mi} = D_{1a} = D_{2a}$ when the two subsystems are isomorphic [14] and is just $D_{mi} = 0$ when the two subsystems satisfy a criterion of dynamical independence. The criterion is that the distance functions for the upper and lower sub-vectors are uncorrelated.

III. To find $I_{2m}(\{\vec{X}_a\},\{\vec{X}_b\},\tau)$, one calculates the generalized correlation integrals $\log C^1_{m,a}(\ell)$, $\log C^1_{m,b}(\ell)$ and $\log C^1_{m,ab}(\ell,\tau)$ where ℓ is
the distance, and a, b and ab represent, respectively, systems a and b
and the composite system. The function $\log C^1_{m,ab}(\ell,\tau)$ is computed directly from the concatenated series.

We have applied our test to two subseries of x_1-components of the
Roessler attractor, each with 4500 elements (N_t). Figure 2 shows a normalized mutual information dimension as a function of the time shift τ between the starts of the series. The shift τ is in units of $T_e/4$. A sample
time $T_s = \pi/25$, a delay time $LT_s = 7T_s$ and embedding dimensions up to
$M = 5$ are used. I_{2m} and D_{mi} decrease slowly as $\tau \to \infty$. Yet D_{mi} is sharply
peaked near $\tau = 0$. This behavior is similar to the mutual information
function for a scalar time series for the Roessler attractor [4], but

with a time scale two orders of magnitude longer. As a check, we have determined the mutual information dimension between the Roessler and Lorenz attractors to be zero.

In conclusion, we have developed a transparent test of the dynamical connectedness of two time series. In our example, we find D_{mi} sensitive to small shifts between the series, but with a persistent tail allowing connectedness to be inferred even for large time shifts.

REFERENCES

1. V. V. Kozlov, M. I. Rabinovich, M. P. Ramazanov, A. M. Reiman and M. M. Sushchik, Correlation dimension of the flow and spatial development of dynamical chaos in a boundary layer, Phys. Lett. A 128:479 (1988).
2. O. E. Roessler, An equation for continuous chaos, Phys. Lett. A 57: 397 (1976); E. N. Lorenz, Deterministic nonperiodic flow, J. Atmos. Sci. 19:130 (1963).
3. F. Heslot, B. Castaing and A. Libchaber, Transitions to turbulence in helium gas, Phys. Rev. A 36:5870 (1987).
4. A. Fraser, Using mutual information to estimate entropy, in: "Dimensions and Entropies in Chaotic Systems," G. Mayer-Kress, ed., Springer-Verlag, Berlin (1986); A. M. Fraser and H. L. Swinney, Independent coordinates for strange attractors from mutual information, Phys. Rev. A 33:1134 (1986).
5. J. Broggi, Evaluation of dimensions and entropies of chaotic systems, J. Opt. Soc. Am. B 5:1020 (1988).
6. N. H. Packard, J. P. Crutchfield, J. D. Farmer and R. S. Shaw, Geometry from a time series, Phys. Rev. Lett. 45:712 (1980); F. Takens, Detecting strange attractors in turbulence, in: "Dynamical Systems and Turbulence," D. A. Rand and L.-S. Young, eds., Springer-Verlag, Berlin (1981).
7. P. Grassberger and I. Procaccia, Measuring the strangeness of strange attractors, Physica D 9:189 (1983).
8. W. A. Brock and C. L. Sayers, Is the business cycle characterized by deterministic chaos?, J. Monetary Econ. 22:71 (1988); N. B. Abraham, L. A. Lugiato and L. M. Narducci, "Overview of instabilities in laser systems," J. Opt. Soc. Am. B 2:7 (1985); W. M. Schaffer, Stretching and folding in lynx fur returns: evidence for a strange attractor in nature?, Am. Nat. 124:798 (1984); A. Brandstäter, J. Swift, H. L. Swinney, A. Wolf, J. D. Farmer, E. Jen and P. J. Crutchfield, Low-dimensional chaos in a hydrodynamic system, Phys. Rev. Lett. 51:1442 (1983).
9. S. Ellner, Estimating attractor dimensions from limited data: a new method, with error estimates, Phys. Lett. A 133:128 (1988).
10. W. Liebert and H. G. Schuster, Proper choice of the time delay for the analysis of chaotic series, unpublished work; K. Pawelzik and H. G. Schuster, Generalized dimensions and entropies from a measured time series, Phys. Rev. A 35:481 (1987).
11. A. M. Albano, J. Muench, C. Schwartz, A. I. Mees and P. E. Rapp, Singular-value decomposition and the Grassberger-Procaccia algorithm, Phys. Rev. A 38:3017 (1988).
12. W. Liebert, K. Pawelzik and H. G. Schuster, Optimal embeddings of chaotic attractors from topological considerations, This Conference (1989).
13. K. Hartt and L. M. Kahn, Dynamically related chaotic variables, unpublished work.
14. D. S. Ornstein, Ergodic theory, randomness, and chaos, Science 243: 182 (1989).

ON PROBLEMS ENCOUNTERED WITH DIMENSION CALCULATIONS

U. Hübner, W. Klische, N.B. Abraham[*], C. O. Weiss

Physikalisch-Technische Bundesanstalt,
D-3300 Braunschweig, Fed. Rep. Germany

[*]Department of Physics,
Bryn Mawr College, Bryn Mawr, PA 19010, USA

INTRODUCTION

In an attempt to quantitatively confirm the presence of Lorenz-like chaos in optically pumped FIR lasers, we have calculated dimensions and entropies (using the algorithm of Grassberger-Procaccia (1983)) from intensity time series obtained from numerically integrating the Lorenz equations and from sampling the output of the laser.

Our contribution deals with technical details of these calculations for the numerical and experimental data. We discuss aspects that need special attention in practical implementations, including the influence of the number of data points, selection of a suitable norm, proper choice of the interpoint spacing of the vector components, selection of a scaling region for the correlation integral, noise, limited resolution, and convergence of dimensions and entropies with embedding dimension. We demonstrate that phase portraits of the data can provide useful insight into the structure of the underlying attractor and we discuss how that structure may affect the correlation integral. Only some selected points of these can be discussed here because of the limited space. Some of these items and a detailed analysis of many different experimental data sets may be found in Hübner et. al. (1989).

THE ALGORITHM

We are analyzing single-variable time series $x(t)$. An E-dimensional phase space is reconstructed using the time delay method by defining the vectors
$$X(t) = (x(t),x(t+\tau),\ldots,x(t+(E-1)\tau))$$
where the time delay is $\tau=n\Delta t$, n being an integer and Δt is the sampling interval. As pointed out by Caputo and Atten (1987) the maximum norm in the algorithm of Grassberger and Procaccia has strong computing advantages and is equivalent (see Takens (1985)) to the Euclidean norm (length of the vector difference). Using the maximum norm the correlation integral is computed with
$$C_E(r) = 2 \left[\text{number of pairs (i,k),} \max_{0 \le j < E} |x_{i+j}-x_{k+j}| \le r \right] /N(N-1)$$

Measures of Complexity and Chaos
Edited by N.B. Abraham *et al.*
Plenum Press, New York

where $i=1,2,\ldots,N-1$, $x_{i+j}=x(t_i+j\tau)$, and $i<k$ ($i=k$ should be excluded). D_2 then follows from the average slopes of the so-called plateau regions of the log-log plot of $C_E(r)$ versus r (discussed later) and the entropy K_2 is evaluated from

$$K_{2,E}(r) = (1/\tau)\ \log_e(C_E(r)/C_{E+1}(r))$$

in the limit $E{\to}\infty$, again averaging over the plateau region.

In order to study the influence of different choices of the norm used to determine the distance between points, we applied the algorithm to a few data sets using the Euclidean norm and the maximum norm. While for each data set both norms delivered the same D_2 values within the uncertainty of the method, the scaling region is wider, and therefore easier to identify, in the case of the Euclidean norm because the peak in the slope curves for large r is much less pronounced. However, the computer time needed in that case is about three times greater than that for the maximum norm. Thus, most of our computations were done using the maximum norm.

THE CHOICE OF τ AND THE MAXIMUM E

The maximum embedding dimension certainly has to be restricted. The question of how to choose E_{max} is closely related to the proper choice of τ. Experience shows that there is a relation $\tau{\cdot}E_c{\approx}T$ between τ, the average period T of the intensity pulses, and the embedding dimension E_c which marks the onset of convergence of the slope curves. It reflects the fact that for Lorenz-like oscillations the structural information needed to reconstruct a point on the attractor is mainly contained in the data sampled during one average period T. Thus choosing an $E>E_c$ adds only little information to that already contained in the components of vectors in embedding dimension $E{\approx}E_c$. To minimize computation time, by minimizing E, a large τ is preferred. On the other hand, τ should be considerably smaller than T to have the data points monitor details of the attractor (see Fig. 1). For example the oscillations in K_2T would not have been resolved if we had used τ comparable to the average oscillation period T. In practice $\tau{\approx}T/7$ was found to be reasonable which yields $E_c{\approx}7$.

If one is only interested in the dimension, E_{max} can be as low as $(2-3)E_c$. Experience shows that to assure convergence of $K_{2,E}$ (comparable to the measurement uncertainty: half the least significant bit of the 8-bit accuracy) one needs embedding dimensions up to $(6-7)E_c$ which means $E_{max}\approx 40$ to 50 for $\tau{\approx}T/7$. This is consistent with studies of the Mackey-Glass system cited by Caputo and Atten (1987). Fig. 1 shows an example of the dimension and entropy convergence. It reveals clearly that τ should not be too large to monitor the oscillatory behavior on the entropy curve. The period of these oscillations when multiplied by n ($=\tau/\Delta t$) equals the average pulsing period of the data itself.

THE SCALING REGION

Fig. 2 shows the slopes of the correlation integral $C(r)$ where the curve parameter is the embedding dimension ($E=1$ to $E=40$ from bottom to top). Fig. 2a) shows results from data which were calculated by numerical

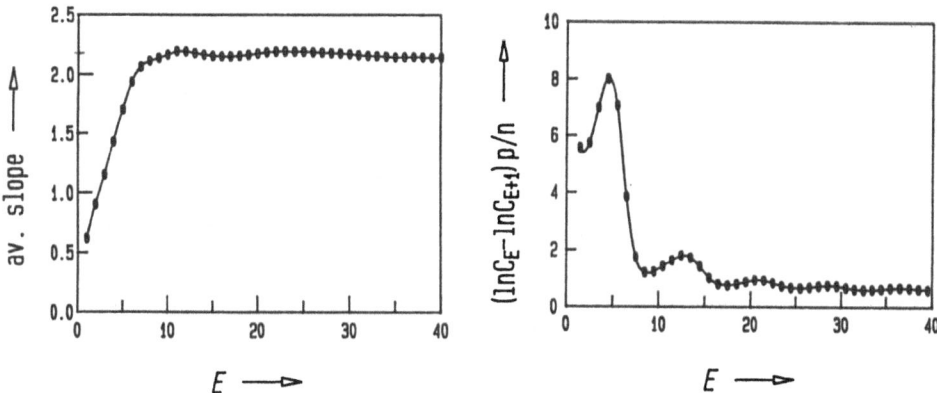

Fig. 1. Correlation dimension D_2 (average slope in the plateau region) and K_2T versus embedding dimension E for our experimental "Lorenz chaos" data set #3. p is the average number (p≈20) of data samples covering the average pulsing period and n (integer; n≈p/7) means the delay between successive components of each vector X. Calculated values (marked by O) were connected by interpolation as a guide to the eye. 25000 8-bit data were used in the calculations.

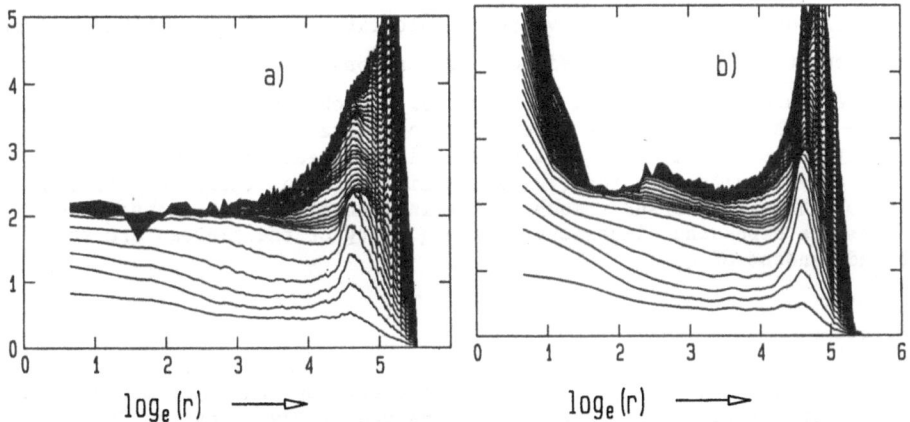

Fig. 2. Slope of log-log plots of the correlation integral versus the distance between points on the embedded attractor, corrected for effects of finite digitizing accuracy of the data (see Möller et. al. (1989)). Parameter of the slope curves is the embedding dimension increasing upward from 1 to 40.
a) Numerically generated data for x^2 (rounded to 8 bit) from solving the detuned Lorenz equations for parameters giving chaotic spirals and slightly detuned (one third of the detuning necessary to get period 3).
b) 8-bit data sampled from intensity pulsations in an FIR laser.

integration of the detuned Lorenz equations. The detuning δ=0.05 (see Zeghlache and Mandel (1985)) was chosen as about one third of that detuning which leads to period three oscillations. This detuning still gives the typical increasing oscillations (spirals) of the "Lorenz type chaos". The other parameters were chosen to be as close as possible to the experiments: R=15, b=0.25, and σ=2. Fig. 2b) was calculated from the experimental data set #3 which shows the closest similarity to our numerical data set.

If a plateau (the "scaling region") exists and converges for increasing embedding dimension it gives the D_2 dimension of the attractor. In both parts of Fig. 2 plateau regions can be identified but the flat region extends to small values of r in a) only. In contrast, noise in the experimental data gives curves rapidly rising as r decreases towards zero. The measured noise level found in signal-to-noise ratio measurements in the experiment can be added to the numerically generated data and corresponds to noise with an rms of approximately one half of the least significant bit of the 8-bit accuracy. The slope curves for the numerically generated data with this added noise are very similar to the slope curves for the experimental data, notably in the rapid rise at low r values.

Plateau identification is also made difficult by peaks caused by structure (density variations) in the attractor which appear for large r. In b), this gives an intermediate peak which obscures the plateau and makes dimension estimates difficult. The region to the right of it continues to change with embedding dimension and should not be used as part of the scaling region because of an artificial decrease in that region along with an enhancement of the last peak caused by the use of the maximum norm.

SPECTRUM OF DIMENSIONS

All topics discussed above for a calculation of D_2 were found to apply to calculations of generalized (q≠2) dimensions D_q. However, for q<2 care must be exercised to discard those vectors which have no neighbors at all within a sphere of radius r. Even with considerably large data sets this rapidly leads to a lack of data points with decreasing q, resulting in acceptable results only for q>0. For positive q, the characteristics of a plot of the slope versus r such as the influence of noise, convergence for successive embedding dimensions or the quality of the plateau are found to be independent of q.

REFERENCES

Caputo, J. G., and Atten, P., 1987, Metric entropy: An experimental means for characterizing and quantifying chaos, Phys. Rev. A, 35:1311
Grassberger, P., and Procaccia, I., 1983, a) Characterization of strange attractors, Phys. Rev. Lett., 50:346
        b) Estimation of the Kolmogorov entropy from a chaotic signal, Phys. Rev. A, 28:2591
Hübner, U., Abraham, N. B., and Weiss, C. O., 1989, Dimensions and entropies of chaotic intensity pulsations in a single mode FIR laser, Phys. Rev. A, manuscript under review.
Möller, M., Mitschke, F., Lange, W., Abraham, N. B., and Hübner, U., Errors from digitizing and noise in estimating attractor dimensions, Physics Letters, 1989, to be published
Zeghlache, H., and Mandel, P., 1985, Influence of detuning on the properties of laser equations, J. Opt. Soc. Am. B, 2:18

SYSTEMATIC ERRORS IN ESTIMATING DIMENSIONS FROM

EXPERIMENTAL DATA

W. Lange and M. Möller

Institut für Angewandte Physik der
Westfälischen Wilhelms-Universität Münster,
D-4400 Münster, Federal Republic of Germany

I. INTRODUCTION

The necessity of characterizing chaotic dynamics quantitatively by assigning to them measures like the dimension of the underlying attractor or the entropy production of the system has generally been accepted. The corresponding methods are practiced now by many experimentalists. Most of them have adopted the method by Grassberger and Procaccia [1] that requires only a single-variable time series by making use of the embedding technique originally proposed by Takens [2]. Though the validity of the method is beyond any doubt, its practical application presents problems, since it relies on assumptions which are not generally fulfilled in the experiment.

First of all the method relies on a very large set of data points, while the actual number of data points in the experiment is usually very restricted. Even if a large number of data is available, it is hard to make use of it due to limitations in computer time; in most experiments the duration of the evaluation of the data exceeds the duration of data acquisition by far and thus is the limiting factor in obtaining information from experiments. Experimentalists often try to get along with very restricted data sets and inevitably run into problems with the best choice for sampling intervals in data acquisition and for the time intervals between vector components in the embedding procedure (see e.g. [3]). These problems are interrelated with systematic errors arising from a limited time span or a limited number of vectors [4, 5]. Very recently a rule on the minimal number of data points became available [6] and it turned out that quite often the actual number of data points evaluated in experiments was below the lower bound. A number of new procedures (see e.g. [7]) and of improved algorithms (see e.g. [8]) have been proposed recently and the experimentalist urgently asks for advice on the tractability, universal applicability and reliability of the proposals.

Measures of Complexity and Chaos
Edited by N.B. Abraham *et al.*
Plenum Press, New York

137

A second problem arises from the inevitable presence of noise in experimental data. It is generally believed that it affects the correlation integral $C(\varepsilon)$ only on length scales of the order of magnitude of the standard deviation of the noise [9]. In the ideal case the slope of a log-log plot of $C(\varepsilon)$ has a small "noise length scale" region (small ε) where the slope increases with embedding dimension and an intermediate range of "proper scaling"; the third region of large ε is without any use due to the finite size of the attractor whose dimension is to be determined. In practice the intermediate range of "proper scaling" may be very short and the experimentalist will tend to extend it somewhat to small values of ε; he tries to justify his approach by the apparent "nearly converging" dependence of the slope on the embedding dimension but has to envisage the possibility of an overestimation of the dimension. This problem will be discussed in some detail later on.

The most obvious method of getting rid of noise problems in experimental physics is the reduction of the observation bandwidth to the band of the signal containing useful information. The influence of - possibly unwanted - spectral filtering on estimated values of dimensions and entropies has been discussed recently in the literature [10, 11, 12, 14] and will shortly be reviewed here.

The role of round-off errors in computer simulations of dynamic systems has been discussed in the literature (see e.g. [13] and references therein). In experiments the effect of finite precision, i.e. of digitizing, can be much more severe, since the resolution of the instruments may be as low as 8 bit, which implies an even lower effective resolution of the experimental data in practice. This can lead to severe systematic errors [11], which will be discussed here on the basis of a very recent paper [13].

II. FILTERING OF CHAOTIC SIGNALS

It has been shown by Badii and Politi [10] that the Lyapunov dimension determined from a time series can be increased if the time series is passed through a low pass filter. The argument is quite simple. If $x(t)$ is the measured component of the state vector X defined in the N-dimensional phase space, then the action of a first order low-pass filter can simply be described by the additional equation of evolution

$$\dot{z} = -\eta z + x \qquad (1)$$

where z is the output signal on which dimensional analysis is performed. The Lyapunov dimension is defined by

$$D_L = j + \sum_{k=1}^{j} \lambda_k / |\lambda_{j+1}| , \qquad (2)$$

where the λ_k's are Lyapunov exponents $\lambda_1 \geq \lambda_2 \geq ... \geq \lambda_N$, and j is the largest integer for which the sum is non-negative. The addition of Eq. 1 yields an additional Lyapunov exponent $-\eta$ but does not affect the previous ones. Therefore D_L remains unaffected as long as η is larger than $|\lambda_{j+1}|$, but changes

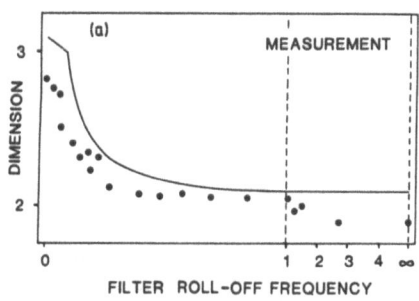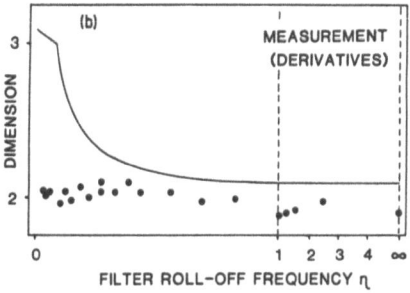

Fig. 1. D_2 values calculated from experimental low-pass filtered signals as a function of filter roll-off frequency η. a) evaluation of the filter output, b) evaluation of the derivative of the filter output

when η becomes smaller. This way $D_L(\eta)$ becomes a piecewise differentiable function. It has been argued that the first "corner point" in the curve $D_L(\eta)$ (see Fig. 1) can be related to a first-order phase transition [14]. The approach can be generalized for higher-order filters.

The behaviour of the dimension D_2 determined from experimental time series which had been filtered numerically has been examined [14]; it was found to show the type of behaviour calculated for D_L.

In a more experimental approach the dimension D_2 of a time series was determined [12] which originated from a well-studied electronic chaotic circuit [15] and had passed a real low pass filter. Results are reproduced in Fig. 1a.

Obviously the value of the dimension should stay unchanged if $\dot{x}(t)$ instead of $x(t)$ is evaluated. On the other hand it was found in [12] that the value of D_2 obtained from the experimental data did __not__ depend on the filter roll-off frequency if the filter output was differentiated before data processing (see Fig. 1b). As a consequence it was proposed in [12] to compare the values of D_2 obtained from the signal and its derivative. If they differ this should indicate the possible influence of a bandwidth limitation in the signal channel. In practice, however, problems with the signal-to-noise ration will often prevent the application of differentiation.

Other means of identifying the possible influence of bandwidth limitations have been proposed very recently [16].

It is an interesting question how the dimensions and entropies are influenced by methods of spectral filtering which cannot be described by differential equations. Work on this question is in progress [17].

Finally it should be noted that the influence of bandwidth limitations, though a possible source of systematic errors, can be useful in obtaining additional information on the chaotic system: In [14] it has been proposed to use the η-dependence of D_L as a means of estimating Lyapunov exponents.

III. DIGITIZING ERRORS

In the analysis of slowly evolving chaotic systems, precision does not present a problem, since 16-bit analogue-to-digital converters are readily available; their resolution can be expected to be much better than the noise level. It takes, however, a long time to obtain the large number of data points needed for a careful analysis and it becomes extremely difficult to keep all parameters of the system constant during data acquisition. In fields like optical chaos typical signal frequencies are in the 10 MHz range and experiments can be completed within milliseconds. Unfortunately, the resolution of commercially available instruments offering a sampling rate of at least 100 MHz seems to be restricted to 8 bit at present. In a recent paper [11] it was shown empirically that considerable errors in the determination of D_2 can result from this low resolution.

Let us call the estimated value of D_2 obtained in an analysis Δ. It was found recently [13] that the values of Δ obtained with different resolution can empirically be described by

$$\Delta = D_2 \left(1 - k\frac{p}{\bar{\varepsilon}}\right);\tag{3}$$

here p is one half of the least significant digit and $\bar{\varepsilon}$ represents an average length scale derived by $\bar{\varepsilon} = (\varepsilon_{min}\varepsilon_{max})^{1/2}$ from the values ε_{min} and ε_{max} between which the slope of $\log(C(\varepsilon))$ versus $\log(\varepsilon)$ was evaluated. The quantity k is of order unity. The formula was checked with the Hénon map (see Fig. 2a), with the Roessler attractor (see Fig. 2b), with results from the Lorenz equations and with numerical values sampled from 3-torus and 5-torus attractors (generated using 3 and 5 incommensurate frequencies, respectively). With respect to arguments making Eq. 3 plausible the reader is referred to [13]. A more rigorous treatment, however, is still desirable.

It should be noted that the results shown in Fig. 2 were obtained by the

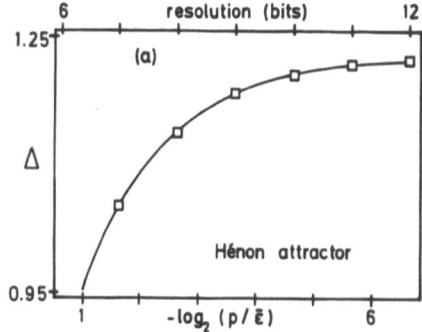 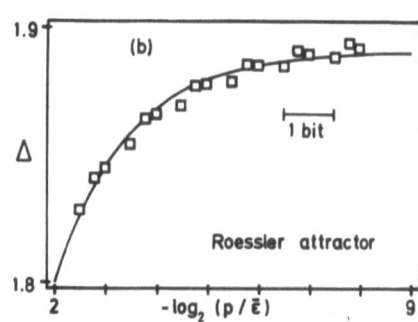

Fig. 2. Slopes of log-log plots of the correlation integral (estimates of D_2) calculated for data rounded to different resolutions, for a) the Hénon attractor, b) the Roessler attractor. In a) one scaling region and in b) three slightly different scaling regions were evaluated (digitizing accuracy p, average length scale $\bar{\varepsilon}$)

use of the maximum norm in the calculations. The arguments presented in [13] show that the Euclidean norm should give slightly modified results. Up to now it is not known how other ways of determining D_2 react to digitizing errors. The considerations presented in Sect. V show that the results depend strongly on details of the procedures used.

IV. COMPARISON OF NOISE AND DIGITIZING ERRORS

The effect of digitizing on experimental data can in many respects be regarded as adding white noise to the signal (see e.g. [18]). On the other hand the presence of noise is expected to cause an overestimate of D_2 (see Sect. I). The authors of [13] tried to find the reason for the different behaviour of stochastic noise and "digitizing noise". An apparent difference is the different amplitude statistics: stochastic noise usually has a more or less Gaussian distribution, while the changes in the data introduced by digitizing might be regarded as a type of (purely additive) noise with a rectangular distribution. In order to test this hypothesis numerical experiments were performed with respect to the effect of adding noise of different probability distributions: (1) rectangular distribution between $-\sigma$ and $+\sigma$; (2) Gaussian distribution with standard deviation σ; (3) Gaussian distribution with standard deviation σ truncated to the range of -2σ to $+2\sigma$.

Results in the case of the Roessler attractor are shown in Fig. 3. It turns out that the values of D_2 can reasonably be approximated by

$$\Delta = D_2 \left(1 + K \left(\frac{\delta}{\varepsilon}\right)^2\right)$$

(4)

here K is a positive factor of order unity and δ is a measure for the standard deviation of the noise; again arguments can be given, why this dependence is to be expected. So in contrast to Eq. 3 there is a second order dimension overestimate. It can be concluded that the shape of the amplitude distribution is of minor importance; the difference lies deeper and is related to the fact that digitizing is a *deterministic* process.

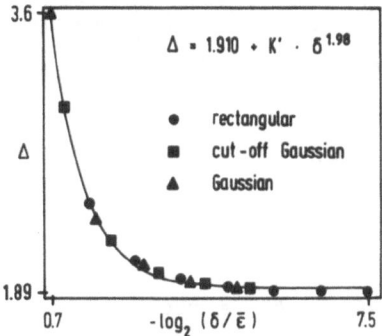

Fig. 3. Slopes of the log-log plots of the correlation integral for noisy data from the Roessler attractor. Noise of three different amplitude statistics with noise amplitudes δ was added before the evaluation. The solid line shows a least-squares fit to Eq. (5).

V. ERROR REDUCTION

For practical purposes the most important question is whether the errors can be avoided or reduced. While the bandwidth limitation discussed in the beginning can be avoided in most cases, the digitizing error and the presence of noise are more severe. Fortunately they can be reduced by changing the procedure of determining Δ.

Power Law Fit

As well the digitizing error as the influence of noise increase with decreasing ε. Simultaneously $C(\varepsilon)$ decreases, and therefore the relative error of $C(\varepsilon)$ increases drastically. The commonly used evaluation of the slope of the log-log plot of $C(\varepsilon)$ vs. ε would be appropriate, if the relative errors did not depend on ε. Without any doubt this procedure is very convenient in the search of the proper scaling region to be evaluated. For the determination of Δ, however, a least squares fit of a power law relation $C(\varepsilon){\sim}\varepsilon^{\Delta}$ without taking any logarithms should be more appropriate. This simple idea was tested on data from different attractors and always resulted in an error reduction. An example is shown in Fig. 4a and b. As a consequence it is **strongly suggested to use the linear data in the fit.**

Use of Correction Formulae

As a consequence of Eq. 3 and Eq. 4 it is reasonable to replace the relation $C(\varepsilon){\sim}\varepsilon^{\Delta}$ in the fitting procedure by a form which takes into account the ε-dependence of the systematic errors. In systems with dominant noise it is sensible to approximate $\Delta = d(\log C_{exp}(\varepsilon))/d(\log(\varepsilon))$ as it varies with different noise amplitudes δ by

$$\Delta = \Delta' \cdot \left(1 + K\left(\frac{\delta}{\varepsilon}\right)^2\right) \quad , \tag{5}$$

where Δ' is an improved approximation of D_2 (cf. Eq. 4) and $C_{exp}(\varepsilon)$ is the experimentally obtained value of the correlation integral. Eq. 5 is equivalent to approximating $C_{exp}(\varepsilon)$ by a corrected function that should be employed instead of the simple power law in the least square fit procedure:

$$C(\varepsilon) \sim \varepsilon^{\Delta'} \cdot e^{-\frac{1}{2}\Lambda(\delta/\varepsilon)^2} \quad . \tag{6}$$

with $\Lambda = K \cdot \Delta' > 0$. Correspondingly it may be preferable to approximate $C_{exp}(\varepsilon)$ by

$$C(\varepsilon) \sim \varepsilon^{\Delta'} \cdot e^{\Lambda(p/\varepsilon)} \tag{7}$$

if the digitizing error is dominant. In the case of the noise error being dominant, Δ' determined after Eq. 6 is a much better approximation to D_2 than Δ in all cases regarded so far. For an example see Fig.4a. Therefore the use of Eq. 6 may be very useful, especially if the scaling region is severely limited by noise. On the other hand the use of Eq. 7 in the case dominated by digitizing noise worked well in Hénon or Lorenz data, but it did not work satisfactorily in several other cases (Roessler, 3-torus, 5-torus).

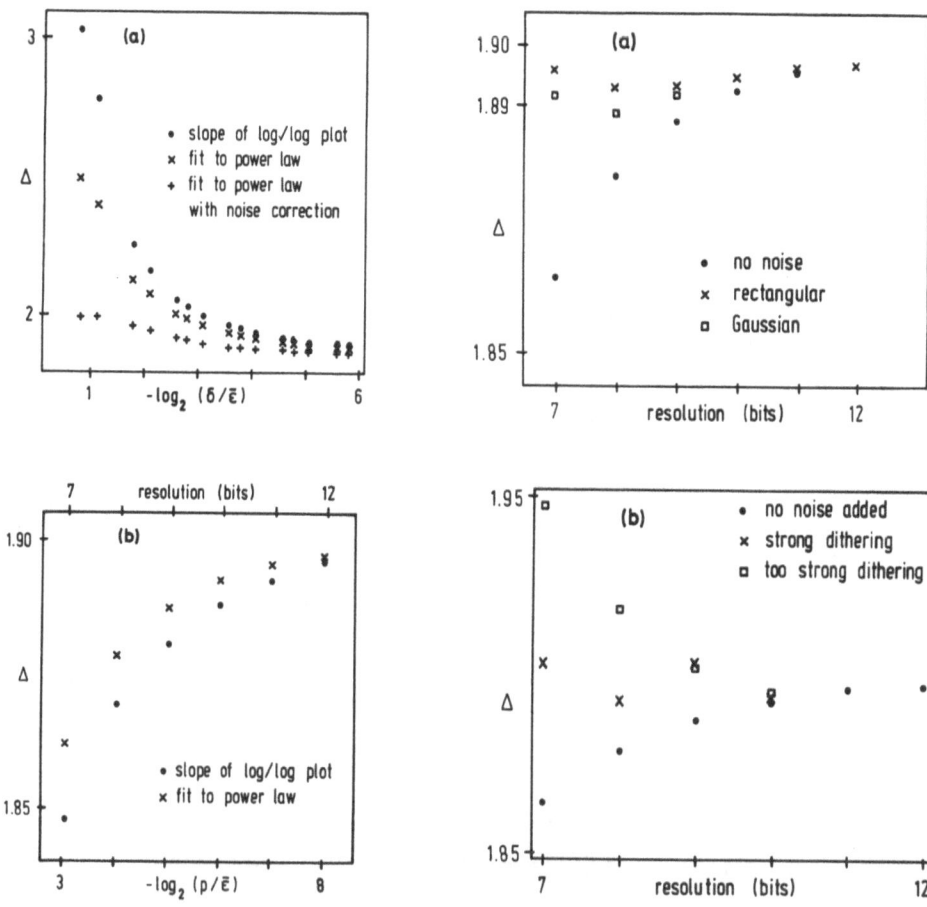

Fig. 4. D_2 estimates from a) noisy, b) digitized data from the Roessler attractor obtained by various evaluation procedures.

Fig. 5. Dimension estimates showing reduction of error from digitized data a) by adding weak noise and using power law fit, b) by adding strong noise and using noise correction formula fit

Artificial Addition of Noise

In the field of engineering it is a standard procedure to avoid systematic errors introduced by digitizing by adding a high frequency signal (random, pseudorandom or deterministic) with an amplitude in the order of the digitizing precision to the signal before digitizing [18]. The procedure is called "dithering". It was simulated by adding random numbers to data computed from the Roessler equations before "digitizing" (rounding off). Different distributions were used. Fig. 5a shows results obtained with weak dithering and Fig. 5b with stronger dithering and application of the correction formula Eq. 6. Generally an improvement can be expected.

VI. K_2-ENTROPY

It may be interesting to ask for the influence of bandwidth limitations and digitizing errors on the K_2-entropy that can be evaluated from the same correlation integrals as D_2.

In [12] it was found that K_2 was not influenced by the low-pass filter. A theoretical prediction of this result is not known to the authors. K_2 also seems to be fairly insensitive to the effect of digitizing. As a demonstration, in Fig. 6 the dependences of D_2 and K_2 - evaluated from the same data sets of the Hénon attractor - on digitizing resolution are compared.

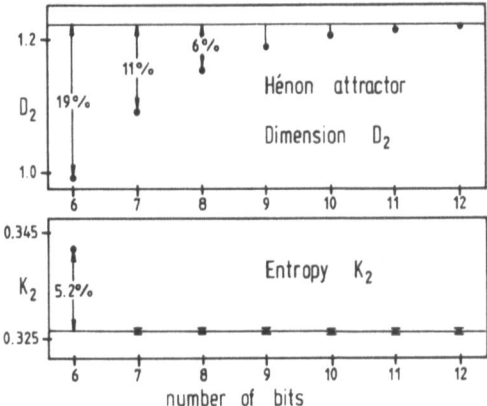

Fig. 6. D_2 and K_2 values in dependence of the digitizing resolution.

VII. CONCLUSION

In this contribution three possible sources of systematic errors are discussed which easily arise in deriving dimensions from experimental data. The origins are limited bandwidth of the signal channel, noise and bad digitizing precision. Means of avoiding or reducing these errors are available.

ACKNOWLEDGEMENTS

This contribution is mainly based on results obtained by F. Mitschke, N.B. Abraham and the authors. The work was supported by the Deutsche Forschungsgemeinschaft and the Commission of the European Community.

REFERENCES

1. P. Grassberger and I. Procaccia,"Characteriziation of Strange Attractors", Phys. Rev. Lett. 50, 346 (1983); "Measuring the Strangeness of Strange Attractors", Physica 9D, 189 (1983); "Estimation of the Kolmogorov entropy

from a chaotic signal", Phys. Rev. A $\underline{28}$, 2591 (1983); "Dimensions and Entropies of Strange Attractors from a Fluctuating Dynamics Approach", Physica 13D, 34 (1984)

2. F. Takens, "Detecting Strange Attractors in Turbulence", in *Dynamical Systems and Turbulence*, Vol.898 of *Lecture Notes in Mathematics*, ed. by D.A. Rand and L.S. Young (Springer, Berlin; 1981)

3. J.G. Caputo, "Determination of Attractor Dimension and Entropy for Various Flows: An Experimentalists Viewpoint", in [5], p.180

4. R. Badii and A. Politi, Phys. Rev. A $\underline{35}$, 1288 (1987); J.G. Caputo and P. Atten, "Metric entropy: An experimental means for characterizing and quantifying chaos", Phys. Rev. A $\underline{35}$, 1311 (1987);

5. *Dimensions and Entropies in Dynamical Systems*, ed. by G. Mayer-Kress, (Springer, Berlin 1986)

6. L.A. Smith, "Intrinsic Limits on Dimension Calculations" Phys. Lett. A133, 283 (1988)

7. A.M. Albano, J. Muench, C. Schwartz, A.I. Mees and P.E. Rapp, "Singular-value decomposition and the Grassberger–Procaccia algorithm", Phys. Rev. A $\underline{38}$, 3017 (1988), R. Badii and A. Politi, J. Stat. Phys. $\underline{40}$, 725 (1985), G. Broggi, J. Opt. Soc. Am. $\underline{B5}$, 1020 (1988)

8. R.L. Somorjai, M.K. Ali, "An efficient algorithm for estimating dimensionalities", Can. J. Chem. 66, 979 (1988)

9. A. Ben Mizrachi, I. Procaccia, and P. Grassberger, "Characterization of experimental (noisy) strange attractors", Phys. Rev. A $\underline{29}$, 975 (1984).

10. R. Badii and A. Politi, "On the Fractal Dimension of Filtered Chaotic Signals", in [5], p.67

11. F. Mitschke, M. Möller, W. Lange, "On Systematic Errors in Characterizing Chaos", in *Optical Bistability IV*, ed. by W. Firth, N. Peyghambarian and A. Tallet, (les Editions Physique, Paris, 1988), p. C2-397 [reprinted from J. Phys. Colloq. 49, suppl.6, C2-397 (1988)].

12. F. Mitschke, M. Möller, W. Lange, "Measuring Filtered Chaotic Signals", Phys. Rev. A, $\underline{37}$, 4518 (1988)

13. M. Möller, W. Lange, F. Mitschke, N.B. Abraham, U. Hübner, "Errors from Digitizing and Noise in Estimating Attractor Dimensions", Phys. Lett. A (to appear)

14. R. Badii, G. Broggi, B. Derighetti, M. Ravani, S. Ciliberto, A. Politi and M.A. Rubio, "Dimension Increase in Filtered Chaotic Signals", Phys. Rev. Lett $\underline{60}$, 979 (1988).

15. F. Mitschke, N. Flüggen, "Chaotic Behaviour of a Hybrid Optical Bistable System without a Time Delay", Appl. Phys. B35, 59 (1984)

16. O. Chennaoui, W. Liebert, K. Pawelzik, H.G. Schuster, "Filterinversion bei chaotischen Zeitreihen", Verhandl. DPG (VI) 24, 4(1989), DY 12-54

17. F. Mitschke (private communication)

18. N.S. Jayant and P. Noll, *Digital Coding of Waveforms*, (Prentice-Hall, Englewood Cliffs, NJ, 1984).

ANALYZING PERIODIC SADDLES IN EXPERIMENTAL STRANGE ATTRACTORS

Daniel P. Lathrop[a] and Eric J. Kostelich[b,a]

[a]Center for Nonlinear Dynamics and Department of Physics
University of Texas
Austin, Texas 78712

[b]Institute for Physical Science and Technology
University of Maryland
College Park, Maryland 20742

Abstract

This paper discusses a way to locate and analyze the periodic orbits in an attractor reconstructed from time series data, and the technique is applied to data from an experiment on the Belousov-Zhabotinskii chemical reaction. The topological entropy of the attractor is estimated by approximating the dynamics with a subshift of finite type. The Lyapunov exponents are computed from the data using a method suggested by Eckmann and Ruelle and agree well with the estimated topological entropy and information dimension.

INTRODUCTION

There has been much recent progress in the analysis of data from experiments whose dynamical behavior can be characterized as low dimensional deterministic chaos. The basic idea is to examine the structure and dynamics of a phase space attractor reconstructed from the experimental time series using the time delay embedding method [1]. Reasonably robust procedures have been developed for estimating attractor dimension [2, 3] and Lyapunov exponents [4, 5], and for reducing noise in chaotic experimental data [6, 7]. In addition, it is often possible to locate the periodic saddle orbits in an attractor [8]. This paper uses data from an experiment on the Belousov-Zhabotinskii reaction to illustrate how these approaches can be used together to quantify much of the dynamical behavior [9, 10, 11].

SADDLE ORBITS

Periodic saddle orbits determine the structure and dynamics on typical strange attractors [12, 13]. Periodic points are dense in Axiom A attractors, and formulas relating

Measures of Complexity and Chaos
Edited by N.B. Abraham *et al.*
Plenum Press, New York

147

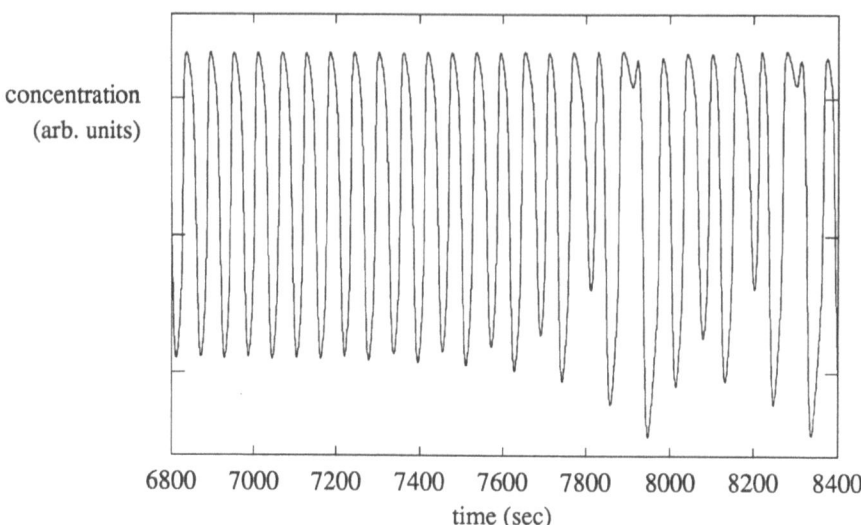

concentration
(arb. units)

6800　7000　7200　7400　7600　7800　8000　8200　8400

time (sec)

Figure 1. Time series of the experimental data. A burst occurs around $t = 7900$ sec.

their eigenvalues to the Lyapunov exponents, natural measure, and fractal dimension can be obtained rigorously in the Axiom A case [14]. In many cases, it is straightforward to locate the saddle orbits of low period in attractors reconstructed from experimental data. An example from a Belousov-Zhabotinskii chemical reaction is given in [8].

In many experiments with chaotic dynamics (*i.e.*, small changes to initial conditions grow exponentially with time), the expansion rate is relatively small. In other words, although trajectories are gradually pushed away from each saddle orbit, they remain nearby long enough to wind around it several times. In these cases it is possible to find unstable periodic orbits by looking for recurrent trajectories as follows.

The experimental time series $\{s_i\}_{i=1}^{n}$ is used to form a vector series $x_i = (s_i, s_{i+\tau}, \ldots, s_{i+(d-1)\tau})$, where d is the embedding dimension and τ is the time delay. (In our example, we have chosen $d = 4$ and $\tau = 23$, based on an information theoretic criterion given in [15].) For each x_i in the series, we find the smallest $j > i$ such that $\|x_i - x_j\| < \epsilon$, where ϵ is a small positive number. (For convenience, the original time series is scaled to the unit interval. In this paper we set $\epsilon = 0.003$.) If such a j exists, we set $m = j - i$ and call x_i an (m, ϵ) *recurrent point*. The number m is the *recurrence time*; in this paper m is expressed in multiples of the time interval between successive measurements.

We consider data from a Belousov-Zhabotinskii chemical reaction for a set of the experimental parameters just past a crisis [10]. The time series consists of 32,768 measurements of the concentration of one of the chemical species in the reactor taken at intervals of 1 sec., a portion of which is illustrated in Fig. 1. The data consist mostly of relaxation oscillations whose relative extrema are about 29 sec. apart, interspersed with occasional bursts.

Figure 2 shows a histogram of the recurrence times tabulated from the data with $\epsilon = 0.003$. There are many points whose recurrence time is approximately 58 sec., corresponding to a period one saddle orbit. All but one of the other recurrence times are clustered in small intervals around multiples of 58 sec. The exceptional peak in the histogram (for $m = 391$) corresponds to the intermittent bursts in the data.

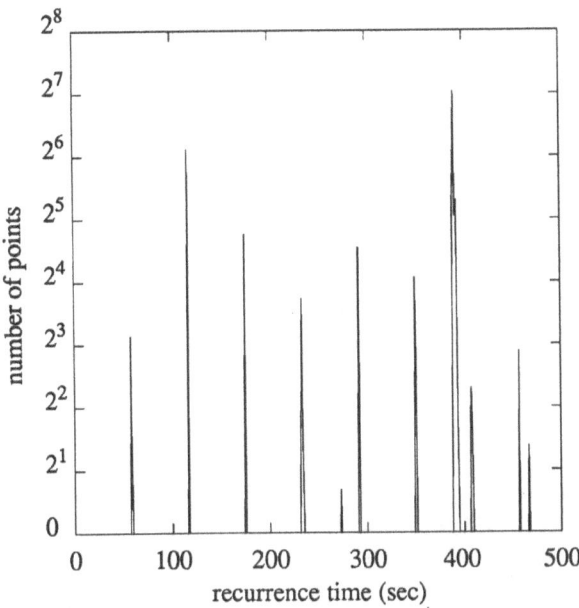

Figure 2. Histogram of the recurrence times computed from the data.

The approximate location of the saddle orbits can be discerned by examining sequences of recurrence times. For each attractor point x_i, we record its recurrence time m_i (if any). We examine the (x_i, m_i) pairs to find sequences where each m_i is the same. One such sequence corresponds to the trajectory shown in Fig. 3(b), near the period one saddle. A graphical analysis shows that the trajectory moves around the saddle in gradually widening spirals until it is kicked away in an "intermittent burst." The burst appears to correspond to a brief approach of the trajectory to another periodic orbit (Fig. 3(c)) which may be responsible for the observed crisis [10]. After the burst, the trajectory returns to a region close to the period one saddle.

The histogram in Fig. 2 suggests that there are orbits of period 2, 3, and higher. A closer inspection reveals that these recurrent points (except those associated with the burst) are an artifact of the "spiraling out" of trajectories as they leave the neighborhood of the period one saddle orbit. This can be seen readily from the return map illustrated in Fig. 4. The period one saddle orbit corresponds to the fixed point marked FP on the graph. Points on a trajectory initially near FP have recurrence times around 58 sec. As the trajectory moves away from FP, it intersects the return plane on alternating sides of FP in the interval marked F. This means that there are points whose recurrence time is 2×58 sec. because the trajectory goes around twice before it returns to within ϵ of its previous intersection on same side of FP. The existence of other multiples of the recurrence times can be explained similarly.

Only two distinct recurrent trajectories are identifiable in this data set. Thus it is not possible to calculate the eigenvalues associated with the saddle orbits and use formulas relating the eigenvalues to the Lyapunov exponents and natural measure [13, 14]. Nevertheless, the identification of saddle orbits is useful for understanding the basic

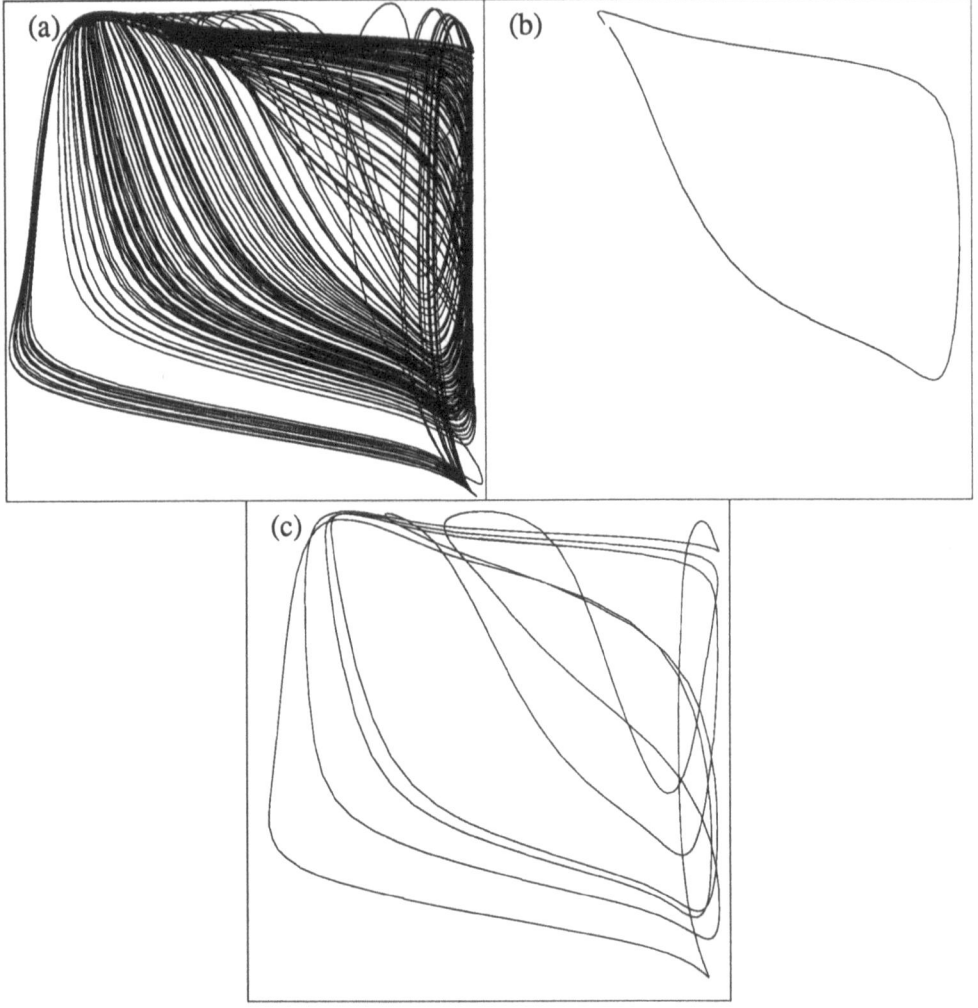

Figure 3. (*a*) The reconstructed attractor. (*b*) An orbit near the period 1 saddle. (*c*) A trajectory corresponding to an intermittent burst.

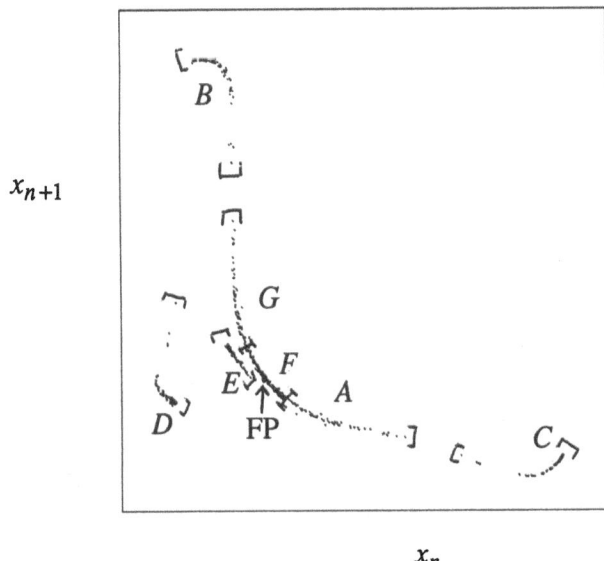

x_{n+1}

x_n

Figure 4. The next amplitude map. The plot is of the the nth relative minimum versus the $(n+1)$st.

structure of the attractor. In this case, the period one saddle and another periodic orbit associated with an intermittent burst appear to govern the dynamics. More generally, one often is interested in how particular periodic orbits change with a parameter. For example, a crisis can occur when the stable and unstable manifolds of one or more unstable periodic saddles collide [16]. The ability to identify periodic saddles can yield a good qualitative understanding of the dynamics associated with a crisis.

Symbolic Dynamics

The use of symbolic dynamics has been a fruitful approach to understanding the behavior of various nonlinear systems [17, 4, 18]. Although the return map shown in Fig. 4 is not a single-valued function, it can be used to find a suitable partition of the phase space in order to estimate the topological entropy.

By graphical inspection, one can identify a sequence of regions (marked A through G in Fig. 4) which are visited by subsequent intersections of the trajectory through the return plane. For example, if x_n is in region A, then x_{n+1} is in B. Likewise, points in B map to those in C and so on. Each time a point is plotted on the return plane, we write a letter (A–G) for the corresponding interval. This approximation of the dynamics is called a *subshift of finite type*, where the allowable sequences are summarized by a *transition matrix* [19], as follows:

$$M = \begin{array}{c|ccccccc}
 & A & B & C & D & E & F & G \\
\hline
A & 0 & 1 & 0 & 0 & 0 & 0 & 0 \\
B & 0 & 0 & 1 & 0 & 0 & 0 & 0 \\
C & 0 & 0 & 0 & 1 & 0 & 0 & 0 \\
D & 0 & 0 & 0 & 0 & 1 & 0 & 0 \\
E & 0 & 0 & 0 & 0 & 0 & 1 & 0 \\
F & 0 & 0 & 0 & 0 & 0 & 1 & 1 \\
G & 1 & 0 & 0 & 0 & 0 & 1 & 0
\end{array}. \tag{1}$$

The nonzero entries correspond to allowable pairs of letters. For example, the first row of the matrix means that A must be followed by B. The points in every other interval map to the interval labeled by the next letter in the alphabet, except for F (where points map either to G or back to F) and G (where points map either to F or A).

If we suppose that the dynamics on the attractor can be approximated in this way, then we can estimate the topological entropy h_t, which counts the number of possible orbits of period p as p becomes large. Let N_p be the number of periodic points for the pth iterate of the return map. Then

$$h_t = \lim_{p \to \infty} \frac{1}{p} \log_2 N_p. \tag{2}$$

In this experiment, it is possible only to identify periodic orbits of low period. Hence we cannot apply Eq. 2 directly. Instead, we can estimate h_t from the transition matrix M. The number of periodic orbits of period p is given by $N_p = $ trace M^p [19] (the trace is the sum of the diagonal entries of M^p). Since the trace is dominated by the largest eigenvalue λ of M for large p, we find that $h_t = \log_2 \lambda$. A numerical calculation shows that $\lambda \approx 1.653$ so that $h_t \approx 0.73$ bits/orbit. (An "orbit" here is defined as 58 time steps, the approximate period of the period one saddle.)

This estimate of the topological entropy probably is too high, because the intervals A–G do not form a *generating partition*. That is, the interval F does not map completely into itself, and points in G do not map completely onto F. It is difficult to assess how accurately the transition matrix models the dynamics, because the length of the data set limits our investigation to the allowable pairs of symbols. If the experimental data set were longer, then it would be possible to investigate longer symbol sequences (e.g., a third return map) and determine more precisely which saddle orbits of low period are present.

Lyapunov Exponents

Several methods have been proposed to estimate the Lyapunov exponents of attractors reconstructed from experimental data [4, 5, 20, 21]. We have used an algorithm based on the approach of Kamphorst et al. [20], in which a tangent map is computed at each point on the observed attractor using linear least squares. Using an embedding dimension of 4 and a time delay of 23, we find that the largest Lyapunov exponent λ_1 is approximately 0.57 bits/orbit. This estimate varies by less than 3% over 10 runs with different sets of algorithm parameters (*e.g.*, the number of points to include in the least squares and their distance from the reference point). We note that the topological entropy is an upper bound for the value of the positive Lyapunov exponent [19]; our value of λ_1 is in reasonable agreement with the estimate of the topological entropy ($h_t \approx 0.73$ bits/orbit) obtained from the transition matrix.

An examination of the return map in Fig. 4 shows how trajectories are pulled into the region near the fixed point. Points in interval D map onto E and then onto F. The presence of this apparent contracting structure (or *stable foliation*) suggests that it may be possible to estimate the largest negative Lyapunov exponent (in this case λ_3) from the data. The median value of λ_3 obtained from the 10 trials above is -8.35 bits/orbit (the variance is about 20%). Although the error bars are only approximate, the Kaplan-Yorke formula for the Lyapunov dimension [22] yields $2 + |\lambda_1/\lambda_3| = 2.07$, which agrees well with the value of 2.10 for the information dimension obtained using the nearest neighbor algorithm of Badii and Politi [3].

CONCLUSION

The time delay embedding method provides a powerful set of tools for the analysis of data from experiments whose behavior can be described as low dimensional deterministic chaos. The identification of periodic saddle orbits can provide a qualitative understanding of the dynamics, particularly in the cases of attractor crises. Return maps can be used to approximate the dynamics by a subshift of finite type; the resulting estimate of the topological entropy agrees reasonably well with the positive Lyapunov exponent. In addition, it may be possible to estimate the largest negative Lyapunov exponent in cases where some contracting structure on the attractor can be discerned.

We thank N. Kreisberg for access to his experimental data, and R. F. Williams and P. Cvitanović for helpful discussions. This research is supported in part by the DARPA Applied and Computational Mathematics Program and the Department of Energy Office of Basic Energy Sciences.

References

[1] F. Takens, in *Dynamical Systems and Turbulence*, ed. by D. A. Rand and L.-S. Young, *Springer Lecture Notes in Mathematics*, Vol. 898 (New York: Springer-Verlag, 1981), p. 366.

[2] P. Grassberger and I. Procaccia, *Physica D* **9** (1983), 189.

[3] R. Badii and A. Politi, *J. Stat. Phys.* **40** (1985), 725.

[4] J.-P. Eckmann and D. Ruelle, *Rev. Mod. Phys.* **57** (1985), 617.

[5] A. Wolf, J. B. Swift, H. L. Swinney, and J. A. Vastano, *Physica D* **16** (1985), 285.

[6] E. J. Kostelich and J. A. Yorke, *Phys. Rev. A* **38** (1988), 1649.

[7] E. J. Kostelich and J. A. Yorke, submitted to *Physica D*.

[8] D. P. Lathrop and E. J. Kostelich, *Phys. Rev. A*, in press.

[9] J. C. Roux, R. H. Simoyi, and H. L. Swinney, *Physica D* **8** (1983), 257.

[10] P. Richetti, P. De Kepper, J. C. Roux, and H. L. Swinney, *J. Stat. Phys.* **48** (1987), 977.

[11] F. Argoul, A. Arneodo, P. Richetti, J. C. Roux, and H. L. Swinney, *Acc. Chem. Res.* **86** (1987), 119.

[12] D. Auerbach, P. Cvitanović, J.-P. Eckmann, G. H. Gunaratne, and I. Procaccia, *Phys. Rev. Lett.* **58** (1987), 2387.

[13] P. Cvitanović, *Phys. Rev. Lett.* **61** (1988), 2729.

[14] C. Grebogi, E. Ott, and J. A. Yorke, *Phys. Rev. A* **36** (1988), 3522.

[15] A. M. Fraser and H. L. Swinney *Phys. Rev. A* **33** (1986), 1134.

[16] C. Grebogi, E. Ott and J. A. Yorke, *Physica D* **7** (1983), 187.

[17] P. Collet and J.-P. Eckmann, *Iterated Maps on the Interval as Dynamical Systems* (Boston: Birkhäuser, 1981).

[18] P. Cvitanović, G. Gunaratne, and I. Procaccia, *Phys. Rev. A* **38** (1988), 1503.

[19] R. L. Devaney, *An Introduction to Chaotic Dynamical Systems* (Menlo Park, CA: The Benjamin/Cummings Publishing Co., Inc., 1986).

[20] J.-P. Eckmann, S. O. Kamphorst, D. Ruelle and S. Ciliberto, *Phys. Rev. A* **34** (1986), 4971.

[21] M. Sano and Y. Sawada, *Phys. Rev. Lett.* **55** (1985), 1082.

[22] J. L. Kaplan and J. A. Yorke, "Chaotic Behavior of Multidimensional Difference Equations," in *Functional Differential Equations and Approximations of Fixed Points*, ed. by H. O. Peitgen and H. O. Walther, Springer Lecture Notes in Mathematics Vol. 730 (New York: Springer-Verlag, 1979).

TIME EVOLUTION OF LOCAL COMPLEXITY MEASURES AND APERIODIC

PERTURBATIONS OF NONLINEAR DYNAMICAL SYSTEMS

Gottfried Mayer-Kress

Mathematics Department, University of California at Santa Cruz

Department of Chemical Engineering, Princeton University

Center for Nonlinear Studies, Los Alamos National Laboratory

MS-B258, Los Alamos, NM 87545

Alfred Hübler

Center for Complex Systems Research, Department of Physics,

Beckman Institute, University of Illinois at Urbana-Champaign

Abstract

We discuss numerical algorithms for estimating dimensional complexity of observed time-series with special emphasis on biological and medical applications. Factors which enter the procedure are discussed and applied to local estimates of pointwise dimensions or crowding indices. We illustrate the concepts with the help of experimental time-series obtained from speech signals. The temporal evolution of the crowding index shows oscillations which can be correlated with properties of the time-series. We compare the time evolution of the dimensional complexity parameter with the original time-series and also with recurrence plots of the embedded time series.

Besides the analysis of spontaneous activity of biological systems it is often more useful to study event related potentials. We have generalized our analysis code in a way that attractors can also be reconstructed from such non contiguous signals. Finally we discuss the possibility of nonlinear, aperiodic stimulation of nonlinear and chaotic systems as a method for very selective excitations of specific nonlinear modes. We discuss possible applications of this method to habituation phenomena and diagnostic use in connection with event-related potentials.

In: "Quantitative Measures of Complex Dynamical Systems", N.B. Abraham (Ed.), Plenum, New York

Measures of Complexity and Chaos
Edited by N.B. Abraham *et al.*
Plenum Press, New York

155

Introduction

It appears that in the field of nonlinear dynamical analysis we are still in the beginning stages of a methodological evolution, particularly in time series analysis. In the calculation of fractal dimensions (d) from time series, many different algorithms and methods have been proposed which efficiently produce converging results in simple cases ($d < 3$). But for more complicated sets the different methods seem to have problems in producing reliable dimension values with a realistic number of data points. (See [16] for an overview.)

Another basic problem in dimensional estimates arises with the interpretation of an observed dimension value [18]. One of the most common confusions arise through a sloppy mixing of purely mathematical concepts and observable quantities. Like many mathematical objects, fractal dimensions are defined through a limiting process which involves both infinite observation time and also infinite precision. The same statement holds for the definition of the Fourier spectrum of a signal. In an experimental context, especially for living systems, the observation of a given dimension value corresponds more to the detection of a (fairly broad) peak in the power spectrum and not to the identification of a finite set of sharp frequency lines. In the same way, as spectro-grams show the time evolution of frequency components in a signal (today widely used in clinical monitoring, although mathematically not well defined) the method of estimating the pointwise dimension (or crowding index) as a function of time might yield relevant information about the evolution of the dominant complexity of the signal. An observed increase in the dimensional complexity could correspond to an increase in the number of independent oscillating subsystems contributing to the signal. In the linear analogue, this number could also be observed by counting the number of independent frequencies in the spectrum. On the other hand, an enhanced "synergetic self-organization" (see [10]) could synchronize some of the subsystems and thereby reduce the observed dimensionality of a time-signal.

Below we give a brief overview of the concepts of dynamical dimension estimates and then describe the methods we have used to obtain an unbiased estimate of dominating dimensional complexity parameters and the detailed structure of the scaling properties of reconstructed attractors.

Local Gauge functions and pointwise dimensions

Assume we are measuring a single variable discrete time-series $x(t_m) = x_m$. Then we can reconstruct vectors $\vec{x}_m$ in a n-dimensional state space through time delay coordinates:

$$\vec{x}_m = (x_m, x_{m-k}, x_{m-2k}, \dots, x_{m-(n-1)k}),$$

where m runs from $(n-1)k+1$ to the number n_{data} of data points and k is the time delay. In figs. (1, 2) we plot the time series x_m of speech signals from two similar but different words.

Here and in the following figures we want to illustrate our method with the help of the two speech signals of figs. (1, 2). The time delay for the reconstruction should be chosen in a way that the coordinates of $\vec{x}_m$ are maximally independent. We use the concept of mutual information content [8] to determine the optimal delay time.

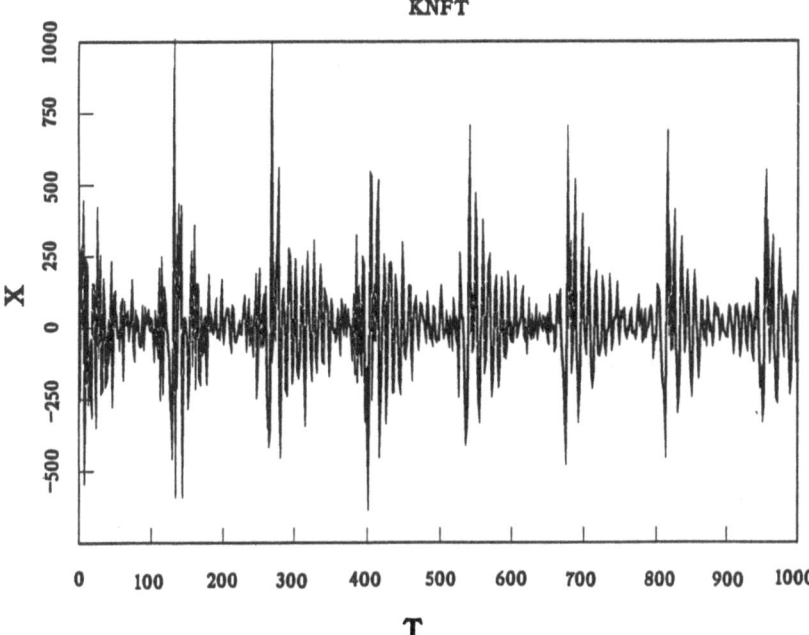

Figure 1. Time series from a segment of a speech signal of the word "know". The modulation of the amplitude corresponds to the dynamics of the vocal chords. (Time measured in sampling units, sampling rate at 12 kHz.)

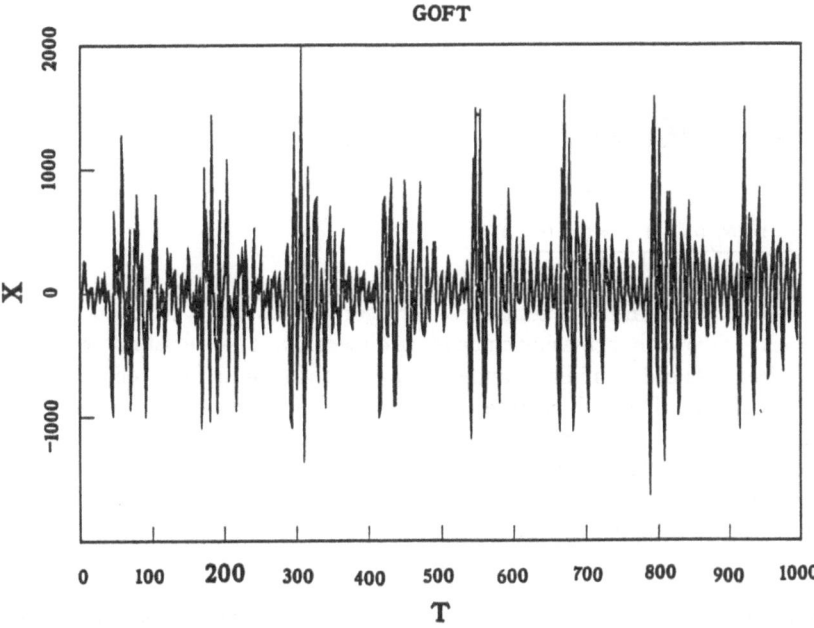

Figure 2. Time series from a segment of a speech signal of the word "go".(Time measured in sampling units, sampling rate at 12 kHz.)

From the data vectors $\vec{x}_m$, we select a subset of n_{ref} equally spaced (in time) reference vectors $\vec{\xi}_j = \vec{x}_{j\nu}$, where $\nu = [\frac{n_{data}}{n_{ref}}]$ [1]. For each of the n-dimensional reference vectors, we determine the local gauge function

$$N_{\vec{\xi}_j}(r) = \sum_{i=1}^{n_{data}} \Theta(r - \|\vec{x}_i - \vec{\xi}_j\|),$$

which counts the number of data points in a neighborhood of $\vec{\xi}_j$ of size r (Θ is the Heavyside step function). In a log-log-representation this function typically exhibits a scaling region

$$I_{\vec{\xi}_j} = [r_{min}(\vec{\xi}_j), r_{max}(\vec{\xi}_j)]$$

over which a slope can be defined. This means that for $r \in I_{\vec{\xi}_j}$ we have

$$N_{\vec{\xi}_j}(r) = c(\vec{\xi}_j)\, r^{d_{\vec{\xi}_j}}$$

where $c(\vec{\xi}_j)$ is a position dependent scale factor. This slope is then interpreted as the pointwise dimension $d_{\vec{\xi}_j}$ of the system at point $\vec{\xi}_j$ (see e.g. [9], [12], [16], [17], [18], [19], [25]). We denote by:

$$\alpha(\vec{\xi}_j) = \frac{r_{max}(\vec{\xi}_j)}{r_{min}(\vec{\xi}_j)}$$

the scaling range of $N_{\vec{\xi}_j}(r)$. If we are restricted to scaling behavior in the region $I_{\vec{\xi}_j}$ then we estimate the dominant dimension at location $\vec{\xi}_j$ from :

$$N_{\vec{\xi}_j}(r_{min}) = c(\vec{\xi}_j)\, r_{min}^{d_{\vec{\xi}_j}}$$

and

$$N_{\vec{\xi}_j}(r_{max}) = c(\vec{\xi}_j)\, r_{max}^{d_{\vec{\xi}_j}}$$

as:

$$d_{\vec{\xi}_j} = \frac{\log(N_{\vec{\xi}_j}(r_{max})) - \log(N_{\vec{\xi}_j}(r_{min}))}{\log(\alpha(\vec{\xi}_j))}.$$

The accuracy of dimension estimates increases with α. For a fixed number of data-points α goes to zero with increasing embedding dimension. There exist different estimates for the requirements of the minimal number n_{data} of data points for a reliable estimate of a given dimension D. They all indicate an exponential dependence on the number of data points n_{data}. In [18] we give arguments for the requirement $n_{data} \geq 10^D$. Nevertheless we think that especially in bio-medical applications, where an exact value of the dimension is not interesting, it might be useful for diagnostic purposes to compare significant changes in the "dimensional complexity parameter" ([17], [19]) even for considerably smaller data sets.

Because of the non-uniformity properties ([2], [20]) of typical attractors, we expect that the location of the scaling region $I_{\vec{\xi}_j}$ (i.e. the value of x_{min}, say) as well as the scaling range α depends on the reference point $\vec{\xi}_j$. In the limit of infinite resolution and number of data points we can obtain the dimension spectra for the multi-fractal set [21].

---

[1] Here $[\cdot]$ denotes Gauss brackets.

Their actual computation is however limited to very low (even compared to the regular dimension estimates) dimensional systems (see e.g. [23]). In the widely used Grassberger-Procaccia algorithm [9] for dimension estimation, the average value of $N_{\vec{\xi}_j}(r)$ is computed over all reference points $\vec{\xi}_j$ at a fixed value of r, i.e. for the dimension estimate one uses $\langle N_{\vec{\xi}_j}(r) \rangle_{\vec{\xi}_j}$, which gives a bad approximation for stongly variing $c(\vec{\xi}_j)$ and small values of $\alpha(\vec{\xi}_j)$. For the estimation of the information dimension [5] we encounter similar problems since we basically average $\langle \log N_{\vec{\xi}_j}(r) \rangle_{\vec{\xi}_j}$ over all reference points $\vec{\xi}_j$.

For the case of finite scaling regions, this technique introduces systematic errors and a bias to lower values [12]. That can be avoided by computing the pointwise dimension individually for each of the reference points and then taking the average. Better accuracy is achieved if, instead of keeping the neighborhood size r fixed, one averages with a fixed number of data points in a neighborhood of a reference points over the different neighborhood sizes, i.e. one computes for given $N_{\vec{\xi}_j}$ the averages $\langle r_{min}(\vec{\xi}_j) \rangle_{\vec{\xi}_j}$ and $\langle r_{max}(\vec{\xi}_j) \rangle_{\vec{\xi}_j}$.

Strictly speaking, the dimension is determined through the scaling behavior at infinitesimally small distances, but, in a practical sense, this scaling is neither feasible nor physically meaningful. The relevant dynamics is not always that observed at very small scales where noise becomes dominant, data are limited, or the statistics becomes weak. In contrast, we want to study the dominant dimension (better: the "dimensional complexity" [17]) which in a sense, describes the scaling behavior with the least irregularity (i.e. the best fit of a straight line in a log-log plot) over the largest range [18]. We think that this method corresponds to a more systematic procedure for extracting a dimension value than from a visual inspection of the dimension curves.

In order to minimize the bias in dimension estimates, we introduced an algorithm which determines the fit-range, goodness of fit (GF), and the estimated dimension automatically for each reference point and for each embedding dimension [19]. In this way we try to make sure that the same criteria are applied to each of the data sets so that results become reproducible and comparable. The main assumption made in the "dominant" dimension estimate is that the relevant scales at each location on the attractor can be found by searching (for a given minimal scaling range $\alpha = \frac{r_{max}}{r_{min}}$ over which the attractor scales) for that value of $\log r$ providing an optimal fit of a linear segment of length $\log \alpha$ to the log of the gauge function. In order to get well defined conditions and make best use of the local scaling properties, we expand the scaling region in the next step until the specified goodness of fit is reached. We use linear interpolation to compensate for the discontinuities due to the finite coarseness of our sequence of distance values r_k.

We determine the scaling range by dividing the logarithm of the normalized signal into 64 bins [2]. The overall range R of the distances is determined by the signal to noise ratio, given as $R = 2^{\frac{64}{S}}, S \geq 1$. For biological signals we typically take $S = 8$, i.e. a resolution of $R = 8$ bits. That means we normalize the signal into a range between $r_{k,min} = 2^{\frac{-27}{4}}$ and $r_{k,max} = 2$. We obtain a logarithmic scale x for distances r through the transformation:

$$r = 2^{\frac{(x-55)}{S}}, \text{ or } x = S \times \log_2 r + 55.$$

A scaling factor of $\alpha = 2$ therefore corresponds to fitting the logarithm of the local gauge function through a set of 8 points. Furthermore we request that the fit range is such that the number of points $N_{\vec{\xi}_j}(r)$ contained in each neighborhood lies between certain bounds in order to avoid artefacts from too large or too small neighborhoods.

---

[2]This allows fast arithmetic on Cray computers

(For the speech data we have $4 \leq N_{\vec{\xi}_j}(r) \leq \frac{n_{data}}{2}$). In our fitting procedure we determine a dimension estimate for each value of the embedding dimension. To avoid a discontinuous jump in the dimension estimate as a function of the embedding dimension which results from a change in the fitting interval, we monitor the dependence of the fitting interval, selected by our algorithm, on the embedding dimension. In most cases the discontinuity in the dimension values is due to this shift in the fit region [19].

In figs. (3, 4) we have plotted the minimal distance $r_{min}(\vec{\xi}_j)$ (on a logarithmic scale) of the scaling range $I_{\vec{\xi}_j}$ as a function of the time index $T = j$. We observe that the location of the scaling region also oscillates with roughly the same frequency as the modulation of the signal. From the interpretation of the signal as relaxation oscillation, mentioned below, one would expect large dimensions in sparse regions with large amplitude. This seems to be the case in the data of fig. (2). The opposite seems to be the case for the other data file. Points which lie close to the boundary of the attractor will typically contribute a smaller effective dimension value due to "edge effects" [18].

We have to be aware that the number of data points of these two files is by far too small to allow any reliable estimate of the true dimension values. The question here is, how we can interpret and understand relative changes of the apparent dimension or dimensional complexity parameter. As described in [18] there exist many geometrical effects at finite scales, which can influence the observed dimension values. Thus more

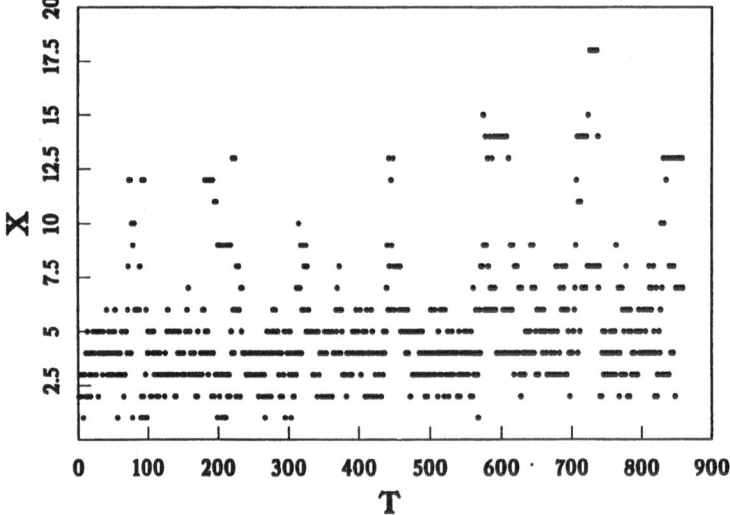

Figure 3. Location x of fit range as a function of time for the data file of fig. (1). The value of x corresponds to a logarithmic representation of $r_{min}(\vec{\xi}_j)$ as used for the definition of $\alpha(\vec{\xi}_j)$ (i.e. we have $x = 8 \times \log_2(r_{min}(\vec{\xi}_j)) + 55$.

work is needed to understand the origins of the oscillations of the apparent pointwise dimensions.

Non-Contiguous Signals

The main limitation for accurate dimension analysis of experimental time series is often given by the amount of available data. Some authors try to generate more data points

160

through higher sampling rates, but that method only introduces systematically low estimates [17]. A better way of increasing the number of data points is given by a repetition of the experiment, i.e. by taking an ensemble average over the attractor. In a biological context this is for example the case in experiments with evoked potentials (see however the discussion of habituation below). In a sequence of contiguous data segments we can embed each individual segment, and if the system is in a well prepared state, then all embedded segments of vector trajectories will belong to the same attractor (or basin of attraction in the case of transient signals). During the dimension computation we have to be cautious not to use reconstructed vectors whose components belong to different segments of the scalar signal. Other than that, the reconstruction and dimension estimation is identical to the procedure described above [19].

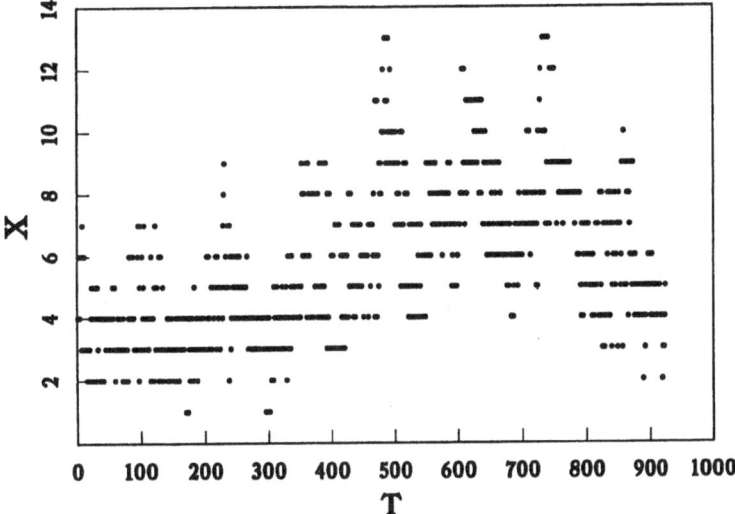

Figure 4: Same as in fig. (3) for the data of fig. (2).

Dynamics of Dimensional Complexity

Since our reference points are sampled at equal time intervals $\nu = \left[\frac{n_{data}}{n_{ref}}\right]$, we obtain a sequence of dimension values that reflects the temporal ordering on the attractor although the dynamics itself is chaotic and recurrences are quasi random. To understand this phenomenon, assume that we observe an intermittently chaotic signal, i.e. regular phases are interrupted at irregular intervals by chaotic bursts. If we choose a reference point at a location which lies in a burst interval, then the data vectors from the regular phase can appear at larger distances than the points from other bursts and therefore contribute less to the local dimension at this reference point. Therefore the value of the pointwise dimension might appear to be higher than for reference points taken during the laminar phase. This effect is not due to nonstationary of the system but it stems from the strong nonuniformity of the dynamics. In the strict mathematical sense [3], reference points with different values for the pointwise dimension are located at singularities of the invariant measure of the attractor (see [21] for details of singularity spectra and $f(\alpha)$ curves). Thus the time dependence of the pointwise dimension would indicate the temporal order in which these singularities are encountered by the system.

This temporal information is completely lost in a direct calculation of the Grassberger-

[3]In the limit of infinitely high resolution.

Procaccia (GP) dimension as well as in the $f(\alpha)$ curves. Our method is also superior to repeated calculation of the (GP)- dimensions on subsets of the time series because in that case one is considerably more limited in the number of relevant data points available for the estimate. It is possible through our method to localize the specific regions on a reconstructed attractor responsible for significant changes in the apparent local dimensionality.

In figs. 5, 6 we plot the pointwise dimension obtained in this way as a function of the reference point for the time series shown in figs. 1, 2.

We can see that the modulation in the amplitude of the signal is correlated with the oscillation in the pointwise dimension [15]. This might be related to the mechanism by which speech is generated: The closing vocal chords interrupt the airflow and thereby generating a delta puls with a white spectrum, i.e. it contains an infinity of frequency modes. Most of them are not resonant with the vocal tract and therefore decay rapidly. This induces a fast decay in the dimensional complexity observed in figs. 5, 6. The decay rate of the dimensional complexity parameter should be characteristic for the resonator and could possibly be used for speech and/or speaker recognition. But because of the extremely small data set and because of possible influences of geometrical effects (see e.g. [18]), it is not clear if this effect has not other (geometrical) origins.

Today there exist several methods for reconstructing dynamical equations from a time series (see e.g. [6], [13]). From the reconstructed equations it is possible to forecast the future behavior of the system. The prediction error E depends on both the dimension of the reconstructed system D as well as on its sensitivity to initial conditions, expressed through its largest Lyapunov exponent λ_1. Farmer and Sidorowich [6] have derived a quantitative expression for the expected prediction error E: If the number of available data points is given by n_{data}, then the error for forecasting the system a time T into the future is estimated as:

$$E \sim n_{data}^{\frac{-1}{D}} \times e^{\lambda_1 T}$$

For the speech signal we would expect that it's dynamical entropy as well as the Lyapunov exponents are relatively constant. Therefore we would expect an oscillation in the prediction error if estimated from these signals according to the number of active modes at any given time. This has been observed with the help of prediction algorithms based on neural networks [14].

The method of estimating the point wise dimension at a sequence of time instances is equivalent to probing the attractor at different geometrical locations. We think that this information is very helpful in associating changes in the complexity of the dynamics with geometrical features of the reconstructed data set; it offers stronger insights into the characteristics of the system.

Recurrence Plots

The time dependent pointwise dimension or crowding index give us information about the scaling behavior in a n dimensional neighborhood of the reconstructed system at a given point in time. As mentioned above, we are only interested in the contribution of those points of the system, which are not temporarily too close, but revisit the neighborhood after some elapsed recurrence time. From the dimensional complexity alone we cannot

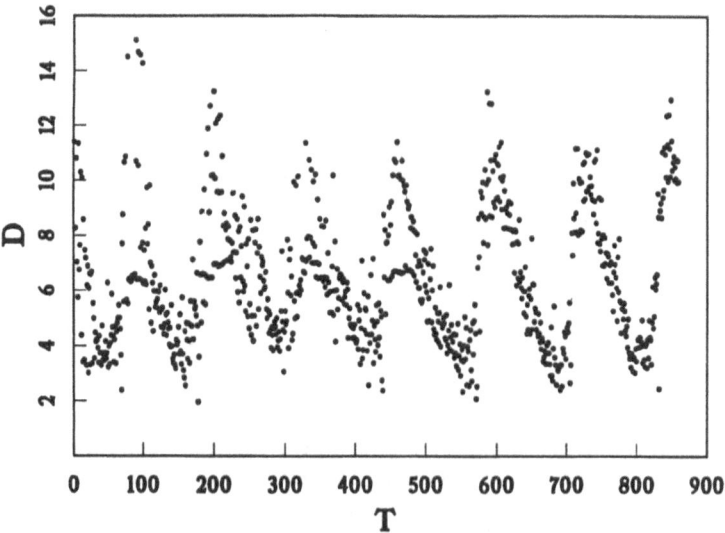

Figure 5. Time series of pointwise dimension values for the speech signal of fig. 1 from the word "know". From the $n_{data} = 1000$ skalar data points we reconstruct a vector time series of embedding dimension $n = 20$ with a time delay of $k = 3$. Under these conditions we obtain from the $n_{data} = 1000$ skalar data points $n_{vec} = 934$ vector data points $\vec{x}_m$ in a 20- dimensional embedding space. Out of those we choose the first $n_{ref} = 930$ vectors as reference vectors $\vec{\xi}_j$. To avoid strongly correlated pairs of points which are temporally very close we don't count vectors $\vec{x}_m$ in a neighborhood of $\vec{\xi}_j$ whenever $| m - j\nu | \leq 4$ (note that in our case $\nu = [\frac{n_{data}}{n_{ref}}] = 1$). For the minimal fit range we require $\alpha \geq \sqrt{2}$. We reject reference points for which the goodness GF of fit at this value of α is larger than $GF \geq 0.2$. For this data file we have to reject 71 reference vectors, which are all located in areas of high pointwise dimension.

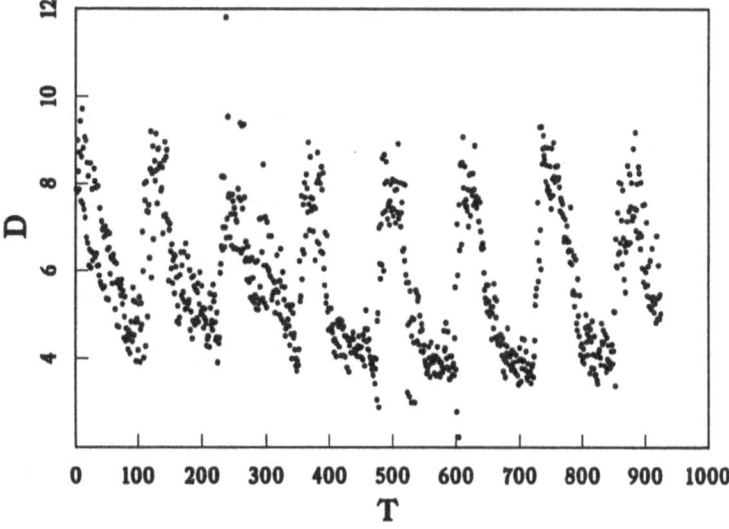

Figure 6. Same as in fig. 5 for the data of fig. 2 (for the word "go"). The numerical parameters are identical. But for this file we only have to reject 5 reference vectors because of insufficient goodness of fit.

deduce any information about the time in which those recurrences take place. It would be possible, as for example in the above mentioned case of an intermittent signal, that a large number of recurrences are generated through some localized small scale dynamics of the system (laminar phase). Then, after a turbulent burst it could take the system a very long time before it visits the same neighborhood again. In other, more regular types of attractors, the recurrence could occur in periodic time intervals.

For some applications it might be interesting to obtain quantitative information about the dynamics of those recurrences. A very elegant way of representing this information was introduced in [3]. We have modified the original method slightly for computational and visualization purposes.

From a (reconstructed) vector time series $\vec{x}_m, m = 1, \ldots, n_{data}$ we compute the distances $\Delta_{m,l}$ between each of the vectors $\vec{x}_m, m = 1, \ldots, [\frac{n_{data}}{2}]$ and the vector $\vec{x}_{m+l}$ shifted in time by an amount of l, for $l = 1, \ldots, [\frac{n_{data}}{2}]$. We now can define a threshold distance $\epsilon > 0$ and define the recurrence times $T_R(m, \epsilon)$ through the condition:

$$\Delta_{m,T_R(m,\epsilon)} \leq \epsilon.$$

In the graphical representation of these functions we observe the periodic structures of the signal in the recurrences (see figs. 7, 8). We obtain more complete information about recurrences at different distances ϵ by plotting the graph of $\Delta_{m,l}$ either in a three dimensional representation or with the help of color coding. Note that each of these reconstructed vectors can be represented as a pattern in the time-series, and therefore this method might be helpful in the context of pattern analysis of scalar signals.

Aperiodic Perturbations of Nonlinear Dynamical Systems

Besides passive analysis and reconstruction of physiological signals the paradigm of nonlinear dynamics and chaos also might hava strong input in more active diagnostic methods. This approach has been successfully applied by A. Hübler and collaborators [13] in a general context. The goal is to reconstruct a model for a given nonlinear system in order to perturb it in an aperiodic way for optimal response or in order to control the dynamics. The tuning parameters in this case are not the frequencies as in traditional spectroscopy, but intrinsic model parameters. We would like to propose such a diagnostic approach in the context of habituation.

The basic idea of active methods is to characterize a system, by determining its response to specific driving forces. The output of the active method is usually the numerical value of the parameters and the dimension of a differential equation or a map which models the dynamics of the system. Active methods, like nonlinear resonance spectroscopy are superior to passive methods because of three reasons: A measurement of the parameters of a system by an active method can be done in this region of the state space of the system, which is of most biological interest in contrast to passive methods which have only access to those regions of the state space which are close to the attractors. Second, the model can be used to control the dynamics of the system even when it is very complex. With the help of variation principles one can calculate from the model the perturbative force of minimal amplitude necessary to reach a certain goal dynamics. Typically this perturbation force is aperiodic but very soft which might be an advantage for many reasons.[4] Third, the active method is especially powerful in those cases, when the experimental system is a set of identical, weakly coupled oscillators behaving incoherently. Brain dynamics might fall into this class. If one gets without a perturbation a compound signal of all oscillators, small and complicated due to interference, a strong response emerges at resonance: every

Figure 7. Recurrence plot $\Delta_{m,T_R(m,\epsilon)}$ for the data of fig. 1 and a threshold value of $\epsilon = .445$. In horizontal direction we have the time index m, in vertical direction we have the time shift l between the two vectors, whose separation is computed. Periodic structures in the signal is represented by horizontal lines. Vertical lines indicate clustering properties of the reconstructed signal.

single oscillator is forced to coherence, synchronized by the aperiodic perturbation.

The following example illustrates the method. The goal is to calculate a perturbation of small amplitude which breaks selectively the bond between a side branch and the main chain of a large molecule.[5] The main chain consists of 14 atoms of unit mass which interact with their nearest neighbours according to the potential

$$V(r) = \frac{1}{2r^2} - \frac{1}{r}$$

where r is the distance to the next neighbours. The position of the first and the last atom is fixed. Close to the fifth atom (B) is an additional atom (A) wich has the same next neighbour interaction with B. We simulate the two dimensional, undamped classical dynamics of this molecule and investigate its response. We assume that B has a negative charge and atom A a positive charge and try to break the bond between these atoms by an oscillating electric field. Due to the electric field forces act on atom B and atom A with the same magnitude but opposite direction. The goal is to find a perturbation which is as small as possible but provides a large energy transfer. Recently it has been show how to solve such a variation problem [13]: We simulate the dynamics of a molecule which has the same structure but where the bond between atom B and atom A is undamped, i.e. there acts an additional force

---

[4]Soft means that they are of small amplitude and have no steep gradients.
[5]Here represented as a chain of coupled oscillators.

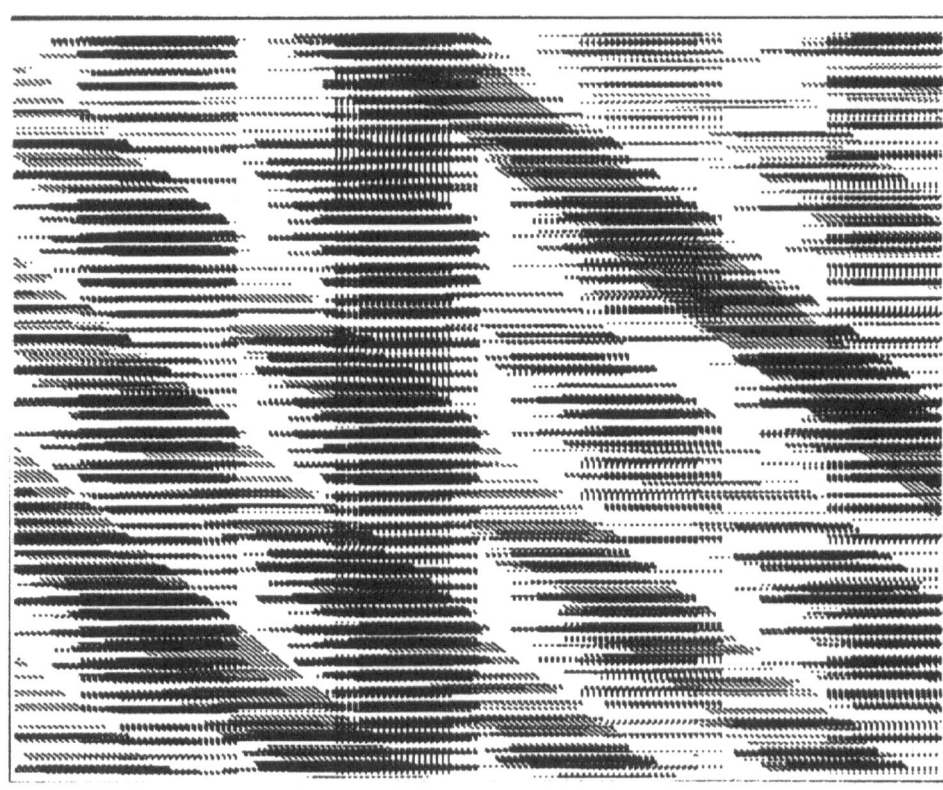

Figure 8. Same as in fig. 7, but for the data of fig. 2.

$$\vec{F}_A = -\eta \left(\vec{\dot{x}}_A - \vec{\dot{x}}_B \right) = -\vec{F}_B,$$

where $\eta = -.05$ and $\vec{\dot{x}}_A$ is the velocity of atom A and $\vec{\dot{x}}_B$ the velocity of atom B. $\vec{F}_A$ and $\vec{F}_B$ are the optimal aperiodic driving forces.

In order to calculate $\vec{F}_A$ and $\vec{F}_B$ we start from an initial condition where all atoms of the main chain are on a straight line and where the side branch is perpendicular to the main chain. The distance of all atoms of the main chain is at the equilibrium position whereas the distance of the side branch A is 0.1% larger than the equilibrium position. The simulation stops when the energy E_s of the bond between atom B and atom A becomes positive (in the numerical simulation $E_s > 0.01$).

Now we apply the resulting driving forces to a molecule where the atoms of the main chain are initially on a straight line and where the side brach is initially perpendicular to the main chain. Further all distances including the side branch are at the equilibrium position.

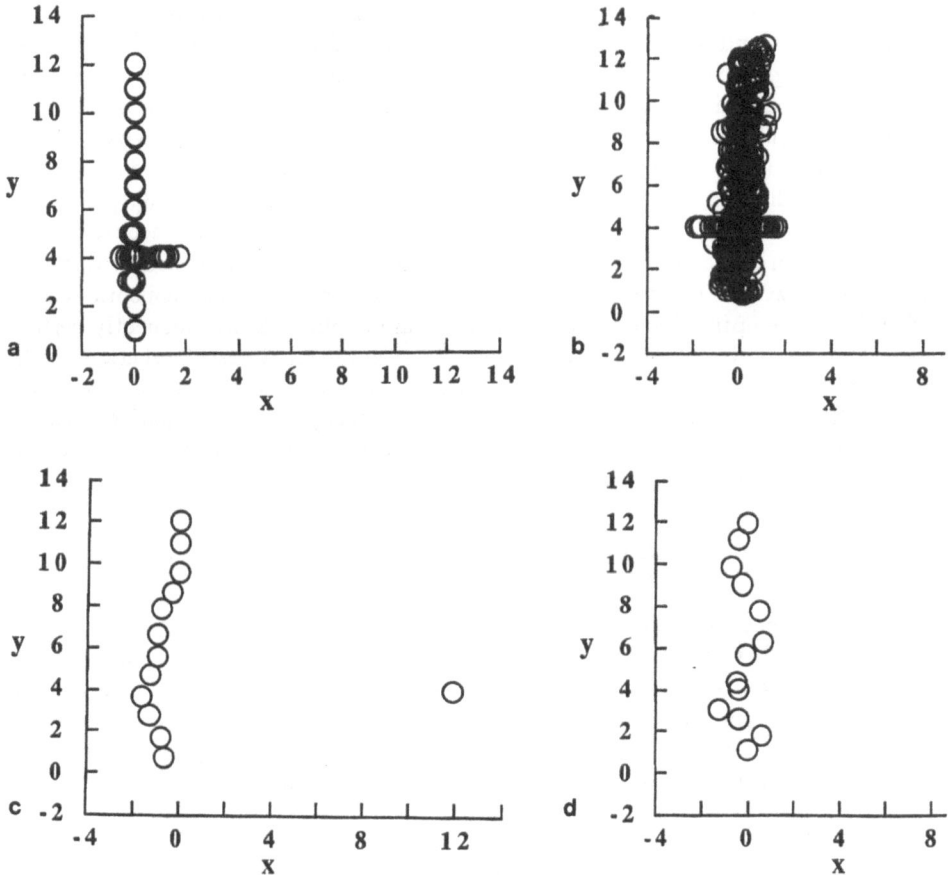

Figure 9. Stimulation of a molecule with sinusoidal (a,b) and optimal aperiodic driving forces(c,d). (a) and (d) show intermediate states, (b) and (d) show the final states.

From numerical simulations we find that the bond is broken and that nearly 100% of the transferred energy is used to break the bond. Only a small portion of the energy is stored as thermal excitation in the main chain, due to the recoil (see fig.(9)).

The result is completely different if a sinusoidal stimulation of the same maximal amplitude is applied for the same period of time: the bond is not broken but a lot of energy is stored in the main chain.

This indicates that optimal aperiodic perturbations can be used for a selective bond breaking. If the side branch of the molecule which has to be stimulated has the interaction potential $V(r) = \frac{c^4}{r^2} - \frac{c^3}{r}$ where c is a parameter, but if the original driving force is used, the bond is not broken if $|c - 1| > .005$. This effect can be used to determine c and other parameters of the equation of motion of the molecule. Since the basic frequency is independent of c sinusoidal perturbations are not selective.

This example shows that it is possible to control the dynamics of a complex system if the differential equation is known. Further it illustrates that it is possible to find the parameters of these equations even if the molecule is initially at a rest. Passive methods would fail in this case.

A clinical application of this method might be the control of epileptic seizures. During the seizure smooth and regular waves dominate the dynamics of the EEG [11] in contrast to the normal highly complex dynamics (see e.g. [17]). If these waves can be modeled by ordinary differential equations, partial differential equations or discrete maps, then the method described above should allow us to compute correcting forces which should reduce the amplitude of these waves and possibly stop the seizure. As stressed before, the necessary aperiodic perturbations typically are of very small amplitude [13], [24]. This effect appears to be in agreement with the well known fact that periodic acoustical and optical stimuli can induce epileptic seizures, i.e. in this terminology they couple very well to the periodic wave states. It is thus conceivable that the intensity of an epileptic seizure is reduced if these optimal aperiodic stimulations are applied. Since frequently certain acoustical and/or optical illusions precede epileptic seizures[22], a possible therapy might include the transformation of the derived aperiodic stimulus into a visual or acoustical image (see [1] for an example of such a mapping). It seems possible that patients can be trained to recall these acoustical or visual patterns at or before onset of the seizure and thus limit or suppress its occurrence.

Habituation and Chaos

Habituation is a phenomenon which occurs as a response to a regular perturbation of an organism by visual or acoustical stimuli. If the stimulus is new it evokes a strong response, if the same stimulus is repeated periodically, then the response goes rapidly to zero. In the context of dynamical systems we can interpret the stimulus as a perturbation which takes the system away from the "spontaneous" attractor, the evoked potential would be a signal of a transient trajectory as it relaxes back to the attractor.

The situation becomes different when we repeat the stimulus periodically: Now we can no longer speak of transients and relaxations (except in cases where the period between successive stimuli is much longer than intrinsic time scales of the system). We now have a driven dynamical system, which can develop completely new and qualitatively different attractors. If we now make the assumption that we are dealing with a complex and adaptive system, then we can see habituation as an adaptation of our brain to the new environment in which a periodic stimulus is present, without any informational value and content. Therefore a reasonable target for adaptation would be to "predict" the occurrence of the next stimulus. If the prediction is correct, the signal does not contain new information and therefore the response might as well be omitted.

It has been observed [7] that the evoked response remains relatively high if the stimulus is presented randomly instead of periodically. Each stimulus is now a surprise, unpredictable and therefore contains new information, to which the system has to respond.

Interestingly enough there exist pathologies in which the habituation is almost completely absent: every new stimulus evokes a strong response, even when they are completely periodic. We can interpret this as a reduction of adaptive performance of the system or a loss of predictive capabilities.

If we now come back to chaotic dynamical systems, then we see that in the field of nonlinear dynamics ther exist several methods to quantify the degree of dynamical complexity and chaos [16]. We can now think of experiments in which we replace the periodic (or stochastic, for that purpose) sequence of stimuli by one which is generated by a chaotic dynamical system with a well defined and known degree of temporal complexity. There are basically two quantities discussed in the chaos literature for the quantification of chaotic signals: the first one is its dimensional complexity (or fractal, or correlation dimension) D, the second one is given by the Kolmogorov entropy K of the signal and also measured through Lyapunov exponents λ_i. In their paper on prediction and noise reduction [6] Farmer and Sidorowich derive an expression for the prediction error E of an predictive algorithm based on dynamical systems. In this expression the predictability of a signal depends both on the dimension as well as on the entropy of the signal.

If we know construct a sequence of stimuli with the help of a chaotic dynamical system, then we have complete control over all these factors. It would be interesting to test if it is possible to measure the habituation time of a given patient as a function of the predictability of the sequence of stimuli. Thus we could possibly obtain a diagnostic tool for the adaptive capabilities of certain brain functions.

We might also add that instead of synthetic, say, acoustical stimuli we could also think of natural stimuli such as pieces of music, rhythms etc. Once we have a measure for the complexity of the musical patterns, say, we also might have some way of quantifying the degree of complexity in a piece of music with which a listener feels comfortable, is able to recognize, can reproduce etc.

We also suspect that this concept of chaotic stimulation also might have applications in cardiac pacemakers and perhaps even in the stimuli offered in acupuncture. (There exist reports according to which the analgesic efficiency of electrical stimulation of skin regions under acupuncture is greatly enhanced if the frequency spectrum of the stimulus has $1/f$ shape. This would be consistent with the above speculation since many chaotic systems have $1/f$ type spectra.

Acknowledgements

One of us (gmk) would like to thank D. Ruelle, Y. Kevrekidis, A. Lapedes, Y.C. Lee, G. Papcun for stimulating discussions and G. Papcun for providing the speech data. This work was partially supported by NSF grant EET-8717787.

References

[1] J. Cowan, in "Pattern Formation by Dynamic Systems and Pattern Recognition", H. Haken (Ed.), Springer Series in Synergetics Vol. **5**, Springer Verlag, Berlin Heidelberg, New York, 1978

[2] U. Dressler, G. Mayer-Kress, W. Lauterborn, "Local Divergence Rates in Nonlinear Dynamical Systems", in preparation

[3] J.P. Eckmann, S. Oliffson Kamphorst, D. Ruelle, Europhysics Letters, **4**,973-977, (1987)

[4] Th. Eisenhammer, A. Hübler, G. Mayer-Kress, P. Milonni "Aperiodic resonant excitation of classical and quantum anharmonic oscillators", in preparation

[5] J.D. Farmer, E. Ott, J. Yorke, "The Dimension of Chaotic Attractors", Physica **7D**, 153, (1983)

[6] J.D. Farmer, J. Sidorowich, "Exploiting Chaos to Predict the Future and Reduce Noise", Reviews of Modern Physics, (1989)

[7] E. Flynn, private communication

[8] A.M. Fraser, H.L. Swinney, Phys. Rev. **A 33**, 1134-1140, (1986)

[9] P.Grassberger, I. Procaccia, "Measuring the Strangeness of Strange Attractors", Physica **9D**, 189, (1983)

[10] H. Haken, "Advanced Synergetics", Springer, Berlin 1983

[11] H. Haken, A. Fuchs, in: "Neural and Synergetic Computers", H. Haken (Ed.), Springer Series in Synergetics Vol. **42**, Springer Verlag, Berlin, Heidelberg, New-York, 1988

[12] J. Holzfuss, G. Mayer-Kress, "An Approach to Error-Estimation in the Application of Dimension Algorithms", in: "Dimensions and Entropies in Chaotic Systems", G. Mayer-Kress (ed.), Springer Series in Synergetics Vol. **32**, Springer Verlag, Berlin etc., 1986

[13] A. Hübler, E. Lüscher, "Resonant Stimulation and Control of Nonlinear Oscillators", Naturwissenschaften 76, 67(1989), A. Hübler, E. Lüscher, "Resonant Stimulation of Complex Systems", to appear in Helv.Phys.Acta **61**

[14] A. Lapedes, private communication

[15] Y.C. Lee, G. Mayer-Kress, G. Papcun, unpublished results

[16] G. Mayer-Kress, (ed.), "Dimensions and Entropies in Chaotic Systems", Springer Series in Synergetics, Vol. 32, Springer-Verlag Berlin, Heidelberg 1986

[17] G. Mayer-Kress, S.P. Layne, "Dimensionality of the Human Electroencephalogram", Proc. of the New York Academy of Sciences conf. "Perspectives in Biological Dynamics and Theoretical Medicine", A.S. Mandell, S. Koslow (eds.) Annals of the New York Academy of Sciences, Vol. **504**, New York, 1987

[18] G. Mayer-Kress, "Application of Dimension Algorithms to Experimental Chaos", in: "Directions in Chaos", Hao Bai-lin (Ed.), World Scientific Publishing Company, Singapore, 1987

[19] G. Mayer-Kress, F. E. Yates, L. Benton, M. Keidel, W. Tirsch, S.J. Pöppl, K. Geist, "Dimensional Analysis of Nonlinear Oscillations in Brain, Heart and Muscle", Mathematical Biosciences **90**, 155-182, 1988

[20] J. Nicolis, G. Mayer-Kress, G. Haubs, "Non-Uniform Chaotic Dynamics with Implications to Information Processing", Z.Naturforsch. **38a**, 1157-1169 (1983)

[21] I. Procaccia, "Characterization of Fractal Measures as Interwoven Sets of Singularities", in G. Mayer-Kress, (ed.), "Dimensions and Entropies in Chaotic Systems", Springer Series in Synergetics, Vol. 32, Springer-Verlag Berlin, Heidelberg 1986

[22] O. Sacks, "The Man who Mistook his Wife for a Hat", Harper & Row, New York 1987

[23] K. Srinivasan, this volume

[24] C. Wagner, W. Stelzel, A. Hübler, E. Lüscher, "Resonante Steuerung nichtlinearer Schwinger", Helv.Phys.Acta **61**, 228, (1988)

[25] J. P. Zbilut, G. Mayer-Kress, K. Geist, "Dimensional Analysis of Heart Rate Variability in Heart Transplant Recipients", Mathematical Biosciences, **90**, 49-70, (1988)

ANALYSIS OF LOCAL SPACE/TIME STATISTICS AND
DIMENSIONS OF ATTRACTORS USING SINGULAR VALUE DECOMPOSITION
AND INFORMATION THEORETIC CRITERIA

A. Passamante, T. Hediger, and Mary Eileen Farrell

Naval Air Development Center

Warminster, Pennsylvania 18974, USA

An algorithm to estimate the average local intrinsic dimension (<LID>) of an attractor using information theoretic criteria is explored in this work. Using noisy sample data the <LID> is computed from an eigenanalysis of local attractor regions, indicating the local orthogonal directions along which the data is clustered. Singular value decomposition (SVD) is used to calculate the eigenvalues as well as to determine the rank of the local phase space data matrix. The <LID> algorithm requires the separation of signal eigenvalues, i.e. the dominant eigenvalues, from the noise eigenvalues for which thresholding mechanisms based on two information theoretic criteria are used. The two information theoretic criteria which we consider for signal/noise separation are the Akaike Information Criterion (AIC) and the Minimum Description Length (MDL) of Rissanen and Schwarz. Several test cases are presented. Results are then compared to the correlation dimension, or fractal dimension, as computed by the Grassberger-Procaccia method. The MDL separation technique produces results in good agreement with the correlation dimension in the range of signal-to-noise ratio (SNR) between 5 to 12 dB. Also, <LID> is calculated in two different ways, as either a spatial average, $<LID>_s$, or as a temporal average, $<LID>_t$. $<LID>_s$ is computed by averaging LID values over randomly selected local regions on the attractor. $<LID>_t$ is computed by first generating consecutive data sets and then averaging LID values over local regions which are restrained to be in approximately the same local vicinity of the attractor. In this way we may compare LID statistics over time and space separately. In addition, we consider the effects of the sampling time, correlation time, window size, and resolution on the <LID> results.

I. INTRODUCTION

The identification of chaotic (or low dimension) characteristics in noisy data samples is a common problem in many scientific areas of investigation. A popular measure of attractor dimension is the correlation dimension which may be estimated using the Grassberger-Procaccia algorithm (GPA).[1] Many conventional fractal dimension (FD) estimators, like the GPA correlation dimension, are very sensitive to the

Measures of Complexity and Chaos
Edited by N.B. Abraham *et al.*
Plenum Press, New York

presence of additive noise. In the method discussed here we take a different approach and analyze moderately noisy experimental data in the phase space using an estimate of the number of orthogonal directions along which the data is distributed. This approach has the advantage of collectively utilizing data to mitigate the effects of the noise. This work extends the work in refs. (2) and (3). The method delineated in ref. (2) for finding the LID required the calculation (via singular value decomposition (SVD)) of the local eigenvalues (EV's) which were subsequently tested by comparing them to an arbitrary threshold to separate the "true" signal (or dominant) EV's from the noise EV's. However, finding appropriate thresholds, in general, is not a straight-forward task. This paper discusses two criteria for signal/noise EV separation based on information theory, the AIC and MDL. Direct thresholds are also considered for comparisons.

II. Theory

A. Local Intrinsic Dimension

We start this approach with noisy time varying data samples (x_i). Points x_i are assumed to consist of a true deterministic signal and white Gaussian noise. The zero mean data is first embedded in high dimension (r) space by forming vectors:

$$\{\mathbf{x}_1 = (x_1, x_2, \ldots x_r), \ldots \mathbf{x}_n = (x_n, x_{n+1}, \ldots x_{n+r-1})\} \qquad (2.1)$$

where typically r = 10 or 20 and here n=20000.

The algorithm randomly selects arbitrary points on the attractor to serve as local centers around which a fixed small number (q) of nearest neighbors will be processed. The essential idea is to randomly cover the entire attractor with local regions, each having its own characteristic spatial distribution of data points extending into various orthogonal dimensions[4,5,6] The number of these local orthogonal, or intrinsic, dimensions is the LID. It is determined by the rank of the local data matrix X formed from the q nearest neighbors (after subtracting out the local center) of each local region, $X = (\mathbf{x}_1, \mathbf{x}_2, \ldots \mathbf{x}_q)^t$, where the superscript t denotes transposition. Values of q range between 30 and 60. We prefer using SVD to find the rank since it is a very numerically efficient method. In general, the output of the SVD contains r singular values, some signal and some noise, and the EV's are the squares of the singular values. Then we apply a thresholding mechanism to separate signal EV's from noise EV's. The number of signal EV's is the number which we call the rank of X. Since this rank will vary all over the attractor we then average each local rank value to find the <LID>.

The <LID> may also be calculated as a time average [<LID>$_t$] at approximately the same region on the attractor. The expectation for <LID>$_t$ is taken over consecutive data sets in time. Results for both space and time averages are presented and compared.

B. Information Theoretic Criteria

In this paper we present two signal/noise separation methods which are formulated on the basis of information theoretic criteria. The first criterion is called the Minimum Description Length (MDL) introduced by Schwarz and Rissanen.[7,8] The second approach is due to Akaike and is called the Akaike Information Criterion (AIC).[9] Akaike has shown that, to within second order, a minimum of the Kullback-Leibler information criterion is acquired when the AIC is also a minimum where:

$$AIC = -2 \log(\text{likelihood}) + 2(\text{number of free parameters in } \Theta) \quad (2.2)$$

where Θ is the parameter vector. The rank of the signal covariance matrix will be denoted as k which also determines the number of free parameters in Θ. The approaches taken by Schwarz[7] and Rissanen,[8] also from information and coding type arguments, yield in the large sample limit, the same criterion called the Minimum Description Length (MDL):

$$MDL = -\log(\text{likelihood}) + (\text{number of free parameters in } \Theta) \log(q)/2 \quad (2.3)$$

For real data the AIC criterion takes on the following form[10,11]:

$$AIC(k) = -2 \log \left\{ \left(\left[\prod_{i=k+1}^{r} 1_i^{1/(r-k)} \right] \Big/ \left[(r-k)^{-1} \sum_{i=k+1}^{r} 1_i \right] \right)^{(r-k)q} \right\}$$

$$+ 2 \left[kr - k^2/2 + k/2 + 1 \right]. \quad (2.4)$$

The MDL criterion, similarly modified by McKee and Rao[11] for real data, is

$$MDL(k) = -\log \left\{ \left(\left[\prod_{i=k+1}^{r} 1_i^{1/(r-k)} \right] \Big/ \left[(r-k)^{-1} \sum_{i=k+1}^{r} 1_i \right] \right)^{(r-k)q} \right\}$$

$$+ \left[kr - k^2/2 + k/2 + 1 \right] \log(q) /2. \quad (2.5)$$

One iterates over several values of k (i.e., k=0, . . . r-1) and chooses that k which minimizes the AIC and MDL. Using MDL the dominant singular values (k) provide a reasonable estimate of the LID which is the local signal distribution of the attractor data within the embedding space.

III. **NUMERICAL RESULTS**

To demonstrate this analysis method we have applied it to several standard test cases of interest, in particular, to the Lorenz system, the Henon map, the Duffing-Ueda attractor,[12] the Rossler attractor, and cases of multiple sinusoids with additive noise over a range of SNR's. All SNR's are measured at the system input. SNR is defined as:

SNR (dB) = $10 \log_{10}$(signal power/noise power)

The input time data window is defined as $t_w = (r-1)t_s$ where t_s is the interpoint sampling time. In choosing an optimum window size we follow the work of Albano et al.[13] which suggests that an appropriate time window is about two to three times the correlation time.

Table 1 presents <LID>'s for various cases where averaging here is done over both space and time. We find reasonably good agreement between the estimated FD of the two-torus, eight-torus, Lorenz, Duffing, and Rossler attractors and their expected values of 2.00, 8.00, 2.05, 2.70, and 2.00 for SNR = 10 dB. These criteria are seen to be inappropriate for SNR's greater than about 12 dB since they will then overestimate the number of dominant singular values. MDL provides reasonable estimates of <LID> whereas AIC tends to greatly overestimate each LID. MDL is the criterion of choice and when applied to the discrete Henon map somewhat overestimate the magnitude of its FD, which should be near 1.2. This difference may be attributed in part to the complex structure of the Henon attractor and in part to the fixed uncontrollable sampling rate for a given map and a given set of parameters.

Table 1. <LID>[spatial/temporal average] via MDL and GPA [a],[b],[c]

	SNR (dB)	q=40 MDL	$\underline{a}$ A	q=50 MDL	$\underline{a}$ A	d_{ij}^{GPA}	d^{GPA}
Lorenz	5	2.34	0.21	2.20	0.22	7.51	nap
	10	2.05	0.15	2.10	0.15	7.21	2.70
	15	1.74	0.10	1.91	0.10	7.09	2.20
Henon	5	1.67	0.33	1.64	0.33	7.57	nap
	10	1.73	0.23	1.76	0.23	6.77	2.80
	15	2.74	0.17	2.71	0.18	4.89	1.90
two-torus	5	1.48	0.25	1.52	0.26	9.81	3.60
	10	2.23	0.18	2.42	0.19	6.45	2.90
	15	3.13	0.14	3.06	0.14	5.00	2.00
eight-torus	5	9.16	0.46	9.08	0.48	5.37	5.00
	10	8.04	0.48	8.56	0.49	4.84	4.50
	15	9.04	0.46	9.20	0.48	5.06	4.40
Duffing	5	2.02	0.37	2.07	0.38	11.10	nap
	10	3.00	0.28	3.09	0.29	9.49	3.54
	15	4.60	0.22	4.47	0.23	6.36	2.92
Rossler	5	1.70	0.34	1.70	0.35	12.40	nap
	10	1.81	0.17	1.64	0.18	7.53	2.08
	15	2.70	0.15	2.83	0.16	9.31	1.88

[a] $\underline{a}$/A = mean local hypersphere radius / full attractor radius.

[b] FD values d_{ij}^{GPA} are calculated using Eq. (3.1) for q_i = 40 and q_j = 50; d^{GPA} is estimated from the slope of the GPA log-log curves.

[c] nap means no apparent plateau in slope value

Table 2. <LID> (using Fixed Thresholding criterion) for Infinite SNR [a]

		D_1	D_{10}	D_{15}	D_{20}	D_{30}
Lorenz	$<LID>_s$	2.36	2.12	2.08	2.04	2.04
	$<LID>_t$	2.00	2.00	2.00	2.00	2.00
	$<\Delta>_s$	0.05	0.04	0.06	0.04	0.03
	$<\Delta>_t$	0.45	0.25	0.21	0.18	0.11
Henon	$<LID>_s$	2.32	1.76	1.72	1.64	1.52
	$<LID>_t$	2.80	2.00	1.90	1.88	1.80
	$<\Delta>_s$	0.07	0.06	0.06	0.06	0.07
	$<\Delta>_t$	0.53	0.57	0.54	0.59	0.47
two-torus	$<LID>_s$	2.30	2.00	2.00	2.00	2.00
	$<LID>_t$	2.60	2.00	2.00	2.00	2.00
	$<\Delta>_s$	0.10	0.00	0.00	0.00	0.00
	$<\Delta>_t$	0.49	0.00	0.00	0.00	0.00
eight-torus	$<LID>_s$	7.82	6.64	6.17	5.73	4.87
	$<LID>_t$	8.00	6.90	6.80	5.88	5.00
	$<\Delta>_s$	0.12	0.09	0.11	0.10	0.11
	$<\Delta>_t$	0.87	0.74	0.62	0.60	0.42
Duffing	$<LID>_s$	4.04	3.00	3.00	2.96	2.81
	$<LID>_t$	4.40	3.00	3.00	2.96	2.92
	$<\Delta>_s$	0.06	0.00	0.01	0.02	0.08
	$<\Delta>_t$	0.29	0.00	0.02	0.04	0.18
Rossler	$<LID>_s$	2.12	2.04	2.02	2.00	2.00
	$<LID>_t$	2.00	2.00	2.00	2.00	2.00
	$<\Delta>_s$	0.03	0.02	0.02	0.01	0.00
	$<\Delta>_t$	0.33	0.16	0.05	0.02	0.00

[a] $<\Delta>_s$ and $<\Delta>_t$ are spatial and temporal standard deviations of <LID>'s.

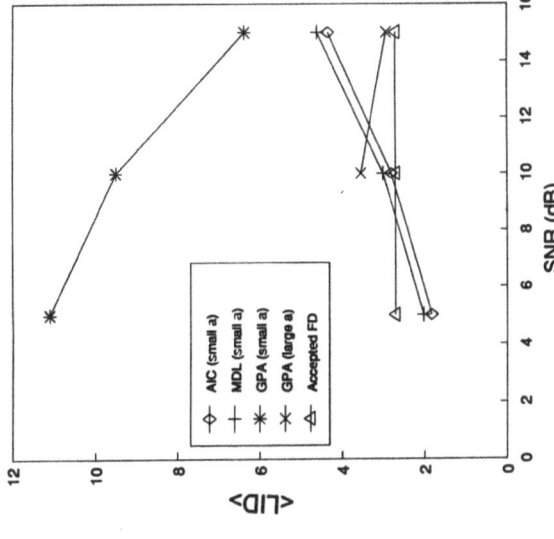

Fig. 1. Plot of $\langle LID \rangle_s$ and $\langle LID \rangle_t$ for two typical attractors at infinite SNR where $\langle LID \rangle_s$ and $\langle LID \rangle_t$ represent spatial and temporal averages, respectively. In the cases shown here thresholding is done using the D_{30} criterion. In the spatial averaging case the index represents the time data set index, while in the temporal averaging case it denotes the randomly chosen spatial region. The true FD of the Lorenz and Duffing attractors are 2.05 and 2.70, respectively.

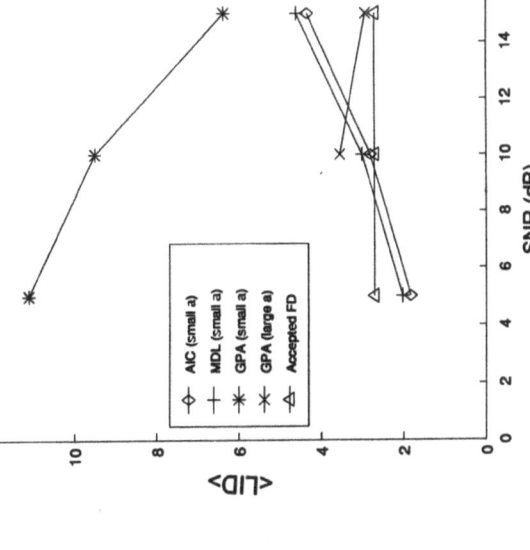

Fig. 2. Shown here for the Duffing attractor we see a comparison of $\langle LID \rangle$ determined using the MDL and AIC criteria with that obtained using the GPA method. Here $\langle LID \rangle$ represents a combined spatial/temporal average of LID's over 25 randomly positioned local regions for each of 25 consecutive data sets (see Table 1). Here the embedding dimension is $r=20$ and the number of samples in each local region is $q=40$. The GPA method at small length scales produces good results only if the SNR is greater than about 20 dB, however, even large (a) GPA results show a significant variation in slopes.

177

For infinite SNR, the arbitrary threshold method used by Passamante et al.[2] gives good results for the <LID> of selected low-dimensional attractors (see Table 2), where D_1 represents those eigenvalues exceeding 1% of the highest eigenvalue, D_{10} represents those exceeding 10% of the highest eigenvalue, et sim., D_{15}, D_{20}, and D_{30}. We should note, however, that this arbitrary thresholding technique produces acceptable results only at D_1 with high-dimensional signals, such as the eight-torus. The FD of the Lorenz, Henon, two-torus, eight-torus, Duffing, and Rossler attractors are 2.05, 1.2, 2.00, 8.00, 2.70, and 2.00, respectively. The spatial and temporal standard deviations of <LID>$_s$ and <LID>$_t$, <Δ>$_s$ and <Δ>$_t$, respectively show that spatial averaging is somewhat superior to temporal averaging for calculation of <LID>. <LID>$_s$ and <LID>$_t$ for the Lorenz and Duffing attractors are plotted in Fig. 1. Notice that the <LID>$_t$ computed over time are indicative of the local attractor structure. In Fig. 1 the Lorenz attractor exhibits a relatively low variance compared to the Duffing attractor due to the Lorenz's benign structure.

All LID results may be compared to the correlation dimension given by the GPA algorithm, which is defined by the slope of the log(c) vs log(a) plot, where c is the correlation summation,[1] and which is calculated by:

$$d_{ij}^{\,GPA} = (\log_{10} q_i - \log_{10} q_j) \,/\, (\log_{10} <a_i> - \log_{10} <a_j>), \qquad (3.1)$$

where q is the number of data points within each local region and <a> is the corresponding mean radius of the local region. In Table 1 $d_{ij}^{\,GPA}$ is calculated for q_i=50, q_j=40 while d^{GPA} is estimated from the slope of the log(c) vs log(r) curves using the GPA algorithm over the whole range of attractor radius. The apparent discrepancy between the $d_{ij}^{\,GPA}$ and the d^{GPA} values in Table 1 is due to the fact that the <LID> technique, from which the <a> used in Eq. (3.1) are generated, is applicable only in the region of small (a) and in the low to moderate SNR domain, whereas the GPA algorithm works best in the high SNR domain at any (a). The results for $d_{ij}^{\,GPA}$, d^{GPA}, and MDL for the Duffing attractor are plotted in Fig. 2. This plot illustrates the fact that, at small length scales, the GPA method produces good results only in the high SNR domain, here greater than about 20 dB, while MDL produces reasonable results in the 5-12 dB SNR range. The results for d^{GPA} obtained using the GPA algorithm over the whole range of attractor radius were estimated from marginal plateaus having no clear convergence.

We also investigate the effect of resolution on determination of the <LID>$_s$. Because data sets have a finite length, there is an inevitable trade-off between resolution and statistical sampling errors. Resolution may be increased by decimating the time series data since this spreads out the power spectrum of the data in the frequency domain. This is illustrated in Figs. 3(a) and 3(b) for the Duffing attractor using decimation factors (DF) of 10 and 4, respectively. A comparison of these figures shows us that a high DF corresponds to increased resolution. Tables 3 and 4 show the variation of <LID>$_s$ with DF for the Lorenz, Duffing, and Rossler attractors. The results for infinite SNR given in Table 3 are seen to be quite insensitive to choice of the DF. However, this is not the case for finite SNR as may be seen from Table 4.

Next we investigate the effect of filtering on determination of the <LID>$_s$. Sometimes an increased SNR is desired, perhaps by filtering the experimental data. For a band-pass filtered swath of the Duffing

Table 3. <LID> Resolution Results for SNR=∞ (DF = decimation factor)

	Criterion	DF=2	DF=4	DF=6	DF=8	DF=10
Duffing	D_1	3.24	3.24	3.40	3.96	3.96
	D_{10}	2.96	3.00	3.00	3.00	3.00
	D_{15}	2.72	2.96	2.96	2.96	3.00
	D_{20}	2.68	2.96	2.92	2.88	3.00
	D_{30}	2.48	2.84	2.84	2.84	2.76
Rossler	D_1	2.16	2.16	2.04	2.12	2.16
	D_{10}	2.04	2.04	2.00	2.04	2.04
	D_{15}	2.04	2.04	2.00	2.04	2.04
	D_{20}	2.04	2.04	2.00	2.04	2.00
	D_{30}	2.04	2.04	2.00	2.00	2.00

Table 4. <LID> Resolution Results for SNR=10 dB (DF = decimation factor)

	Criterion	DF=2	DF=4	DF=6	DF=8	DF=10
Duffing	MDL	1.96	3.28	3.44	3.76	2.68
Rossler	MDL	1.96	2.12	2.36	2.32	2.24

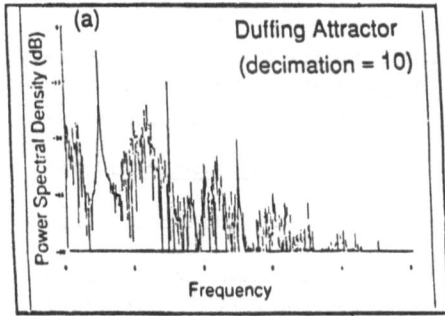 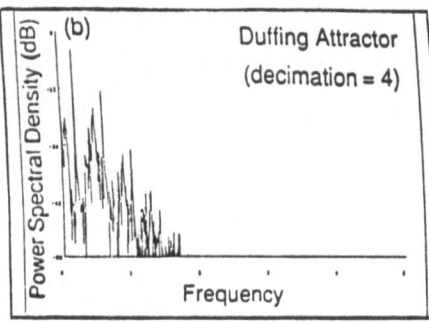

Fig. 3. (a) Power Spectral Density of the Duffing-Ueda attractor $\ddot{x} + b\dot{x} + x^3 = B\cos(wt)$ for parameters b=0.05, B=7.5, and w=1.0 for a decimation factor of 10; (b) Power Spectral Density of the same Duffing attractor for a decimation factor of 4.

attractor an $<LID>_s$ of $D_{30}=3.2$ is obtained for $q=40$, which is higher than the D_{30} value of 2.8 obtained without filtering. In a separate test, we translated the swath down to base band (by multiplying it by a complex exponential) and then low-pass filtered it in order to remove the high frequency components of the signal. We find that the process of translation does not change the value of $<LID>_s$ $[D_{30}=3.2]$.

IV. CONCLUSIONS

We have shown that the MDL method, coupled with SVD, can provide reasonable estimates of the dimension of both high and low dimensional attractors having SNR's between 5 and 12 dB. AIC produced values of $<LID>$ which are unacceptably high. An MDL type criterion cannot be applied to signals with SNR of 12 dB and over as it will inherently overestimate the number of dominant singular values due to the signal. If the SNR is low to moderate, the technique presented here gives good $<LID>$ results. Since these information criteria are not suitable for SNR's above 12 dB, other methods of signal/noise separation are being investigated for the high SNR case. In addition, we find that spatial and temporal averaging of LID values to obtain $<LID>_s$ and $<LID>_t$ produce comparable results, with the spatial average, in general, being closer to the true FD.

REFERENCES

1. P. Grassberger and I. Procaccia, Measuring the Strangeness of Strange Attractors, Physica 9D:189 (1983).
2. A. Passamante, T. Hediger, and M. Gollub, Fractal dimension and local intrinsic dimension, Phys. Rev. A 39:3640 (1989).
3. T. Hediger, A. Passamante, and Mary Eileen Farrell, Estimation of fractal and local dimensions of attractors using SVD and information theoretic criteria, submitted Phys. Rev. A, May, 1989.
4. D. S. Broomhead, R. Jones, and Gregory P. King, Topological dimension and local coordinates from time series data, J. Phys. A 20:L563 (1987).
5. E. R. Pike, Singular System Analysis of Time Series Data, in: "Computational Systems: Natural and Artificial", H. Haken, ed., Springer-Verlag, Berlin (1986).
6. K. Fukunaga and D. R. Olsen, An Algorithm for Finding Intrinsic Dimensionality of Data, IEEE Transactions on Computers C-20:176 (1971).
7. G. Schwarz, Estimating the dimension of a model, Ann. Stat. 6:461 (1978).
8. J. Rissanen, Modeling by shortest data description, Automatica 14:465 (1978).
9. H. Akaike, A New Look at the Statistical Model Identification, IEEE Transactions on Automatic Control AC-19:716 (1974).
10. M. Wax and T. Kailath, Detection of Signals by Information Theoretic Criteria, IEEE Transactions on Acoustics, Speech, and Signal Processing ASSP-33:387 (1985).
11. B. W. McKee and S. S. Rao, An Application of the Minimum Description Length Criterion to Signal Detection and Model Order Selection, Proceedings of the IASTED International Symposium on Applied Control, Filtering and Signal Processing, Geneva, 15-18 June 1987.
12. Y. Ueda, Randomly Transitional Phenomena in the System Governed by Duffing's Equation, J. Stat. Phys. 20:181 (1979).
13. A. M. Albano, J. Muench, C. Schwartz, A. I. Mees, and P. E. Rapp, Singular-value decomposition and the Grassberger-Procaccia algorithm, Phys. Rev. A 38:3017 (1988).

ENTROPY AND CORRELATION TIME

IN A MULTIMODE DYE LASER

Michael G. Raymer

Department of Physics and Chemical Physics Institute
University of Oregon
Eugene, OR 97403

In a recent paper the results of an experimental study of a multimode cw dye laser were reported.[1] It was found that above a certain pumping level the intensities of the cavity modes became unstable and began to fluctuate nonperiodically. From time series of mode intensities the intensity correlation time τ_c and order-2 Kolmogorov entropy K_2 were determined. The correlation time was defined as the width of the peak of the monotonically decreasing autocorrelation function of intensity, and was found to decrease with increasing pump level. This can be understood by arguments based on coherent mode coupling.[1] The entropy was determined using the algorithm of Grassberger and Proccacia,[2] and was found to increase with increasing pump level.

The purpose of this paper is to emphasize an interesting relation observed to hold between the measured correlation time and entropy, that is $\tau_c \cong 1/K_2$. This relation appears plausible, as it is reasonable to associate with $1/K_2$ a time-scale for destruction of information. We are not aware of any discussion of the conditions under which it will be true generally.

ACKNOWLEDGEMENT

This work was supported by the U. S. Department of Energy, Basic Energy Sciences.

REFERENCES

1. I. McMackin, C. Radzewicz, M. Beck, and M. G. Raymer, Phys. Rev. A 38, 820 (1988).

2. P. Grassberger and I. Procaccia, Physica 9D, 189 (1983).

Measures of Complexity and Chaos
Edited by N.B. Abraham *et al.*
Plenum Press, New York

181

DIMENSION CALCULATION PRECISION WITH FINITE DATA SETS

Chera L. Sayers

Department of Economics
University of Houston
Houston, TX 77204-5582

A. Introduction

The study of nonlinear dynamics has become quite popular recently. Originating in the natural sciences, the topic has now spread to numerous diverse fields such as brain research, optics, meteorology, economics, finance, statistics, mathematics, natural resources and agriculture. The interest in nonlinear dynamics was sparked by the discovery in the natural sciences of processes characterized by deterministic chaos; that is, highly complex behavior that is generated by relatively simple nonlinear functions. Observed time series generated by chaotic processes appear to be random by standard linear time series methods.

While empirical studies in the natural sciences are usually characterized by large data sets numbering in the tens of thousands, data sets in recent applications in various fields often number less that one thousand observations. As a result, statistical procedures designed to be implemented in the case of large data sets may not necessarily be appropriate in the case of small, finite data sets.

The popular Grassberger and Procaccia (1983) algorithm utilized in estimating the correlation dimension is an example of one such procedure under which finite data sets induce biased estimates. While no formal statistical distribution theory exists for the notion of the correlation dimension, researchers have attempted to examine the behavior of this estimating procedure under various circumstances. Abraham, et al. (1986) and Albano, et al. (1987) examine the behavior of the correlation dimension estimate for finite data sets. Theiler (1988) examines the statistical precision of the correlation dimension for finite data samples, as dependent upon the tolerance distance. Moller, et al (1989) examine systematic errors in the correlation integral and the correlation dimension estimate as a result of noise in the data series and as a result of finite precision errors.

Ramsey and Yuan (1989a, 1989b) examine the bias and error bars associated with correlation dimension estimation and postulate empirical relationships for the dimension estimate and the associated variance. In particular, the statistical properties of the correlation dimension depend directly upon

Measures of Complexity and Chaos
Edited by N.B. Abraham *et al.*
Plenum Press, New York

sample size and embedding dimension. Ramsey, Sayers and Rothman (1988) have recently extended this work to re-examine specific studies utilizing the notion of correlation dimension in the field of economics.

B. The Correlation Dimension, Estimation and Results

In order to utilize the Grassberger-Procaccia algorithm, let the ordered sequence $\{X_t\}$, $t=1,\ldots N$, represent the observed time series. Then, for a given embedding dimension, d, create a sequence of d-histories,

(1) $\{(x_t, x_{t+T}, \ldots, x_{t+(d-1)T})\}$,

where T stands for the time delay parameter. The sample correlation integral is given by,

(2) $C^N_r = \{\Sigma_{i,j} \ \theta(r - |X_i - X_j|)\}/N^2$, $r>0$, $X_i = (x_i, x_{i+1T}, \ldots, x_{i+(d-1)T})$,
where $\theta(.)$ is the Heaviside function which maps positive arguments into one and non-positive arguments into zero. $\theta(.)$ counts the number of points which are within the tolerance distance, r, from each other. Grassberger and Procaccia show,

(3) $\lim C^N_r \rightarrow C$, as $N \rightarrow \infty$, and $r \rightarrow 0^+$, and $d\ln(C)/d\ln(r)=D_2$.

Thus, an estimate of the correlation dimension can be obtained by a plot of $\ln(C)$ versus $\ln(r)$ and identifying a scaling region of r such that the power law, $C^N_r \approx r^{D2}$ holds. Utilizing the traditional regression model yields,

(4) $\ln(C) = a + b \ln(r) + u$,

where b is an estimate of the correlation dimension, dc, and u is an independently and identically distributed normal error term distributed independently of r. Ramsey and Yuan (1989a,1989b) and Ramsey, Sayers and Rothman (1988) present an alternative random coefficient regression model of the following form,

(5) $\ln(C) = a(e|N,ed) + b(e|N,ed)\ln(r) + u$,

where the variance of u is very small relative to the variance of e for small N and large embedding dimension, ed. The term e represents an experimental error term whose distribution is dependent on the sample size, N and the embedding dimension, ed. Empirically, e appears to be distributed independently of u, but u itself is autocorrelated. Ramsey and Yuan demonstrate that the conditional mean of the estimate of dc takes the following form,

(6) $\ln(k+dc) = \alpha_1 + \alpha_2 N^{\alpha_3} + \alpha_4 N^{\alpha_5}[\exp(\alpha_6/ed^{\alpha_7})-1.0]$.

For random variables that scale monotonically in ed, $k=0$, $\alpha_1=0$ and both α_2 and α_3 are positive. If one has an attractor with a given dimension dc, then $k=1$, $\alpha_1>0$ and $\alpha_3<0$; one would expect α_3 to be -1.0 for attractors with low dimension. For both random and attractor generated data, α_6 should be negative and α_7 should be positive. α_5 seems to be negative for attractors and zero for random variables. α_4 depends upon the units of measurement chosen for ed and the relative weight of the ed effect to sample size.

In addition, (6) implies that the usual regression formula for the variance of the slope estimator, b, is not valid. Empirically, the usual estimate of b is greatly biased downward. Ramsey and Yuan document that the true equation is,

(7) $\ln(\text{std}) = \beta_1 + \beta_2 \ln(N) + \beta_3 \ln(ed) + \beta_4 (ed/N)$.

In empirical studies, the standard deviation of the correlation dimension estimate increases in embedding dimension and decreases in sample size. However, the relative effect of ed is much greater than that for N. Thus, small increases in ed must be offset by proportionately larger increases in N, given that the goal is a constant variance.

C. Conclusion

In summary, results from Ramsey and Yuan (1989a,1989b) and Ramsey, Sayers and Rothman (1988) indicate that dimension estimates, in general, are upward biased for attractors and downward biased for random noise. This bias effect increases with embedding dimension and decreases with sample size. Due to relevant bias effects, this implies that, for example, a correlation dimension estimate of 0.214 could imply an actual correlation dimension value of as high as 1.68.

Results from empirical studies indicate that the standard deviation expressed in (7) declines roughly according to the square root of sample size, but the increase in standard deviation due to ed varies from a low of the square root of ed to a high of a power of 16. Thus, the actual variance of dimension estimates can be as high as 64 times larger than that estimated by the usual least squares approach. The variance of the dimension estimate declines with sample size and increases very rapidly with embedding dimension.

Acknowledgement

Much of the above draws from joint research on nonlinear dynamics with James B. Ramsey and Philip Rothman of New York University. My intellectual debt to James B. Ramsey is especially great.

References

Abraham, N.B., A.M. Albano, B. Das, G. de Guzman, S. Yong, R.S. Gioggia, G.P. Puccioni and J.R. Tredicce (1986): "Calculating the Dimension of Attractors From Small Data Sets," Physics Letters A, 114, 217-221.

Albano, A.M., A.I. Mees, G.C. de Guzman and P.E. Rapp (1987): "Data Requirements for Reliable Estimation of Correlation Dimensions," in A.V. Holden (ed.), Chaotic Biological Systems, Pergammon Press, Elmsford, New York.

Grassberger, P. and I. Procaccia (1983): Measuring the Strangeness of Strange Attractors," Physica, 9D, 189-208.

Moller, M., W. Lange, F. Mitschke, N.B. Abraham and U. Hubner (1989), "Errors From Digitizing and Noise in Estimating Attractor Dimensions," Physics Letters A, January 1989.

Ramsey, J.B. and H. Yuan (1989a): "Bias and Error Bars in Dimension Calculations and Their Evaluation in Some Simple Models," Physics Letters A, 134, 287-297.

Ramsey, J.B. and H. Yuan (1989b): "The Statistical Properties of Dimension Calculations Using Small Data Sets," C.V. Starr Center for Applied Economics, New York University, submitted to <u>Nonlinearity</u>.

Ramsey, J.B., C.L. Sayers and P. Rothman (1988): "The Statistical Properties of Dimension Calculations Using Small Data Sets: Some Economic Applications," C.V. Starr Center for Applied Economics, New York University, forthcoming, <u>International Economic Review</u>.

Theiler, J. (1988): "Statistical Precision of Dimension Estimators," Institute for Nonlinear Science, University of California at San Diego.

CHAOS IN CHILDHOOD EPIDEMICS

W. M. Schaffer* and L. F. Olsen**

* Department of Ecology and Evolutionary Biology
The University of Arizona
Tucson, AZ 85721 USA

** Institute of Biochemistry
Odense University
Odense, DENMARK

Case For Chaos

Whereas childhood infections such as chickenpox evidence effectively periodic dynamics, case reports for diseases such as measles and rubella fluctuate more erratically [1]. Elsewhere [2,3], it has been suggested that the observed fluctuations correspond to low dimensional chaos of the sort which arises in epidemiological models [5] of the SEIR variety, i.e., differential equations, subject to periodic forcing.[1] The basis for this assertion is as follows:

1. Reconstructed phase portraits, Poincaré sections, and return maps constructed from actual data are remarkably similar to those obtained from the models (Fig. 1).

2. For measles, both the models and the data yield correlation dimensions of about 2.5. A comparable value is obtained for the Lyapunov dimension as computed directly from the differential equations.

3. Both the models and the data yield estimated positive Lyapunov exponents of between 0.4 and 0.5 bpy.

More recently, it was shown [5] that Monte Carlo simulations, which allow for stochastic forcing as a consequence of finite population size, exhibit similar correspondences with the data both for large, first world cities with $N > 200,000$ (Fig. 2) and also for small islands with $N < 50,000$.

Biennial Cycles Plus Noise

An alternative interpretation [6,7] is that the observed dynamics correspond to a twice periodic orbit (alternating high and low years) perturbed by noise and corrupted by observational error. The argument for this hypothesis is as follows:

1. The differential equations exhibit period-doubling to chaos as one increases the intensity of seasonal forcing.

2. If one assumes a sinusoidal forcing function - an idealization at best - it is claimed [6] that

---

1. In childhood epidemics, seasonal forcing results from enhanced transmission rates in schools.

Measures of Complexity and Chaos
Edited by N.B. Abraham *et al.*
Plenum Press, New York

187

the magnitude of this parameter is insufficient to place the system in the chaotic region. Inasmuch as the differential equations exhibit a very sharp transition from essentially biennial, albeit chaotic, dynamics to highly irregular chaos, we contend that it is probably impossible to resolve this question by estimating a parameter. In particular, we point to the difficulty in obtaining accurate estimates,[2] and the aforementioned fact that the assumption of sinusoidal forcing is almost certainly inaccurate. An alternative approach is outlined below.

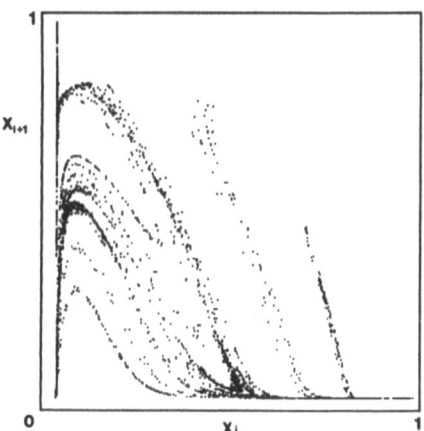

Chaos *vs.* Periodicity Plus Noise

Our analysis rests on the assumption that despite their inadequacies, the differential equations (periodically forced SEIR model) exhibit a range of dynamical possibilities that reasonably approximate what one can expect to observe in nature, though not necessarily for the same parameter values. In part, this assertion is based on simulations [5] in which we varied the form of the forcing function to more nearly correspond to the on-off pattern of enhanced contagion which results from crowding in the classroom. Such changes affect the parameter values at which bifurcations occur, but not the overall sequence of the most important transitions. On the other hand, real world infections are almost certainly subject to the effects of finite population size which manifest themselves as stochastic forcing. Our approach therefore has been to incorporate such effects while at the same time seeking to determine the range of dynamical behaviors in model epidemics which best fit the data.

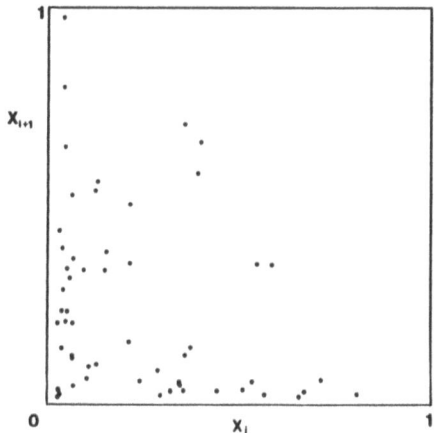

Figure 1. Return maps for measles epidemics imagined and real. **Top.** The map computed from the SEIR model. **Bottom.** City of Baltimore, 1900-1963.

To this end, we have performed Monte Carlo experiments in which we varied both the population size and the magnitude of seasonal forcing. For each combination, we computed 10 fifty year replicates and for each replicate, the correlation dimension as described in [5]. The results are shown in Figure 3. In particular, note the following:

1. For $N > 250,000$, the correlation dimension (solid curves) is essentially independent of population size. Furthermore, the dimension effectively discriminates among the various values of B_1, which is our index of seasonality.

---

2. In the ODEs, the transition to large amplitude chaos, occurs at about $B_1 = 0.28$, B_1 being the parameter which indexes seasonality. Schwartz [6] suggests a maximum reasonable value for this parameter is 0.2, which would place the system well in the region of a two-year cycle. On the other hand, the figure provided by London and Yorke [1] suggests a value of 0.25-0.30, which, in terms of the present discussion, could correspond to anything.

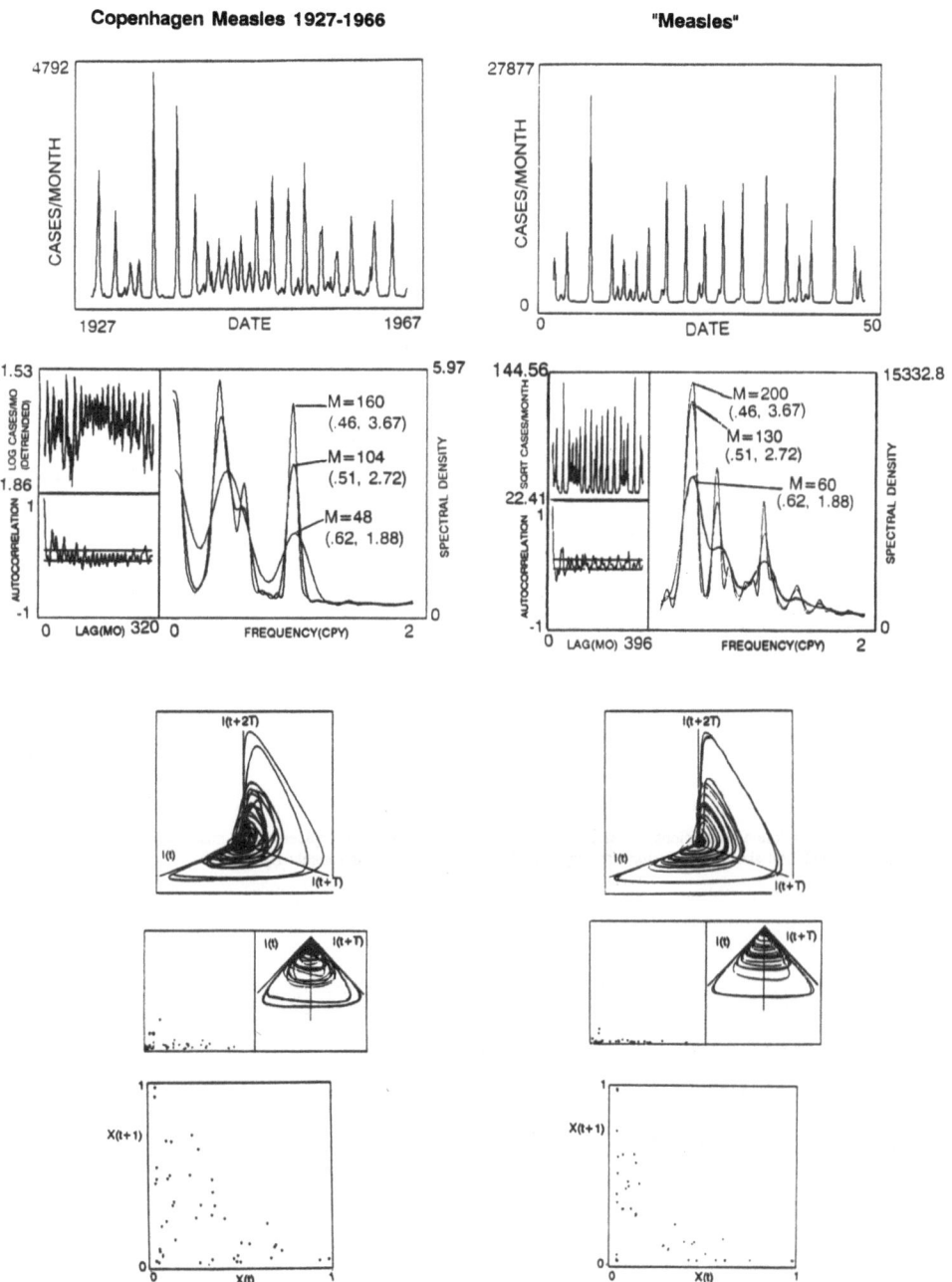

Figure 2. Measles in Copenhagen, Denmark and a Monte-Carlo simulation for N = 1,000,000. From top to bottom, we show time series, power spectra, reconstructed phase portraits, Poincaré sections and return maps.

2. Measles epidemics in first world cities (circles) with populations exceeding this figure, in every case but one, have correlation dimensions which require a value of B_1 in excess of 0.28, in which case the simulations are unquestionably chaotic.

3. The calculations suggest that for small, semi-isolated populations, the correlation dimension might possibly serve as an independent estimator of immigration rates of infective individuals.

The foregoing results complement previous estimates [8] of Lyapunov exponents and correlation dimensions for the differential equations subject to noise. In sum, epidemics of measles in large first world cities are consistent with SEIR-type chaos and inconsistent with periodic behavior in the presence of noise.

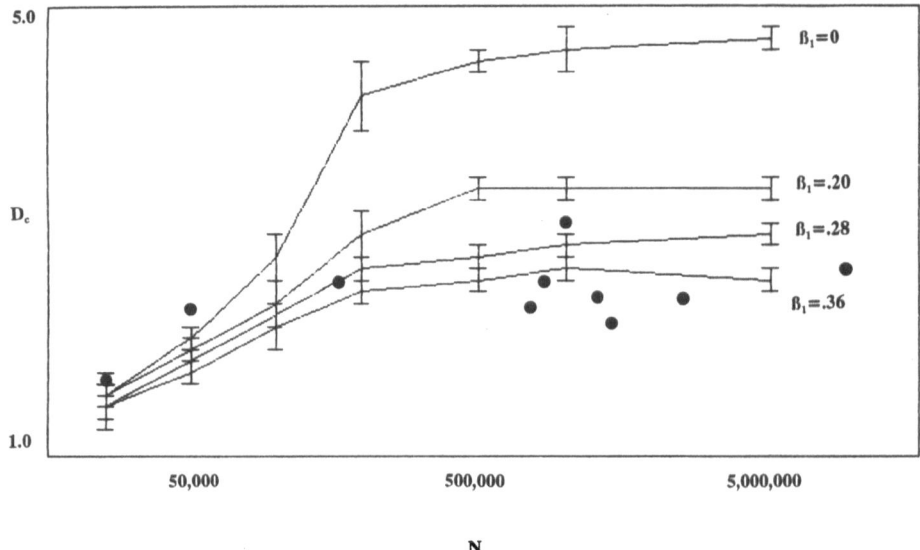

Figure 3. Correlation dimensions for simulated (lines) and actual (circles) epidemics of measles plotted against population size. The transition to large amplitude chaos occurs at roughly $B_1 = 0.28$.

References

1. W. P. London and J.A. Yorke. *Am. J. Epidemiol.* **98**, 453 (1973).

2. W. M. Schaffer and M. Kot. *J. Theor. Biol.* **112**, 403 (1985).

3. W. M. Schaffer. *IMA J. Math. Appl. Med. Biol.* **2**, 221 (1985).

4. K. Dietz. *Lect. Notes Biomath.* **11**, 1, (1976).

5. L. F. Olsen, G.L. Truty and W. M. Schaffer. *Theor. Pop. Biol.* **33**, 344 (1988).

6. I. B. Schwartz. *Proc. 12th IMACS World Congr. '88 on Scient. Comput.* (In press).

7. J. L. Aron and I. B. Schwartz. *J. Theor. Biol.* **110**, 665 (1984).

8. W. M. Schaffer, L. F. Olsen, G. L. Truty and S. L. Fulmer. **In**, S. Krasner. *The Ubiquity of Chaos.* AAAS. (1989).

MEASUREMENTS OF $f(\alpha)$ FOR MULTIFRACTAL

ATTRACTORS IN DRIVEN DIODE RESONATOR SYSTEMS

Z. Su, R. W. Rollins
and E. R. Hunt

Department of Physics and Astronomy
Ohio University
Athens, Ohio 45701, USA

SUMMARY

A single diode resonator system is a nonlinear circuit consisting of a series combination of a resistance, inductance and a p-n junction diode. Driven single diode resonator systems experimentally show the period-doubling route to chaos, while driven coupled diode resonator systems show the quasi-periodic route to chaos. We report $f(\alpha)$ spectra[1,2] calculated from experimental data[3,4] and model calculations[5] for driven diode resonator systems at transitions to chaos. The method used to obtain $f(\alpha)$ has no "free" adjustable parameters.

The experimental system of two coupled diode resonators is driven to exhibit quasiperiodicity with a winding number approximated by the continued fraction <4111...> to an accuracy of one part in 10^6. The accumulation point for the period doublings is found to a precision of about 0.1% for the single diode resonator system. The numerical simulations are based on an extension of the Hunt and Rollins[6,7] model of the diode (ideal diode modified to include a fixed reverse capacitance and a finite reverse recovery time). The extended model allows the reverse recovery time to depend on several previous cycles. The numerical simulations show remarkable global agreement with measurements on the corresponding experimental system.

The $f(\alpha)$ spectra obtained are compared with those calculated for the circle map (for the quasiperiodic route) and the logistic map (for the period-doubling route). For the quasiperiodic route, the $f(\alpha)$ spectra are also reported for sub- and super-critical orbits. Good agreement with theory is found in all cases, providing experimental evidence for the universality of scaling properties at criticality within certain classes of nonlinear dynamical systems.

REFERENCES

1. T. C. Halsey, M. H. Jensen, L. P. Kadanoff, I. Procaccia and
    B. I. Shraiman, Fractal measures and their singularities: The
    characterization of strange sets, Phys. Rev. A 33:1141 (1986).

Measures of Complexity and Chaos
Edited by N.B. Abraham et al.
Plenum Press, New York

191

2. M. H. Jensen, L. P. Kadanoff, A. Libchaber, I. Procaccia, and
    J. Stavans, Global Universality at the onset of chaos; Results of a
    forced Rayleigh-Bénard experiment, Phys. Rev. Lett. 55:2798 (1985).

3. Z. Su, R. W. Rollins and E. R. Hunt, Measurements of $f(\alpha)$ spectra of
    attractors at transitions to chaos in driven diode resonator
    systems, Phys. Rev. A 36:3515 (1987).

4. Z. Su, R. W. Rollins and E. R. Hunt, Universal properties at the
    onset of chaos in diode resonator systems, Phys. Rev. A, in press.

5. Z. Su, R. W. Rollins and E. R. Hunt, Simulation and characterization
    of strange attractors in driven diode resonator systems, Phys. Rev.
    A, in press.

6. R. W. Rollins and E. R. Hunt, Exactly solvable model of a physical
    system exhibiting universal chaotic behavior, Phys. Rev. Lett.
    49:1295 (1982).

7. E. R. Hunt and R. W. Rollins, Exactly solvable model of a physical
    system exhibiting multidimensional chaotic behavior, Phys. Rev. A
    29:1000 (1984).

"IS THERE A STRANGE ATTRACTOR IN A FLUIDIZED BED ?"

S. W. Tam and M. K. Devine

Chemical Technology Division
Argonne National Laboratory
9700 S. Cass Avenue
Argonne, IL   60439

INTRODUCTION

Fluidized beds have long been used as multiphase reactors in the chemical processing industries and as fossil-fuel combustion beds in the energy production industries.[1] They operate on the principle of fluidization in which a fluid is allowed to flow through a solid phase (usually a system of particles of varying sizes) at a velocity above a minimum value U_{mf} at which the drag exceeds the net particle weights. At that stage the particles "fluidize," which is a state of motion in many ways similar in a reverse manner to sedimentation. The process provides efficient mixing among the phases in the system, resulting in good mass/heat transfer characteristics.

Depending on the extent to which the fluid flow velocity V exceeds U_{mf}, the fluidized bed can be in different flow regimes. In the regimes of interest (including the so-called bubbling, slugging and turbulent regimes), the fluidized-bed flow parameters tend to fluctuate in time. The flow parameters include particle velocities, pressure, and porosities. The presence of porosities (or voidage) to different extents is a common characteristic of these flow regimes. Figure 1 is a plot of a typical time series data set for the pressure fluctuation of a fluidized bed.[2] Both the qualitative features of the data and the fact that fluidization processes can be described by the nonlinear equations of multi-phase flows prompt the following question. Do these fluidization data exhibit only stochastic features, or do they contain chaotic characteristics as well? To address this question we have employed a combination of nonlinear dynamical techniques on a relatively large data set for fluidized bed pressure fluctuation.

THE APPROACH

The calculational approach that we have used is the Grassberger-Procaccia correlation integral [GP] method.[3] However, since we are analyzing actual experimental data, noise inevitably arises from a variety of sources. We have utilized a noise reduction technique called singular spectrum decomposition (SVD), which is a well-known method in signal processing.[4] In the present analysis, SVD has been

Measures of Complexity and Chaos
Edited by N.B. Abraham *et al.*
Plenum Press, New York

used only for processing the experimental data for subsequent analysis by the GP algorithm. No attempt has been made to extract directly with SVD any dimensional information on the possible attractor underlying the experimental time series data. Such a combined GP/SVD approach has been shown to be useful in analyzing model-generated data, including the Lorenz and the Rössler systems.[5] Here we have integrated this composite approach with a mutual information analysis and applied them to an experimental time series data set for fluidized bed pressure fluctuation.[2]

The pressure data were obtained at a sampling rate of 500 digitizations per second with a 10 bit A/D converter.[2] The total data set represents an hour of continuous run.[2] In the present analysis several subsets of the data of length between 15,000 and 50,000 points have been utilized.

Given by scalar time series x(t) one can construct d-dimensional vectors by the embedding technique with the following form:[3,7]

$$(x(t),x(t + T),...,x(t + (d - 1)T)) \text{ ---------- (1)}$$

where d is the embedding dimension, and T is the time delay. The geometrical structure generated with these vectors has the same dimensional characteristics such as the correlation dimension (see discussion below) of the attractor generated from the dynamics underlying the original scalar time series.

For practical purposes one needs to choose an appropriate value for T, the time delay. We have utilized the method of mutual information to obtain T.[6] The mutual information $I(x(0),y(T))$ measures the degree of predictability on a variable y at time T given a measurement of another variable x at time zero. The general shape of I(T) has a large central peak around T=0, which arises from near term-correlation. Longer range temporal correlation exhibits itself in oscillatory behavior in I for large T. It has been suggested that the "optimum" T is the time delay at which the first minimum of I occurs.[8]

We have calculated I for the pressure fluctuation data as a function of time delay T. The results are displayed in Fig. 2. The first minimum occurs at a T value of about 53, i.e., a characteristic time of approximately 1/10 of a second. This is the time delay used in subsequent analysis.

The correlation integral (CI) is defined as: $C(l) = \lim_{N \to \infty} 1/N^2 \{$numbers of pairs of points on the attractor whose distance is less than l$\}$

Here N is the number of d-dimensional vectors constructed out of the scalar time series. For some range of l called the scaling region, C(l) scales as l^D where D is the correlation dimension.[3] A saturation of D as the embedding dimension d increases would be an indication that the experimental signal has a non-random component.[3] Before the pressure data were subjected to a CI analysis, they were first processed by SVD[4]. Basically SVD involves the construction of a trajectory matrix A out of the N - d + 1 dimensional vectors generated from the time series. Performing SVD on A gives the usual singular spectrum and a new matrix A', which geometrically is just a rotated version of A. The extent to which the singular values decrease toward zero is an indication of the noise level present with the data. Figure 3 shows the normalized eigen-spectrum (which is just

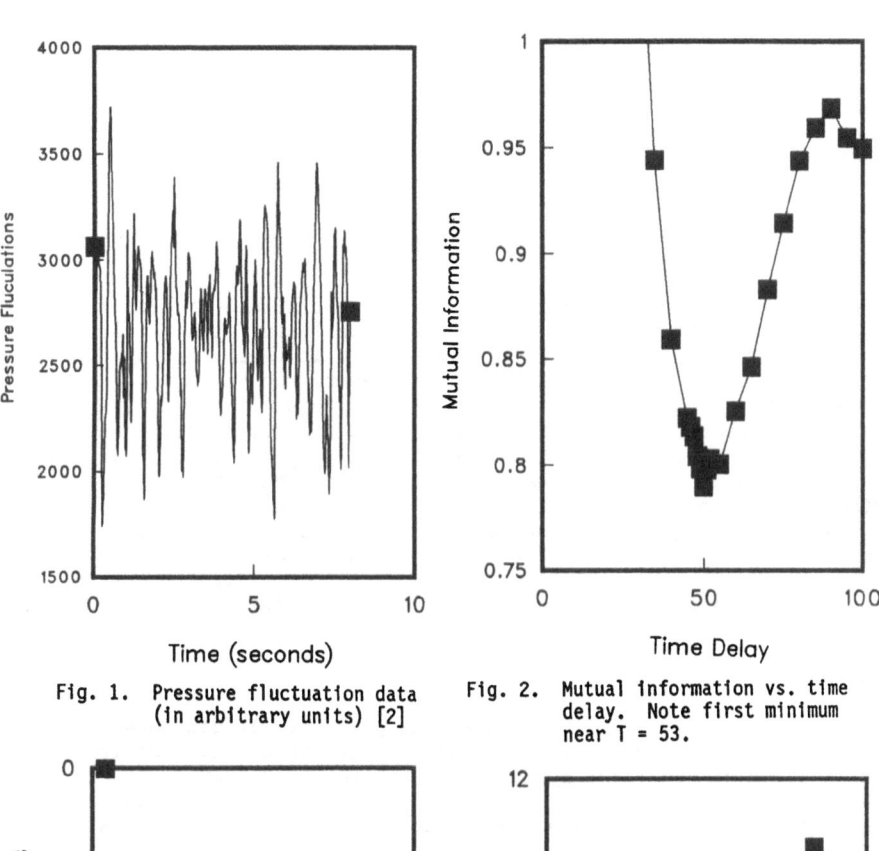

Fig. 1. Pressure fluctuation data
(in arbitrary units) [2]

Fig. 2. Mutual information vs. time
delay. Note first minimum
near T = 53.

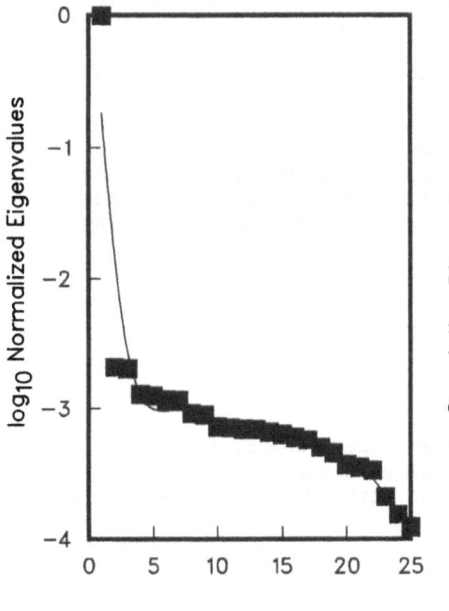

Fig. 3. Normalized eigenvalue vs.
embedding dimension.

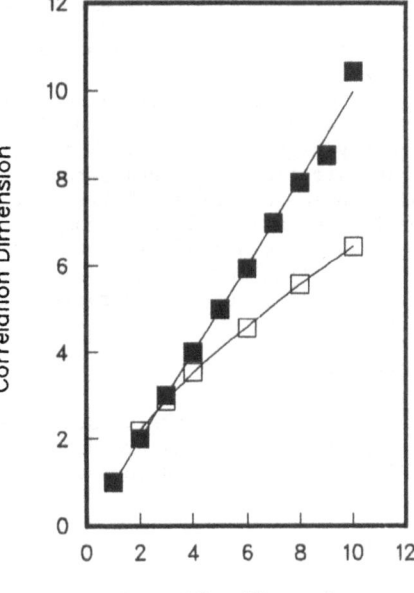

Fig. 4. Correlation dimension vs.
embedding dimensions.
Empty square, pressure
data; filled square,
random number.

the square of the singular values) of the pressure fluctuation data as a function of the embedding dimension d up to 25. One can see a rapid initial decline in the eigenvalues, follow by a more gradual decline. For the higher embedding dimensions the matrices A (or A′) are increasingly contaminated by noise. We have thus chosen to consider submatrices of A′ up to d = 10. These subsets of the rotated matrix A′ are expected to be less affected by noise than the original A. However, it must be pointed out that a rigorous general criterion for choosing the upper bound to the embedding dimension with SVD using experimental data is still an open issue.[9]

The CI has been calculated from the subsets of A′ and the correlation dimension D extracted from a local slope analysis. The resulting D as a function of d is displayed in Fig. 4. It is clear that up to d = 10 there is yet no sign of saturation. However, the value(s) of D that have been obtained are significantly below those from a truly stochastic process such as arising from a "good" random number generator. The present results seem to indicate the presence of a nonrandom component in the process underlying the time series data. Nevertheless, the signal-to-noise ratio (S/N) and the rapidly shrinking scaling region with increasing d rule out an unambiguous determination of the saturation regime and, hence, the evaluation of the dimension of the corresponding attractor. If an attractor exists for the present data, it does not have a low-dimensional structure (i.e., with D in the range of 2-3).

CONCLUSION

The present analysis on the time series data for fluidized-bed pressure fluctuation highlighted several critical issues arising from the application of nonlinear dynamical techniques to the analysis of systems of 'real world' interest (including technological, climatical, and medical/biological). These issues include the problem of noise, the presence of moderate-to-high dimensional attractors, and the constraint of data size limitation. The smallness of the data set size and the presence of noise would invalidate embedding of data to arbitrarily high dimensions. This in turn, would severely limit the possible detection of moderate-to-high dimensional attractors. These factors provide a serious challenge to the effectiveness of applying the available data analysis techniques to chaotic time series. Nevertheless, useful information such as the presence of nonrandom processes and the lower bound on the relevant dimensions of the possible attractor can be extracted, as the present analysis has shown.

ACKNOWLEDGMENT

The authors would like to acknowledge D. Dent and R. La Nauze of CSIRO for making the pressure data available to them and J. Stringer of EPRI for stimulating discussions. This work is partially supported by EPRI.

REFERENCES

[1] D. Geldart (eds.) Gas Fluidization Technology, (John Wiley & Sons Ltd. U.K.) 1986.
[2] D. Dent, private communication 1989.
[3] P. Grassberger and I. Procaccia, Physica D 9 (1983) 189.
[4] D. S. Broomhead and G. P. King, Physica D 20 (1986) 217.
[5] A. M. Albano, J. Muench, C. Schwartz, A. I. Mees, and P. E. Rapp, Phys. Rev. A. 38 (1988) 3017.

[6] A. M. Frazer and H. L. Swinney, Phys. Rev A 33 (1986) 1135.

[7] F. Takens, in Dynamical Systems of Turbulence, D. A. Rand and
     L.-S. Young (eds.) (Springer, Berlin) 1981.

[8] R. S. Shaw, The Dripping Faucet as a Model Chaotic System Aerial
     Press, Santa Cruz, CA (1985).

[9] A. Passamante, T. Hediger, and M. E. Farrell, preprint, 1989.

STATISTICAL ERROR IN DIMENSION ESTIMATORS

James Theiler

Institute for Nonlinear Science
University of California at San Diego

Present address: MIT Lincoln Laboratory, L-244
Lexington, MA 02173-0073
e-mail: jt@ll.ll.mit.edu

ABSTRACT: The statistical error in an estimate of the correlation dimension based on a sample of N points (*e.g.*, from a time series) generically scales as $O(1/\sqrt{N})$. The coefficient of the $1/\sqrt{N}$ term can be related to the variance over the attractor of the pointwise mass function. In many cases, the coefficient is small, and for very particular examples the coefficient is zero. In the latter case, anomalously precise $O(1/N)$ scaling is obtained. These results are shown to have practical implications for computing the dimension of an attractor from a time series.

INTRODUCTION: Though computing the dimension of fractal sets has achieved widespread popularity, it is difficult to characterize the accuracy and reliability of these computational estimates [1]. In this paper, we set aside the issue of systematic errors, noting that they can be important [2], and focus on the statistical error. We comment that although systematic errors are in many cases more significant than statistical error, the two are qualitatively different in their behavior, and an understanding of both is necessary for quantifying the goodness of a dimension estimate. In particular, we note that although most systematic errors can be reduced by decreasing the distance scale R, statistical error *increases* with decreasing R. To minimize the total error in a dimension estimate, one must trade-off the geometrical effects of large R against the statistical effects of small N.

We will take a different approach than in Ref. [3] and measure the precison of a computational estimate directly as the variation in estimated values over many trials with equivalent but independent datasets. For a dynamical system, such independent datasets might arise from an ensemble of distinct initial conditions.

First, two naïve models will be introduced, the independent distances hypothesis (IDH), and the average pointwise mass model (APM). Neither is "correct," but each one gives intuitive insight into a different aspect of statistical error as it arises in dimension estimations. Next, a model will be derived which is exact for uncorrelated random data of arbitrary distribution (uniform, gaussian, *etc.*) and in arbitrary embedding dimension. Finally, we will show how these results can be applied to the practical estimation of correlation dimension.

CORRELATION ALGORITHM: The problem is to compute the dimension of a fractal attractor A from a finite set of points $\{x_1, \dots, x_N\}$, with $x_i \in A \subset \mathbf{R}^m$ sampling the attractor. Often these points arise from an embedding of a one-dimensional time series [4]. An effective and very popular algorithm for computing dimension was developed by Grassberger and Procaccia [5] and Takens [6], and is based on the distances between pairs of points. This algorithm computes $\hat{C}(N, r)$, a finite-N estimate of the correlation intgral $C(r)$ whose scaling with r defines the correlation dimension.

The correlation integral estimator $\hat{C}(N, r)$ is given by the fraction of distances $r_{ij} = \|x_i - x_j\|$ less than the cutoff distance r. We can express this estimator in terms of an average of pointwise mass

Measures of Complexity and Chaos
Edited by N.B. Abraham *et al.*
Plenum Press, New York

199

function estimators $\hat{B}_{x_i}(N, r)$, which are given by the fraction of points that are within r of the reference point x_i.

$$\hat{C}(N, r) = \frac{1}{N} \sum_{i=1}^{N} \hat{B}_{x_i}(N, r) \tag{1}$$

The correlation dimension ν expresses the scaling $C(r) \sim r^{\nu}$; or more formally, $\nu = \lim_{r \to 0} \frac{\log C(r)}{\log r}$. Stuck with finite N, we of course cannot apply this formula directly. One option is to choose a specific $r = R$; then estimate

$$\hat{\nu}(N, R) = \frac{\log \hat{C}(N, R)}{\log R}. \tag{2}$$

In practice, we choose R as small as we dare, thereby reducing systematic bias. However, with N fixed, statistical precision usually degrades as R decreases, since fewer distances are being counted. Quantifying this trade-off should yield good values to use for R. In the meantime, we will for notational ease omit explicit reference to the parameter R.

ANALYTICAL MODELS: Although we will later give an exact equation, we first introduce two naïve models: the Independent Distances Hypothesis (IDH), and the Average Pointwise Mass (APM) model. Both make simplifying assumptions which provide intuitive insight into the sources of statistical error. Both provide, as well, useful lower bounds on the statistical precision.

Independent Distances Hypothesis (IDH): From the N points in a given sample, $D = N(N-1)/2$ distances are computed. These distances are obviously related to and constrained by each other (there are, for example, a huge variety of triangle inequalities among them). We emphasize that this is the case even when the N points $\{x_1, x_2, \ldots, x_N\}$ are independent of each other. The Independent Distances Hypothesis ignores these relations and constraints, and asserts that all D distances are fully independent. In this case the central limit theorem applies, and $\sigma_C^2 = \overline{C^2} - \overline{C}^2$ can be shown to be given by

$$\sigma_C = \sqrt{\frac{1}{D}[C - C^2]} = \sqrt{\frac{2C(1-C)}{N(N-1)}} \approx \frac{\sqrt{2C(1-C)}}{N}. \tag{3}$$

We comment that IDH is not generally valid, since the assumption of independence of distances was unjustified; however, IDH does provide a lower bound on the statistical precision, and in the regime of small N, it is this error which dominates.

Average Pointwise Mass (APM) Model: Recall Eq. (1) which defines the correlation integral in terms of the average of pointwise mass functions. In the APM model, we replace each $B_{x_i}(N)$ by its $N \to \infty$ limit. In other words, each $B_{x_i}(N)$ is replaced by the actual pointwise mass for which the finite N value is an estimator. Then,

$$\hat{C}_{APM}(N) = \frac{1}{N} \sum_{i=1}^{N} B_{x_i}. \tag{4}$$

Since we have replaced the pointwise mass function estimators with their more precise counterparts, we expect that $\hat{C}_{APM}(N)$ will be a better estimator than $\hat{C}(N)$. In particular, we expect σ_{APM} will be a lower bound on the actual statistical error. We will further assume that the points $\{x_1, \ldots, x_N\}$ are independent (though of course the distances between them must satisfy all the usual constraints: that is, we are treating this as a system with N degrees of freedom, not N^2 as in the IDH). With distinct points independent, we can write $\overline{B_{x_i} B_{x_j}} = \overline{B_{x_i}} \cdot \overline{B_{x_j}} + \delta_{ij} \left[\overline{B_{x_i}^2} - \overline{B_{x_i}}^2 \right] = \langle B \rangle^2 + \delta_{ij} \left[\langle B^2 \rangle - \langle B \rangle^2 \right]$. Introducing B_{rms} as the root-mean-square fluctuation of the pointwise mass function,

$$B_{rms} = \sqrt{\langle [B - \langle B \rangle]^2 \rangle} = \sqrt{\langle B^2 \rangle - \langle B \rangle^2}, \tag{5}$$

we can compute the expectation values of $\hat{C}_{APM}(N)$ and $\hat{C}_{APM}(N)^2$, and from these we can compute $\sigma_{APM}^2 = \overline{C_{APM}^2} - \overline{C_{APM}}^2$, which gives

$$\sigma_{APM} = \frac{B_{rms}}{\sqrt{N}} \tag{6}$$

Although this is a lower bound, we see that in this model, the error scales as $O(1/\sqrt{N})$, which for large N is much larger than the $O(1/N)$ scaling given by the IDH model, at least in the asymptotic regime $N \to \infty$. This model further tells us that the coefficient of the $1/\sqrt{N}$ depends on the rms variation of the pointwise mass function B_x. Actually, as we will later see, this model underestimates the coefficient by a factor of two; however it is qualitatively correct in identifying the *source* of the $1/\sqrt{N}$ statistical error, namely the variation in the pointwise mass function.

Exact: It is possible to derive an equation which is exact for independent and identically distributed (IID) time series of arbitrary distribution: uniform, gaussian, or fractal. Without going into detail (see [7]), the result we obtain is

$$\sigma_C(N) = \sqrt{\frac{4B_{rms}^2}{N} + \frac{2\left[C(1-C) + B_{rms}^2\right]}{N(N-1)}}. \tag{7}$$

Notice that the leading behavior of σ_C is exactly twice what was predicted by the more naïve APM model. Note also that for the special case $B_{rms} = 0$ (or whenever $B_{rms} \ll \sqrt{C(1-C)/2N}$), the IDH gives the correct answer.

IMPLICATIONS FOR DIMENSION ALGORITHMS:

No advantage for $N_{ref} \ll N$: A popular (for example, see [1,3,8]) alternative method for computing correlation dimension involves choosing relatively few $N_{ref} \ll N$ reference points $\{y_1, \ldots, y_{N_{ref}}\}$, and computing (at much less cost) the correlation integral defined by

$$C(N_{ref}, N) = \frac{1}{N_{ref}} \sum_{i=1}^{N_{ref}} B_{y_i}(N). \tag{8}$$

The problem with this approach is that there are only N_{ref} samples to average the pointwise mass functions. Using the APM model to derive a lower bound on the statistical error, we find

$$\sigma_C > \frac{B_{rms}}{\sqrt{N_{ref}}}. \tag{9}$$

This is not very much precision for the money. Indeed, let $M = O(N_{ref}N)$ be the money (computer time) required to compute the $N_{ref}N$ distances. Let $N_{ref} = sN$, with $s \ll 1$. Then the error is given by

$$\sigma_C = O(1/\sqrt{N_{ref}}) = O((sM)^{-1/4}) \tag{10}$$

and we can decrease the error, keeping M fixed, simply by increasing s, the ratio of reference points to total points. There is no advantage, from the point of view of statistical error, in taking $N_{ref} \ll N$. It is more efficient to take $N_{ref} = N$, which corresponds to the original formulation of the correlation integral.

Optimal R: Though previous sections have omitted reference to the parameter R, we now consider the issue of how best to choose the distance scale R for which dimension will be computed. The choice involves a trade-off between accuracy and precision. The former is improved by decreasing R and the latter is improved by increasing R. A natural goal is to minimize the sum of systematic and statistical error, treating this as the total error of the dimension estimate. This properly requires explicit expressions for both kinds of error in terms of R, and a priori estimates of systematic error are by definition difficult. (After all, if you really did know the systematic error, you could simply "subtract it off," and be done with it.) Nonetheless, we can make some assumptions about how systematic errors generically scale with R and from these obtain at least qualitative understanding of the tradeoff between systematic and statistical error.

We begin by expressing our main equation, Eq. (7) as the square of the relative error:

$$\left(\frac{\sigma_C}{C}\right)^2 = \frac{4B_{rms}^2(R)}{NC(R)^2} + \frac{2C(R)[1 - C(R)] + 2B_{rms}^2}{C(R)^2 N(N-1)} \tag{11}$$

We are interested in the behavior of this error in the limit of large N and small R. We note first that the correlation integral $C(R)$ scales as R^{D_2}, and that B_{rms} scales as R^{D_3} where D_2 and D_3 are the generalized dimensions that were introduced in [9,10]. We write $\nu = D_2$ and $\beta = D_2 - D_3$, noting that $0 \leq \beta \leq \nu$, and that for uniform fractals, $\beta = 0$. Thus we have two dominant terms in the above equation: the first (APM) scales as $N^{-1}R^{-2\beta}$ and the second (IDH) scales as $N^{-2}R^{-\nu}$.

To the two terms above we add a heuristic term to account for the systematic error. We will assume that the statistical error scales independently of N and with length scale as R^s. We take $s > 0$ to indicate that the systematic error increases with increasing R. Typically, as in the case of the edge effect, we have $s = 1$, although examples can be found with $s > 1$ as well [2].

Thus, the dominant error for a given value of R is given by $\sigma_C/C \sim \max\left[N^{-1/2}R^{-\beta}, N^{-1}R^{-\nu/2}, R^s\right]$. The optimal R will be that value where the systematic and statistical errors are approximately equal. This gives

$$R_{opt} \sim \max\left[N^{-2/(2s+\nu)}, N^{-1/(2s+\beta)}\right]. \tag{12}$$

Define ε to be the total relative error σ_C/C at optimal R; we will use this miminal error in the correlation integral $C(R)$ as an indication of the the the best estimate of fractal dimension that we can make taking into account both systematic and statistical error. We have $\varepsilon = \sigma_C/C \sim R_{opt}^s \sim \max\left[N^{-2s/(2s+\nu)}, N^{-s/(2s+\beta)}\right]$. This equation suggests that the number of points N needed to estimate the correlation integral to a given tolerance ε scales with that tolerance according to

$$N \sim (1/\varepsilon)^\theta \quad \text{with} \quad \theta = \max\left[1 + \frac{\nu}{2s}, 2 + \frac{\beta}{s}\right]. \tag{13}$$

ACKNOWLEDGEMENTS: I am happy to acknowledge useful discussions with Henry Abarbanel, W. D. Dechert, Hermann Riecke, Harry Luithardt, and Bette Korber. This work was supported by DAPRA-URI contract N00014-86K-0758.

REFERENCES

[1] *It is not difficult to develop an algorithm that will yield numbers that can be called dimension, but it is far more difficult to be confident that those numbers truly represent the dynamics of the system.* A. Brandstater and H.L. Swinney, "Strange attractors in weakly-turbulent Couette-Taylor flow," Phys. Rev. **A 25** (1987) 2207.

[2] J. Theiler, "Quantifying chaos: practical estimation of the correlation dimension," Ph.D. Thesis, Caltech (1988).

[3] J. Holzfuss and G. Mayer-Kress, "An approach to error estimation in the application of dimension algorithms," in *Dimensions and Entropies in Chaotic Systems*, ed. G. Mayer-Kress (Springer-Verlag, Berlin, 1986).

[4] N.H.Packard, J.P.Crutchfield, J.D.Farmer, and R.S.Shaw, "Geometry from a time series," Phys. Rev. Lett. **45** (1980) 712.

[5] P. Grassberger and I. Procaccia, "Characterizing strange attractors," Phys. Rev. Lett. **50** (1983) 346.

[6] F. Takens, "Invariants related to dimension and entropy," in *Atas do 13º* (colóqkio brasiliero do matemática, Rio de Janeiro, 1983).

[7] J. Theiler, "Statistical precision of dimension estimators," preprint.

[8] J. G. Caputo and P. Atten, "Metric entropy: An experimental means for characterizing and quantifying chaos," Phys. Rev. **A. 35** (1987) 1311.

[9] H. G. E. Hentschel and I. Procaccia, "The Infinite Number of Generalized Dimensions of Fractals and Strange Attractors," Physica **8D** (1983) 435.

[10] P. Grassberger, "Generalized dimensions of strange attractors," Phys. Lett. A **97** (1983) 227.

DYNAMICAL COMPLEXITY OF STRANGE SETS

Ditza Auerbach

Dept. of Chemical Physics, Weizmann Institute
Rehovot, Israel 76100

Systems governed by deterministic dynamics may exhibit unpredictable time evolution, thus appearing *chaotic*. Such phenomena are ubiquitous in nature, ranging from turbulent fluid flows to heart arrythmia. The unpredictability of chaotic systems arises from the abundancy of trajectories that exist for slight changes of the initial conditions. The entropy provides a measure of the multitude of possible time evolutions a system may exhibit, but does not provide a quantification of the ease or difficulty with which the set of all possible motions can be organized and encoded. Many of the proposed definitions [1] for dynamical complexity reduce to entropy related quantities such as the Kolmogorov entropy, and as such are measures of randomness. In this communication a measure of complexity unrelated to the entropy is introduced in order to quantify the difficulty in organizing the possible motions of a chaotic system. Other proposed topological definitions of complexity [2], such as the *algorithmic complexity* usually diverge for a generic chaotic system.

A description of a strange set which is invariant under the dynamics requires a hierarchical encoding of the possible trajectories on the limiting set. The unstable periodic orbits underlying the strange set can provide a good characterization of it. They are dense on the attractor (proven for hyperbolic systems) and are arranged hierarchically according to their length. Typically, the number of periodic points grows exponentially with the period length [3]. The scaling structure of a chaotic attractor can be deduced from the unstable periodic orbits and their eigenvalues [4,5]. In order to probe increasingly fine scaling structure, successively longer periodic orbits must be used. The presence of a particular periodic orbit as well as the size of its eigenvalues are invariant under smooth coordinate transformations, making them the relavent quantities to be extracted from data. In fact, numerical procedures for obtaining the unstable periodic orbits from a chaotic time series have recently been developed [6].

A real chaotic trajectory can be considered to be a walk in phase space amongst the unstable periodic orbits; it approaches a periodic orbit along its stable manifold, only to be subsequently thrown away from it along its unstable manifold onto the stable manifold of another periodic orbit. The difficulty in characterizing the set of allowed periodic orbits to arbitrary length is equivalent to the effort necessary to reconstruct the possible trajectories on the attractor. A simple system, characterized by zero complexity, will be one in which low order data, only the short periodic orbits, are sufficient in order to predict the allowed periodic orbits to arbitrary length. For a system with non-zero complexity, the low order data are insufficient in order to

Measures of Complexity and Chaos
Edited by N.B. Abraham *et al.*
Plenum Press, New York

characterize all the long orbits correctly. In the following, this insufficiency will be exactly quantified.

In the discussion we consider only systems for which a symbolic representation consisting of a finite alphabet $\{s_1, s_2, \ldots, s_k\}$ exists for the dynamics. Each periodic orbit of period n is uniquely represented by a string of n consecutive symbols from the alphabet. For a 1-hump map in one dimension good symbolic dynamics is provided using the two symbols $\{0, 1\}$, with the partitioning point given by the critical point [7]. Consider the quadratic map at the parameter value corresponding to fully developed chaos

$$x' = 1 - ax^2 \qquad \text{with } a = 2 \qquad (1)$$

There are exactly two fixed points denoted by 0 and 1 and four period two points denoted by 01, 10, 00 and 11, the last two of which are the fixed points repeated twice. For this choice of parameter values all the periodic points corresponding to all combinations of 0's and 1's exist. The topological entropy is maximal, $\log 2$, and the allowed periodic orbits can be arranged on a complete binary tree with symbols 0 and 1 at the top of the tree. All allowed periodic orbits can be derived by taking all possible combinations of the two *building blocks* 0 and 1. The system has zero *grammatical complexity* since all the possible varied evolutions of the system can be reconstructed using a finite number of building blocks (0 and 1). Note, that in the notion of grammatical complexity, only the topology or the grammar is considered without taking into account the varying probabilities associated with the different periodic orbits. Generalizations of the grammatical complexity to include metric properties are reported elsewhere [8].

As the parameter of the quadratic map (1) is lowered below $a = 2$, periodic orbits are pruned off the strange set by inverse tangent bifurcations and inverse period doubling bifurcations. Not all the orbits on the binary tree are allowed, rendering the tree incomplete. Knowledge of the basic two symbols of the tree, 0 and 1, is insufficient in order to predict the entire set of allowed periodic orbits. For certain parameter values, it may be possible to construct a **complete** n-ary tree (fig. 1) which includes only the allowed periodic orbits. It would then be possible to predict the entire set of unstable periodic orbits from the n building blocks on the top level of the tree. Again, such a system should be considered simple, possessing zero grammatical complexity, due to the fact that the strange set could be decoded from a finite amount of data: the n orbits composing the building blocks of the tree.

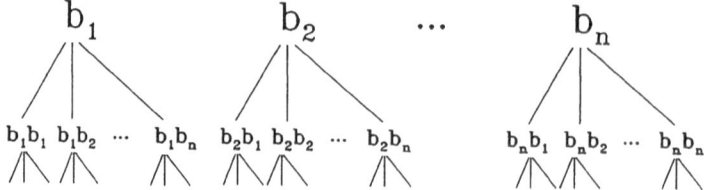

Figure 1. *The first two levels of a complete n-ary tree are shown with the building blocks forming the top level.*

In general, for generic parameter values of (1) in the chaotic regime, it is impossible to arrange all the periodic orbits on a finitely branched complete tree with n building blocks $\{b_1, b_2, \ldots, b_n\}$, where each b_i is constructed from the alphabet symbols $\{s_1, s_2, \ldots, s_k\}$. If one is only interested in orbits up to any given period, a finitely branched tree is sufficient, but as one proceeds to describe orbits of longer periods, additional building blocks of increasing lengths must be added to the top level of the tree. In much the same way as the irrationality of a real number is given by how well it is approximated by rationals (continued fractions), the grammatical complexity of a set is defined by how well it is approximated by complete finitely branched trees. The rate at which the branching of the tree grows with increasing length of allocated periodic orbits, or equivalently with increasing length of the building blocks, is a measure of the asymptotic complexity of the dynamical system's grammar. The number of building blocks upto a given length grows at most linearly with the maximum length of the building blocks [8], so that the grammatical complexity is defined as:

$$C_0 = \lim_{l \to \infty} \frac{n(l)}{l} \quad , \tag{2}$$

where $n(l)$ is the total number of building blocks up to length l. For cases where the limit does not exist, lim is replaced by lim sup. It can be shown [8] that C_0 satisfies the inequality $0 \leq C_0 \leq 2$.

A complete finitely branched tree can be constructed for any parameter of the quadratic map for which a Markov partition exists, i.e, a partition in which each element is mapped **onto** a union of other elements. For concreteness consider the parameter value where the critical point is mapped in 4 steps onto the unstable period 2 orbit (fig. 2(a)). The attractor of the system is known to be chaotic [7]. The unstable periodic orbits underlying this set can be organized on a complete ternary tree which is obtained by representing the dynamics between the Markov partition elements by a graph (fig. 2(b)) with nodes denoting the partition elements $\{A, B, C, D, E\}$. Directed links from a given node denote the possible preimages of that partition element. For parameter values lower than that of fully developed chaos, some points on the attractor may have only one preimage rather than two. The links are labeled by 0 or 1 depending on whether the preimage is obtained by backwards iteration of the left or right branch of the map respectively. All possible trajectories (backwards in time) on the attractor can be reconstructed from the graph of fig. 2(b), simply by choosing an initial node and traversing the graph according to the directed links.

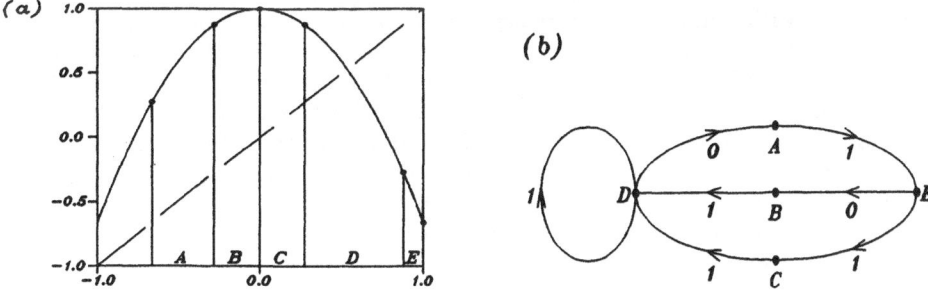

figure 2. The quadratic map (1) at a = 1.6612..., where the critical point is mapped onto the unstable period 2 orbit in 4 iterations (a). The critical point and its iterates (dark circles) form a Markov partition covering the strange attractor, with transitions represented by the graph (b). A directed link from a partition element to its preimage is labeled by 0 (1) if obtained from the left (right) branch of the map.

The allowed periodic orbits of period n, say, can be read off the graph as the paths which start at a given node (the start node) and return to it after traversing exactly n links. One obtains the building blocks by considering solely the periodic orbits which visit the start node only at the end of the traversal cycle. Assigning node D as the start node, the complete ternary tree of periodic orbits is composed of the building blocks 1, 0101, and 1101. A different choice of the start node may lead to a different complete tree. In general, a building block may be a family of periodic orbits rather than just one orbit. An acceptable block has the form of a regular expression of a regular language [9], consisting of concatenations of the alphabet symbols 0 and 1, along with the operations + (exclusive or) and * (arbitrary repetitions), with the restriction that * cannot enclose the entire building block. The form of a regular expression is best understood by considering an example such as:

$$1111^*(00 + 01)000^* \tag{3}$$

This family of orbits can be described as all those which begin with at least three 1's, followed by either 00 or 01, and then ending with a sequence of at least two zeros. From the point of view of complexity, it is reasonable to group periodic orbits as regular expressions, since one would assume the existence of all members of the family (3) from knowledge of finitely many of the short members. The complete tree of periodic orbits corresponding to the attractor of fig. 2(a) is described by the regular expression $(1 + 0101 + 1101)^*$ which is composed exclusively of the building blocks of the tree.

On a complete tree such as that of fig. 1 each periodic orbit appears at least once (several cyclic permutations may appear). Periodic orbits that never visit the start node of fig. 2(b) are apparently missing from the tree. The fact that they are allowed orbits can be simply deduced from the tree itself since their preimages are allocated on the complete tree. Therefore, it is unnecessary to consider the fact that they are only indirectly present on the tree in any quantification of the grammatical complexity.

In order to deduce the grammatical complexity of a set whose periodic orbits cannot be arranged on a complete finitely branched tree, one approximates it hierarchically by sets of zero complexity. As an example, consider the quadratic map (1) at the parameter value $a = a_\infty = 1.7103989\ldots$, for which the itinerary of the critical point is aperiodic and can be constructed from the inflation rule $0 \to 10, 1 \to 110$. One considers a sequence of parameter values $a = a_i$ of (1) , where the critical point is superstable with period i and itinerary which approaches the aperiodic itinerary of $a = a_\infty$. Although at these superstable parameter values the attractor is simply periodic, a strange invariant set which is dense with unstable periodic orbits exists but is repelling. One deduces the complexity of the attractor at $a = a_\infty$ by approaching it with the grammar of the repellers at parameter values corresponding to increasingly longer superstable orbits. The points of the superstable orbit at the parameter value a_i generate a Markov partition, from which a finite graph can be constructed, and subsequently the building blocks of the repeller can be extracted. Their number grows with the length of the superstable orbit. The infinite set of building blocks at $a = a_\infty$ can be deduced from the hierarchical approach along the repellers corresponding to the superstable parameter values. The result of this procedure [8] leads to $C_0 = w_G$ where w_G is the golden mean. Fig. 3 is a plot of the number of building blocks of length less than l found in a time series of 10^6 iterations obtained numerically at $a = a_\infty$. As can be seen from the straight line fit with slope $C_0 = 0.618 (\approx w_G)$, all the building blocks up to a length of approximately 50 are present in this time series.

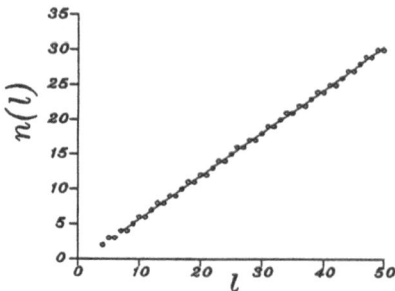

figure 3. The growth of total number $n(l)$ of building blocks upto length l is shown for the map (1) at $a = a_\infty$. Only the building blocks appearing in a time series of length 10^6 are counted.

The above discussion of the grammatical complexity can be directly applied to higher dimensional systems, provided that they are endowed with good symbolic dynamics. When one takes into account the natural measure of the dynamical system over the building blocks, a whole spectrum of generalized complexities C_q can be defined and computed from a time series [8]. The generalized complexities quantify the non-uniformity of the natural measure on the building blocks, analogously to the way the spectrum D_q [10] measures the multifractality of a fractal distribution.

Acknowledgements

I would like to thank R. Badii for introducing me to the problem of quantifying complexity in dynamical systems. Since then, I have benefitted from collaboration with I. Procaccia and illuminating discussions with J.-P. Eckmann and C. Langton.

References

[1]   A. N. Kolmogorov, Probl. Inform. Transm. **1**, 1 (1965); G. Chaitin, J. Assoc. Comp. Math. **13**, 547 (1966); A. Lempel and J. Ziv, IEEE Trans. Inform. Theory **22**, 75 (1976).

[2]   S. Wolfram, Commun. Math. Phys. **96**, 15 (1984); P. Grassberger, Inter. Jour. Theo. Phys. **25**, 939 (1986); J. Crutchfield and K. Young, Phys. Rev. Lett. **63**, 109 (1989).

[3]   J.-P. Eckmann and D. Ruelle, Rev. Mod. Phys. **57**, 617 (1985).

[4]   D. Auerbach, B. O'Shaughnessy, and I. Procaccia, Phys. Rev. A **37**, 2234 (1988).

[5]   G. K. Gunaratne and I. Procaccia, Phys. Rev. Lett. **59**, 1377 (1987).

[6]   D. Auerbach, P. Cvitanovic, J.-P. Eckmann, G. Guneratne and I. Procaccia, Phys. Rev. Lett. **58**, 2387 (1987).

[7]   P. Collet and J.-P. Eckmann, *Iterated Maps on the Interval as Dynamical Systems* (Birkhauser, Boston, 1980).

[8]   D. Auerbach and I. Procaccia, *in preparation.*

[9]   J. E. Hopcroft and J. D. Ullman, *Introduction to Automata Theory, Language and Computation* (Addison-Wesley 1979).

[10] H. G. E. Hentschel and I. Procaccia, Physica **8D**, 435 (1983).

CHARACTERIZATION OF COMPLEX SYSTEMS BY APERIODIC DRIVING FORCES

Daniel Bensen, Michael Welge , Alfred Hübler, Norman Packard

Center for Complex Systems Research, Department of Physics
Beckman Institute, 405 N Mathews Ave, Urbana, IL 61801

The response of a complex system is usually very complicated if it is perturbed by a sinu-siodal driving force. We show, however, that for every complex system there is a special aperiodic driving force which produces a simple response. This special driving force is related to a certain nonlinear differential equation. We propose to use the parameters of this differential equation to describe the complexity of the system.

INTRODUCTION

Generalized dimensions, entropies, Lyapounov exponents[1] and approximations of the flow vector field[2,3] are used to describe the periodic and chaotic dynamics of nonlinear experimental systems. In addition to the passive observation of a nonlinear oscillator and the description of the measured time series using statistical quantities. it is possible to characterize a nonlinear oscillator by an active method, namely by determining its response to specific driving forces[4]. The output of the active method is usually the numerical values of the parameters and the dimension of a differential equation or a map which models the dynamics of the system. These parameters can be used in order to define classes of complexity[5]. Active methods, like nonlinear resonance spectroscopy, are superior to passive methods because of three reasons: First, a measurement of parameters of a system by an active method can be done in the region of the state space of the system that is of the most physical interest, in contrast to passive methods. which have access only to those regions of the state space which are close to the attractors. Second, the model can be used to control the dynamics of complex systems[6,7]. Third. the active method is especially superior in those cases where the the experimental system is a set of identical, weakly coupled oscillators behaving incoherently. If one gets without a driving force a compound signal of all oscillators, small and complicated due to interference, a strong response emerges at resonace. At resonance every single oscillator is forced into coherent oscillation, sychronized by the driving force. In the next section we show that the response of a nonlinear oscillator to certain aperiodic driving forces can be serveral orders of magnitude larger than the response to sinusoidal perturbations and how to use this response in order to find a model of the dynamics.

Measures of Complexity and Chaos
Edited by N.B. Abraham *et al.*
Plenum Press, New York

RESONANT PERTURBATIONS OF NONLINEAR OSCILLATORS

In order to investigate resonant driving forces we consider in the following a damped or conservative oscillation in a nonlinear Potential $V(y)$:

$$\ddot{y} + \eta\dot{y} + \frac{\partial V(y,\vec{p})}{\partial y} = F(t) \tag{1}$$

where $\eta \geq 0$ is a friction constant, $\vec{p}$ are the parameters of the potential, and $F(t)$ is a driving force. In order to calculate resonant driving forces, we integrate a goal equation[7]:

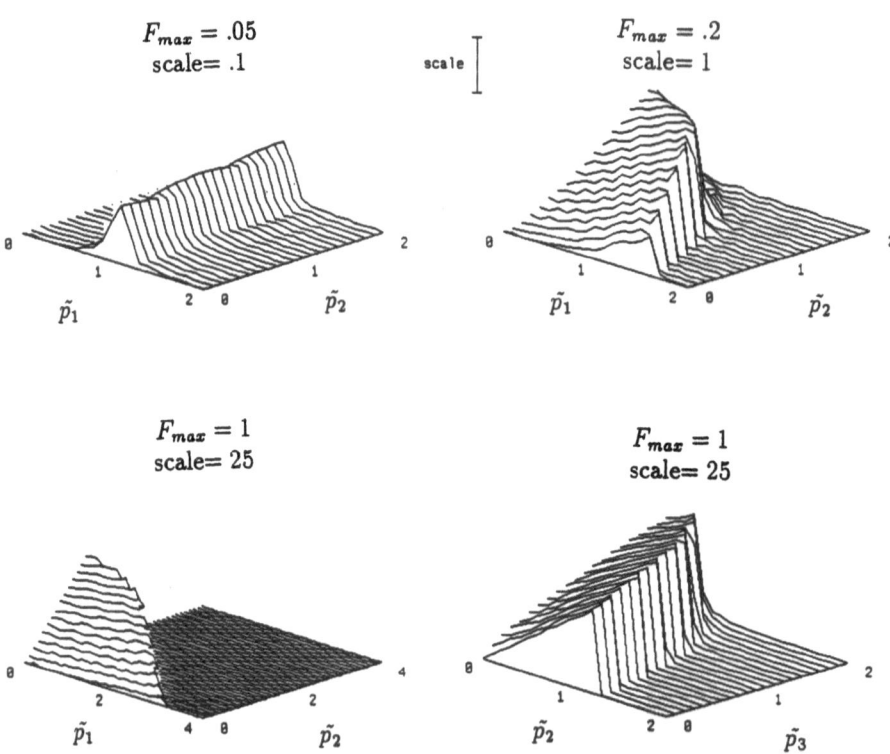

Fig. 1 illustrates the final energy $E(T = 50) = .5\dot{y}^2 + V(y,\vec{p})$ versus the parameters of the model of the potential $V(x,\vec{p}) = .5\tilde{p}_1 x^2 + .25\tilde{p}_2 x^4 + .01\tilde{p}_3 x^6$, where $p_1 = p_2 = p_3 = 1$ are the parameters of the experimental system and where the magnitude of the driving force has various values. The initial conditions are $y = \dot{y} = 0$ and $x = \dot{x} \approx .02$. At resonance, i.e. for $\tilde{p}_1 = \tilde{p}_2 = \tilde{p}_3 = 1$ the final energy is more than one order of magnitude larger than the largest final energy due to a sinusoidal perturbation with the same amplitude ($F_{max} = 1$).

$$\ddot{x} + \tilde{\eta}(t)\dot{x} + \frac{\partial V(x,\vec{p})}{\partial x} = 0 \tag{2}$$

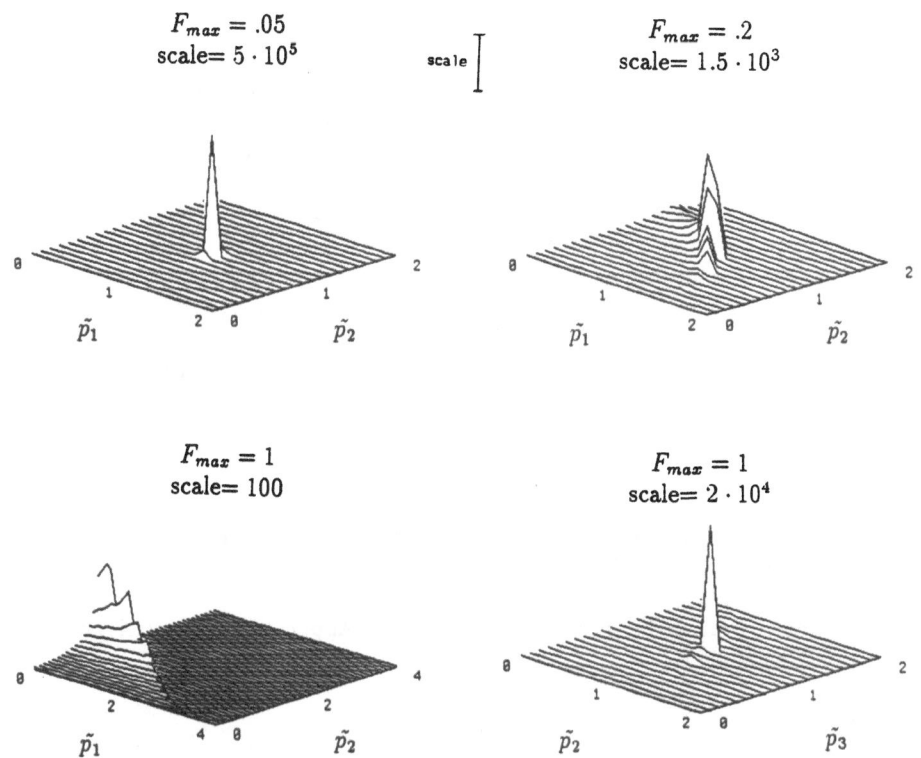

$F_{max} = .05$
scale= $5 \cdot 10^5$

scale

$F_{max} = .2$
scale= $1.5 \cdot 10^3$

$\tilde{p}_1$ $\tilde{p}_2$ $\tilde{p}_1$ $\tilde{p}_2$

$F_{max} = 1$
scale= 100

$F_{max} = 1$
scale= $2 \cdot 10^4$

$\tilde{p}_1$ $\tilde{p}_2$ $\tilde{p}_2$ $\tilde{p}_3$

Fig. 2 illustrates the ratio between the transfered and the reflected energy versus the parameters of the model of the potential $V\left(x,\vec{\tilde{p}}\right) = .5\tilde{p}_1 x^2 + .25\tilde{p}_2 x^4 + .01\tilde{p}_3 x^6$, where $p_1 = p_2 = p_3 = 1$ are the parameters of the experimental system and where the magnitude of the driving force has various values. The initial conditions are $y = \dot{y} = 0$ and $x = \dot{x} \approx .02$.

where $\tilde{\eta}(t) = \eta - \frac{F_{max}}{(2E)^{.5}}$. F_{max} estimates the amplitude of the driving force defined by $F(t) = (\eta - \tilde{\eta}(t))\,\dot{x}$. Due to the special choice of $\tilde{\eta}(t)$ the amplitude of the driving force is approximately constant. If $\vec{\tilde{p}} = \vec{p}$, $y(t) = x(t)$ is a solution of Eq. (1). For a large variety of systems this solution is stable[6]. For $y(t) = x(t)$ the energy transfer $\Delta E = \int_0^T F(t)\,\dot{y}\,dt$ is positive for all T and the reflected energy $E_r = \int_0^T F(t)\,\dot{y}\,\Theta\,(-F(t)\,\dot{y})\,dt$ is zero, which is its smallest value. Θ is the Heavyside step function. For $\vec{\tilde{p}} \neq \vec{p}$ the reflected energy is non zero and the energy transfer is much smaller (see Fig.1 and Fig.2).

NONLINEAR RESONANCE SPECTROSCOPY

The largest response emerges when the parameters of the model coincide with the true parameters of the system (Figs. 1,2). Therefore the parameters of the experimental system can be found by a systematic search for the largest response. This method is equivalent to the usual linear resonance spectroscopy for $\tilde{p}_2 = 0$. In the linear case the driving forces are sinusoidal perturbations and the corresponding resonance curves result. They represent a cut through the shoulder of the nonlinear resonance peak in the background. For large driving forces the maximum amplitude of the linear resonace curves is usually serveral orders of magnitude smaller than the maximum amplitude of the nonlinear resonance curve. A general problem of finding a model of a nonlinear system is the large number of parameters which might be necessary. A systematic scan in the high dimensional parameter space in order to find the maximum response is often too time consuming. Therfore we propose the following systematic search: When a small driving force is applied, the response is sensitive only to the linear parameter of the model (see Fig. 1a). Therefore a small driving force can be used to determine this parameter. If a moderate driving force is applied the response is sensitive to the parameter of the next order, but is still insensitive to higher order terms. By a systematic increase of the magnitude of the aperiodic driving force the coefficients of a Taylor series of the flow vector field of the model can be determined step by step.

ACKNOWLEDGEMENT

This work was supported in part by the grant N00014-88-K-0293 from the ONR and by the grant NSF-PHY 86-58062

* permanent address: National Center for Supercomputer Applications, University of Illinois at Urbana-Champaign

REFERENCES

1. see e.g. J.P. Eckmann and D. Ruelle. Ergodic Theory of Chaos and Strange Attractors, Rev.Mod.Phys. 57:617(1985)
2. J. Cremers and A. Hübler, Construction of Differential Equations from Experimental Data, Z.Naturforsch. 42a:797(1987), J.P. Crutchfield and B.S. McNamara, Equations of Motions from Data Series, Complex Systems 1:417(1987)
3. J.D. Farmer and J.J. Sidorowich, Predicting Chaotic Time Series, Phys.Rev.Lett. 59:845(1987)
4. G. Reiser, A. Hübler, and E. Lüscher. Algorithm for the Determination of the resonances of Anharmonic Damped Oscillators, Z.Naturforsch 42a:803(1987)
5. R. Thom, "Structural Stability and Morphogenisis",W.A. Benjamin, Reading, Mass. 1975
6. A. Hübler and E. Lüscher, Resonant Stimulation and Control of Nonlinear Oscillators, Naturwissenschaften 76:67(1989)
7. A. Hübler, Adaptive Control of Chaotic Systems, to appear in Helv.Phys.Acta 61(1989)

STABILIZATION OF PROLIFIC POPULATIONS THROUGH MIGRATION AND LONG-LIVED PROPAGULES

Byers, R.E. and R.I.C. Hansell

Department of Zoology,
University of Toronto,
25 Harbord St.,
Toronto, Ontario, M5S 1A1

Abstract

A spatial analogue of the logistic model is constructed, with sub-populations located on a 10 by 10 grid, with terms representing long lived propagules such as seeds left in the seed bank by plants or eggs left by many insects and other animals, and with terms representing migration between sub-populations. Both long lived propagules and migration proved sufficient, individually, to stabilize the population when 'r' is set much greater than 4. The usual sequence of bifurcation to chaos is seen as 'r' is increased to 4, but the degree of chaos diminishes greatly as 'r' is increased beyond 4, ultimately becoming locked in a period 3 cycle. As would be expected, low migration rates allowed each sub-population to operate as an independent oscillator while high migration rates synchronized the sub-populations. Estimation of the correlation dimension is not feasible for this model since the combination of the memory induced by the persistent propagules with the spatial distribution results in very high dimensional behaviour.

Introduction

Schaffer (1981) showed that complex systems can be decomposed into relatively independent subsystems, for the purpose of modelling, and derived the dependance of the parameters of the model of a subsystem on the dynamics of those parts of the system which are excluded from the model. The criteria Schaffer used was that each subsystem was comprised of species which operate on similar time scales, with each subsystem in the ecosystem operating on different time scales. Other criteria, such as the degree of connectance between components of the system might also be used. In each case, the parameters of the estimated model will be functions of the ecosystem containing the modelled population. Baleen whales feeding on euphasiids may change their abundance so little during several generations of euphasiids that there is little point in modelling their dynamics in a model of euphasiid population dynamics, but the survival parameter that is appropriate for the euphasiid model will depend on whether the whales are common or scarce during the time for which the model is to apply. There is a strong connection between the whales and euphasiids, but the great disparity between the time scales on which the two organisms operate allows the study of them separately. Similarly, although birds of prey may, in a particular area, depend on trees for nest sites, the connection between birds of prey and trees is so weak that they can be studied independently: the dynamics of the birds of prey are much more likely to depend on the abundance of song birds and small mammals each year than a change in the number of trees, barring annihilation of the forest.

The simplest model which might be examined, in the process of decoupling an ecosystem into subsystems, is the logistic model. The diversity of behaviours which the logistic model can show, ranging from period one fixed points to chaos, is now well known, and it may be necessary to understand the dynamics of this simple model before a more complicated model can be understood. However, there are a variety of population processes which may alter the dynamics

Measures of Complexity and Chaos
Edited by N.B. Abraham *et al.*
Plenum Press, New York

of the population, and there are a number of problems with a direct application of the logistic model to a population. First, the logistic model shows a series of period doubling bifurcations as the population growth rate increases up to 400%. However, a growth rate of 400% is not particularly fast since all it requires is the production of four offspring that survive to reproduce, if the mother dies, or three if she doesn't. Many small mammals and birds produce more than four offspring each year, sometimes breeding more than once each year, and may live to breed for several years. With such species, the proportion of the population which doesn't breed, or the death rate of both juveniles and adults would have to be very large in order to yield a net rate of increase less than 400%. And there are many species which are much more prolific than mammals and birds. Many species of fish, for example, may produce from several hundred thousand to several dozen million offspring (e.g. 0.8-2.4 million for atlantic sturgeon, 60 million for a 300 kg bluefin tuna) each time they reproduce, and live to reproduce dozens of times (e.g. bluefin tuna may live 38 years, and atlantic sturgeon may live 60 years) (Scott and Scott 1988). For a species that produces a million offspring in a year, 99.9994% of the offspring must die in order to reduce the net annual growth rate to 400% (assuming a sex ratio of 1:1), and it is difficult to imagine a predator, or a set of predators, being efficient enough to annihilate such a large proportion of, the young produced. Trees, similarly are very prolific and long lived. For prolific, long lived species, very high mortality rates are expected and observed, but it is questionable as to whether or not the mortality rates are high enough to reduce population growth rates low enough to prevent these populations from showing chaos.

A second problem in applying the logistic model is that many species produce propagules that remain inert for a considerable period of time before producing an offspring. In plants, this phenomenon is described as a seed bank (Harper 1977). Many zooplankton produce resting eggs. Protists may use spores. These inert propagules are used to survive temporarily inhospitable environments such as dense forest cover, in the case of shade intolerant plants that contribute to the seed bank, or winter, in the case of zooplankton. They may also serve to preserve a prolific species which has increased beyond the capability of the environment to support it: i.e. the inevitable crash in the logistic model when r > 4.

Finally, the logistic model ignores spatial aspects of population dynamics. In a now classic, simple experiment, Huffaker (1958) found that the predatory mite *Typhlodromus occidentalis* usually annihilated the phytophageous mite *Eotetranychus sexmaculatus*. The principal feature of his experimental system was that the phytophageous mites were fed on oranges arranged on a grid intermixed with rubber balls. When few oranges were used, the system quickly crashed, the prey being annihilated. However, when a large number of oranges were used in conjunction with a variety of barriers to the dispersal of the mites, the system persisted. With a single population of each of the predator and prey, the prey is annihilated quickly, but with a large number of populations, connected with limited migration, local populations go extinct but the total system persists. The spatial aspect of the population biology of these species, including dispersal, stabilized an otherwise unstable system.

In this study, the logistic model is extended to allow consideration of these three issues. The effects of persistent propagules and migration on the dynamics of a population is examined. In particular, the ability of each of these to preserve a population with a r > 4, and the degree of synchrony between populations, is examined.

Method

A spatial logistic model is constructed consisted of 100 populations set up on a 10 by 10 grid on a 2d-torus. This is a two stage model in that given a population size at time t, the number of persistent propagules is calculated along with the number of emigrants and then the emigrants are equally divided between the eight nearest neighbors of the population. Finally, the population size at time $t+1$ is obtained by applying the logistic map 'rX(1-X)' to the sum of the population size at time t and the immigrants to the population, and the result (which is set to zero if it becomes negative) is added to a contribution from the pool of persistent propagules that existed at time $t+1$. The sequence of events within a time period (between t and $t+1$) is 1) the production of propagules, 2) emigrants leave, 3) immigrants arrive, and 4) a portion of the persistent propagules produce new adults. Most of the simulations were run for 1000 iterations, although a few were run for up to 100,000 iterations. The longer simulations were done primarily to verify the stability of spatially distributed populations with extremely large values of 'r' that appeared to persist in the short simulations.

Results

Figure 1 provides the bifurcation diagram for this spatial model. This was obtained from the frequency distribution of population sizes observed across the torus for several iterations after 2000 iterations. This procedure is based on the observation that in the absence of synchrony the frequency distribution of population sizes across space is the same as that through time, and in the presence of synchrony the frequency distribution through time is the same as that through space averaged over a number of iterations equal to the period of the attractor (e.g. compare part i and ii in figures 2 through 6). At no time was synchrony observed with long period or chaotic attractors. Figures 2 through 6 provide frequency distributions for a variety of parameter values. Such distributions provide an indicator of the complexity of the system, with the simplest systems having all of the populations having the same size at the same time and the most complicated having a different size for each population. An examination of these distributions also allows detection of the period three orbit for large values of r (e.g. for r = 7 in figure 2) since three distinct frequency distributions are obtained which are observed successively.

Both persistent propagules and migration stabilized the spatially distributed population when 'r' was made arbitrarily large. This held for all levels of production of persistent propagules, but only for small amounts of migration: migration too small to synchronize components of the spatially distributed population. Small levels of migration allow each component of the spatially distributed population to act as independent oscillators while large levels of migration synchronized them, so that they act as a single population which is not spatially distributed (e.g. compare figure 2 with figures 3 and 4). The production of persistent propagules substantially delays the onset of synchrony, for a given migration level, and increases the amount of migration required to generate synchrony at a given time (e.g. compare figure 2 with figure 3 and 6). The value of 'r' has the same effect (e.g. compare figure 4a with figure 4b). Synchrony has not been observed for emigration levels less than 20% with r greater than 3.2.

The usual sequence of bifurcation to chaos is observed (figure 1), and then the chaos diminishes to ultimately yield a period three cycle. The distribution of the sizes of the components of the spatially distributed population at a given time is the same as the distribution of population sizes at a randomly selected site through time, until synchrony appears at which time the spatial frequency distribution must be averaged over several iterations.

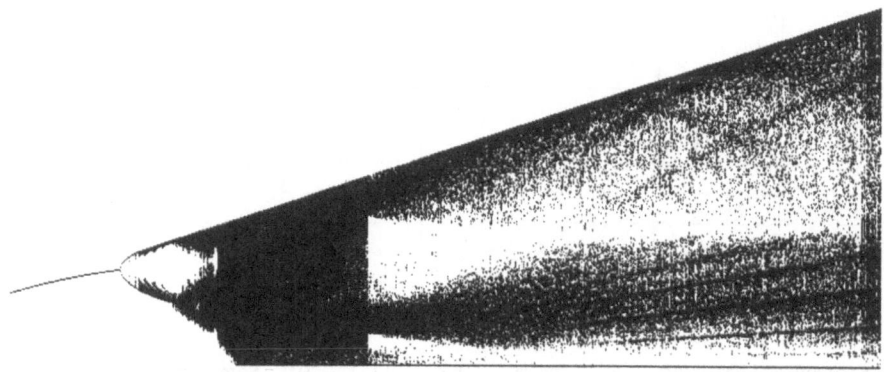

Figure 1. A bifurcation diagram for the spatial logistic model with the production of persistent propagules.
$2 < r < 10, 0 < x < 3$. The data is taken at t = 2000. After the second bifurcation, transients take much more than 2000 iterations to decay. However, the usual pattern is apparent, so most of the transients observed may be due to the presence of semi-stable attractors whose stability is enhanced by the inclusion of spatial aspects of the model.

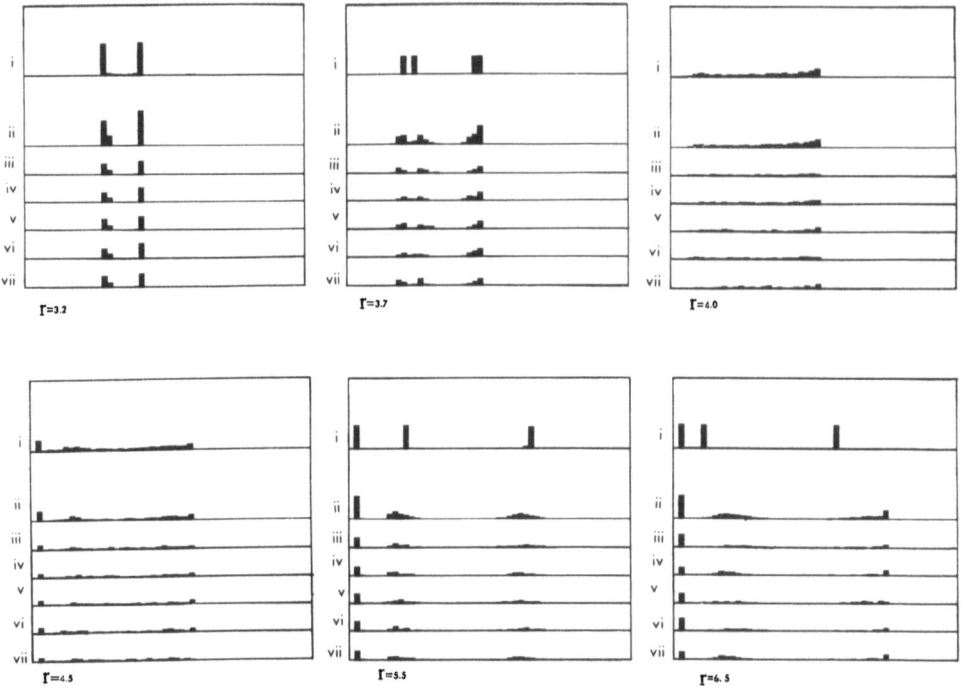

Figure 2. Frequency distribution i) at a given site through time, ii) across all sites through time for the last 10 iterations, iii)-vii) across all sites at t-4 through to t. $0 < x < 2$, $e = 0.1$, $f = 0.1$, kmax = 1000.

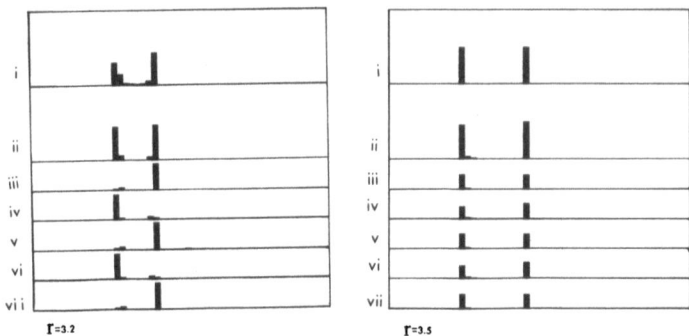

Figure 3. Frequency distributions, as in figure 2, for $e = 0.1$, $f = 0.0$. Some degree of synchrony is apparent for $r = 3.2$, but not for higher values of r.

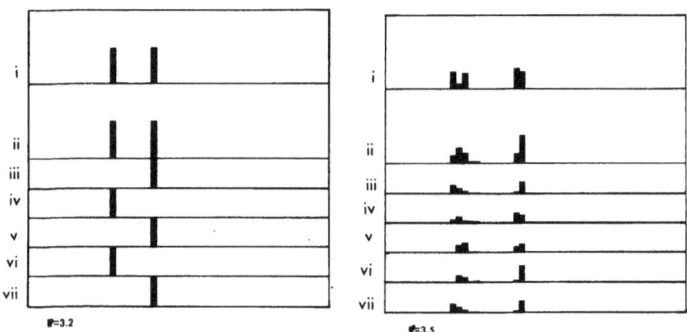

Figure 4. Frequency distributions, as in figure 2, for e = 0.2, f = 0.0. Complete synchrony is apparent for r = 3.2, but not for higher values of r.

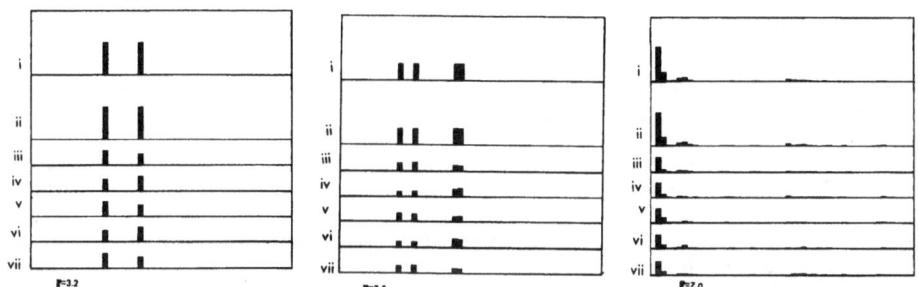

Figure 5. Frequency distributions, as in figure 2, for e = 0.0, f = 0.1.

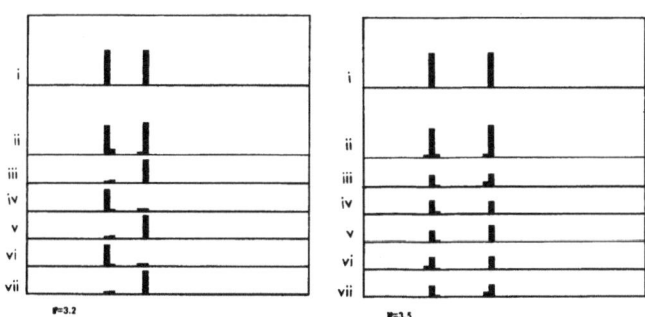

Figure 6. Frequency distributions, as in figure 2, for e = 0.1, f = 0.1, kmax = 5000. Some synchrony is apparent for r = 3.2.

Discussion

Numerical estimation of the dimension of this system is not feasible since it will be very high. For example, Politi (this issue) showed that the dimension estimated for a series of coupled oscillators on a 1 by 20 grid will approximate the embedding dimension for feasible embedding dimensions. However, the frequency distribution of populations sizes across space can be used to assess the complexity of the system.

When r < 4, there is little effect of either being spatially distributed, with migration between components of the population, or of the production of persistent propagules. The behaviour of each component of the population behaves much like the simple logistic model. The only important observation to make with small values of 'r' is that small amounts of migration allowed each component of the population to operate approximately as an independent oscillator, while large amounts of migration synchronized the components of the population.

When r > 4, both the production of persistent propagules and small amounts of migration stabilized the population. As would be expected, local extinctions occurred, but immigration and local production of new individuals from the persistent propagules re-established the population locally whenever local extinction occurred. The only time this fails is when persistent propagules are not produced and migration is large enough to synchronize the components of the system, in which case, all of the components of the population ultimately go extinct at the same time, leaving no source of new individuals to re-establish the population either locally or globally.

References

Harper, J.L. 1977. Population Biology of Plants. Academic Press, London.

Huffaker, C.B. 1958. Experimental studies on predation: Dispersion factors and predator-prey oscillations. Hilgardia 27:343-383.

Schaffer, W.M. 1981 Ecological abstraction: the consequences of reduced dimensionality in ecological models; Ecological monographs 51(4):383-401.

Scott, W.B. and M.G. Scott. 1988. Atlantic Fishes of Canada. Canadian Bulletin of Fisheries and Aquatic Sciences No. 219.

COMPLEX BEHAVIOUR OF SYSTEMS DUE TO SEMI-STABLE ATTRACTORS:
ATTRACTORS THAT HAVE BEEN DESTABILIZED BUT WHICH STILL
TEMPORARILY DOMINATE THE DYNAMICS OF A SYSTEM

Byers[1], R.E., R.I.C. Hansell[1], and N. Madras[2]

1. Department of Zoology,
University of Toronto,
25 Harbord St.,
Toronto, Ontario, M5S 1A1

2. Department of Mathematics,
York University,
4700 Keele St.,
Downsview, Ontario, M3J 1P3

Abstract

Semi-stable attractors are provisionally defined as invariant sets around which volumes are contracting. Three examples are referred to: one is extensively documented by Grebogi, et al. (1986), a second is generated by an age-structured model, and the last is seen in the logistic model. The first example is described by Grebogi as a chaotic transient. However, this transient lasts in excess of 80,000 iterations, which, for real ecological systems in which each iteration represents a year, is longer than many systems can survive. The second looks very much like a saddle point, and indeed, saddle points may be semi-stable attractors, but only if the absolute value of the determinant of the Jacobian at the saddle point is less than 1. In the last example, the trajectories don't stay near the semi-stable attractor very long, but they generate interesting dynamics when 'r' is permitted to vary, and illustrate the relation between semi-stable attractors and the stable attractors they're derived from.

Introduction

Simple models, like the discrete logistic model, show a wide variety of stable behaviours, ranging from simple equilibria to chaos. In such simple models, transient behaviour disappears very quickly. However, in complicated model systems, some transient behaviours may take a considerable period of time to disappear. For example, Grebogi, et al (1986, Battelino, et al. 1988) describe 'chaotic transients' that 'seems to occur often' and take 'extraordinarily long times' to converge to the relevant periodic attractor (see also Carroll et al 1987, 1988). The extraordinarily long times Grebogi, et al., refer to are between 20,000 and 80,000 iterations (see e.g. figure 7 in Battelino et al, 1988). If similarly long times are considered for an ecological system, and each iteration represented a year, these transients could exist for periods longer than most temperate and polar terrestrial ecosystems have existed. In such a case, the transient is the only behaviour that is observed, and may be more persistent than the system itself.

Long lived transients, which are not chaotic, occur also in matrix models of population dynamics. In the matrix model considered in this paper, these 'transients' last from 200 to 1000 iterations (figure 1). In this case, the trajectory takes 10-20 iterations to reach a period one fixed point, and then no change is apparent for several hundred iterations on average, and then it moves quickly to a period two attractor. The important feature of this is that the qualitative behaviour, until the trajectory begins to move away from the period one fixed point, is indistinguishable

Measures of Complexity and Chaos
Edited by N.B. Abraham *et al.*
Plenum Press, New York

219

from the behaviour shown when, with a small change in one parameter in the model, the trajectory approaches, and remains at, an asymptotically stable period one fixed point. That the transient in this model looks like an attractor that existed in a similar position with slightly different parameter values is what it has in common with the chaotic transients described by Grebogi, et al. (1986).

This creates a problem, if the transient exists for decades or centuries, for a field biologist who has to examine an ecosystem and determine whether or not it is stable, e.g. for the purpose of exploitation. It is unlikely that estimation of parameter values from data collected from an ecosystem will yield precision better than two or three significant figures, even with enormous (and expensive) datasets. Nor is it likely that the system will be observed for more than a few years. In such circumstances, how can an ecologist distinguish between a stable attractor and a 'transient'? When an ecosystem has apparently not changed greatly for a number of years (or decades or centuries), and then it changes rapidly to some other, apparently stable, state, it is usually assumed that some natural or anthropogenic catastrophe or pressure has caused the change. However, as the transients examined here show, there may have been no external change causing the observed behaviour. The observed behaviour may just be a long lived transient, and a search for a causal mechanism, such as over-exploitation or pollution of the system, or climatic change, may be futile.

Semi-stable Attractors

There are three reasons for defining the transients considered here to be semi-stable attractors. First, they derive from an attractor with a small change in a parameter. What has happened, in the case of the matrix population model for example, is that an equilibrium has been transformed into a saddle point as the parameter has passed through the catastrophe set. However, not all saddle points will display this behaviour. What is required is that the rate of change in the unstable manifold is substantially smaller than that in the stable manifolds: small enough that the absolute value of the determinant of the Jacobian is less than one. There is thus an intimate connection between semi-stable attractors and attractors in that the semi-stable attractor is a destabilized attractor.

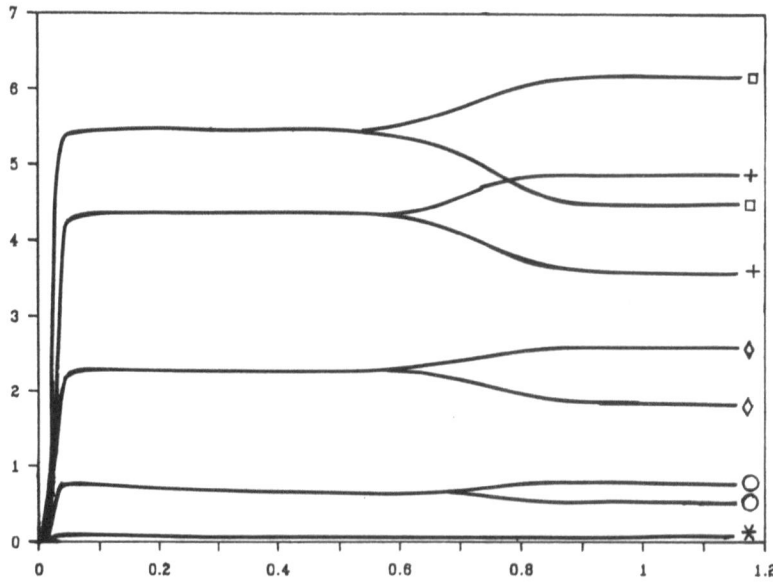

Figure 1. Age specific abundances through time. (Population size vs time.) After the bifurcation, after about 500 iterations, each age group oscillates between the two populations sizes indicated for it

Legend: □-Age group 1, +-Age group 2, ◇-Age group 3, O-Age group 4, *-Age group 5

Second, semi-stable attractors dominate the dynamics of the system in question for an extended period of time and in doing so produce behaviours qualitatively indistinguishable from that produced by the related attractor. For this reason, semi-stable attractors might be detectable as such only when 1) the system changes its behaviour after an extended period of stable behaviour under circumstances in which the only known change is internal to the system that has changed, 2) when a very large number of equivalent system have been observed in a given state and a small proportion of them change for no apparent reason, or 3) if unusually precise estimates of parameter values for a model estimated for the system suggest that the 'stable' behaviour observed is unstable.

Finally, the transients of interest here are qualitatively different from transient behaviours usually considered. In the simplest case, transients decay exponentially toward an attractor, and the only relationship between the transient and an attractor is that the trajectory tends toward the attractor. In marked contrast, the transients of interest here are the remains of a destabilized attractor and, for extended periods of time, show no decay toward an attractor. It is useful to consider stable behaviours encountered in ecosystems to represent attractors, with the understanding that they may be either attractors or semi-stable attractors.

A semi-stable attractor (Byers, et al., in prep.) can be defined as an invariant set for which there are open sets A and B, both containing the invariant set, for which set B has a volume smaller than that of set A, and for which all initial conditions contained in set A at time 't' are part of trajectories that are all in set B at time 't + t_1' (i.e. set B contains the image of set A after t_1 units of time), and if the invariant set is an attractor under some small change in one or more of the parameters of the system. Observe that if B is contained in A, then it follows that the invariant set is a stable attractor (see, for example, the definition of attractors in Eckmann and Ruelle 1985). With our definition, set A does not necessarily contain all of set B, and so escape from the semi-stable attractor is possible through that part of set B which is not in set A. The requirement for a shrinking volume near the semi-stable attractor justifies, in an intuitive way, the name attractor. It is also clear that the principle effect of the parameter change that destabilizes an attractor is that the basin of attraction is distorted so that set B is no longer contained in set A, in multi-dimensional systems (e.g. figure 2).

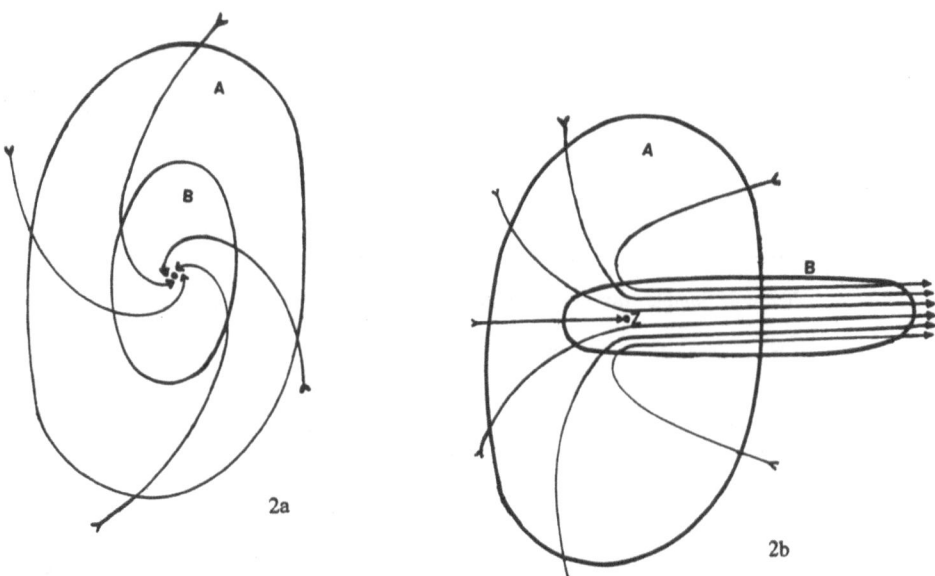

2a

2b

Figure 2. a) Stable attractor. Trajectories in set A at time t_1 are in the smaller set B at some later time t_2, and never sray. b) Semi-stable attractor. Trajectories in set A at time t_1 are in the smaller set B at some later time t_2, and may eventually escape from the attractor through that part of B which is not in A (z is a fixed point).

Implications for non-autonomous systems

The existence of semi-stable attractors can also generate interesting behaviours in non-autonomous systems when the parameter which is a function of time is that which destabilizes the attractor of a system, and the nature of its variation through time takes it through the catastrophe set repeatedly. In figure 3', sample results of simulations using the logistic model in which 'r' is a periodic function of time: the upper half of this figure represents the attractor that would be stable at time 't' if 'r' was to be fixed at the value of r(t) (thus times at which r(t) = 4 are identifiable when the attractor fills the interval). The value of 'r' in this figure varies from values which would allow a period one fixed point as an attractor up to 4; r being a simple sine function of time with a suitable period (150 iterations). Although the period one fixed point is unstable for most of the time, the trajectory is constrained into chasing the period one fixed point. This trajectory is asymptotically stable: all trajectories converge to this trajectory regardless of the initial conditions. Transients decay in a chaotic fashion in the first few dozen iterations, and then the trajectory tends to follow the period one fixed point. This occurs because the trajectory moves to increasing values of 'x' when it is below the current period one fixed point and to decreasing values of 'x' otherwise: as a result, the trajectory is constantly being pushed into the path of the fixed point that would be obtained for r = r(t). The rate at which the trajectory can move away from the period one fixed point is minuscule because the period one fixed point is either an attractor or a semi-stable weak attractor most of the values of 'r' (the trajectory is either attracted toward the current period one fixed point or repelled only very slowly).

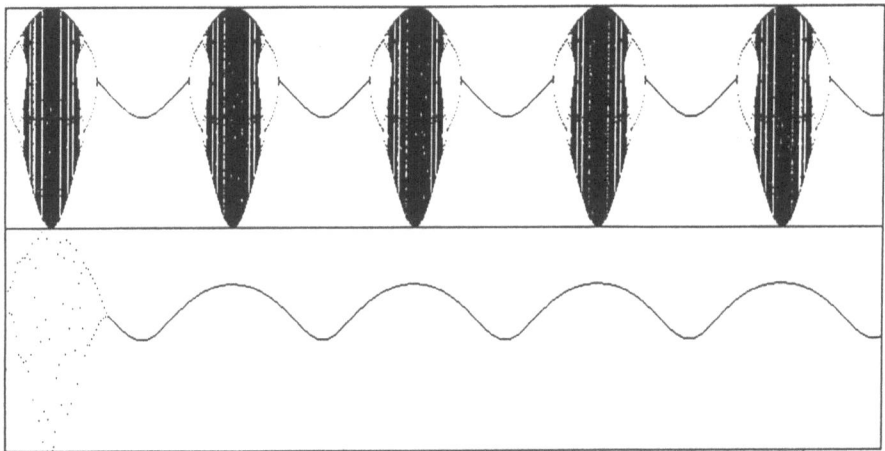

Figure 3. Sample trajectory and stable states vs. time for the model

r = 4 - a + a * sin(2 * pi * t/tau)

x(t + 1) = r * x(t) * (1.0 - x(t))

for a = 1 and tau = 150.

All simulations using these values converge quickly to the trajectories shown. For each simulation shown, there are two things plotted. In the top half of each plot, the attractor that would be obtained if r was forced to be constant with the value of r(t) at time t is plotted vs t. The regions of solid black indicate that r(t) is high enough to produce chaos. In the bottom half of each plot, the trajectory is plotted. Both halves range from 0 to 1 as a function of time, for times 0 through 720.

222

Discussion

Semi-stable attractors provide a means of describing transient behaviour which is qualitatively indistinguishable from that produced by an attractor in the short term, and for explaining both why this transient behaviour looks like that produced by an attractor and why a system moves away from an apparently stable state for no apparent reason. For example, they provide a way of considering the development of forests, for which rates of change may be reduced from noticeable change over a period of years, in early successional stages (e.g., in temperate regions, development from abandoned field to coniferous forest), to noticeable change over a period of decades (from coniferous to deciduous forests), to almost imperceptible change over periods of centuries. Old growth forests are expected to be vulnerable to local perturbation, such as wind-throw (trees blown over by the wind) or death of old trees, which lead to a restarting of the successional process, creating a mosaic of meadow, coniferous forest, mixed coniferous-deciduous forest, immature deciduous forest and old growth, deciduous forest. Such a mosaic may be considered to be a result of perturbations by wind, disease, insect pests, or whatever it was that killed the old trees. However, these perturbations would have little or no effect in a young deciduous forest: the reason that they can have an effect in old forests is because the old trees are much more vulnerable than their young counterparts. The result is that the old growth forest is vulnerable to perturbations it was not vulnerable to earlier in its history: the forest has, in effect, been destabilized by age. The behaviour of the forest as a young deciduous forest looks very much like an attractor and can last for centuries, yet the forest will change when the forest is old.

References

Battelino, P.M., C. Grebogi, E. Ott, J.A. Yorke and E.D. Yorke. 1988. Multiple coexisting attractors, basin boundaries and basic sets. Physica D 32:296-305.

Byers, R.E., R.I.C. Hansell, and N. Madras. in prep. Stability-like Properties of Population Models.

Carroll, T.L., L.M. Pecora, and F.J. Rachford. 1987. Chaotic transients and Multiple attractors in spin-wave experiments. Physical Review Letters 59(25):2891-2894.

Carroll, T.L., F.J. Rachford, and L.M. Pecora. 1987. Occurrence of chaotic transients during transitions between quasiperiodic states un yttrium iron garnet. Physical Review B 38(4):2938-2940.

Eckmann, J.-P. and D. Ruelle. 1985. Ergodic theory of chaos and strange attractors; Reviews of Modern Physics 57:617-656.

Grebogi, C., E. Ott, and J.A. Yorke. 1986. Critical exponent of chaotic transients in nonlinear dynamical systems; Physical Review Letters 57(11):1284-1287.

UNIVERSAL PROPERTIES
OF THE RESONANCE CURVE OF COMPLEX SYSTEMS

Kenneth Chang, Alfred Hübler, Norman Packard

Center for Complex Systems Research, Department of Physics
Beckman Institute, 405 N. Mathews Avenue, Urbana, IL 61801

The dynamics of a large variety of complex systems are confined to a low-dimensional manifold. We show that the resonance curve of those systems has a universal shape. The parameters of the resonance curve can be used to characterize a complex system.

INTRODUCTION

Resonance spectroscopy has proved very successful in many fields of physics. However nonlinear oscillators usually respond to a purely periodic perturbation in a complex and often chaotic manner[1] due to the amplitude-frequency coupling. This chaotic response is small and difficult to characterize.[2] It is also not resonant, because the driving force and the oscillator are out of phase.[3] Recently, for oscillators with well-defined energies, a method has been proposed to calculate a driving force which is in phase with the oscillator velocity at all amplitudes.[4] Usually these driving forces are aperiodic and are called resonant driving forces because the reflected energy and the reaction power are zero.[5] Based on these ideas, we introduce a more general definition of resonance that is also valid for systems without a well-defined energy function. We apply this concept in calculating resonant driving forces for the chaotic dynamics of an logistic map.

GENERAL RESONANCE SPECTROSCOPY

Consider a nonlinear oscillator of type

$$\ddot{y} + \eta_1 \dot{y} + \frac{dV(y,\mathbf{p}_1)}{dy} = F(t) \tag{1}$$

where η is a friction constant, $\mathbf{p}_1$ is a set of parameters of the nonlinear potential V and F is a driving force which is independent from y. Let us assume that the amplitude y at time $t = 0$ is in the vicinity of a minimum of the potential. Starting with a model of the dynamics of the unperturbed experimental system

$$\ddot{x} + \eta_2 \dot{x} + \frac{dV(x,\mathbf{p}_2)}{dx} = 0 \tag{2}$$

Measures of Complexity and Chaos
Edited by N.B. Abraham *et al.*
Plenum Press, New York

one calculates a resonant driving force by integrating the goal equation[5]

$$\ddot{z} + \eta_2 \dot{z} + \frac{dV(z,p_2)}{dz} = \eta_3 \dot{z} \tag{3}$$

which differs from the dynamics of the model by just an additional friction term. Usually η_3 is larger than η_2. The driving force results from

$$F(t) = \eta_3 \dot{z} \tag{4}$$

A driving force is called resonant if the reaction power is zero, i.e. the transferred energy $P = F\dot{y}$ is either always positive or always negative. If the parameters of the model coincide with the parameters of the system $y(t) = z(t)$ is a solution of Eq. (1). In many cases this solution is a stable solution with a large basin of attraction.[6] If the initial conditions lie within the basin of attraction, the driving force given by Eq. (4) is a resonant driving force. The basic idea of this procedure is to find a model which can predict the experimental system's response for any perturbation of physical interest and to use this model to force the system into a certain dynamics. Usually this goal dynamics is calculated by a variation principle, e.g. to get a large energy transfer. Of course, from an experimental point of view it is interesting to search for a model with perturbations that produce a large energy transfer in order to get a good signal to noise ratio, but any other type of perturbation which produces a large response, such as a large frequency shift, might be as useful.

A common feature of all these methods is the search for a model where the difference between all the goal dynamics of physical interest and the response of the system becomes as small as possible, i.e. the quantity $R = \overline{(y(t) - z(t))}^{-1}$ should be as large as possible. The average is taken over time and over all types of goal dynamics which are of physical interest and which are in the basin of attraction of the solution $y(t) = z(t)$ if the model would be correct. A systematic search for the maximum value of R is called general resonance spectroscopy since it can be applied to systems where no well-defined energy function exists. In the next section, we apply this concept to the dynamics of a logistic map and several other maps and show that the corresponding resonance curves are sharp and have a similar, simple shape.

RESONANCE CURVES OF SOME MAPS

We investigated discrete maps of the form

$$y_{n+1} = f(y_n, c_e) + F_n \tag{5}$$

where c_e is the parameter of the map, where F_n is a time dependent driving force and where $n = 0, 1, 2....$ We use a map with a single unknown parameter c_m in order to model the dynamics of the unperturbed map

$$x_{n+1} = f(x_n, c_m) \tag{6}$$

As the goal dynamics, we use a map which provides chaotic and periodic dynamics

$$z_{n+1} = g(x_n, c_g) \tag{7}$$

and which satisfies the condition $\left| \frac{df(z_n, c_m)}{dz_n} \right| < 1$ in order to ensure the stability of $y(t) = z(t)$. The deviation between y_0 and x_0 is about 1%. The driving force is given by

$F_n = -f(z_n,c_m) + g(z_n,c_g)$. Fig. 1 shows the resonance curves of several systems. The resonance curves have a sharp peak at the exact value of the parameter.

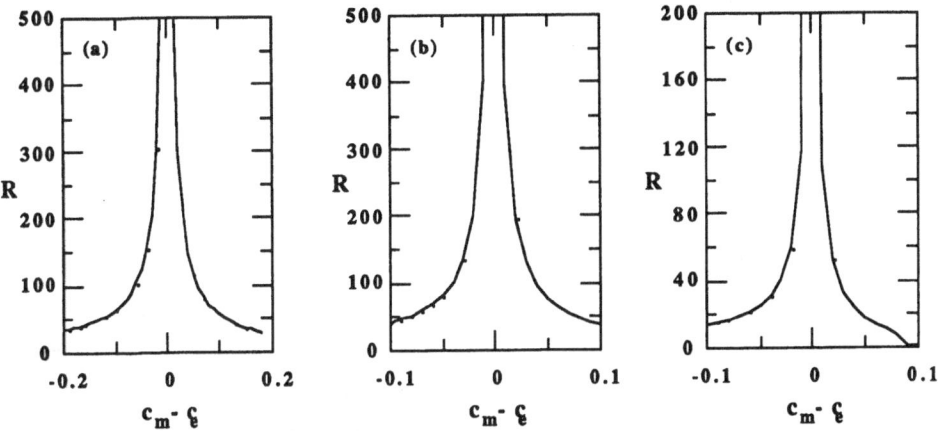

Fig. 1. R versus $c_m - c_e$ for (a) $f = c_e y_n(1 - y_n)$, where $c_e = 3.8$, $g(z_n,c_g) = f(4(z_n - \frac{3}{8}),c_g)$, $c_g = 3.7$; (b) $f = c_e(\exp(-2(y_n - \frac{1}{2})^2) - \exp(\frac{1}{2}))$, where $c_e = 1.9$, $g(z_n,c_g) = f(2z_n - \frac{1}{2},c_g)$, $c_g = 1.8$; and (c) $f = c_e \sin(\pi x)$ where $c_e = 0.8$, $g(z_n,c_g) = f(5z_n - 2,c_g)$, $c_g = 0.6$.

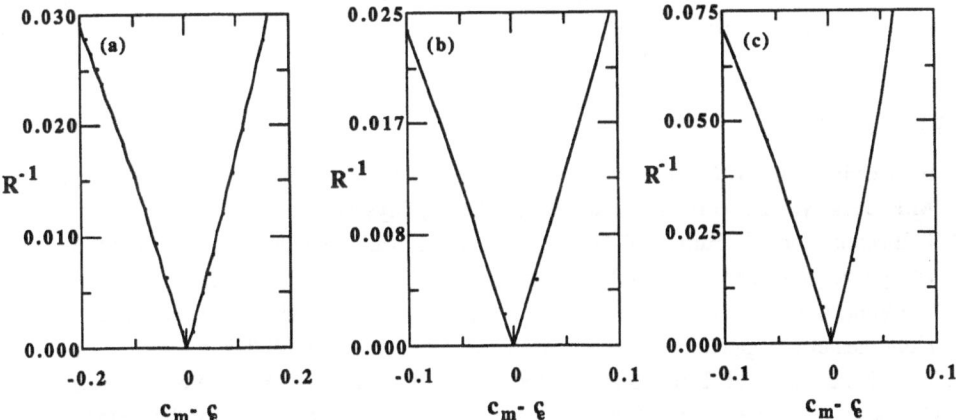

Fig. 2. R^{-1} versus $c_m - c_e$ for the same systems as in Fig. 1.

Close to the peak of the resonance curve, R^{-1} versus the parameter of the model can be estimated by a straight line. (Fig.2) This shows that the resonance curves can be estimated by a hyperbola in this parameter region. The slope of the straight line is a measure of the width of the resonance curve. Fig. 3 shows that the slope depends on c_g. When the goal dynamics is in a periodic window, the resonance curve is sharper than for neighboring value for c_g which generate a chaotic dynamics.

Acknowledgement

This work was supported in part by the grant N00014-88-K-0293 from the ONR and by the grant NSF-PHY 86-58062

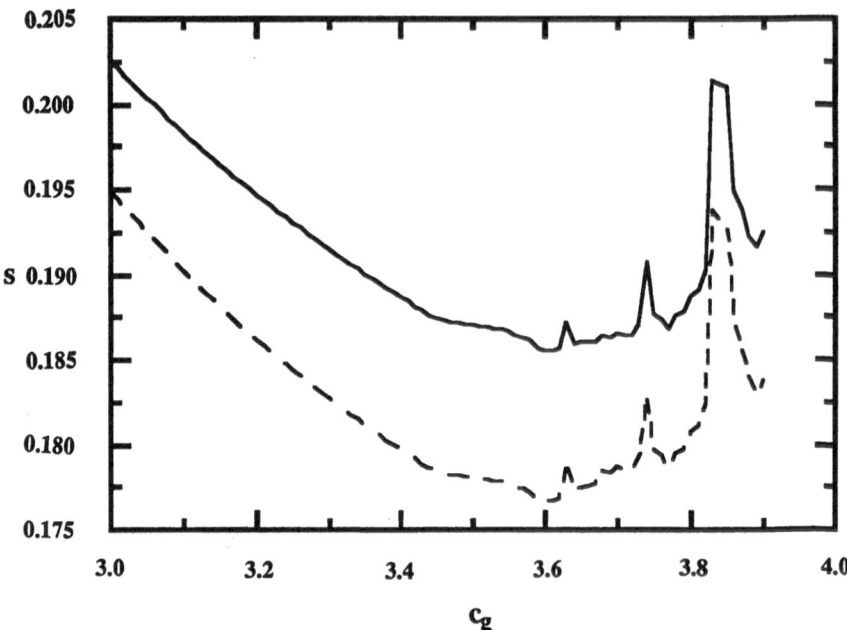

Fig. 3. The slope of a linear approximation of R^{-1} (see Fig. 2) versus c_g for the logistic map $f = 3y_n(1 - y_n)$. The solid line represents the magnitude of the slope above c_e, the broken line represents the slope below c_e.

REFERENCES

1. B.A. Huberman and J.P. Crutchfield, Phys. Rev. Lett. 43:1743(1979); D.D. Humieres, M.R.Beasley, B.A. Huberman, and A. Libchaber, Phys.Rev.A 26:3483(1982)
2. T. Eisenhammer, A. Hübler, T. Geisel, E. Lüscher, Scaling behavior of the maximum energy exchange between coupled anharmonic oscillators, to be published; T. Eisenhammer, T. Hecht, A. Hübler, E. Lüscher, Skalengesetze für den maximalen Energieaustausch nichtlinearer gekoppelter Systeme, Naturwissenschaften 74:336(1987)
3. T.F. Hueter and R.H. Bolt, "Sonics", John Wiley & Sons, New York 1966, 5th ed., p. 20
4. G. Reiser, A. Hübler, and E. Lüscher, Algorithm for the Determination of the Resonances of Anharmonic Damped Oscillators, Z.Naturforsch 42a:803(1987)
5. A. Hübler and E. Lüscher, Resonant Stimulation and Control of Nonlinear Oscillators, Naturwissenschaften 76:67(1989)
6. E.A. Jackson and A. Hübler, Periodic Entrainment of Chaotic Logistic Map Dynamics, to be published

THE EFFECTS OF EXTERNAL NOISE ON COMPLEXITY IN TWO

DIMENSIONAL DRIVEN DAMPED DYNAMICAL SYSTEM

Fang Jinqing

Center for Studies in Statistical Mechanics
University of Texas at Austin, Austin. TX78712

We consider the following Langevin equation with two dimensional driven damped dynamical system (2-D DDDS)[1]

$$\dot{X}_1 = X_2$$

$$\dot{X}_2 = A\cos(2\pi X_3) + BX_1 + CX_1^3 - (DX_1^2 - E)X_2 + F_w \qquad (1)$$

$$\dot{X}_3 = \omega/2\pi$$

where F_w is Gaussian white noise which has the properties

$$<F_w(t)> = 0 \qquad (2)$$

$$<F_w(t)\,F_w(s)> = 2\,D_F\,\delta(t-s) \qquad (3)$$

where D_F is diffusion coefficient, δ the Dirac delta function. Eq.(2) and (3) completely determines all of its statistical feature.

The following main results of the effects of white noise on the complexity of the system (1) are obtained by computer simulation.

(1) The effects of external white noise on the complexity can be represented by the generalized winding number diagram[2] and the corresponding largest Lyapunov exponent as a function of the ratio of D_F to A and of D_F to D, as shown on Fig.1 for two groups of control parameters. Clearly, the interplay between the white noise and the nonlinearities of the system can lead to the mode locking for the system with the right parameter space and a alternating periodic-chaotic sequence with D_F/D occurs in such parameter space as the corresponding largest Lyapunov exponents are both of nonpositive and positive values with increasing D_F/D.

(2) External white noise in the right part of parameter space makes a symmetry breaking of the pattern in the system, i.e., it changes shape of the patterns in the phase space. Fig.2 makes some comparisons of the phase portraits between the absence of and the presence of diffusion coefficient by keeping the other parameters fixed, say, A = B = C = 1, D = 0.075, E = 0.270975 and ω = 2.015. Apparently, the pattern do not change at all when white noise is of

Measures of Complexity and Chaos
Edited by N.B. Abraham *et al.*
Plenum Press, New York

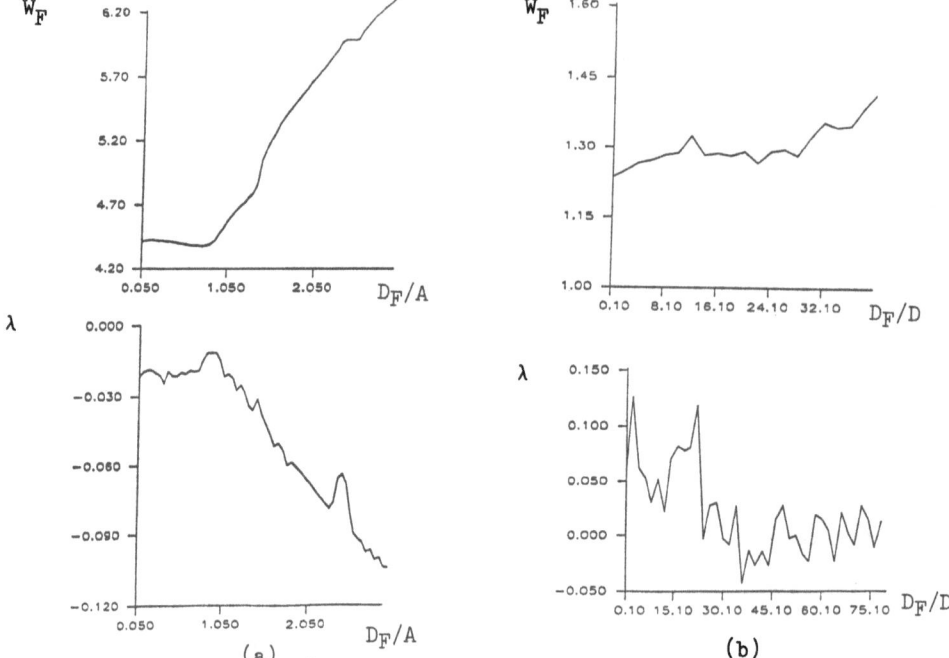

Fig.1 Generalized winding number W_F and the largest Lyapunov exponent λ as a function of D_F/A and of D_F/D.
(a) $B = C = 1$, $D = 0.0075$, $E = 0.270975$, $\omega = 2.015$;
(b) $A = 0.01$, $B = C = 1$, $E = 0.05$, $\omega = 2.015$.

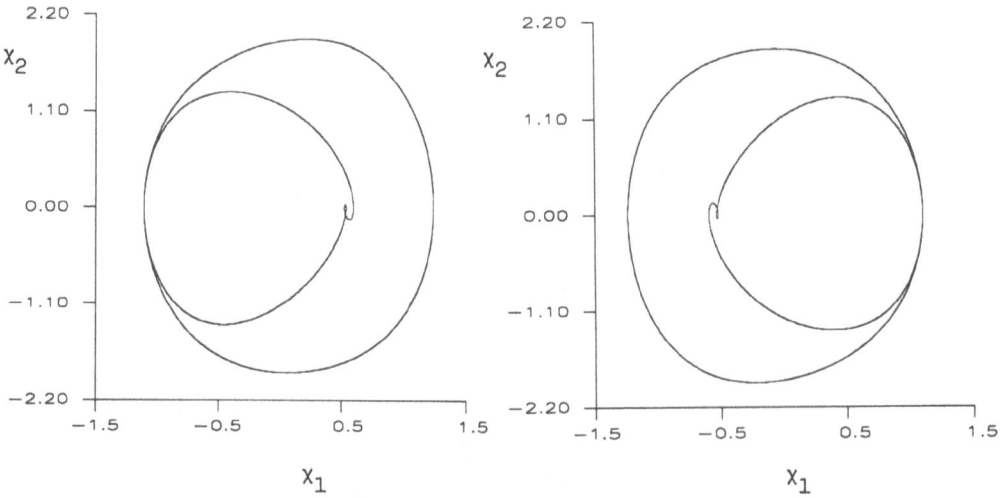

Fig.2 Phase portrait between the absence of and the present of white noise for $A = B = C = 1$, $D = 0.075$, $E = 0.270975$ and $\omega = 2.015$.
(a) $D_F = 0$; (b) $D_F = 1.5$.

value of small (say, $D_F = 0.15$), however, the pattern is changed into antisymmetry and the element of the Farey tree from (2,2,3) to (2,3) if only the diffusion coefficient is increased enough large, say, up to 1.5. However, the element of the Farey tree comes back to (2, 2, 3) again if the diffusion coeffecient is increased even further (say, $D_F/A = 15$). This phenomenon can appear alternatively when the ratio of diffusion coefficient D_F to A is increased continuously, as shown in Table 1. We see that the S - AS sequence go to a limit cycle, where the letters S and AS denote a "symmetry " shape of the pattern and the corresponding " antisymmetry " shape of the pattern above, respectively.

(3) Under the right part of parameter space, the presence of the white noise may play a role of organization and stabilization of the dynamical system otherwise chaotic state. The original chaotic state in the absence of white noise ($D_F = 0$) jumps into a limit cycle at once when the effect of white noise is considered. The amplitude of the oscillation with a unique frequency in the presence of white noise is much larger than anyone of the original oscillations in the absence of white noise. Physically, this fact shows that a strong enough white noise can compensate the dissipation processes and plays a role of amplification of signal of the system with the right parameters.

Table 1. Symmetry-Antisymmetry sequence as a function
of the ratio of the D_F/A
$A = B = C = 1, D = 0.1, E = 0.25, \omega = 2.015$

D_F/A	0	0.35	0.45	0.53	0.58	0.68	...	5.0	...15.0	24.5	26.5	27.0
pattern	S	AS	S	AS	S	AS	...	AS	... AS	S	AS	P=1

(4) The effect of the white noise with different intensity may lead to an alternating periodic- chaotic sequence, as shown on Fig.3 for A = B = C = 1, D = 0.075, E = 0.270975 and $\omega = 2.2$. we have obtained that $D_F = 23.98, 24.03, 24.031$ and 24.05 are four transition points from period-5 to chaos then alternating from a new chaos to a new period and the like.

(5) The effect of the white noise can also be illustrated by looking at the power spectra, $S(\omega)$ at fixed values of ω, B, C and E/D while the noise intensity is increased. Fig.4 shows such a sequence for A = 2.5, B = 1, C = D = 5, E = 0 and ω = 2.256, in which Fig.4(a) corresponds to $S(\omega)$ in the absence of the white noise exhibits a period-3. In the presence of the white noise a transition takes place into a chaotic regime characterized by broad band noise with subharmonic content of periodicity P = 6. As the noise intensity is increased even further, a new bifurcation occurs from which a new chaotic state appears.

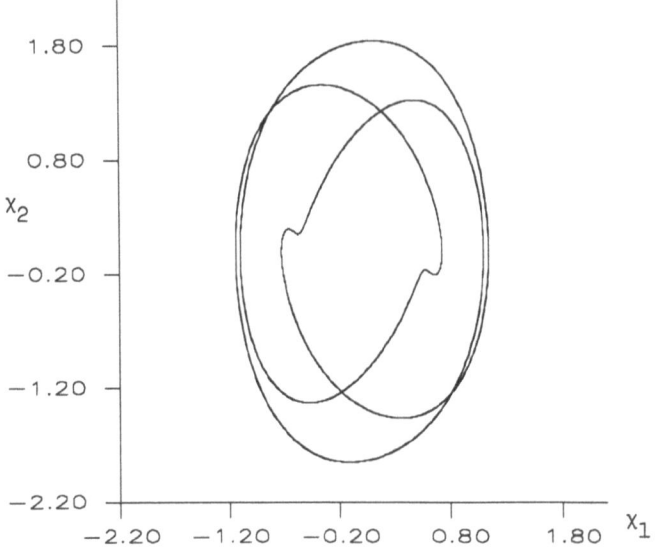

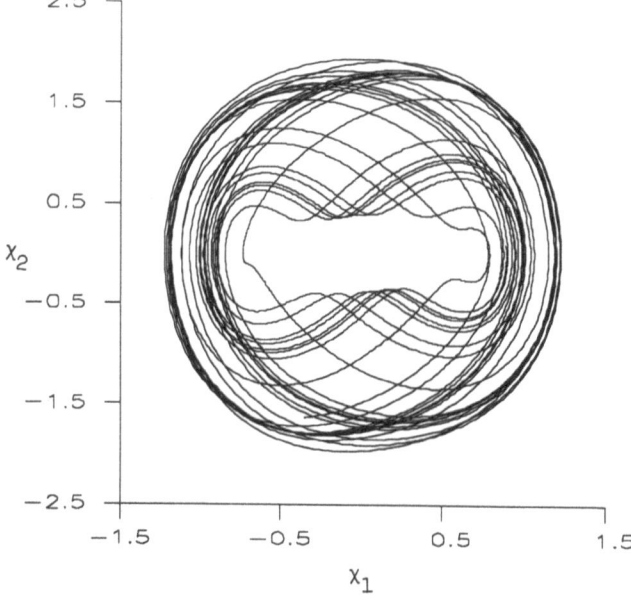

Fig.3 Phase portrait of
alternating periodic-
chaotic states with white
noise intensity
$A = B = C = 1$, $D = 0.075$
$E = 0.270975$ and $\omega = 2.2$
(a) $D_F = 23.98$, $P = 5$;

(b) $D_F = 24.03$, chaos.

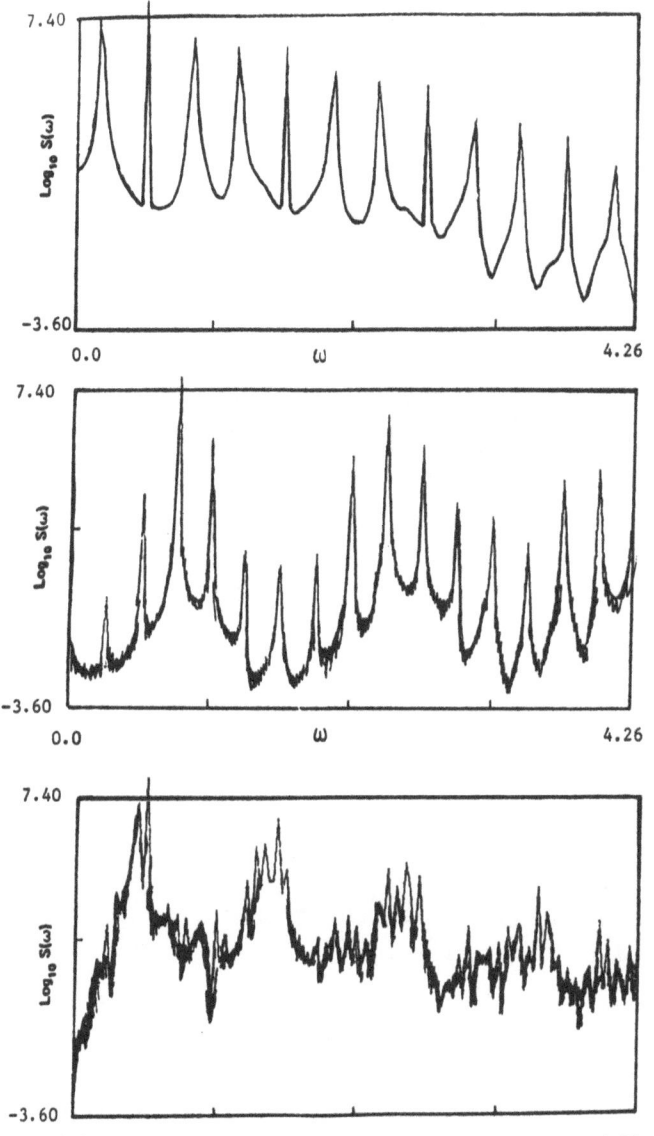

Fig.4 The power spectra at increasing values of the ratio of D_F/A for $A = 2.5$, $B = 1$, $C = 5$, $E = 0$, $\omega = 2.256$.
(a) $D_F/A = 0$;
(b) $D_F/A = 1.5$;
(c) $D_F/A = 9.75$.

(6) As a matter of fact, the results above have therefore shown that the interplay of the white noise with the nonlinearity of the system can induce nonequilibrium transition which depends on the ratio of the diffusion coefficient to the coefficient of nonlinearity of the system, say, D_F/A and $D_{F/D}$. In other words, there are some threshold values in the parameter space, say, $D_{F/A} = 23.98$, 24.03 and 24.31 (also to see Fig.3 and Tab.1) for which the system undergoes a nonequilibrium transition with D_F/A or D_F/D. Thus this can be modified by varying diffusive coefficient as well as the other parameters.

ACKNOWLEDGEMENTS

The author is grateful to Prof. I. Prigogine and Prof. L. E. Reichl of Ilya Prigogine Center for Studies in Statistical Mechanics, The University of Texas at Austin for their hospitality and support.

REFERENCES

1. Fang Jinqing, Generalized Farey Organization and The Effects of White Noise on Complexity in 2-D DDDS, to be submitted to Phys. Rev. A, 1989.
2. U. Parlitz and W. Lanterbom, Period-doubling Cascades and Devil's Staircases of the Driven Van Der Pol Oscillator, Phys. Rev. A , 36:1428(1987).

CHAOS ON A CATASTROPHE MANIFOLD

S. T. Gaito and G. P. King

Nonlinear Systems Laboratory, Mathematics Institute,
University of Warwick, Coventry CV4 7AL, United Kingdom

Because a detailed account of this work is in preparation [4,5] we will present only a summary in these proceedings.

1 A catastrophe with chaos

The work reported here grew out of an attempt to develop a global understanding of the bifurcations and chaotic dynamics in a bi-stable chaotic oscillator [10,3]. Our theoretical work follows Zeeman's programme of incorporating non-trivial dynamics into Catastrophy theory modelling by allowing control parameters to have a state-dependent component [11]. We note that this is similar in spirit to the approach adopted by King and Swinney who treated a state parameter as if it were a control parameter in their experimental investigation of the stability of wavy Taylor vortices [6].

The system that we have studied is an electronic oscillator of the van der Pol-Duffing type with an extra degree of freedom (that is, it is three-dimensional). Our oscillator is a variant of the one introduced by Shinriki et al [10] and investigated experimentally and theoretically by Freire et al [3]. (We note that it is essentially the same system studied by Matsumoto et al [7].) The parameter space can be divided up according to how many different basins of attraction co-exist. Our oscillator has either one, two or three co-existing attractors. In this work we restrict ourselves to the following series of bifurcations from a unique stationary state (the trivial solution) as a parameter is varied:

1. ... pitchfork bifurcation to two non-trivial equilibria (there are now two basins of attraction);

2. ... hopf bifurcation about each of the non-trivial equilibria;

3. ... period doubling cascade to chaos about each of the non-trivial equilibria (the system trajectory is confined to one of the basins);

4. ... merging of the basins of attraction of the two non-trivial equilibria (the system trajectory explores both basins - usually chaotically).

In the experiments the pitchfork bifurcation is not observed, so one must consider the unfolding of the pitchfork bifurcation - which brings the cusp catastrophe and hence catastrophe theory thinking into the problem. Physically, the reason for the cusp catastrophe is that the negative resistor always has some bias. We can understand this in the following way. Assume that the current through the negative resistor is related to the voltage across it by a relation of the form

$$I(V) = bV^3 + aV + \mu,$$

where b is positive. A pitchfork bifurcation occurs when $a = \mu = 0$. Most models of the negative resistor commonly used in electrical networks (e.g., [10,3,7]) assume that $\mu = 0$. This is incorrect. In future work [4,5] we show the consequences of varying this parameter.

Measures of Complexity and Chaos
Edited by N.B. Abraham *et al.*
Plenum Press, New York

The catastrophe theory interpretation of this problem allows us to reconsider the concept of a 'potential' in a dynamical system. A helpful analogy for the physical scientist is the motion of a particle in a double-well potential subjected to a stochastic force (see, for example,[8]). In this analogy the stochastic fluctuations that give the particle enough energy to cross over the potential barrier separating the two wells is determined by the deterministic chaos. Analogous to the temperature is the amplitude of the orbit, and the difference in the forward and reverse probability of transition between the two potential wells is determined by the difference in the depth of the wells (i.e., the bias).

Although it is clear that catastrophe manifolds (especially the cusp catastrophe) are often useful to characterize and model the static behaviour of a system, it is not obvious how they can be used to help understand the complex dynamics in a chaotic system. However in [4] we show that the unfolding of the pitchfork in the steady problem is reflected in the 'unfolding' of the underlying bi-modal return map for the chaotic system. This result was based on intuitive guess work and subsequently confirmed experimentally. In order to test our ideas we had to go beyond the simple rules used by experimentalists to locate the return interval for a single basin attractor. We found it useful to consider the concept of a knot-holder.

2 Knot-holders

Knot-holders were introduced by Birman and Williams and expanded by Holmes [1]. Knot-holders encode the intuitive structure of the attractor since they are a mathematical approximation to the actual 'thick' surfaces observed in most low-dimensional experimental strange attractors. We argue that the intuitive information contained in the knot-holder is useful in constructing interval return maps.

For the simple attractors (such as the Rössler attractor [9]) and for their knot-holders (as studied by Birman, Williams, and Holmes), the location of an appropriate Poincare section and hence return map interval is obvious. In the two basin case the situation is not as simple. Therefore one has to consider how the return map interval should be chosen. In making our choice we have followed two basic principles: (1) all closed solutions should pass through the return map interval, and (2) the resulting return map should be as simple as possible.

For a single basin attractor the best place for the return map interval is just after the branch line. In a two basin attractor there are two different branch lines to choose from. If only one of the two branch lines are chosen for the return map interval the resulting return map could have an infinite number of monotonic sections corresponding to orbits cycling an arbitrary number of times through the other basin. By using a return map interval consisting of two pieces, one for each basin, the resulting return map will not have an infinite number of monotonic sections.

With a return map interval consisting of two pieces there are 2^2 ways to orient the pieces to produce a return map.

3 Results

In figs 1 and 2 we show Poincare surface of sections and return maps for two values of the control parameters. Fig 1 is actually a superposition of data from two experiments (but same parameter values) illustrating the bi-stable nature of the chaotic attractors. (Note that the solid line in fig 1a separates the data obtained from the two experiments.) Fig 2 shows the case after the basins of the two attractors have merged. The surface of section was chosen so that the fixed-points of the dynamical system were always contained in it. The line on which the fixed-points must lie is indicated by the dashed line in figs 1a and 2a. The phase portrait was reconstructed using the singular systems approach developed in [2]. In the figures c_1 and c_3 are singular vectors obtained from the singular value decomposition of the trajectory matrix.

In the regime of two independent basins of attraction, the knot-holder for our bi-stable oscillator has two copies of the 'henon' knot-holder (more familiarly known as the Rössler attractor) which yields a continuous uni-modal return map for each basin. In the regime where the two basins of attraction have merged the situation is much more complex.

Our study of the experimental data has suggested the knot-holder shown in fig 3. Our empirically constructed knot-holder consists of two copies of the single basin ('henon') knot-holders joined by four filaments; one filament cut off from each of the original branches. Two filaments (B+, D+) from the S+ basin cross over to the S- basin and join that basin along the branch line. Similarly the two filaments (B-, D-) from the S- basin cross over to the S+ basin and join that basin along the branch line.

236

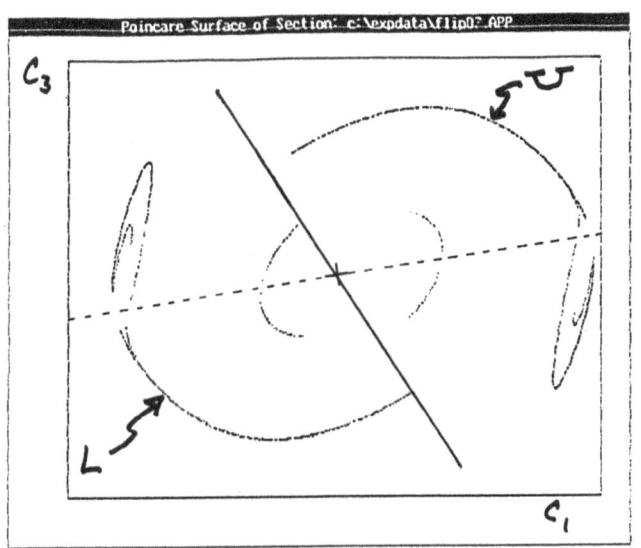

Fig. 1a

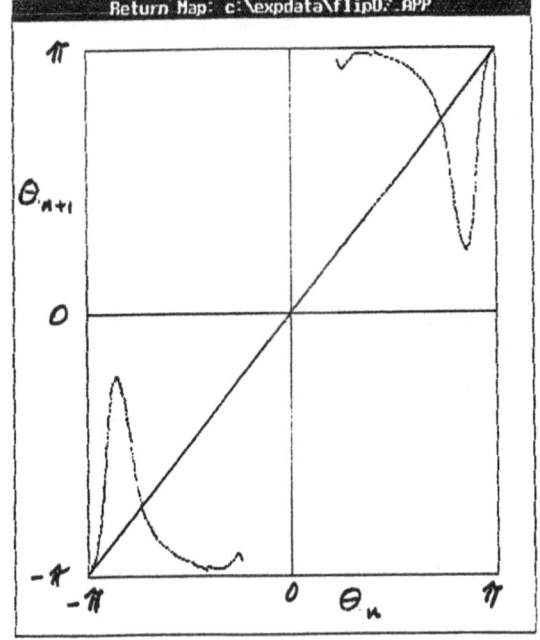

Fig. 1b

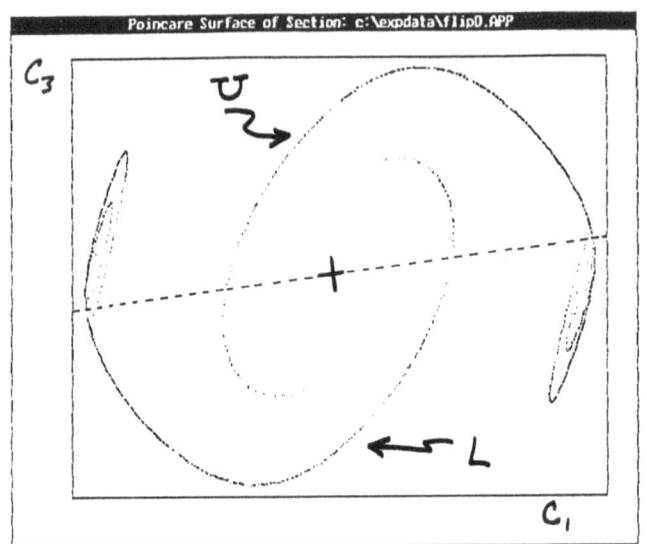

Fig. 2a

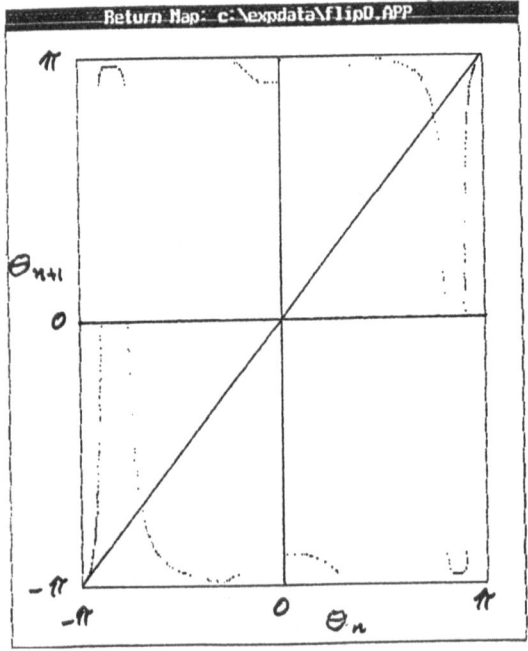

Fig. 2b

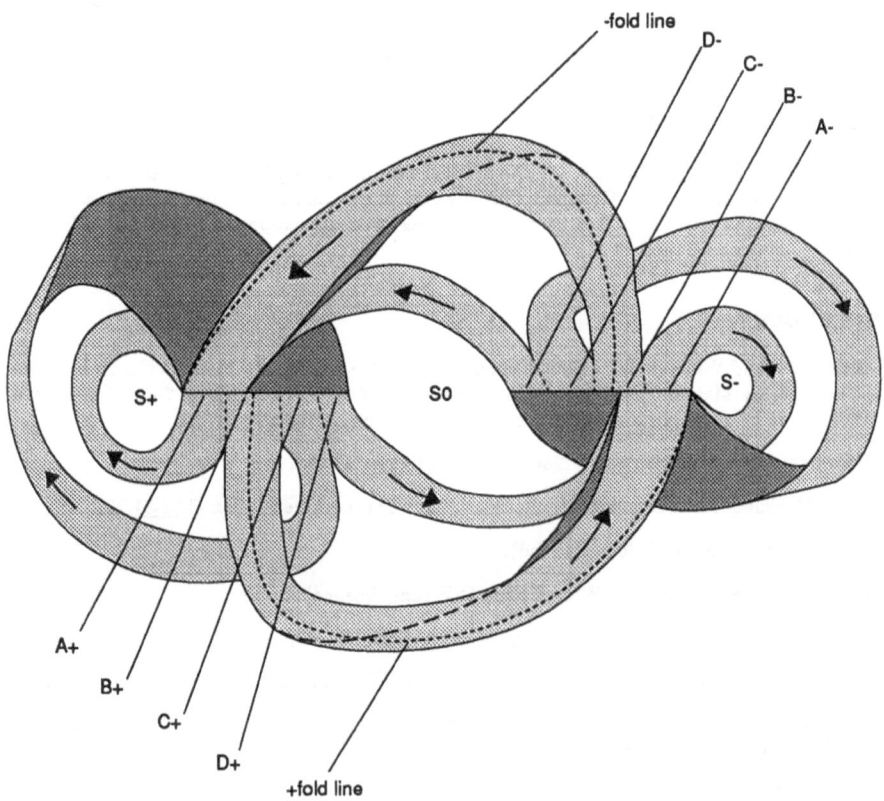

-fold line

D-

C-

B-

A-

S+

S0

S-

A+

B+

C+

D+

+fold line

Fig. 3

We have chosen to use a two piece return map interval, with each piece cutting the knot-holder (attractor) just after the S+ and S- branch lines. Since our two intervals lay roughly in a circle on the Poincare section, we choose to parameterise the return map interval with an angle in the following way. We locate our origin (approximately) at the stationary saddle as shown in figs 1a and 2a, and then rotate our coordinate system so as to identify the "x-axis" with the 'fixed-point line'. Only the set of points indicated by U and L in figs 1a and 2a were used to construct the return maps. The set U gives the return map in the interval $(0, -\pi)$ and the set L gives the return map in the interval $(0, \pi)$. The angles are measured in a way that ensures that the fixed points at each end of the attractor are at $\pm\pi$.

4 Conclusions

The knot-holder encodes information about the closed periodic orbits. It also indicates where best to choose the return map interval. Both the return maps and the knot-holders give us this information with different ways to read it off the object. Knot-holders 'should' deform nicer than return maps as the parameters change as *they* are the geometric object that is changing.

Acknowledgements

We thank Mark Roberts, Ian Stewart, and Christopher Zeeman for helpful discussions and Barry Joyce for his help on the experiments. One of us (STG) would like to thank the Canadian National Science and Engineering Research Council for support. This research was supported by the UK Science and Engineering Research Council.

References

[1] Birman J S and Williams R F (1983) Topology 22 pp 47-82; Birman J S and Williams R F (1983) Contemporary Mathematics 20 pp 1-60; Holmes P (1986) Physica 21D pp 7-41; Holmes P and Williams R F (1985) Arch. Rat. Mech. and Annal. 90 pp 115-194.

[2] Broomhead D S and King G P (1986) Physica 20D pp 217-236.

[3] Freire E, Franquelo L G and Aracil J (1984) *IEEE Trans. Circuits Syst.* CAS-31 237.

[4] Gaito S T and King G P (unpublished).

[5] Gomes G, Ashwin P, Swift J W, and King G P (unpublished).

[6] King G P and Swinney H L (1983) Phys. Rev. A27 pp 1240-3.

[7] Matsumoto T, Chua L O, and Komuro M (1985) *IEEE Trans. Circuits Syst.* CAS-32 798.

[8] Procaccia I and Ross J (1978) J. Chem. Phys. 67 pp 5558-5564; Procaccia I and Ross J (1978) J. Chem. Phys. 67 pp 5565-5571.

[9] Rösler O E (1979) in *Structural Stability in Physics* ed W Güttinger and H Eikemeier (Berlin-Heidelberg-N.Y. Springer-Verlag) pp 290-309.

[10] Shinriki M, Yamamoto M and Mori S (1981) *Proc. IEEE* 69 pp 394-395.

[11] Zeeman E C (1972) in *Towards a Theoretical Biology, Vol 4* ed C H Waddington (Edinburgh University Press) 8-67; Zeeman E C (1976) *Bull. Inst. Math. and Appl.* 12 207-214.

TOPOLOGICAL FREQUENCIES IN DYNAMICAL SYSTEMS

R. Gilmore, G. Mindlin, H. G. Solari

Department of Physics and Atmospheric Sciences,
Drexel University
Philadelphia PA 19104.

INTRODUCTION

In this work the presence of a seemingly anomalous peak at a non 2^n frequency in the power spectrum of the period doubling attractor at the onset of chaos is discussed . Using tools from the theory of knots we study the behaviour of the topological invariant responsible for that effect in a wide class of systems.

BACKGROUND AND MOTIVATION

Fourier analysis has been a widely used tool in dynamical systems, in particular in experimental systems. The presence of a wide band Fourier spectrum is a well known feature of a chaotic solution of a deterministic system. Among the factors that affect the shape of the spectrum, the systematic behaviour of a topological invariant was discussed in a previous work (1). That topological invariant is essentially the linking number between an orbit and its period doubled daughter, and the effect of its systematic behaviour is the presence of seemingly anomalous peaks in the spectrum.

In the study of dynamical systems, much attention has been given to how the orbits are linked or knotted because these characteristics are topological invariants; to change the knot type means that the orbit intersects itself, contradicting uniqueness of solution. But the search for topological invariants has been carried out mainly numerically until Birman, Williams and Holmes introduced the template construction or knot-holder (2,3).

If a system is ruled by a map such that for certain values of a control parameter there is a hyperbolic invariant set we can identify the points with the same future (the effect of that identification is a collapse along the stable manifolds) and the remarkable result by Birman and Williams is that knots of orbits and links between orbits survive in the collapse. The advantage of this representation is that we have 'almost' reduced our problem to a one dimensional one. The bi-infinite sequences that can be associated to each point become infinite sequences (now we are able to order them), and the knot information survived . Now is trivial to compute the linking number between two orbits (2).

TOPOLOGICAL FREQUENCIES

In a dissipative system undergoing period doubling, the attracting set of a chaotic solution is knotted in a complicated way. The fourier spectrum of such a solution will have

Measures of Complexity and Chaos
Edited by N.B. Abraham *et al.*
Plenum Press, New York

both information about the particular features of the system, and about the frequency components needed for the intrinsic construction of the knot. For a system

$$dx_1/dt = f1(\lambda, x_1, x_2, x_3)$$

$$dx_2/dt = f2(\lambda, x_1, x_2, x_3)$$

$$dx_3/dt = f3(\lambda, x_1, x_2, x_3)$$

with x_1, x_2, x_3 reals and λ a real parameter controling the period doubling cascade, we can linearize the vector field around a stable solution and construct the evolution operator of an initial condition as

$$U(t,0) = Te^{\int_0^t Df}$$

where T is the time order operator and Df is the jacobian matrix. As t evolves, the eigenvectors of U(t,0) will rotate around the periodic solution. Now, when l is a value of the control parameter such that the periodic solution is about to loose stability, the eigenvalue of U(T,0) will be -1. That means that the eigenvectors of the operators U(t,0), with t in (0,T) will rotate around the periodic solution an odd number of times. Let's call this number the torsion number. This number has been numerically and analytically computed, and follows a strong systematic in a period doubling cascade . Note that if x_n is a solution of period T_n and x_{n+1} the period doubled solution, the torsion number associated to x_n is just the linking number between x_n and x_{n+1}.

Now, how does this systematic behaviour of the torsion number affect the shape of the spectrum? If we define the sequence G={ $g_1, g_2, ..., g_n, ...$} where g_i is the torsion number for the period T_n divided by T_n, x(t) defined as

$$x(t) = \sum_i a_i \cos(g_i * t + phase_i)$$

will contain the information needed to construct the knot. The coefficients a_i and the phases can be choosen such that a_i/a_{i+1} goes to Feigenbaum's alpha and such that for any arbitrary map of this flow the visit orders are respected.

If we take the fourier transform of x(t), it's clear that if there is an accumulation value for $\{g_n\}$, there will be in the spectrum a peak at that value. In reference 1 an experiment is reported in which the peak is very clear. It comes out from the previous discussion, that a main issue in order to predict the position of the peak is how to compute the linking number between a solution and its period doubled, to obtain the sequence G. Here is where the template construction helps. One only has to locate the orbits in the template, and the twists of the different manifolds indicate how the orbits wrap around each other.

A very important point should be stressed here. The sequence G previously introduced was defined for attracting orbits. All the orbits that can be located in a template are orbits of the hyperbolic invariant set, where no attractor exists; every orbit is stable in one direction and unstable in the other. Now is where the fact that the knot is a topological invariant plays a crucial role. Being an invariant, the way in which the orbits are linked in the process of the period doubling will remain the same as we change the parameters of our system towards the situation in which there is a complete hyperbolic invariant set.

By using these knot theory methods, the results in reference 1 can be generalized to situations in which the dynamics is given by three-symbols sequences; it must only be remembered that two branches topologically equivalent in terms of maps may not be equivalent in terms of flows. For example in the annulus map, one manifold introduces a 0 twist and the other 2 twists. See figure below.

Fig.1. Template for the annulus map.

CONCLUSION

Knot theory methods have been very useful in bifurcation theory. Holmes exploited knot invariants to predict changes in the order of different sequences of bifurcations as dissipation in the system is changed . Also, methods of the theory of knots have been succesful in identifying the dynamics behind a system (4). In this work, we use it to predict the location of seemingly anomalous peaks in the spectrum of the period doubling chaotic solution for arbitrary systems. We show how to deal with cases in which more than one orientation preserving branches of the map are present,having different topological continuations in the flow.

Knot theory methods are not yet systematically used in the study of dynamical systems. As soon as numerology is obtained for topological numbers, self similarities and closure relations appear.We think that the algebraic structure behind that is worth exploring and remains an open problem.

REFERENCES

1.   Gonzalez D., Magnasco M., Mindlin G., Romanelli L., Larrondo H: A universal departure from the classical period doubling spectrum. Physica D (in press).

2. Birman J.S., Williams R.F: Knotted periodic orbits in dynamical systems-I:Lorenz's equations. Topology Vol.22 N1,47, 1983.

3.   Holmes P.: Knotted periodic orbits in suspensions of annulus maps. Proc.R Soc.Lond. A 441,351,1987.

4. Solari H.G, Gilmore R.: Relative rotation rates for driven dynamical systems. Physical Review A. Vol 37. N8,3096,1988.

PHASE TRANSITIONS INDUCED BY DETERMINISTIC DELAYED FORCES

M. LE BERRE, E. RESSAYRE AND A. TALLLET

Laboratoire de Photophysique Moléculaire, Bât. 213

Université Paris-Sud, 91405 Orsay, France

The frontier between noise and deterministic chaos was recently shown[1] to be free, in the sense that the very simple deterministic retarded equation

$$\frac{dx}{dt} + \frac{\partial V}{\partial x}(x(t)) = \sin [Ax (t - d)] \tag{1}$$

with $\partial V/\partial x = x$ was shown to behave as a <u>linear Langevin</u> equation with Gaussian noise, in the limit $A \to \infty$. In deterministic systems the statistics of the feedback term cannot be stated a priori, it follows from internal properties of the equation. Let us point out that the Gaussian behavior of x results from the analytical form choosen for the feedback f(x) in the sense that the periodic character of f(x) is responsible for the short memory effects in f[ax(t)]. On the contrary, in the case of Mackey-Glass equation[2], the feedback will never get short memory as the parameter A increases, because the corresponding feedback $f(x) = x / 1+x^c$ has only one maximum.

Here we want to investigate the role of the memory time of the feedback term in Eq.(1), for the three cases of $\partial V/\partial x = x$, x^3, and $x(x^2-B^2)$ (double well potential).

In all cases chaotic solutions are obtained for A x d larger than few units, they are numerically obtained by a Runge-Kutta method[3] with an adaptative step size developed for retarded equations.

I.   LANGEVIN LIKE BEHAVIOR

When A is larger than about 15 the "Langevin-like behavior" of Eq.(1) is illustrated in Figs 1 and reported in Ref. 4. *As for Gaussian noise it is found that the distribution probability of x is given by $P(x) \propto \exp - V(x) / D$ in the limit of large A.*

Measures of Complexity and Chaos
Edited by N.B. Abraham *et al.*
Plenum Press, New York

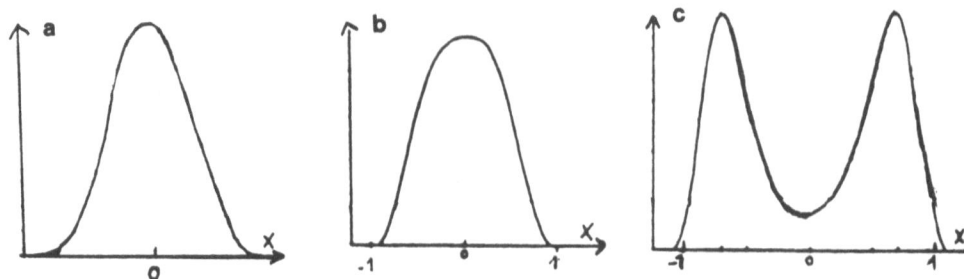

Fig. 1 Stationary probability distribution for x (Eq. 1).
(a) $\partial V/\partial x = x$, $A=14.4$, $d=5$; (b) $\partial V/\partial x = x^3$, $A=13.3$, $d=2.5$;
(c) $\partial V/\partial x = x(x^2-0.49)$, $A=20$, $d=1$.

This property results in all cases from the fact that the correlation function $\Gamma_f(\tau)$ of sin Ax(t) tends to behave like a δ-function as A increases.

The statistical properties of the feedback (which is not Gaussian except in the linear case), and its short range memory effects are also studied. A striking property of $\Gamma_f(\tau)$ is that it has always the same shape as in the linear case [1], with a central peak width equal to $2/A$ and with energy, or diffusion coefficient, $D = \int_0^\infty \Gamma_f(\tau)d\tau$ only depending on A.

In the case of non-Gaussian colored feedback (neither Gaussian, nor exponentially correlated) with short memory $(A \geqslant 20)$ the statistics of x fits fairly well the expression proposed for diffusion of oscillators driven by coloured noise[5] of correlation time τ_f, but here one must take $\tau_f = 5/A$, i.e. five time larger than for a coloured noise with the same energy.

II. PHASE TRANSITIONS INDUCED BY DELAYED FEEDBACK

For small delay $(d \propto 1)$, noticeable correlations appear between x(t) and sin Ax(t-d) because of the retarded structure of Eq.(1). A generalized Fokker-Planck equation can be derived. The probability distribution is shown to obey the following expansion

$$dP/dt = - \sum_n (- \partial/\partial x)^n \{ D_n P(x) \} \frac{1}{n!} \qquad (2)$$

as for ordinary systems, but here the coefficients D_n are conditional averages for the given set (x_t, x_{t-d}). For $A > 10$, the central peak width of $\Gamma_f(\tau)$ is still much smaller than unity, then Eq.(2) can be truncated after the first two terms, which gives the stationary probability distribution

$$P(x) = \frac{N}{D_2(x)} \exp - \int_0^x D_1(x') / D_2(x') \, dx' \qquad (3)$$

with

$$D_1(x) = - \partial V/\partial x + < \sin Ax(t-d) >_x \qquad (4)$$

and $D_2(x) = \frac{1}{\tau} \lim_{\tau \to 0} \iint_t^{t+\tau} dt_1 dt_2 < \sin Ax(t_1) \sin Ax(t_2)_{x(t+d)} >.$

The additional drift term $\sin Ax(t-d)$ is responsible for the new peaks in the probability distribution as illustrated in Figs. 2 for double well potential, A=10, B=0.7. The left curve corresponds to d=20 ($<\sin Ax(t-d)_x > \propto 0$), while in the right curve the small decay (d=1) induces an additional drift term. The trouble here is that the drift term has to be derived numerically. Nevertheless an analytical expression is proposed, which is shown to be valid around the maxima of P(x), within the rough hypothesis of Gaussian statistics for x. With $c = \Gamma_x(d)$, one obtains

$$< \sin Ax(t-d_x > = c . \exp - \sigma_x^2 (1 - c^2) . \sin cA x,$$

which actually predicts the main changes in the statistics, as illustrated by the dotted line in Figs. 2.

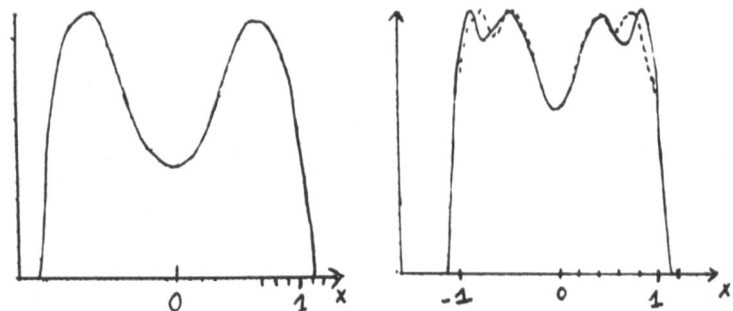

Figs. 2   Probability distribution P(x), numerical results in full line for A=10, left d=20, right d=1. Dotted line $P(x) \propto \exp -D^{-1} \int_0^x D_1(x') dx'$ with $D_1(x)$ in Eq.(4).

III.- LARGE MEMORY EFFECTS

When A decreases further, the central peak width of $\Gamma_f(\tau)$ becomes of order unity, it follows that one must take into acount all the terms in the modified Kramers-Moyal expansion proposed above (Eq.2). The Fokker-Planck equation is no more valid because the coefficients D_n (n ≥ 3) get noticeable values. An analytical derivation was not derived, however numerical results show that the above effects, which <u>has some similarity with noise induced phase transitions</u>[6], <u>persist and even increase</u>. Fig. 3 show this phenomenon for small A.

Let us notice that such splitting of the maximum would never be obtained in the case of linear langevin equation with coloured noise.
In conclusion we have shown that the correlation between the feedback and x(t) is responsible for large changes in the probability distribution of chaotic solutions in Eq.(1). As the delay decreases, the "noise-like distribution" changes continuously for (A > 15) and may have new peaks. This phenomenon could be named "delay induced transition" since it is the signature of the retarded structure of the equation.

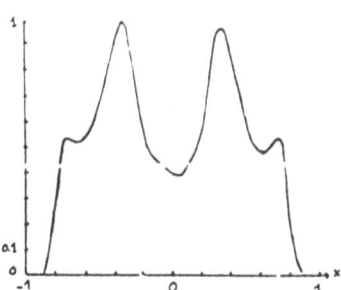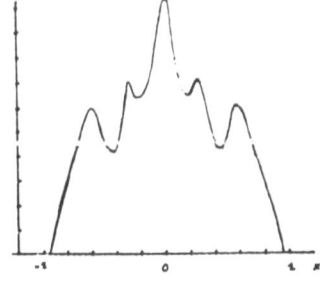

Fig. 3 : Low dimensional chaos : for small A, the probability distribution P(x) is strongly modulated, because of large correlations in the energetic feedback. Left : $\dot{x} + x = \sin 4x(t-5)$; right $\dot{x} + x^3 \sin 6x(t-1)$

REFERENCES

1 a) B. Dorizzi, B. Grammaticos, M. Le Berre, Y. Pommeau, E. Ressayre, A. Tallet, Phys. Rev. A 35, 328, 1987 ; b) M. Le Berre, Y. Pommeau, E. Ressayre, A. Tallet, H.M. Gibbs, D.L. Kaplan, M.J. Rose, in "Far from equilibrium phase transitions", Proc. Sitges, Barcelona 1988, ed. by L. Garrido, Springer Verlag, n° 319, Berlin 1988 ; c) The Gaussian character of Eq.(1) was mentioned by K. Ikeda, O. Akimoto, in Coherence and Quantum Optics V, Rochester, 1983, ed. by L. Mandel and E. Wolf (Plenum N.Y., 1984).
2 M.C. Mackey, L. Glass, Science 197, 287 1977. See also J.D. Farmer, Physica 4D, 366, 1982.
3 E. Hairer, S.P. Norsett, G. Wanner, "Solving ordinary differential equations", Springer Verlag, 1987.
4 M. Le Berre, E. Ressayre, A. Tallet, Proc. of IQEC Conference in Rochester, June 1989.
5 See for example J. Masoliver, B.J. West, H. Lindenberg, Phys. Rev. A 35, 3086, 1987 and references herein.
6 W. Horsthemke, R. Lefever "Noise induced transitions : Theory and applications in Physics, Chemistry and Biology", Springer Verlag, 1984. See also R.R. Grigolini, L.A. Lugiato, R. Mannella and P.V.E. McClintock, M. Meni, M. Pernigo, Phys. Rev. A38, 1966, 1988 and references herein.

MUTUAL INFORMATION FUNCTIONS VERSUS CORRELATION FUNCTIONS IN BINARY SEQUENCES

Wentian Li

Center for Complex Systems Research, Department of Physics
Beckman Institute, 405 N Mathews Ave, Urbana, IL 61801, and
Physics Department, Columbia University, New York, NY 10027

Mutual information is a well known concept used in information theory [1]. Recently, it has been suggested that it can be used in the study of chaotic dynamical systems [2] and for the characterization of spatial complex patterns [3]. Although it has been shown that mutual information is a better quantity than correlation function in the determination of the time delay for the delayed signal in reconstructing the phase space of the chaotic trajectory [4], there is no attempt to systematically compare it with the more frequently used correlation functions. Binary sequences provide an excellent example for this comparative study. Some results are included here, for more details see Ref [5].

INTRODUCTION

Considering a binary sequence $\{x_i\}(i = 1, N)$, with state value $\{a_\alpha\}(\alpha = 0, 1)$, both the correlation function and the mutual information function can be defined using probabilities of state configurations: $p_{\alpha\beta} \equiv \overline{p(x_i = a_\alpha, x_{i+d} = a_\beta)}$ and $p_\alpha \equiv \overline{p(x_i = a_\alpha)}$. They are: (i) for correlation function,

$$\Gamma(d) = \sum_\alpha \sum_\beta a_\alpha a_\beta p_{\alpha\beta} - \left(\sum_\alpha a_\alpha p_\alpha\right)^2 \tag{1}$$

and (ii) for mutual information function:

$$M(d) = \sum_\alpha \sum_\beta p_{\alpha\beta} log \frac{p_{\alpha\beta}}{(p_\alpha)(p_\beta)}. \tag{2}$$

From these definitions, we can see that the main difference between the two functions is that the value of the correlation function is effected by the state values a_α while the mutual information is not. This basic feature makes the mutual information function more suitable for measuring correlation in symbolic sequences, as well as numerical sequences whose symbolic aspect is essential, for example, Ising spin configurations, Markov partition of a chaotic dynamics, galaxy distributions, and so on.

Measures of Complexity and Chaos
Edited by N.B. Abraham *et al.*
Plenum Press, New York

RELATION BETWEEN THE TWO FUNCTIONS

By a straightforward counting of the configurations, the mutual information functions and the correlation functions for simple binary sequences, such as random sequences and periodic sequences, can be easily determined. Both $M(d)$ and $\Gamma(d)$ are zero for random sequences, and both oscillate for periodic sequences (though $M(d)$ is always positive but $\Gamma(d)$ can either positive or negative).

For binary sequences, the mutual information function and the correlation functions can be directly related. To be specific, if $\Gamma(d) = p_{11} - p_1^2$, then mutual information can be shown to be related to the correlation function by the formula:

$$M(d) = \Gamma(d)log\frac{(1 + \frac{\Gamma(d)}{p_1^2})(1 + \frac{\Gamma(d)}{p_0^2})}{(1 - \frac{\Gamma(d)}{p_0 p_1})^2} + p_1^2 log(1 + \frac{\Gamma(d)}{p_1^2})$$

$$+p_0^2 log(1 + \frac{\Gamma(d)}{p_0^2}) + 2p_0 p_1 log(1 - \frac{\Gamma(d)}{p_0 p_1}) \tag{3}$$

If $\Gamma(d)$ decays to zero at longer distances, it can be shown that

$$M(d) \approx \Gamma(d)^2 \frac{1}{2p_0^2 p_1^2}. \tag{4}$$

In particular if a sequence has power law correlation function $1/d^\alpha$, the mutual information function behaves like $1/d^{2\alpha}$, as illustrated in [5] for the sequence generated by the expansion-modification context-free language [6].

BLOCK-TO-BLOCK MUTUAL INFORMATION FUNCTIONS

One can easily extend the definition of mutual information functions between sites to that between blocks. By doing so, the fluctuation in mutual information function is smoothed out. Figure 1 shows an example of both functions of the sequence generated by cellular automaton Rule-110 [7]. It is known that this sequence has underlying periodic background with period 14.

Another example is the simple $\cdots 100100 \cdots$ periodic sequence. The site-to-site mutual information is

$$M(d) = \begin{cases} log(3) - 2/3log(2) & d = 3i \\ log(3) - 4/3log(2) & d \neq 3i \end{cases} \tag{5}$$

But block-to-block mutual information functions become a constant $log(3)$ when the block length is larger than 3.

Though longer block lengths produce smoother curves, they also require a greater amount of storage space and give poor statistics. For binary sequences, a block length of 2 to 6 usually give a good balance.

ESTIMATION OF ERRORS FOR MUTUAL INFORMATION FUNCTIONS DUE TO FINITE LENGTHS

The determination of the mutual information function depends solely on the determination of the probability distributions $p_{\alpha\beta}$. For a finite sequence the counting $c_{\alpha\beta} = Np_{\alpha\beta}$'s fluctuate around the average value $\overline{c_{\alpha\beta}}$. Under the assumption that the fluctuation of $c_{\alpha\beta}$ is proportional to the square root of itself, we can estimate the error of the mutual infor-

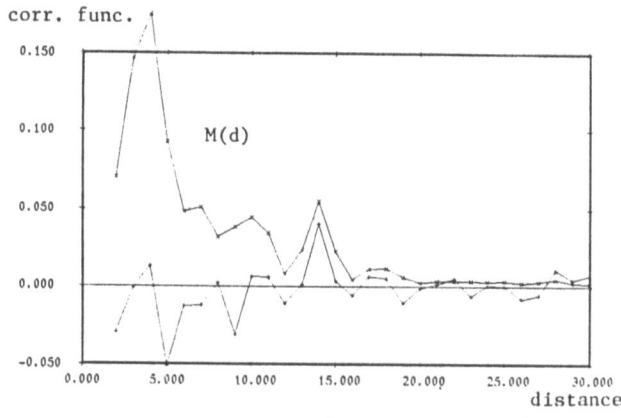

Figure 1. The mutual information function and the correlation function for the sequence generated by cellular automaton Rule-110. The sequence length is 8100, and the block with length 2 is used in the calculation of M(d).

mation function:

$$M(d) - \overline{M(d)} = \frac{1}{N} \sum_{\alpha\beta} \left(\frac{1}{2}(1 + \frac{\overline{c_{\alpha\beta}}}{c_{\alpha}} + \frac{\overline{c_{\alpha\beta}}}{c_{\beta}}) \pm (\sqrt{\frac{\overline{c_{\alpha\beta}}}{c_{\alpha}}} \pm \sqrt{\frac{\overline{c_{\alpha\beta}}}{c_{\beta}}}) \right). \tag{6}$$

Generally speaking, the probability distribution is unknown. When they are known in some cases, such as the equal distribution for random sequences, the formula gives a very simple result. For example, for random sequences the error of the mutual information is roughly:

$$M(d) - \overline{M(d)} \sim 1/\overline{c_{\alpha\beta}}. \tag{7}$$

Intuitively, the higher the counts for each configuration, the less the fluctuation, which is precisely what this formula tells us. Such residue errors are indeed observed in the study of the structure of cellular automata rule spaces using the spatial mutual information [8].

SYMBOLIC 1/f NOISE?

Finally, I want to mention that mutual information function makes the "symbolic 1/f noise" meaningful. It is known that 1/f noises are the signals whose power spectra (Fourier transformation of the correlation function) are inversely proportional to the frequency. 1/f noises have correlations up to many length (or time) scales.

For symbolic sequences such as English texts, the definition of correlation function fails as the tool of measuring correlation. The correlation functions can be used only if the symbolic sequence is converted to a numerical sequence. However, different conversion schemes usually lead to different correlation functions.

On the other hand, there is no such problem with mutual information function. One can imagine that a symbolic sequence with a power law mutual information function is similar to the numerical sequences with 1/f spectra. These symbolic sequences can be called "symbolic 1/f noises". Unfortunately, preliminary analysis of the mutual information function on English texts do not show a good power law over many length scales [9].

ACKNOWLEDGEMENTS

I would like to thank Norman Packard and Tom Meyer for discussions. The research is partially supported by the NSF grant PYS-86-58062.

REFERENCES

1. C.E.Shannon, "The mathematical theory of communication", Bell Syst. Techn. Journal, 27, 379-423 (1948).
2. Rob Shaw, "The dripping faucet as a model chaotic system", (Aerial Press) (1984); and unpublished ideas.
3. Gregory J. Chaitin, "Toward a mathematical definition of 'life' ", The Maximum Entropy Formalism, ed. Levine and Tribus, (MIT Press 1979).
4. A.M. Fraser, H.L. Swinney, "Independent coordinates for strange attractors from mutual information ", Physical Review A, 33, 1134-1140 (1986).
5. Wentian Li, "Mutual information versus correlation functions", (CCSR Tech Report, CCSR, Univ. of Illinois, 1989).
6. Wentian Li, Problems in Complex Systems (Ph.D thesis, Columbia University, 1989); "Context-free languages can give 1/f spectra", (CCSR Tech Report No.10, CCSR Univ. of Illinois, 1988).
7. ed. Stephen Wolfram, *Theory and Application of Cellular Automata*, (World Scientific, 1986).
8. Chris Langton, Norman Packard, Wentian Li, "Bifurcation-like phenomena in cellular automata rule space", (paper in preparation, 1989).
9. Wentian Li, "Correlation analysis of JFK's inaugural speech", (work in progress).

REDUCTION OF COMPLEXITY BY OPTIMAL DRIVING FORCES

Thomas Meyer, Alfred Hübler, Norman Packard

Center for Complex Systems Research, Department of Physics
Beckman Institute, 405 N Mathews Ave, Urbana, IL 61801

In general nonlinear waves are not stable in a chain of finite length. Since they have a finite lifetime, it is important to investigate the production of nonlinear waves, e.g. the production of solitons. A general feature of nonlinear waves is the amplitude frequency coupling, which causes the excitation by sinusoidal driving forces to be very inefficient. The response is usually very complex in addition. We present a method to calculate special aperiodic driving forces, which generates nonlinear waves very efficiently. The response to these driving forces is very simple.

INTRODUCTION

When nonlinear oscillator is perturbed by a sinusoidal force, the response is comparatively small in amplitude[1], and does not fulfil any well defined resonance condition[2], even when the frequency of the driving force coincides with a peak (resonance) in the power spectrum of the unperturbed system[3]. Outside the region of entrainment the response is complicated, in many cases chaotic[4]. In order to obtain large, simple and predictable response, the frequency of the driving force has to be varied in such a way, that it coincides at all amplitudes with the characteristic frequency of the oscillator[5]. Since the characteristic frequencies of nonlinear oscillators usually depends on the amplitude the optimal driving force has to be aperiodic. Recently a method to calculate those optimal driving forces has been presented[6]. We apply this method in order to calculated optimal driving forces for the creation of solitons.

CREATION OF SOLITONS BY APERIODIC DRIVING FORCES

Nonlinear waves and solitons provide good mathematical models in various fields of science[7]. In most experimental systems solitons have a long but finite lifetime. Therefore we investigate the creation of solitons by external perturbations. We assume that the dynamics of the experimental system can modeled by a sine Gordon equation

Measures of Complexity and Chaos
Edited by N.B. Abraham *et al.*
Plenum Press, New York

$$u_{xx} - u_{tt} - sin(u) = F(x, t) \tag{1}$$

where $u(x, t)$ is the field amplitude which depends on space x and time t and where F is an external perturbation which only depends on time and space. In order to calculate resonant driving forces we integrate according to Ref. 6 the following goal dynamics

$$w_{xx} - w_{tt} - Bsin(u) + w_t \Theta \left(|x - 50| - 2.5 \right) = 0 \tag{2}$$

where B is a parameter and where Θ is Heavyside's step function. We take circular or fixed boundaries at $x = 0$ and $x = 100$. The simulation if finished at time T when $|w(x, T)| \geq \pi$. The initial conditions are $w(x, 0) = .0$ and $w(50, 0) = .001$. The driving force results from

$$F(x, t) = -w_t(x, t) \Theta \left(|x - 50| - 2.5 \right) \tag{3}$$

and $F(x, t) = 0$ for $t \geq T$. The basic idea is, that if the structure of Eq. (1) and Eq. (2) are the same, i.e. $B = 1$, $u(x, t) = w(x, t)$ is a special solution of Eq. (1). In this case the energy transfer $P(t) = \int_0^{100} F\dot{u}dx$ is positive for all t i.e. no energy is reflected since F is proportional to w_t. Therefore the coefficient of absorption is 100%, the reaction power is zero and the perturbation is resonant. The special space dependence of F was taken in order to create solitons instead of other nonlinear waves. Fig. 1a shows the result of a numerical simulation of the response of the sin-Gordon system. For the integration we use 100 homogeneously distributed break points. The initial amplitudes of u at these break points are randomly distributed in the interval $[-10^{-5}, 10^{-5}]$ and the initial velocity is set equal zero. Fig. 1a illustrates that nearly all the transferred energy is used for the creation of a soliton antisoliton pair since there are no additional waves in the chain. The situation is completely different if we apply a sinusoidal driving force of the same magnitude for the same period of time and in the same region of the chain. In this case no solitons are created (see. Fig.1b) but a very complicated dynamics results due to the misfit of the driving frequency and the eigen frequency of the system (Fig.2a). This example illustrates that the response of a nonlinear system is usually very complicates whereas the response can be well predictable and simple if special aperiodic driving forces are used, since $u(x, t) = w(x, t)$ and $w(x, t)$ can be calculated in advance for an infinite long period of time.

NONLINEAR RESONANCE SPECTROSCOPY

An essential condition in order to get such a simple response is to have a correct model. Otherwise u differs from w and usually the dynamics is chaotic and an essential part of the energy is reflected. Fig. 2b show the ratio R between the reflected and the absorbed energy versus B. R reaches its maximum value when the parameters of the model and the parameters of the goal dynamics coincide. In this case the response is simple and predictable for an infinite long period of time, while in all other cases including periodic perturbations a very complicated response was found. By a systematic search

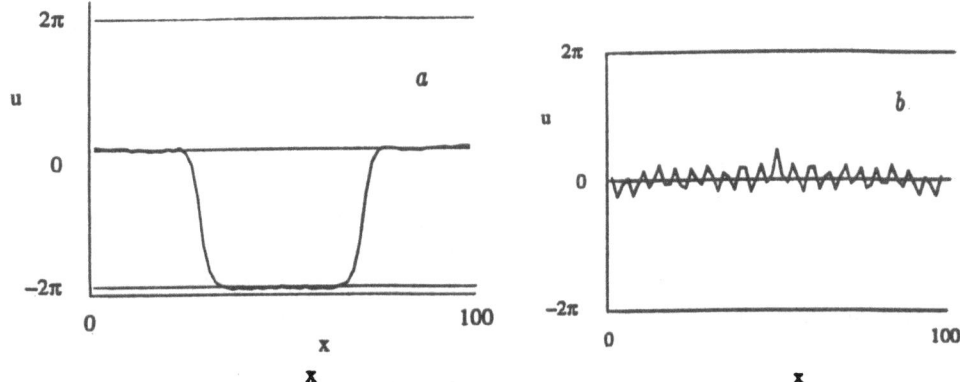

Fig. 1 The field amplitude u versus x after an aperiodic optimal stimulation (a) and after a sinusoidal stimulation (b)

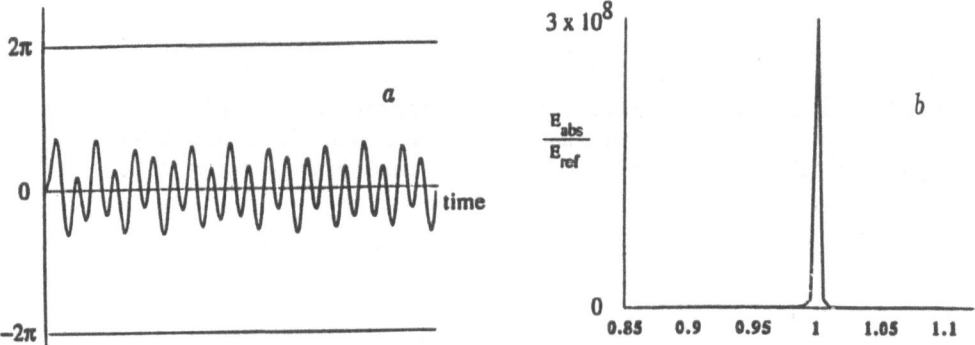

Fig. 2 The field amplitude $u(50, t)$ versus *time* for a sinusoidal perturbation (a) and the ratio between the reflected and the absorbed energy versus the parameter B of the model (b).

for the minimum of the reflected energy as a function of the parameters of the model the correct magnitude of these parameters can be determined.

ACKNOWLEDGEMENT

This work was supported in part by the grant N00014-88-K-0293 from the ONR and by the grant NSF-PHY 86-58062

REFERENCES

1. T. Eisenhammer, A. Hübler, T. Geisel, E. Lüscher, Scaling Behavior of the Maximum Energy Exchange between Coupled Anharmonic Oscillators, to be published; T. Eisenhammer, T. Hecht, A. Hübler, E. Lüscher, Skalengesetze für den maximalen Energieaustausch nichtlinearer gekoppelter Systeme, Naturwissenschaften 74:336(1987)
2. T.F. Hueter and R.H. Bolt,"Sonics",John Wiley & Sons, New York 1966, 5th ed., p.20
3. D. Ruelle, Resonances of Chaotic Dynamical Systems, Phys.Rev.Lett. 56:405(1986); U. Parlitz and W. Lauterborn, Superstructure in the Bifurcation Set of the Duffing Equation, Phys.Lett. 107A:351(1985)
4. B.A. Huberman and J.P. Crutchfield, Chaotic States of Anharmonic Systems in Periodic Fields, Phys.Rev.Lett. 43:1743(1979); D.D. Humieres, M.R. Beasley, B.A. Huberman, and A. Libchaber, Chaotic States and Routes to Chaos in the Forced Pendulum, Phys.Rev.A 26:3483(1982)
5. A. Hübler, E. Lüscher, Resonant stimulation and control of complex systems, Helv.Phys.Acta 61:(1989)
6. A. Hübler, E. Lüscher, Resonant stimulation and control of nonlinear oscillators, Naturwissenschaften 76:67(1989)
7. R.K. Bullough and P.J. Caudrey eds., "Solitons", Springer, Berlin 1980; A.S. Davydov, "Solitons in molecular systems", D.Reindel, Dordrecht (1985); S. Takeno, "Dynamical Problems in Soliton Systems", Springer Series in Synergetics, Springer, Berlin (1985)

SYMBOLIC DYNAMICAL RESOLUTION OF POWER SPECTRA

M.A. Sepúlveda and R. Badii

The Weizmann Institute, Chemical Physics Dept., 76100 Rehovot, Israel,
Fakultät für Physik, Universität Konstanz, 7750 Constance, W.Germany

Power spectra have been for a long time employed as a means for the characterization of experimental time signals. After the discovery of low-dimensional chaotic behaviour in physical systems the analysis of power spectra contributed to the detailed understanding of transitions to chaos by period-doubling and quasiperiodicity [1]. However, their usefulness for the investigation of typical chaos has been questioned, since they are not invariant under smooth coordinate changes [1]. Here we show that power spectra are characterized by the topological and the metric properties of symbolic orbits, together with the actual numerical values of the observable. The former two ingredients are dynamical invariants and affect the spectra much more deeply than the latter one, which is obviously non-invariant.

Dissipative deterministic chaotic systems have power spectra exhibiting a scenario of broadened peaks whose position and height are the effect of the complex stretching and folding mechanism acting on phase-space. Conservative dynamical systems display, additionally, a collection of sharp peaks which are originated by motion in the vicinity of invariant tori [2]. The usage of symbolic dynamics allows one to understand the "grammatical" rules governing the system's evolution by organizing them hierarchically on a logic tree [3]. The major peaks can then be correlated with the levels of the tree, thus providing an importance ordering, and their positions are explained in terms of joint contributions of particularly relevant ("primitive") symbolic sequences. As a result of this unfolding procedure, the dynamics is described as a succession of deterministic paths (blocks of symbols) which appear at random in time, according to measured transition probabilities. The tree is employed to make predictions about the future symbolic time-evolution and to reconstruct a signal statistically and dynamically equivalent to the original one. The comparison between the power spectra of the two signals provides an effective tool for a quantitative evaluation of the achieved accuracy.

Following the approach of ref. [3], we consider an input symbol sequence $S_N \equiv s_1 \ldots s_N$, with $N \approx 10^8$. All periodic subsequences S_n of length n are allocated on the logic tree, respecting all relevant parental relations. The first level of the tree contains "primitive" sequences (i.e., periodic and not decomposable in periodic substrings), which are generally of variable length and constitute the basic components of the original signal S_N. All other sequences on the tree are formed as a concatenation of primitives, respecting the intrinsic dynamical laws of the system. The approach is authentically hierarchical because the degree of understanding about the system which one has gained at a certain step of the procedure is ideally measured by making predictions on the composition of the next level of the tree. These predictions are then compared with the "actual" structure of the tree (i.e., with what is obtained by letting n go to infinity). In ref. [3] the information gain in the learning process was proposed as a measure of the system's complexity. If there exists an $n_0 < +\infty$ such that

Measures of Complexity and Chaos
Edited by N.B. Abraham *et al.*
Plenum Press, New York

257

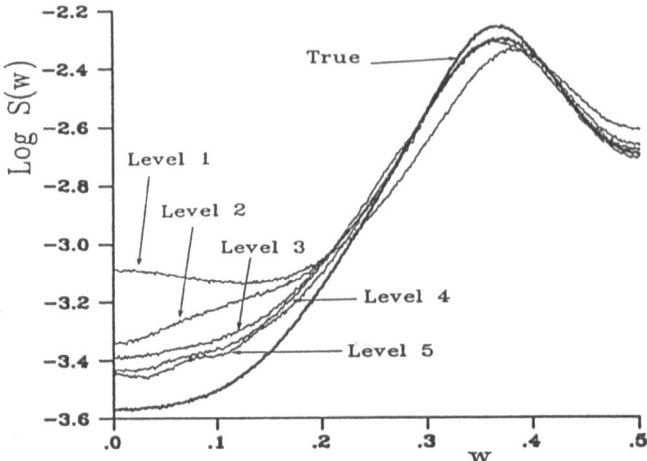

Figure 1. Comparison of the power spectrum $S(\omega)$ of the roof map (thick line) with those corresponding to the successive hierarchical reconstructions obtained from the first five levels of the logic tree (thin lines). Very satisfactory convergence is reached, because of the simplicity of the map. Base-10 logarithms have been used.

for $n > n_0$ all predictions match, the system is simple; if new surprises are encountered at each step, it is complex.

The symbolic signal is first reconstructed by taking a random combination of the sequences allocated at a certain level l of the tree, according to the measured transition probabilities: for example, after sequence 01 has been selected, the block 001 can appear with probability $P(001|01) \equiv P(01001)/P(01)$, the symbol 1 with probability $P(1|01) \equiv P(011)/P(01)$ and so on. The signal obtained with this procedure is then compared with the real symbolic signal generated by the system (roof, logistic or Hénon map[3]). Very good agreement is found in all cases: in particular, all principal peaks are reproduced by considering level $l = 1$, whereas the lower ones are better and better approximated by using sequences at lower levels (2 to 6).

For a complete understanding of the real system's dynamics, however, it is necessary to attach a continuous variable x to the reconstructed symbolic signal. The following method is particularly convenient: each time a block S_n containing n symbols is selected, the previously calculated x-value of the centre of mass of the corresponding phase-space element is taken; the remaining $n-1$ x-values are then simply computed as the centres of mass of the images of such element. This procedure can also been used in the analysis of experimental signals, provided that a satisfactory symbolic dynamical representation can be found. A comparison between the spectrum of the reconstructed "continuous" signal and of the true one is illustrated in figures 1-3, for the roof map described in ref. [3], for the logistic map $x' = 1 - 1.85x^2$ and for the Hénon map at standard parameter values[4]. The agreement is quite good, although a new source of errors is introduced by the procedure. In fact, phase-space elements corresponding to short symbolic sequences are rather large and the substitution of any of their internal points with the centre of mass may yield, in some cases, a rather poor approximation of the true dynamics. The usage of the x-coordinates of the periodic orbit with the same symbolic sequence as the selected one furnishes, in general, a less accurate result: periodic orbits can have, in fact, points lying very close to the border of the associated phase-space element.

An improvement can be obtained by adding to the x-coordinate of the centre of mass a random number between $-\sigma/2$ and $\sigma/2$, where $\sigma = \sqrt{\langle x^2 \rangle - \langle x \rangle^2}$ is also computable from

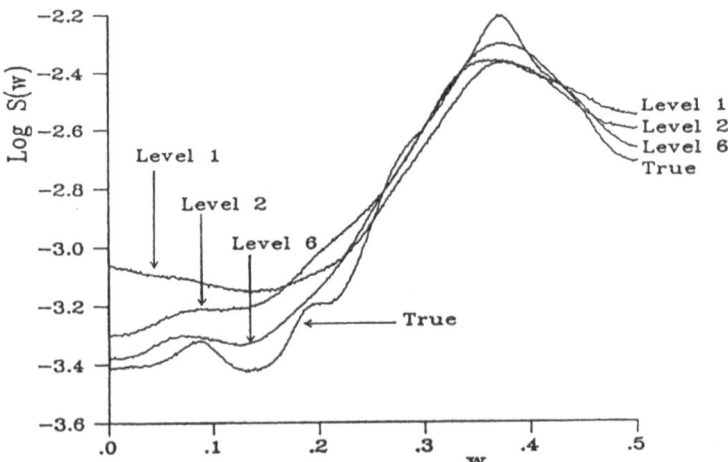

Figure 2. Same as in fig. 1 for the logistic map $x' = 1 - 1.85x^2$. Although the tree is topologically very similar to that of the roof map, worse convergence is observed here because of the nonhyperbolicity of the logistic equation.

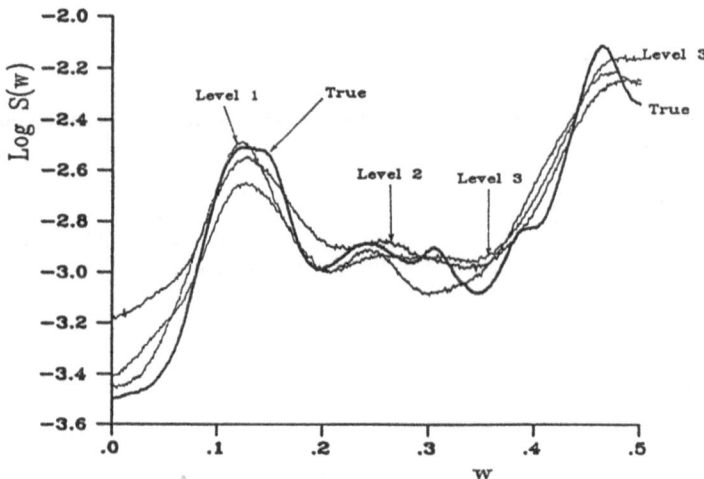

Figure 3. Same as in fig. 1 for the Hénon map at standard parameter values. The accuracy is less satisfactory than in the previous examples, since longer-range memory effects are present. The total probability of the sequences corresponding to the frequencies 1/7 and 1/8 is 2.5 times less than that of sequence 0111 (corresponding to frequency 1/4) and of sequence 01 (peak close to frequency 1/2). A naive model based on the weight of periodic sequences would, therefore, yield a wrong prediction for the spectrum.

the time-series. In this way, the dynamics is split into a macroscopic part, given by the jumps between elements in phase-space, and a microscopic random contribution which represents the unresolved motion within each element of the covering. The presence in the time-signal of deterministic paths of different length (the primitives) accounts for the existence of peaks in the power spectra which correspond to non-integer periods (i.e., to frequencies that are not the inverse of an integer). This apparently puzzling (for discrete-time maps) feature of the spectra is rather general and shows that the signal cannot be interpreted as a mere superposition of periodic orbits: it is rather a mixture of determinism and randomness, as discussed above. Our approach presents many advantages over constant-length procedures, since the intrinsic deterministic rules of the system are fully incorporated: the primitives are, in fact, the longest indecomposable itineraries described by the orbit.

The reconstruction of the time-signal is based on two ingredients which are invariant under smooth coordinate changes (the topology and the metrics of the logic tree) and on one which is not (the values of the continuous variable x). The topological rules account for the complicated folding mechanism acting on phase-space and reflect the most fundamental dynamical properties of the system. They immediately provide a piecewise linear approximation of the equations of motion and can be used to calculate quantities like invariant measure, fractal dimensions and metric entropies [3]. Moreover, different systems can be regrouped into topological "equivalence classes", on the basis of the structure of their logic trees. The metric properties of the system are particularly relevant in the investigation of nonhyperbolic systems, because of the presence of square-root singularities in the invariant measure. The effect of nonhyperbolicity on power spectra can be seen by comparing figures 1 and 2. In fact, the roof and the logistic map (at the chosen parameter values) are topologically described by essentially the same tree, apart from small differences at low levels, whereas their probabilistic properties are rather different. Much slower convergence is observed for the nonhyperbolic logistic map. The method is rather insensitive to changes in the probabilities which only concern prefactors in the mass-size scaling laws. Differences in the exponents (dimensions), instead, lead to poor approximations if a linear model is used, since the square-root singularities in the invariant measure cannot be well reproduced. Construction of fully nonlinear maps from the knowledge of the logic tree would provide much more efficient reconstruction algorithms.

Finally, the accuracy is very weakly depending on the chosen values of the x-coordinate.

Acknowledgements

We benefitted from discussions with E. Pollak, G. Broggi and I. Procaccia. This work has been partly supported by a grant of the Von Humboldt Foundation.

REFERENCES

[1] J.P. Eckmann and D. Ruelle, Rev.Mod.Phys. **57**, 617 (1985).

[2] M.A. Sepúlveda, R. Badii and E. Pollak, to be published.

[3] R. Badii, *Unfolding Complexity in Nonlinear Dynamical Systems*, this issue; *Quantitative Characterization of Complexity and Predictability*, submitted for publication.

[4] M. Hénon, Comm.Math.Phys. **50**, 69 (1976).

RELATIVE ROTATION RATES FOR DRIVEN DYNAMICAL SYSTEMS

Hernán G. Solari and R. Gilmore

Department of Physics and Atmospheric Science
Drexel University
Philadelphia, PA   19104

INTRODUCTION

The results presented here represent a first step in an attempt to construct a topological classification of nonlinear dynamical systems.[1] We consider nonlinear systems which are periodically driven at a single forcing frequency. This allows definition of a period-one Poincaré section and reduction of the flow to a return map. Since nonlinear behavior is typically associated with stretching and folding, the return map undergoes formation of a horseshoe[2] as a function of one or more control parameters. We characterize the topological properties of orbit pairs according to the positions of their returns in the horseshoe. The topological indices which result can then be lifted to characterize orbit pairs in the original flow.

BACKGROUND AND MOTIVATION

The properties of the CO_2 laser with modulated parameters have been studied extensively both experimentally[3-11] and theoretically.[12,13] As a model for this system we use the simple pair of coupled equations

$$du/dt = [z - R \cos(\Omega t)] u$$

$$dz/dt = (1 - \varepsilon_1 z) - (1 + \varepsilon_2 z) u \qquad (1)$$

where u is the laser intensity, z the population inversion, $\Omega = 2\pi/T$ is the driving frequency and ε_1, ε_2 are damping parameters. This system exhibits multistability. Periodic orbits of all periods between 1 and 11 have been seen both experimentally and theoretically, and many additional orbits of period up to 20 have also been seen. Each orbit undergoes a period-doubling cascade accumulating in chaos;[14] the resulting strange attractor then undergoes one or more crises.[15]

It was natural to ask whether orbits of arbitrarily high period could exist, even if their detection was rendered impossible by the small size of the high period basins of attraction in the phase space and the short duration of their existence in the control parameter space. More generally, is it possible for more than one topologically distinct orbit to exist for a given periodicity? In short, we wondered if there were some completeness result for dynamical systems comparable to the completeness theorems of linear mathematics.

Measures of Complexity and Chaos
Edited by N.B. Abraham *et al.*
Plenum Press, New York

The only completeness result in nonlinear analysis relates to the horseshoe construction.[2] The horseshoe is the minimal construction incorporating both stretching and folding. The spectrum of periodic orbits in a completed horseshoe is known.[16] Accordingly, we searched for, and found, a horseshoe in the period-one return map of the laser equations (1).

RELATIVE ROTATION RATES FOR THE HORSESHOE

The reduction of a flow to a map is unique, but the lift of a map to a flow is not. The nonuniqueness lies in the number of complete turns the map may make about an axis orthogonal to its direction of propagation during one period. The lift of a map to a flow is unique up to this global torsion, which is a topological index.

The Relative Rotation Rates[1] of a pair of periodic orbits determine how the orbits are intertwined with each other. If d_{AB} is the difference vector between a point on orbit A of period p_A and a point on orbit B of period p_B, then d_{AB} will evolve in time under a flow and finally return to its initial orientation after p_A*p_B periods. During the course of its forward propagation it will make an integer number, n_{AB}, of turns about its axis of propagation. The ratio, $n_{AB}/(p_A*p_B)$, is the relative rotation rate of the pair of orbits A and B. It describes the average number of turns the orbits make about each other during one period.

We have computed the Relative Rotation Rates for all low period orbits in a zero-torsion flow which develops a horseshoe in its return map. The Relative Rotation Rates for the Newhouse nodes and saddles[17] are shown in Table 1.

COMPARISON WITH LASER MODEL

The Relative Rotation Rates for all periodic orbits of the laser model which have been located were computed. They were then compared with those computed abstractly for the Smale horseshoe return map. There were no discrepancies. Moreover, the Relative Rotation Rates were used as a diagnostic to distinguish the three topologically inequivalent period-five orbits which were located numerically.

CONCLUSION

The table of Relative Rotation Rates for the periodic orbits of the laser is identical to the table of Relative Rotation Rates for the zero torsion lift of the horseshoe return map.

The backbone of a periodically driven nonlinear dynamical system is its set of periodic orbits. We have constructed a topological description of how these orbits are intertwined among each other. This description is presented in terms of a matrix of Relative Rotation Rates which can be determined from the underlying horseshoe in the return map of the flow. The lift of the Relative Rotation Rates from the map to the flow is unique up to global torsion, which is the Relative Rotation Rate of the two period-one orbits of the Smale horseshoe. In this program we replace the description of the flow in terms of (uncountable) sets of differential equations by a topological description in terms of countable indices which are in turn derived from a very modest set of mappings of horseshoe type. The Relative Rotation Rates exhibit systematics under period doubling cascades and provide selection rules which show which bifurcation processes are allowed and which are forbidden at any stage in the development of the horseshoe, for example, as some control parameter is varied. These properties are discussed more fully in Ref. 1.

ACKNOWLEDGMENTS

We wish to thank Prof. J. R. Tredicce, Prof. L. M. Narducci and Dr. E. V. Eschenazi for useful discussions.

Table 1. Relative Rotation Rates for the low period Newhouse orbits in the Smale Horseshoe. All orbits intertwine the stable period one orbit but do not intertwine the unstable period one orbit in the zero torsion lift of the horseshoe. The global torsion is the Relative Rotation Rate between the two period-one orbits. The Relative Rotation Rates shown in this table are the same for the Newhouse saddles and nodes.

Period	1	2[a]	3	4	5	6
1		1/2	1/3	1/4	1/5	1/6
2	1/2		1/3	1/4	1/5	1/6
3	1/3	1/3		1/4	1/5	1/6
4	1/4	1/4	1/4		1/5	1/6
5	1/5	1/5	1/5	1/5		1/6
6	1/6	1/6	1/6	1/6	1/6	

[a] This is the period two orbit bifurcated from the period one orbit. There is no period-two Newhouse orbit in the Smale horseshoe.

REFERENCES

1. H. G. Solari and R. Gilmore, Phys. Rev. **A37**, 3096 (1988).
2. S. Smale, Bull. Am. Math. Soc. **73**, 747 (1967).
3. F. T. Arecchi, R. Meucci, G. Puccioni, and J. Tredicce, Phys. Rev. Lett. **49**,1217 (1982).
4. R. S. Gioggia and N. B. Abraham, Phys. Rev. Lett. **51**, 650 (1983).
5. F. T. Arecchi, G. L. Lippi, G. Puccioni, and J. Tredicce, Opt. Commun. **51**, 308(1984).
6. W. Klische, H. R. Telle, and C. O. Weiss, Opt. Lett. **55**, 561 (1984).
7. J. R. Tredicce, F. T. Arecchi, G. L. Lippi and G. P. Puccioni, J. Opt. Soc. Am. **B2**, 173 (1985).
8. G. P. Puccioni, A. Poggi, W. Gadomski, J. R. Tredicce, and F. T. Arecchi, Phys. Rev. Lett. **55**, 339 (1985).
9. J. R. Tredicce, N. B. Abraham, G. P. Puccioni, and F. T. Arecchi, Optics Commun. **55**, 131 (1985).
10. J. R. Tredicce, F. T. Arecchi, G. P. Puccioni, A. Poggi, and W. Gadomski, Phys. Rev. **A34**, 2073 (1986).
11. T. Midavaine, D. Dangoisse, and P. Glorieux, Phys. Rev. Lett. **55**, 1989 (1986).
12. I. I. Matorin, A. S. Pikovskii, and Ya. I. Khanin, Sov. J. Quantum Electronics **14**, 1401 (1984).
13. H. G. Solari, E. Eschenazi, R. Gilmore, and J. R. Tredicce, Optics Commun. **64**, 49 (1987).
14. M. J. Feigenbaum, J. Stat. Phys. **21**, 669 (1979).
15. C. Grebogi, E. Ott, and J. A. Yorke, Phys. Rev. Lett. **57**, 1284 (1986).
16. R. M. May, Nature **261**, 459 (1976).
17. S. E. Newhouse, *Progress in Mathematics*, Boston: Birkhauser, 1980, Vol 8, p. 1.

STRETCHING FOLDING TWISTING

IN THE DRIVEN DAMPED DUFFING DEVICE

Hernán G. Solari, Xin-Jun Hou and R. Gilmore

Department of Physics and Atmospheric Science
Drexel University
Philadelphia, PA   19104

INTRODUCTION

The results presented here represent another step in our attempt to provide a topological understanding of the behavior of nonlinear dynamical systems. The conditions under which "bifurcation bubbles" exist for a dynamical system has always been a mystery. They occur, for example, in the driven Duffing oscillator. We began by computing the Relative Rotation Rates for the periodic orbits of the Duffing oscillator in order to determine the systematics of "bifurcation bubbles." The systematics which we have discovered have led to an increased understanding of the mechanics responsible for the bifurcation phenomena seen in driven nonlinear oscillators. In particular, these phenomena can be classified as local bifurcation phenomena and global bifurcation phenomena in a natural way.

BACKGROUND AND MOTIVATION

The driven Duffing oscillator

$$dx/dt = y$$

$$dy/dt = -\delta y - x - x^3 + R\cos(\Omega t) \tag{1}$$

has been thoroughly investigated for many years. It is known now that the control parameter plane (R-Ω) is divided into a number of "fingers" in which many periodic orbits, both stable and unstable, coexist. These peninsulae are surrounded by a "sea" in which only one period-one orbit is present. This decomposition is shown schematically in Fig. 1. In sweeping one of the control parameters holding the other constant, we move across these peninsulae and experience bifurcation bubbles. That is, upon entering a peninsula a period doubling cascade is initiated and saddle-node bifurcations occur. They recollapse upon leaving the peninsula. In this way a series of bifurcation bubbles is encountered on sweeping the control parameters.

Since the mechanism responsible for bifurcation bubbles is mysterious, we used the recently developed Relative Rotation Rates[1] as a tool to study these bubbles. A point (set of control parameter values) was chosen within each peninsula. All periodic orbits, both stable

Measures of Complexity and Chaos
Edited by N.B. Abraham *et al.*
Plenum Press, New York

and unstable, which coexist at each control parameter value were then located, identified by their logical name, and the Relative Rotation Rates of each coexisting pair of orbits was computed. These were compared with the Relative Rotation Rates for the corresponding pairs of orbits as determined from the horseshoe return map for the Duffing oscillator. This return is not a cubic S-shaped curve lying on its side, as often suggested in the literature. Rather, it is the second iterate of a Smale horseshoe.

LOCAL AND GLOBAL TORSION

The comparison of the set of Relative Rotation Rates within each peninsula with those of the iterated horseshoe was exact and moreover uniquely identified the global torsion of that peninsula. The global torsion of each peninsula is an odd integer which differs by two between adjacent peninsulae.[2]

The picture that emerged is the following. Within each finger the global torsion is fixed, an odd integer. On moving along the axis of the finger (increasing R) the return map becomes increasingly deformed into a completed horseshoe. While this occurs the standard bifurcation scenario (e.g., logistic map or Henon Heiles phenomena) takes place. We call the bifurcation scenario accompanied by the formation of a horseshoe the *local bifurcation scenario*.

On moving to the border between the peninsula and the sea, the horseshoe must "unwind." As it does so, the local bifurcation scenario reverses itself. As the control parameters move into the strait between two adjacent peninsulae, only a single orbit remains, of period-one. The return map "pivots" about this orbit, its global torsion changing by two in the process. As the control parameter enters the next peninsula, the local horseshoe begins the formation process anew. This time all periodic orbit pairs have their Relative Rotation Rates changed by two.

LOCAL BIFURCATION AND GLOBAL BIFURCATION SCENARIO

Two independent forces are at work creating the local and global bifurcation scenarios. One force is the increasing strength of the forcing term, or decreasing damping. This results in increased deformation of the return map until it assumes the shape of a completed horseshoe. This force is responsible for the creation of many periodic orbits in saddle node bifurcations, the period-doubling cascade to accumulation along each initially stable branch, and the crises of the residual strange sets in the coexisting basins.

A second force is responsible for "torquing the return map." This force attempts to increase the Relative Rotation Rates of all periodic orbit pairs. However, since the Relative Rotation Rates are topological indices, they are constant *while they exist*. As a consequence, when the "torque" becomes too great all orbits but one are forced to disappear (by unfolding the horseshoe), the entire return map responds to the "torque" by rotating about the one remaining period-one orbit, and the return map begins bending into a horseshoe again.

This leaves unresolved the source of the "torque." "Torque" occurs naturally in any oscillator, nonlinear or not. A point in the phase space typically rotates under free propagation (e.g., for $dx/dt = y$, $dy/dt = -kx$, the point x,y rotates in phase space). The longer the interval between Poincaré sections, the more the return map rotates. In a driven nonlinear oscillator the spacing between Poincaré sections is determined by the frequency of the forcing term. Therefore the "torque" is inversely related to the forcing frequency, or directly proportional to the period, T $(\Omega T = 2\pi)$.

In a typical nonlinear oscillator, such as the Duffing oscillator, we expect the following qualitative behavior as a function of the forcing term $f(t) = R \cos (2\pi t/T)$. For fixed T the global torsion is fixed (and related to the integer part of T) and the return map becomes increasingly distorted into a horseshoe as R increases. For fixed R, the horseshoe

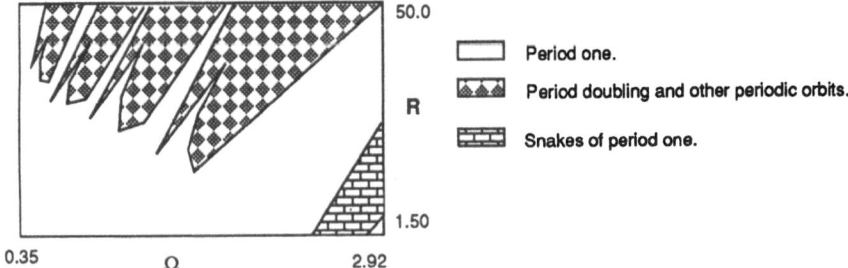

Fig. 1. The control parameter plane of the driven Duffing oscillator (1) is divided into "fingers" of constant odd integer global torsion surrounded by a sea in which only a single periodic orbit, of period one, exists.

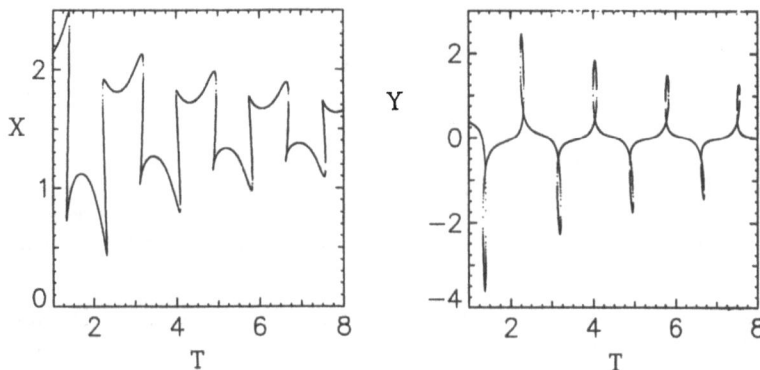

Fig. 2. (a) x- and (b) y- values for the period-one fixed points in the Poincaré section of the driven Duffing oscillator (2) as a function of the period, T, of the forcing term, with R = 5.0 and δ = 0.2.

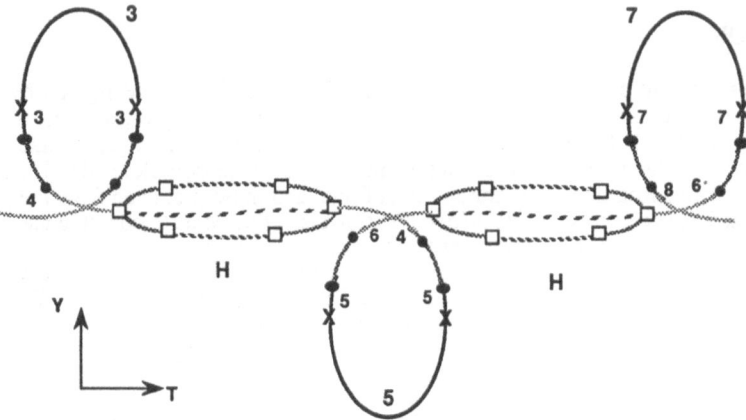

Fig. 3. Schematic blow-up of Fig. 2(b). Proceeding from the left: A single period-one orbit is joined by two others created in a saddle-node bifurcation at X. One new orbit (top) is a regular saddle with constant local torsion (3) until it is annihilated in an inverse saddle-node bifurcation with the initial period-one orbit. The other orbit created in the saddle-node bifurcation remains a stable node for a short while until it becomes a stable focus (at dot). During its existence as a focus its local torsion increases from 3 to 4. It becomes a stable node (at dot 4) until a period-doubling cascade begins (at square). The orbit is then a flip saddle (dashed) with local torsion 4. The side branches represent an asymmetric pair of orbits which undergo a bifurcation bubble until recollapse when the flip saddle disappears (at square). The orbit then becomes a stable node, stable focus (second dot 4) during which the local torsion increases from 4 to 5, a stable node, and finally disappears in an inverse saddle-node bifurcation.

forms and unforms as the control parameter T increases and passes through a succession of bifurcation peninsulae, with the global torsion increasing as successive peninsula are encountered.

In large part, increasing R drives the local bifurcation scenario while increasing T drives the global bifurcation scenario.

QUANTITATIVE PROPERTIES

To confirm this scenario we studied a modified form of the Duffing equation (1). The phase space volume change induced by (1) is $(1/V)dV/dt = -\delta$ so that the volume contration per period is $V(T)/V(0) = e^{-\delta T}$. To keep the contraction of the map constant as the control parameter T is varied, we studied the modified Duffing equation

$$dx/dt = y$$

$$dy/dt = -(\delta/T)y - x - x^3 + R \cos(2\pi t/T) \qquad (2)$$

as a function of the forcing strength, R, and the period, T.

In particular, we searched for fixed points of the period-one return map as a function of T for fixed R. The striking results obtained from this search are shown in Figs. 2a and 2b. For some parameter values there is one fixed point (in the sea between the peninsulae). For others there are three fixed points. The systematic behavior of these fixed points is shown schematically in Fig. 3.

One fixed point, created and destroyed in saddle-node bifurcations, is a regular saddle. The local torsion about this saddle is fixed and is an odd integer. The other two fixed points evolve systematically from a stable node with odd integer local torsion, to a stable focus, during which the local torsion increases by one, to a stable node again and finally to a flip saddle. This process then recurs, in reverse order, with the local torsion again increasing by one during the focus phase of the "perestroika."

CONCLUSION

The systematic behavior of the Duffing oscillator has been investigated by both qualitative and quantitative methods. We used topological techniques, specifically the Relative Rotation Rates, to determine the systematics of the bifurcation peninsulae in the control parameter plane. In doing so, we were able to make complete identifications between the Relative Rotation Rates for all orbits found and those predicted from the return map, the iterated Smale horseshoe. This identification also allowed us to compute the global torsion of the return map, or equivalently, the associated peninsula. The systematic behavior of the global torsion then suggested that two independent forces are at work in molding the bifurcation behavior of nonlinear driven damped dynamical systems. The local bifurcation scenario is governed by the evolution of the return map into some kind of horseshoe. The global bifurcation scenario is driven by the "torque" applied to the return map by some control parameter, typically the period of the forcing term.

These hypotheses were tested quantitatively on a modified form (2) of the Duffing oscillator. These quantitative studies not only failed to reject the qualitative hypotheses, but also revealed new properties of the Duffing oscillator whose presence was suggested by these hypotheses.

REFERENCES

1. H. G. Solari and R. Gilmore, Phys. Rev. **A37**, 3096 (1988).
2. H. G. Solari and R. Gilmore, Phys. Rev. **A38**, 1566 (1988).

CHARACTERIZING CHAOTIC ATTRACTORS UNDERLYING SINGLE MODE LASER EMISSION

BY QUANTITATIVE LASER FIELD PHASE MEASUREMENT

C.O. Weiss, N.B. Abraham

Physikalisch-Technische Bundesanstalt

3300 Braunschweig, F.R. Germany

INTRODUCTION

We have carried out experiments on single-mode laser-pumped gas lasers to show that dynamics of single mode autonomous lasers is at least in this class of lasers not an exception but rather the rule. Lasers emitting from the far infrared to the visible on rotational[1], vibrational[2] and electronic[3] transitions have shown periodic and chaotic single-mode dynamics.

The NH_3-laser used in the far infrared was chosen[4] since it appeared to be an experimentally easy candidate for showing the dynamics of the Lorenz model[5], which was developed for an approximate description of convective Bénard flow and later found to be isomorphic with the equations of the simplest kind of laser: homogeneously broadened, single mode, traveling wave[6] in the plane wave approximation; and which had so far no counterpart in a real experimental system. The laser pumped gas laser is strictly a coherently coupled 3-level system described by nine density matrix equations as opposed to the incoherently pumped two-level (laser-) system described by the Lorenz equations. The limits in which the three-level laser equations can be adiabatically eliminated to reduce to the Lorenz equations have been discussed in[7]. From this discussion it is at least doubtful whether these conditions can be met with the NH_3-lasers. Consequently a large number of theoretical investigations[8] have come to the conclusion that it would be in principle impossible to realize the dynamics of the Lorenz model with this type of laser. None the less the experimental results[9] were all compatible with and actually characteristic for the Lorenz model. The most complete model developed for this type of laser[10] which tried to model as realistically as possible the experiments, permits indeed to conclude that the NH_3 laser dynamics is very similar to that of the Lorenz equations. The two-dimensional bifurcation structure (parameters: resonator tuning and pump strength) is predicted to be identical for the NH_3-laser and the Lorenz model extended for detuning. An important difference, however, was pointed out: while in the Lorenz attractor the field changes irregularly from positive to negative and vice versa, the model for the NH_3-laser yields single sided (but bistable) attractors, that is: the field never changes sign, even though it may be of either sign. In intensity measurements this difference does not show up.

Measures of Complexity and Chaos
Edited by N.B. Abraham *et al.*
Plenum Press, New York

Experiment

In order to test the predictions of the NH$_3$-laser model versus the
Lorenz model, an experiment was carried out in which the <u>laser field</u> can
be measured rather than the usual laser intensity, the latter not containing any information about the sign of the phase of the laser field.

In order to measure the optical phase of a radiation, interference
with a second reference field is needed. If the phase is to be measured
equally precisely for all phase angles two interference signals in phase
quadrature are needed. This is particularly necessary since the phase
angles at the beginning of the measurements are unknown.

The laser used operates on the aR(4.4) transition of isotopic NH$_3$
(^{15}NH$_3$) at a wavelength of 152.7 μm pumped by the 10R(18) line of an isotopic CO$_2$-laser (^{13}CO$_2$) via the aQ(5.4) vibrational transition. We decided
for experimental simplicity to create the necessary reference frequency by
harmonic generation of a mm wave in a micron size GaAs-Schottky barrier
diode. See Fig. 1 for the experimental set-up.

The ^{15}NH$_3$-laser under study uses a ring resonator whose reflectors
are a grating used in zero order reflection for the 153 μm radiation and in
first order to couple in the 10 μm pump radiation; a 33 μm grid constant
copper wire mesh used as the outcoupling element; and a curved gold mirror.

The spatial filter assembly with a diaphragm serves to continuously
attenuate the pump power without changing pump beam geometry or its frequency. Unidirectional emission in the ring is achieved utilizing Doppler

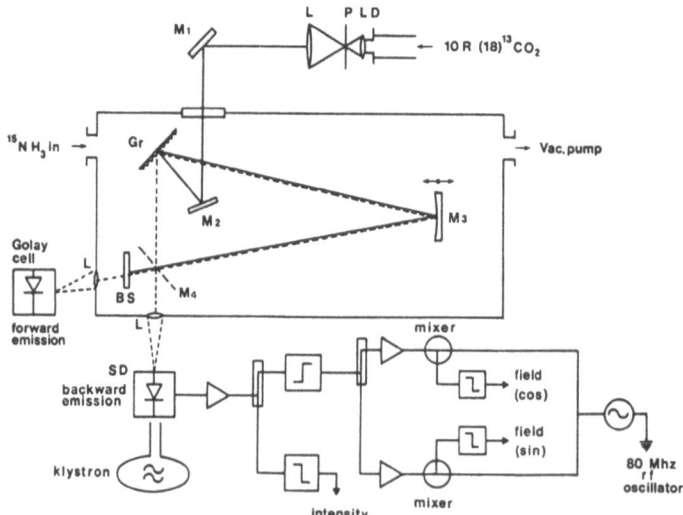

Fig. 1. Experimental set-up showing FIR ring laser; Schottky barrier
diode detector illuminated by klystron microwave source; and
circuits to separate the homodyne and heterodyne portions of
the signal and then to beat the heterodyne signal with an
80 MHz reference oscillator to bring the electric field signal
to a zero-frequency reference frame. (M—mirror; L—lens; BS—CO$_2$
beam stop; Gr—grating; P—aperture of CO$_2$ spatial filter.)

effects on the pump and lasing transition[11]. The backward (ccw) emission was used since the gain line profile is closer to a homogeneously broadened line profile than the forward (ccw) emission line which is possibly split due to the pump AC-Stark effect[12].

To ensure single direction backward emission, the forward emission was continuously monitored by a Golay cell which was sometimes replaced for time resolved monitoring by a second GaAs-Schottky diode.

The GaAs Schottky diode detecting the backward emission was additionally irradiated by a 70 GHz wave from a 4 mm wave klystron. In the diode harmonics of the 70 GHz radiation are generated. The klystron frequency was adjusted so that the frequency of its 27th harmonic was 80 MHz from the laser frequency. The diode then mixes the laser frequency with the harmonic frequency to generate a beat at 80 MHz. The latter is amplified by a low-noise preamplifier to exceed noise levels of the registration electronics.

The output of the diode is first passed through a low pass filter with a cut-off at 10 MHz. This signal contains only information about the laser intensity.

The diode signal is additionally passed through a high pass filter with a low frequency cut-off at 20 MHz. This signal is detected by two mixers which have 80 MHz as their local oscillators supplied to the mixers with 90° phase difference. The two mixer outputs are the quadrature field signals which allow in principle to follow laser field amplitude and phase as a function of time.

The requirements on the phase-stability of the klystron and the laser are extreme. However, it was possible to clearly distinguish the phase changes resulting from frequency detuning of the reference frequency from those resulting from the laser dynamics. The latter occur at a rate of $\pi/100$ ns while the "technical" phase drift was often less than $\pi/100$ µs. The measurements were done by adjusting the frequency of the mm- wave harmonic (the reference frequency) as precisely as possible to generate a beat of 80 MHz. The signals corresponding to laser intensity and to the two quadrature components of the laser field were recorded on digital storage oscilloscopes with recording length of up to 30 000 points. The recorded signals were then evaluated by computer.

Measurements (higher pressure)

The measurements intend to provide quantitative comparisons with the predictions of the Lorenz model (extended for detuning[13]) on the one hand and with those of the full NH_3-laser model.

The spiral shapes of the intensity pulses typical for the Lorenz model, the period-doubling transition to chaos predicted for tuning the laser towards the line center by the extended Lorenz equations[13], and the instability threshold and the abrupt onset of chaos predicted by the Lorenz equations had already been shown experimentally[9].

However, more structure exists in the two-dimensional bifurcation diagram (pump strength, tuning) of the extended Lorenz equations (Fig. 2). Weakly chaotic attractors e.g. "almost periodic" P3 and P5 are seen for small detuning in Fig. 2[13] above the chaos threshold. These were searched for first in the experiment. Fig. 3 shows intensity pulses measured at different detunings of the NH_3-laser resonator from the NH_3 gain line center.

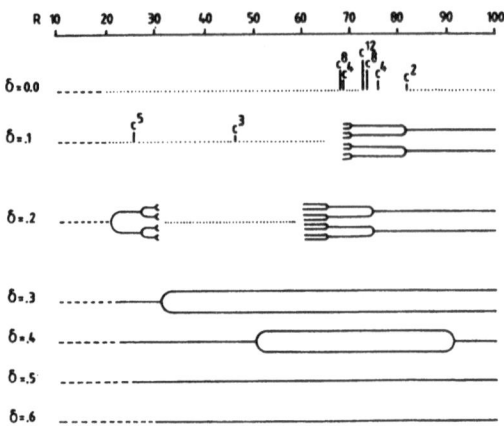

Fig. 2. Two-dimensional bifurcation diagram of the homogeneously broadened laser. R is the pumping strength referred to the laser
threshold (R=1). δ is the laser detuning in units of the homogeneous linewidth. Dashed lines: continuous emission; dotted lines:
chaotic emission; full lines: periodic pulsing. The period doublings are symbolized, C_n: periodic or weakly chaotic emission
with period n.

From the period-doubling cascade P2, P4 and P8 are shown, followed
by chaotic pulsing. For illustration a P12 window in the chaotic range is
also shown. The predicted P3-attractor (whose chaoticity is too weak to
be experimentally distinguishable from a periodic P3 attractor) follows
before the spiral pulsing predicted by the Lorenz equations, which describe only line center tuning, is reached.

Thus the laser intensity shows complete agreement with the extended
Lorenz model. The question of the attractor symmetry e.g. in the sign of
the field was then addressed using field phase measurements.

Fig. 4 shows a recording of Lorenz-like ("spiral-") intensity pulses
with the two quadrature components of the laser field. Keeping in mind
that as mentioned above, a slow phase drift caused by technical reasons
is superimposed on the phase and magnitude changes of the field, already
this recording shows convincingly that at the end of each "spiral" the
field pulse changes sign. This is clear indication of a symmetric
attractor as predicted by the Lorenz equations and contrary to the result
of the NH_3-laser model.

A more quantitative evaluation of the phase evolution as a function
of time is shown in Fig. 5. Again Lorenz-like intensity pulses are shown
together with the quadrature field components. Below, the laser field
phase calculated from the two quadrature field components is shown. It
can be seen that at the end of each intensity "spiral" the phase changes
by π which means a change of sign in the field confirming the symmetric
nature of the attractor. It is noteworth that there is no measurable phase
change during the evolution of the complete "spirals". There are only two
values of the phase differing by π. (Again having in mind that the slow
phase change superimposed over the π-phase jumps are of technical origin.)
This is in remarkable agreement with the predicitons of the Lorenz equations
where a "phase" does not exist since they describe the perfectly tuned
laser with real field variable. The only "phase" in the Lorenz equations

272

is the sign of the field corresponding to π difference in phase. It may be mentioned that the NH₃-laser model predicts on the contrary frequency pulling and pushing during the pulses, clearly in disagreement - as the attractor symmetry - with the experimental findings.

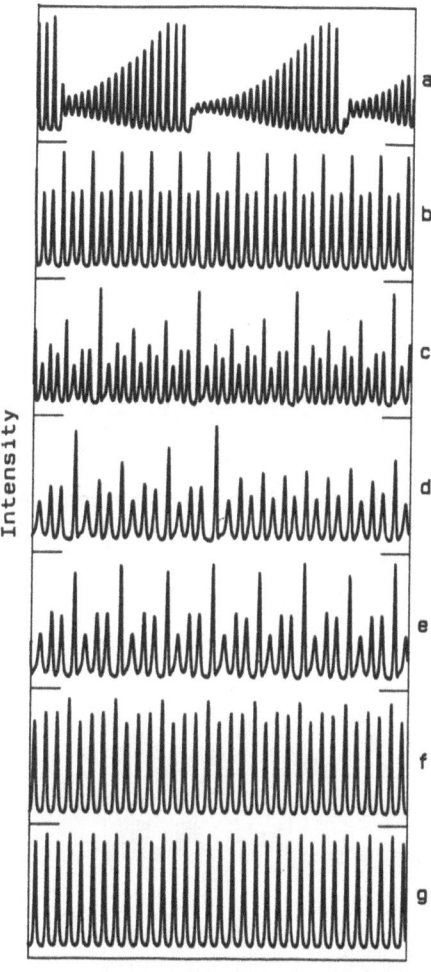

Fig. 3. Intensity signals measured for increasing detuning of
the laser cavity with respect to emission line center:
a) resonant case: Lorenz like chaos
b) small detuning: stable period-3 oscillation
c) period-12 window within the period-doubling chaos
domain d)
d) period-doubling chaos
e) stable period-8 oscillation
f) stable period-4 oscillation
g) stable period-2 oscillation

Fig. 6 illustrates the attractor symmetry again on a P-3 window measured within the chaotic range when the laser is tuned to line center. The change of signs between groups of three pulses is evident and the phase evaluation properly shows phase jumps of π and no phase modulation accompanying the pulses. The attractor is clearly symmetric and frequency pulling is absent as to be expected from the Lorenz equations and contrary to the model of the NH_3-laser. Fig. 7 shows another even more complex example of periodic pulsing measured as expected for the Lorenz model, a P6 symmetric attractor (field).

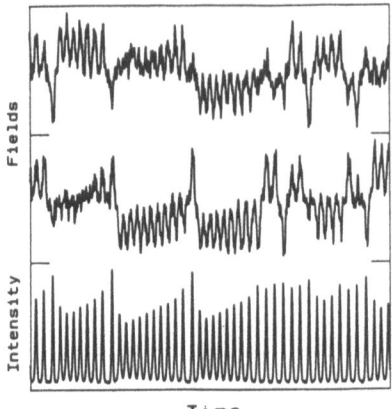

Fig. 4. Upper traces: in-phase and in-quadrature components of the field emitted by the laser without detuning. Lower trace: intensity emitted by the laser.

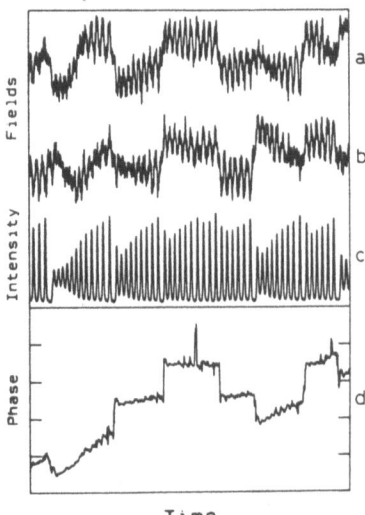

Fig. 5. High pressure (p=9 Pa) chaotic pulsing of 153-μm $^{14}NH_3$ laser emission for resonant tuning. Trace marked "intensity": laser intensity pulses. Pulsing period is 1 μs. Traces marked "fields": in-phase and in-quadrature heterodyne signals measuring the laser field. Trace marked "phase": phase changes of the laser field as a function of time, reconstructed from field traces. One division on the vertical axis corresponds to a phase change of π rad.

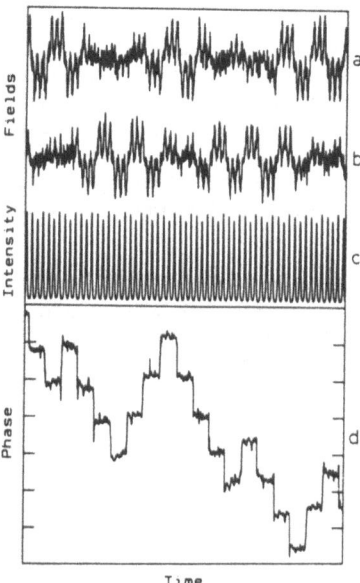

Fig. 6. A periodic window in the high-pressure chaotic range.
Same conditions as Fig. 1 but slightly changed pump
intensity. Traces and units as in Fig. 1. Note phase
changes of π as in Fig. 1 indicating a symmetric
attractor as expected for the Lorenz model.

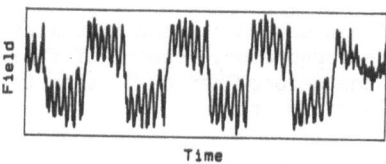

Fig. 7. One of the electric field components for a symmetric
window within the chaotic domain for centrally tuned
cavity: the laser performs regular motion on a
symmetric attractor with six loops on each side.

Fig. 8 shows the intensity and the field quadrature components for
the same pressure as the preceeding measurements but for significant
detuning. Here the laser is simply pulsing periodically. The conditions
correspond to Fig. 3 at a detuning larger than trace g or to Fig. 2 for
a detuning δ of ~ 0.4.

The field recordings clearly show a one-sided attractor i.e. the
field does <u>not</u> change sign. It must, however, be kept in mind that since
this is a detuned case the field phase is meaning fully a continuous
variable. Hence the pulses are possibly accompanied by phase modulation
which can not be deduced from Fig. 8 without quantitative evaluation. In
any case the "one-sided" attractor is in agreement with the predictions
of the extended Lorenz equations.

Pulsing at smaller detuning, corresponding to trace g in Fig. 3, or $\delta \sim 0.3$ m Fig. 2, is shown in Fig. 9. this is the first period doubling, apparently with a one-sided attractor. The prediction of the extended Lorenz equations are shown for this case in Fig. 10 in intensity and phase. The large pulses are accompanied by a large phase excursion and the small pulses by small phase excursions. Note that the large phase excursions approach $\pi/2$ already. The quantitative phase evaluation of Fig. 9 is shown in Fig. 11 which bears a striking resemblence with the theoretical prediction of Fig. 10.

Concluding, one may say that all measurements of the field phase at high pressure $(P(NH_3) = 8$ Pa) are in detailed agreement with the "extended" Lorenz model and contradicting in the attractor symmetry and the frequency pulling those of the NH_3-laser model.

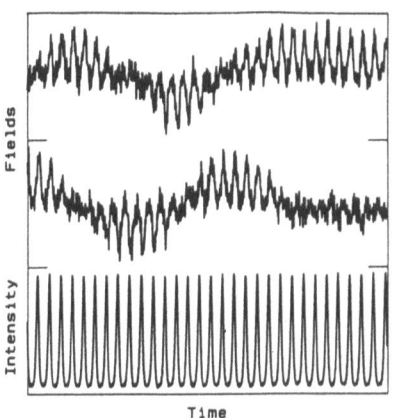

Fig. 8. Electric field components and intensity of a laser with detuned resonator: absence of field sign changes indicates the detuned period-1 state to be described by a single-sided attractor.

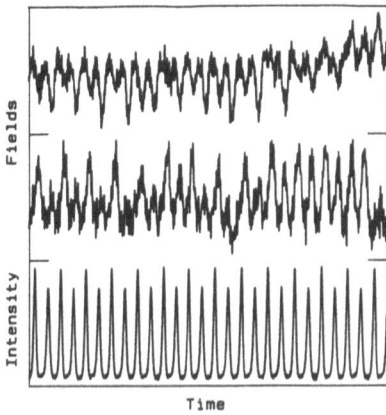

Fig. 9. Similar to Fig. 8, but for less detuning, showing the single-sided detuned period-2 state.

Measurements (lower pressure)

The difference in the predictions of the extended Lorenz equations
and the NH_3-laser model are of course due to 3-level coherences included
in the latter. Coherences can be destroyed by relaxation processes. It
is therefore that we have looked for laser dynamics similar to the
Lorenz model predictions at the highest possible NH_3 working pressure
(highest collisional relaxation rates), permitting dynamic phenomena.
This is close to the limit of "bad cavity"[6], which is a necessary requirement for dynamics in the Lorenz model. It appeared then that three-level
coherence effects would more likely manifest themselves at lower pressures
(smaller relaxation rates). Measurements were therefore also carried out
at lower NH_3 pressure to possibly find similarities with the NH_3-laser
model.

Fig. 12 shows a measurement taken at 5 Pa pressure. The intensity
pulses are "spirals" as in the Lorenz-like cases e.g. Figs. 4 and 5 and
not distinguishable from them. The field traces, however, lack any clear
sign changes in contrast to e.g. Figs. 4, 5. The quantitaive phase evaluation consequently shows no phase jumps of magnitude π. This behavior must
be interpreted as arising from a one-sided attractor as predicted by the
NH_3-laser model.

Consequently it appears that somewhere between 5 Pa and 8 Pa a
"glueing" of the one-sided attractors (which are bistable according to
the model) occurs, to form a double-sided (Lorenz-like) attractor.

A first indication of such "glueing" might be seen in the measurement Fig. 13 and 14 taken at 6.5 Pa. Here we see an intensity pulsing
indicating a "period-2". The field measurements, however, reveal that the
field changes sign after each two pulses (notice here as also in Fig. 6
the perfect symmetry of the pulses on either side - best judged by the
intensity traces).

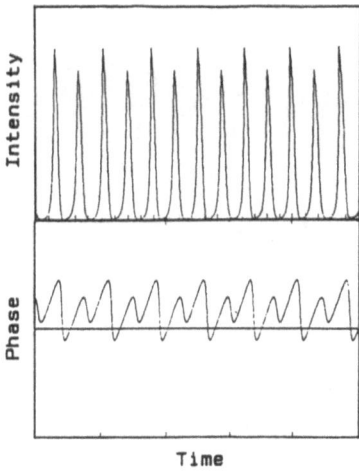

Fig. 10. Intensity and phase for a case corresponding to Fig. 9
calculated from the extended Lorenz equations.

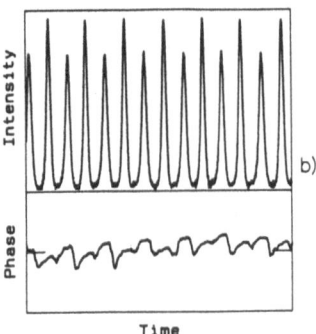

Fig. 11. Intensity and phase
Intensity and phase evaluation
of Fig. 9. Note the similarity
with the theoretical data
of Fig. 10.

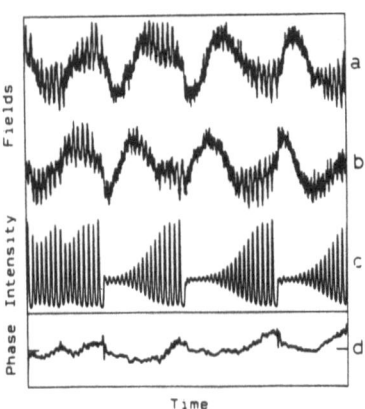

Fig. 12.
Chaotic pulsing of 153 μm
$^{15}NH_3$-laser emission, for
resonant tuning and p=5Pa.
Traces marked as in Fig. 5,
with same time and phase
scales.

The phase evaluation Fig. 14 shows the difference from a symmetric
Lorenz case: the large pulses are accompanied by phase changes of π
(i.e. a change of sign in the laser field) while the small pulses are
accompanied by phase changes of π (i.e. a change of sign in the laser
field) while the small pulses are accompanied by phase changes of ~ π/3.
The occurrence of phase changes other than π shows that the system in
this case has more degrees of freedom than the Lorenz model.

When the laser has one-sided attractors on line center at 5 Pa
pressure, Fig. 13, 14 can be viewed as the "glueing" of two such one-sided
P2 attractors into a double-sided symmetric attractor, which, however,
has still more degrees of freedom than the Lorenz cascade, manifesting
itself in the frequency pushing and -pulling corresponding to the π/3
phase changes.

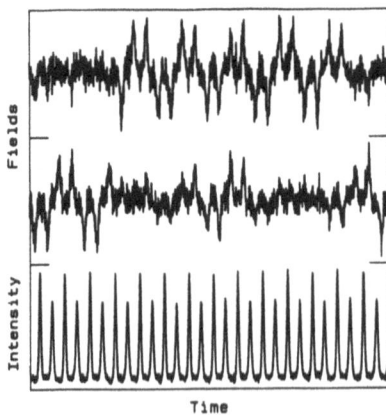

Fig. 13. Electric field components and intensity for central
tuning of the cavity (lower pressure: coherence
effects from pump are present). The laser is shown
to be in a symmetric state performing two loops
on each side of the attractor.

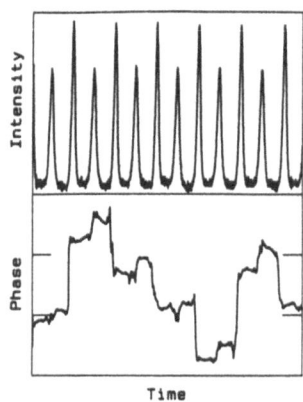

Fig. 14. Intensity and phase for the state given in Fig. 13.

We imagine that as the pressure is increased, and with it the relaxation rates, this allows to adiabatically eliminate degrees of freedom, so that at higher pressure the dynamics reduces to the simple Lorenz model dynamics. We note, concluding, that the techniques of optical field measurement provides a wealth of information inaccessible to simple intensity measurements.

We expect from its utilization, on the NH_3-system only, to map out the bifurcation structure of the range of validity of the (extended) Lorenz equations, and in the range of validity of the NH_3-laser model. Additionally we expect to unravel the details of the transition between these models; all of which will provide exciting opportunities for cross-fertilization between theorists and experimentalists.

More generally we think these examples show that this technique should be generally used in the study of optical instabilities and dynamics, be it in active or passive systems. The cases of optical pattern formation where the phases appears to play a particularly important role[14] would definitely call for optical field measurements.

REFERENCES

1. C.O. Weiss, W. Klische, P.S. Ering, M. Cooper,
    Opt. Comm. 52, 6 (1985)
2. K.J. Siemsen, J. Reid, C.O. Weiss, Opt. Comm. 63,
    415 (1987)
3. T.Q. Wu, C.O. Weiss, Opt. Comm. 61, 337 (1987)
4. C.O. Weiss, W. Klische, Opt. Comm. 51, 47 (1984)
5. E. Lorenz, J. Atm. Sci 20, 130 (1963)
6. H. Haken, Phys. Lett. 53A, 77 (1975)
7. M.A. Dupertuis, R.R.E. Salomaa, M.R. Siegrist,
    Opt. Comm. 57, 410 (1986)
8. See Refs. No. 6 in:
    C.O. Weiss, N.B. Abraham, U. Hübner,
    Phys. Rev. Lett. 61, 1587 (1988)
9. C.O. Weiss, J. Brock, Phys. Rev. Lett. 57, 2804 (1986)
10. F. Laguarta, J. Pujol, R. Vilaseca, R. Corbalan,
    Journal de Physique, Colloque C2, Suppl. No. 6 Tome 49,
    P. 409 (1988)
11. J. Heppner, C.O.Weiss, Appl. Phys. Lett. 33, 590 (1978)

12. J. Heppner, C.O.Weiss, U. Hübner, G. Schinn,
    IEEE Journ. Quant. El. QE-16, 392 (1980)
13. H. Zeghlache, P. Mandel, JOSA B2, 18 (1985)
14. R. Lefever, L.A. Lugiato, W. Kaige, N.B. Abraham,
    P. Mandel, Opt. Comm. 135, 254 (1989)

SHIL'NIKOV CHAOS: HOW TO CHARACTERIZE HOMOCLINIC AND HETEROCLINIC

BEHAVIOUR

F.T. Arecchi, A. Lapucci and R. Meucci

Istituto Nazionale di Ottica, Largo E. Fermi, 6
50125 Firenze, Italy

ABSTRACT

We introduce the concepts of Shil'nikov chaos and competing instabilities in a nonlinear dynamics including at least a saddle focus and a saddle point, a parameter change induces a smooth transition from a homoclinic to a heteroclinic trajectory.
In terms of the return map to a given Poincaré section, the two trajectories have the following characterization. In the homoclinic case, the global behavior is recovered from the local linear dynamics within a unit box around the saddle focus. The heteroclinic case requires the composition of two linearized maps around the two unstable points.
By an exponential transformation the geometrical map yields the return map of the orbital times. This new map represents the most appropriate indicator for experimental situations whenever a symbolic dynamics built on geometric position does not offer a sensitive test. Furthermore the time maps display a large sensitivity to noise. This offers a criterion to discriminate between a simulation (either analog or digital) with a few variables and experiment dealing with the physical variables embedded in the real world and thus acted upon by noise.

1. Theory of Shil'nikov homoclinic chaos[1]

Shil'nikov dynamics corresponds to orbits asymptotic to an unstable saddle focus in at least a 3D space. Limiting to a 3D space let us call $\alpha \pm i\omega$ the pair of complex eigenvalues on the stable ($\alpha < 0$) manifold and $\gamma > 0$ the eigenvalue in the unstable direction orthogonal to the plane.
Let us consider a dynamics where all fixed points are unstable, within a given range of control parameters. We call such situation a regime of competing instabilities[2]. In physical implementations we can adjust[3] the control parameter in order to isolate a non zero set of initial conditions such that all trajectories departing from there approach asymptotically the unstable saddle focus and remain at a finite distance from all other fixed points. In such a case, under the Shil'nikov

Measures of Complexity and Chaos
Edited by N.B. Abraham *et al.*
Plenum Press, New York

condition[1]

$$|\alpha/\gamma| < 1 \qquad (1)$$

the motion becomes chaotic.

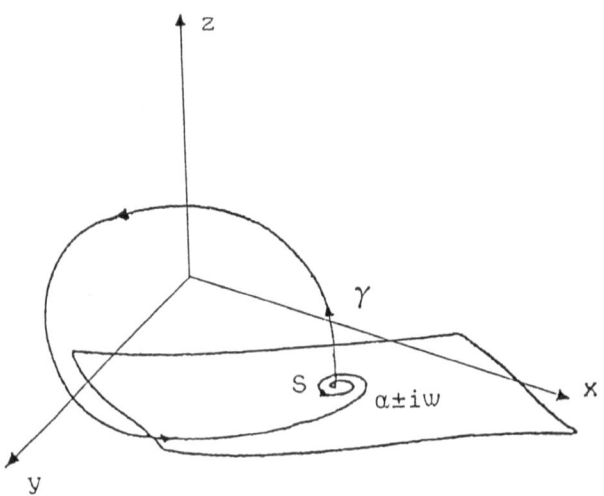

Fig. 1. Schematic rappresentation of a trajectory in Shil'nikov dynamics

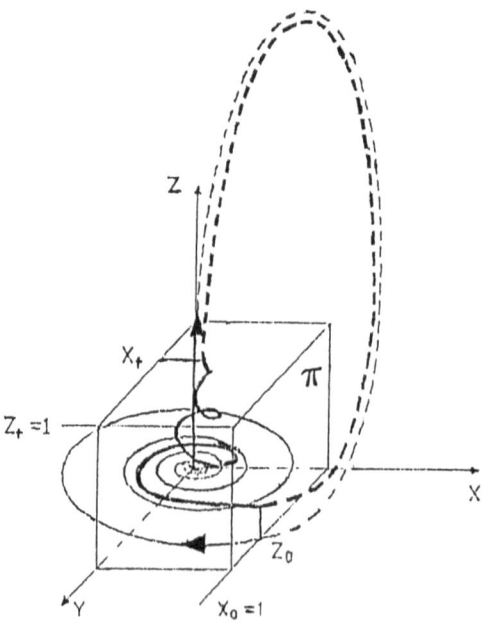

Fig. 2. Construction of unit box leading to the unidimensional map (5)
through the linearization of the flow around the saddle focus.

A single orbit of this type spiralling around an unstable saddle focus S is qualitatively sketched in fig. 1.

With the understanding that the only interesting dynamical features occur around point S we obtain a global description by just studying the linearized dynamics within a small box around S (Fig. 2).

In fig. 2, we orient the three axes along the eigenvectors with x-y coinciding with the stable plane and z being the expanding direction.

We take the Π plane (vertical plane of equation x=1 containing a face of the cube) as the Poincaré section and we calculate the return map for the coordinate z. Starting at t=o at z=o on x=1 (y is irrilevant for the following considerations) the phase point leaves the upper cube side z=1 at time τ such that

$$1 = z_0 \, e^{\gamma \tau} \tag{2}$$

from which it results

$$\tau = -\frac{1}{\gamma} \log z_0 \tag{3}$$

The horizontal coordinate x evolves over the same time as

$$x(\tau) = e^{-\alpha \tau} \cos \omega \tau \tag{4}$$

since the initial condition is x(0)=1. Neglecting a phase shift due to the y position, we constrain the motion external to the box to a rigid translation (see dashed trajectories).

$$x(\tau) \longrightarrow z_1$$

besides an offset ε added at each turn and which may be considered as a second control parameter, the first one being the ratio $|\alpha/\gamma|$.

Using relation (3) and writing z as z_{n+1} and z_0 as z_n we obtain the return map

$$z_{n+1} = z_n^{\alpha/\gamma} \cos\left(\frac{\omega}{\gamma} \log z_n\right) + \varepsilon \tag{5}$$

which describes the homoclinic orbit.

The map (5), even though representing a sensible global description, may provide a poor experimental criterion whenever the z coordinates on the Π plane are clustered in a small region. A lack of experimental sensitivity appears in experimental return maps which do not display the nice features that Eq(5) provides for the theory. Such was the case for the Belouzov Zabotinski reaction[4]. On the other hand, the above behaviour appears rather universal whenever one can isolate a spiral type orbit, as it occurs in Lorenz or Roessler chaos[5].

In dealing with a quantum optical experiment, whose details will be reported on the next section, we introduced a more sensitive dynamical indicator[6,7]. Based on the logarithmic relation between position z on the Π plane and times τ that the orbits take to return to that plane, and assuming that the relevant time is that spent in the box of fig. 2, map(5) transforms via relation(3) into a return map for orbital times. We rescale τ as $T = \delta\tau = -\log z$ and obtain

$$T_{n+1} = -\ln[\exp(-\lambda/\gamma T_n)\cos(\omega/\gamma T_n) + \epsilon] = -\ln[\varphi(T) + \epsilon], \qquad (6)$$

Comparison of Eqs(5) and (6) shows the enhanced sensitivity to fluctuations of the T map with respect to the z map. Indeed, suppose that the offset ϵ from homoclinicity is affected by a small amount of noise. The sensitivities of the two maps to such a noise are given, respectively, by $\partial z/\partial\epsilon = 1$ and

$$\partial T/\partial\epsilon = [\varphi(T) + \epsilon]^{-1}. \qquad (7)$$

This sensitivity factor acts as a lever arm whenever $\varphi(T) + \epsilon$ becomes very small. Note the following: (1) This is not deterministic chaos; in fact, large fluctuations can be expected even for a regular dynamics, implying a fixed point T*, (2) It is not associated with the homoclinicity condition $\epsilon = 0$; in fact, for finite ϵ there may a T* such that $\varphi(T*) + \epsilon = 0$.
Since a homoclinic orbit is the dynamic counterpart of repeated decays out of an unstable state, the result is like repositioning the initial condition in an experiment on a single decay. Here the repetition is automatically provided by the contracting motion asymptotic to the stable manifold. As a consequence, superposed upon the deterministic dynamics (either regular or chaotic), the high sensitivity (Eq(7)) may provide a broadening of the T maps not detectable in the z maps whenever noise in the offset ϵ is present.
In fact, the model description   x=F(x) of a large system in terms of a low-dimensional dynamic variable x is just an ensemble-averaged description, and residual fluctuations on position x must be considered at some initial time, even though the successive evolution is accounted for by a deterministic law. In our case such a fluctuation is a stochastic spread $\delta\epsilon$ on the offset ϵ of the position z .
As shown in Fig. 3, the same amount of $\delta\epsilon$ in Eqs.(5) and (6) leaves the z maps unaltered, while it strongly affects the T maps.
If we specialize the map parameters α , γ, ω , and ϵ to a regular orbit (fixed points both in z and T spaces), introduction of $\delta\epsilon$ does not broaden the z point, while the T point broadens.

For example, the values $\alpha/\gamma = 0.98$, $\omega/\gamma = 2.98$, and $\epsilon = 0.01$ yield one fixed point T* = 5.327, with a sensitivity $\delta T*/\delta\epsilon = 182$ (Fig. 3(c)).
Note that the noise effect reported here has nothing to do with additive noise effects on return maps already described[8]. Indeed, the latter effects refer to the scaling behavior near stationary bifurcations, whereas our data refer to transient fluctuation enhancement, and they do not leave a permanent mark (such as an orbital shift or broadening).
Thus, while Shil'nikov chaos is a deterministic effect described on

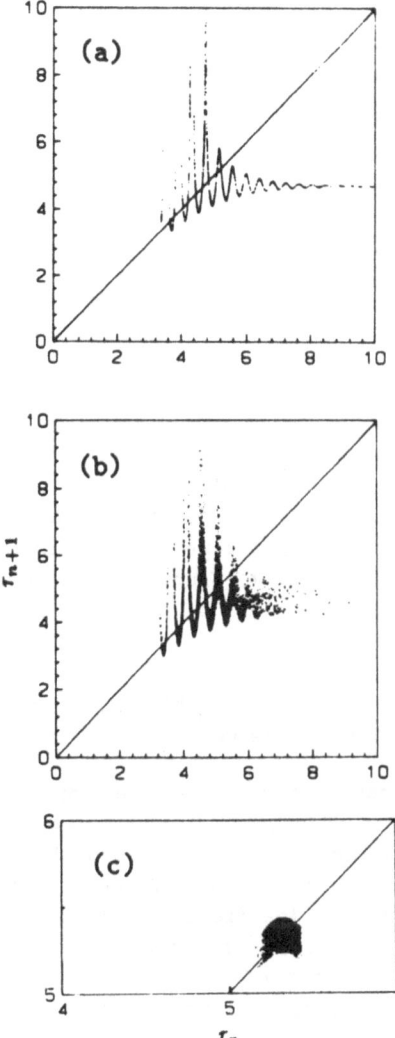

Fig. 3. Numerical iteration maps for Shil'nikov chaos. Parameter values: $\frac{w}{\gamma} = 13.0$, $\alpha/\gamma = 0.986$, $\varepsilon = 0.01$. (a) and (b), T maps without and with noise $\delta\varepsilon = 10^{-2}$, respectively. (c) Stable fixed point of the regular dynamics, broadened by a noise $\delta\varepsilon = 10^{-2}$.

average by the backbone of the z or T maps, the superposed thickening is a noise effect peculiar to T maps and undetectable in z maps. This new effect is a specific indicator of intrinsic fluctuations, and it permists a demarcation line to be drawn between a real-life experiment and a model simulation, from which this second feature is absent.

In order to explore the regular behavior of these closed orbits, we take the fixed part of map(6)

$$T^* = -\ln[\exp(-\lambda/\gamma T^*)\cos(\omega/\gamma T^*) + \epsilon] \tag{8}$$

Eq.(8) gives a stable fixed point, provided Shilnikov condition (i) is violated, that is.

Solving trascendental Eq.(8) and plotting the Poincaré frequency $1/T$ versus the control parameter $\mathcal{E}$, yields two different items, namely

  i) a staircase region implying histeresis cycles
  ii) a logarithmic divergence for small $\mathcal{E}$.

2. Experimental implementation of Shil'nikov dynamics

For pedagogical reasons we have collected in Sec. 1 the main features of Shil'nikov chaos as well as those of a regular (non chaotic) dynamics and here we describe the corresponding experiments. As a matter of fact, things have gone in the reverse order: we first found evidence of spiral type orbits, including large time fluctuations, or regular periods scaling with control parameter as i), ii) of Sec. 1; then we looked in the theoretical literature and found that, using the orbital period as a dynamical indicator more sensitive than Poincaré position, we could nicely describe what was previously treated only at a qualitative level, in terms of a symbolic dynamics coding the number of spirals around the saddle focus[4].

Our experimental setup consists of a single mode CO_2 laser with an intracavity electro-optic modulator. A signal proportional to the laser output intensity is sent back to the electro-optic modulator[9]. Single mode CO_2 lasers have a dynamic behavior described by two coupled differential equations, one for the field amplitude and the other for the population inversion, the fast polarization being adiabatically eliminated from the complete set of Maxwell-Bloch equations. Thus, the presence of feedback introduces a third degree of freedom. When the feedback loop is so fast that it provides a practically instantly adapted loss coefficient, it does not modify the phase-space topology. On other hand, if the time scale of the feedback loop is of the same order as of the other two relevant variables, the system becomes three dimensional. With suitable normalizations such a system is described by three first-order differential equations for the laser intensity $x(t)$, the population inversion $y(t)$, and the modulation voltage $z(t)$ as follows:

$$
\begin{aligned}
x &= -K_0 x \left[1 + \eta\sin^2(z) - y \right], \\
y &= -\gamma_{\parallel}(y + xy - A), \\
z &= -\beta (z - B + r\ x),
\end{aligned} \tag{9}
$$

where $K_0 = (c/L)T$ is the unmodulated cavity-loss parameter, L is the cavity length, T is the effective transmission of the cavity, and γ_N is the population decay rate. The intensity $x(t)$ is normalized to the saturation intensity; the population inversion $y(t)$ is normalized to the threshold inversion $z(t)$ is the modulation voltage normalized to $\pi/V_{\lambda/2}$ with $V_{\lambda/2}$ the $\lambda/2$ modulator voltage; A is the normalized pump parameter; ß is the damping rate of the feedback loop; r is a coupling coefficient between the detected intensity $x(t)$ and the normalized $z(t)$ voltage; B is the bias voltage applied to the electro-optic modulator; and $\eta = (1 - T)/T$. In Fig. 4 we present a schematic view of the trajectory in the three-dimensional space, obtained by a linear stability analysis of the motion around the stationary points, and qualitative connections between the linear manifolds (dashed lines).

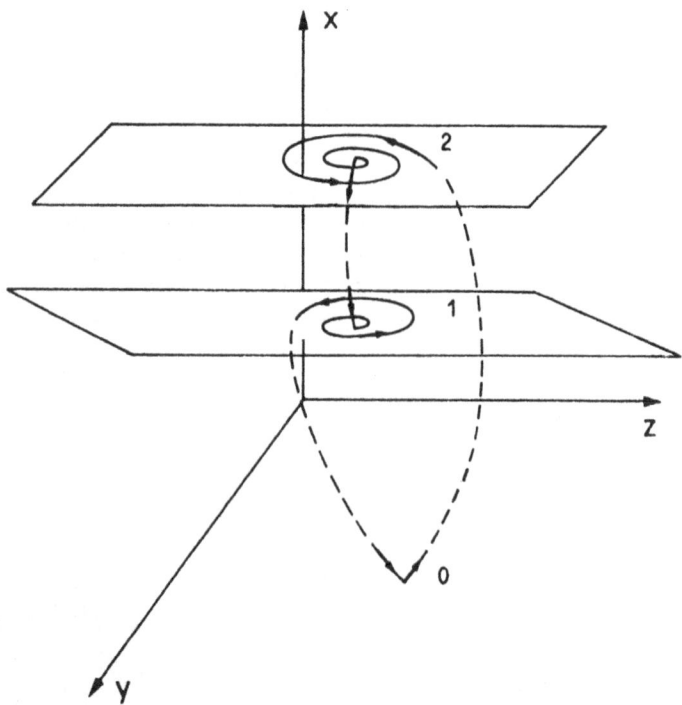

Fig. 4. Schematic view of a trajectory in the phase space when the dynamics are affected by all three unstable stationary points.

From an experimental point of view we are able to visualize $(x - z)$ phase-space projections, obtained by feeding onto a scope the photodetector signal proportional to the laser out-put intensity $x(t)$ and the feedback voltage $z(t)$. This phase-space projection consists of closed orbits visiting successively the neighborhoods of the three unstable stationary points 0,1, and 2.

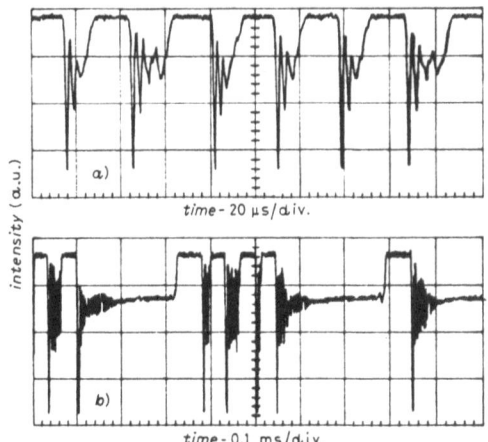

Fig. 5. Time plots of the laser intensity in the regime of Shil'nikov
chaos. a) and b) refer to the same B value (B=0.427), but two
different gains (0.487 and 0.696, respectively) of the feedback
loop. b) shows two long transients corresponding to a large
number of small spirals around the saddle focus.

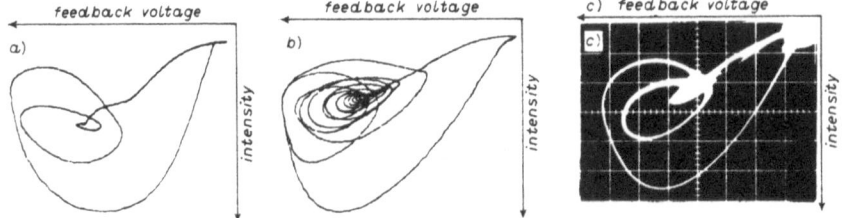

Fig. 6. Phase space projections x–z (laser intensity–feedback voltage).
a) and b) are single orbits obtained by a digitizer, referring to
the same parameters of Fig. 5a) and b), respectively. c) is the
superposition of 30000 orbits of type a).

The local chaos around point 1, established at the end of a subharmonic sequence, has been characterized by standard methods as power spectra and correlation dimension measurements[9].

The competition of the three instabilities in controlling the global features of the motion was described in Ref. 3. There $|\alpha/\gamma|$ was adjusted major then one showing regular behavior and experimental evidence of items i) and ii) of sec. 1. Here we adjust the control parameters in order to have a dominance of the saddle focus, so that the motion consists of a quasi-homoclinic orbit asymptotic to it.

In fig. 5 we report experimental plots of the laser intensity vs. time for two slightly different conditions. Keeping a fixed population inversion (pump 1.5 times above the threshold value) the system has two control parameters, the bias B and the gain r of the amplifier driving the electro-optic modulator.

We have kept B fixed at B=0.380 and increased r from r=0.467 in fig. 5a) to r=0.491 in fig. 5b). Figure 5b) shows clear evidence of a homoclinic orbit in the two long transients, which provide a lengthy permancence in a phase space region of almost constant intensity. This appears more clearly in the corresponding phase space projections (fig. 5a) and b)). For comparison we give in fig. 5c) a photographic exposure (over 1 s) of 30.000 orbits as that of fig. 5a), to show the stability of shape. We see that the first three large oscillations of fig. 5 have strong anharmonic distorsions and display common features over different repetitions, while the small oscillations around the saddle focus display slight differences from pulse to pulse. These latter oscillations are ruled by the linearized dynamics which consists of a contracting spiral exp $-|\alpha|t \cos(\omega t)$ on the stable manifold and of an expansion exp γt along the unstable direction.

Before we discuss comparison with Shil'nikov theory, a crucial question arises: how much of the spread in the return times has to be attributed to point 2 or 0? Indeed, we have quasi-heteroclinic orbits visiting the surroundings of the two unstable points 2 and 0. But in our experimental situation the dynamics can be assimilated to a quasi-homoclinic orbit around point 2, which is thus mainly responsible for the spread in return times. This is easily proved by measuring the spread T in the residence times T_0 around 0 (zero intensity stripes) and the spread T_2 in the residence times T_2 around 2 (complementary stripes, such that $T_0 + T_2$ is the total orbital time). In Fig. 5, which shows typical time sequences used to build the two averaged relative spreads are approximately

$$\langle \Delta T_0 \rangle / \langle T_0 \rangle \sim 14\%, \qquad \langle \Delta T_2 \rangle / \langle T_2 \rangle \sim 80\%,$$

$$\langle \Delta T_0 \rangle / \langle T_0 \rangle \sim 40\%, \qquad \langle \Delta T_2 \rangle / \langle T_2 \rangle \sim 250\%.$$

The comparison shows that point 0 introduces a perturbation around 14% with respect to pure homoclinicity, that is, the orbital regularity is ruled mainly by point 2. Thus a theoretical approach to our experiment in terms of homoclinic chaos appears justified.

We measure the time spacings by setting a threshold circuit near the top of the largest peak of the intensity signal. A time-to-amplitude converter (TAC) yields the sequence T_i of successive time spacings, which is then classified as a statistical distribution by a multichannel pulse height analyser, or stored in a digitizer, so that correlation functions or iteration maps can be sorted out.

The statistical distribution of return times is a broad featureless curve which does not offer cues on the ordering of T_i. On the contrary, the iteration map (T_{i+1} vs. T_i) displays a regular structure (fig. 7a)). To check whether we are in the presence of a one-dimensional (1D) iteration map, and the remaining thickness is due to the observation technique, or the map is more that 1D, we report in fig. 7b) the iteration maps corresponding to three regular situations.

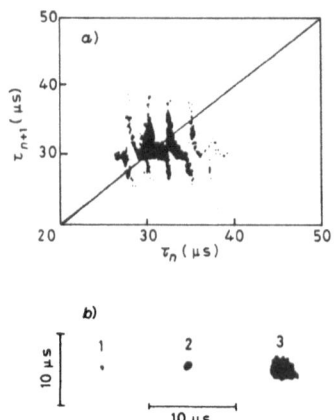

Fig. 7. Experimental iteration maps of the return times. a) refers to r=0.487 and to B=0.350. b) shows the maps corresponding to regular periodic situations, namely, 1) an electronic oscillator, 2) the laser in a regular periodic regime and 3) the laser just at the onset of the instability but still with a regular period.

In the absence of fluctuations in T_i they should be pointlike (the image of a stable fixed point). In fact 1) corresponds to an electronic oscillator and it just shows the resolution of the TAC, 2) corresponds to the laser in a regular periodic regime away from the Shil'nikov instability, and 3) corresponds to the laser on the verge of the instability but still with a regular period. In this last case, the fluctuation associated with the nearby transition shows that, even without chaos in the return time, the close approach to an instability point introduces a fluctuation enhancement, which has no theoretical counterpart in the current treatment of deterministic chaos. To deal with this broadening, the dynamical equations should include a statistical spread in the injection coordinate at the Poincaré section near the saddle focus, to account for the macroscopic character of the experimental system. As it was shown in ref. 10, even though this spread has no relevance on the average dynamics, it contributes a large transient fluctuation whenever the system decays from an unstable point.

3. Toward heteroclinicity: role of points 0 and 1

We have already seen that we can isolate quasi-homoclinic conditions when most of the time fluctuations are due to the residence time around point 2.

An intensity plot seems to approach dangerously the zero-intensity state, however this does not change yet the homoclinic characterization.

More frequently, we have situation where 2 and 1 are very close and the decaying spiral around 2 is followed by a weakly amplifying spiral around 1. Evidence of this is offered by fig. 6b and 5a.

To better emphasize this point we report a new experimental plot (fig. 8a) and a corresponding numerical experiment (fig. 8b).

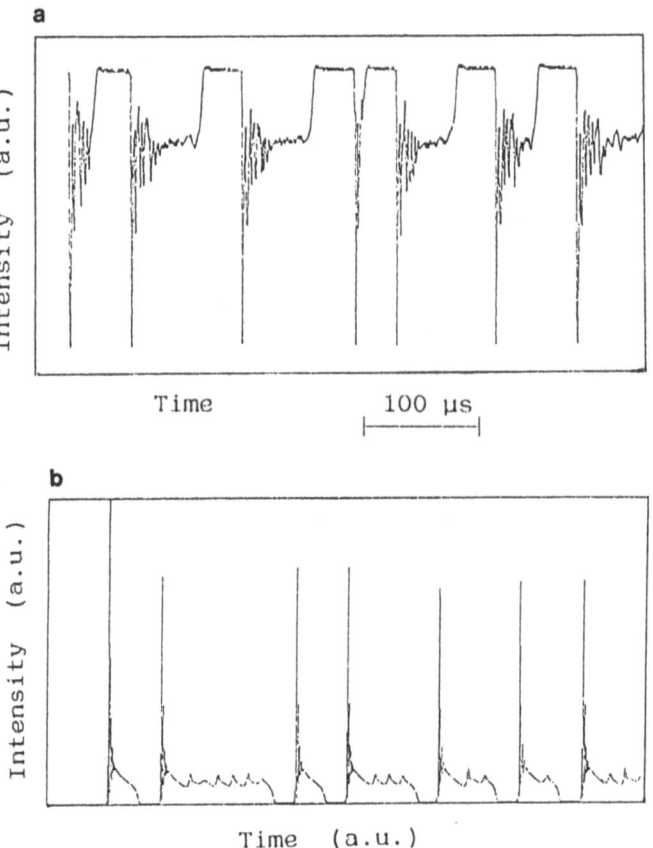

Fig. 8. a) Experimental time evolution of the laser intensity for B=0.433 V=0.558 b) Time evolution in numerical simulation. Parameter values for a 3 level model[13] $K_0=2.0\ 10^7$; $\gamma=6.10^4$; $\gamma'=4.10^6$; $B=3.10^6$; A=.39; $\eta=.2$; B=1.86; r=.603

Fig. 8b shows evidence of a different number of turns around point 1; this effect was already observed by Argoul et al.[4] and renamed "hesitation" by Arimondo et al[11] who have been observing similar phenomena in L.S.A. (Lasers with saturable absorbers). By the way, similar experiments on L.S.A. have been carried also by the group of Lille[12]. Fig. 9 shows the phase space plots of a quasi homoclinic dynamics (rolling only around 2) and a quasi-heteroclinic one (rolling also around 1).

Notice that the numerics of fig. 8b) and 9) were done with a 3 level model.

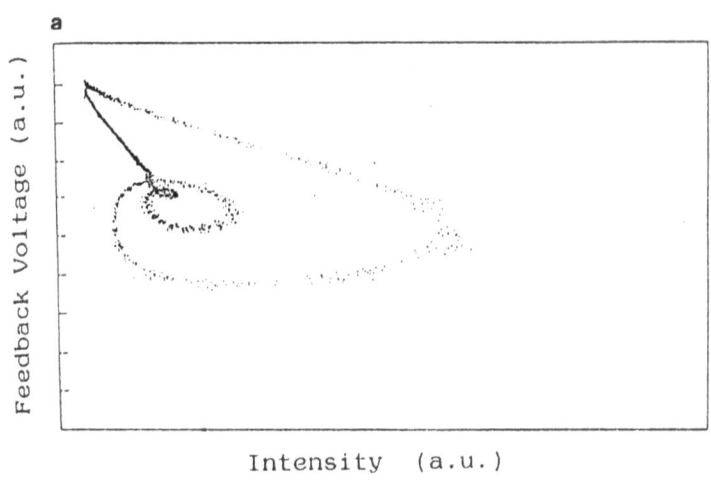

Intensity (a.u.)

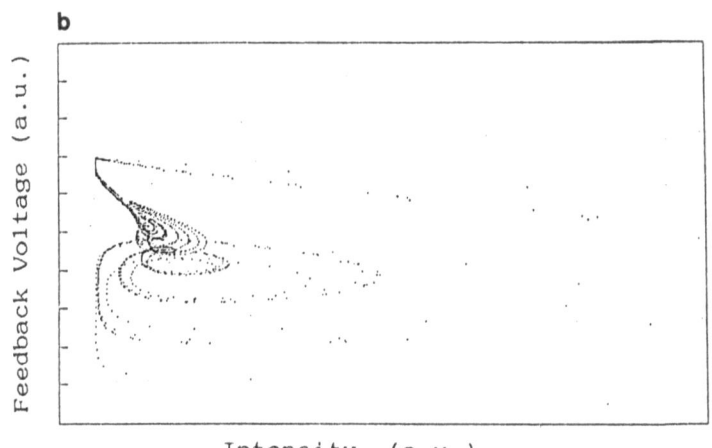

Intensity (a.u.)

Fig. 9. Phase space plots in numerical simulation. Parameters values:
a) $K_0 = 2.10^7$, $\gamma = 6.10^4$, $\gamma' = 6.10^5$, $\beta = 1.3.10^6$, A=3.9, $\eta = .2$, B = 2.0 r = . 5 with the addition of a small amount of noise $\delta B/B=10^{-3}$
b) $K_0 = 2.10^7$, $\delta = 6.10^4$, $\gamma' = 6.10^5$, $\beta = 1.3.10^6$, A = 3.9, $\alpha = .2$ B =1.95, r = .605.

A 3 level model for CO_2 molecular introduced by Shimizu et al[13] was used by the two groups working on L.S.A. (Ref. 11,12 plus private communication). In our case, we have tested both 2 and 3 level models but in our time range we have not found relevant differences. This may be due to the fact that the feedback dynamics goes over time of the order of 10-100 us and this time range is not sensitive to the further decays introduced by the 3 level model.

4. Conclusions

We conclude stressing the universality of the phenomena here reported. The spiralling around foci and associated phenomena should occur in all Lorenz type or Roessler type dynamics. Evidence of it is in the recent report on Lorenz chaos[14] as well as in the above reports on B-Z chemical instabilities[4].

The best assessments are obtained by time maps as said above. Time maps are now being used also by other investigators[15].

The technique of a local linear dynamics to get a global description can be extended to the heteroclinic case, in terms of a composition of two linearized maps around the two unstable points. A preliminary investigation of this kind has lead us to a conjecture on stabilization of orbital times in heteroclinic dynamics which seems relevant for biological clocks. This subject is developed elsewhere[16].

REFERENCES

1. L.P. Shil'nikov, Dokl. Akad. Nauk SSSR **160**, 558 (1965); L.P. Shil'nikov, Mat. Sb. **77**, 119, 461 (1968); **81**, 92, 123 (1970).

2. A. Arneodo, P.H. Coullet, E.A. Spiegel, and C. Tresser, Physica **14D**, 327 (1985).

3. F.T. Arecchi, R. Meucci and W. Gadomski, Phys. Rev. Lett. **58**, 2205 (1987).

4. F. Argoul, A. Arneodo, and P. Richetti, Phys. Lett. **A120**, 269 (1987).

5. P. Glendinning and C. Sparrow, J. Stat. Phys. **35**, 645 (1984); P. Gaspard, R. Kapral and G. Nicolis, J. Stat. Phys. **35**, 697 (1984).

6. F.T. Arecchi, A. Lapucci, R. Meucci, J.A. Roversi, and P. Coullet, i) ELITE Meeting, Torino, Italy, march 1987.
   ii) International Workshop on Instabilities and Chaos in Nonlinear Optical systems, Il Ciocco, Italy, July 1987 paper WC 13-2.
   iii) Europhys. Lett. **6**, 677 (1988).

7. F.T. Arecchi, W. Gadomski, A. Lapucci, R. Meucci, H. Mancini, and J.A. Roversi, JOSA **B5**, 1153 (1988).

8. J.P. Crutchfield, D. Farmer, and B.A. Hubermann, Phys. Rev. **92**, 45 (1982).

9. F.T. Arecchi, W. Gadomski, and R. Meucci, Phys. Rev. A **34**, 1617 (1986).

10. F.T. Arecchi, V. Degiorgio, and B. Querzola, Phys. Rev. Lett. **19**, 168 (1967);
    F.T. Arecchi, and A. Politi, and L. Ulivi, Nuovo Cimento **71B**, 119, (1982).

11. D. Hennequin, F. De Tomasi, B. Zambon, and E. Arimondo, Phys. Rev. A **37**, 243 (1988).

12. D. Dangoisse, A. Bekkali, F. Papoff, and P. Glorieux, Europhys. Lett. **6**, 335 (1988).

13. M. Tachikawa, K. Tanii, and T. Shimizu, J.O.S.A. **B 5**, 1077 (1988).

14. C.O. Weiss, N.B. Abraham, and U. Hubner, Phys. Rev. Lett. **61**, 1587, (1988).

15. F. Papoff, A. Fioretti, E. Arimondo, and N.B. Abraham, "Return time and distribution in the laser with saturable absorber", this same Conference

16. F.T. Arecchi, A. Lapucci, and R. Meucci, to be published.

TIME SERIES NEAR CODIMENSION TWO GLOBAL BIFURCATIONS

Paul Glendinning

Department of Applied Mathematics and Theoretical Physics

Silver Street, Cambridge CB3 9EW, U.K.

INTRODUCTION

For continuous maps of the interval the transition to chaos (at least, in the sense of positive topological entropy) is always via the generation of periodic orbits of with periods which are powers of two. Discontinuous maps of the interval allow considerably more freedom in the choice of a route to chaos, think of circle maps as maps of the interval, for example. Discontinuous maps of the interval arise naturally in the study of certain classes of flows, and here we wish to outline some of the phenomena associated with such maps and the corresponding flows they are intended to model. In particular, we shall concentrate on two sequences of bifurcations of periodic orbits. In the first sequence, described in section 3, there are bifurcations from period p_n to period p_n+1, ($p_0=1$) accumulating at some value of the parameter in the problem. The second sequence (section 4) is more complicated; it is defined by $p_{4(n+1)}=4p_{4n}+2$, with $p_0=1$, $p_{4n+1}=p_{4n}+1$, $p_{4n+2}=2p_{4n+1}$ and $p_{4n+3}=p_{4n+2}-1$ (so $p_{4(n+1)}=2p_{4n+3}$). There is a fundamental difference between these two sequences: the first is not really part of a transition to chaos whilst the accumulation point of the second sequence in parameter space is a new codimension two point on the boundary of chaos (although the precise sequence itself is not codimension two). This is only one example of the possible anharmonic routes to chaos which can arise in the classes of differential equations defined in the next section. Unfortunately, I know of no examples to date.

The work described below ia a précis of a series of papers[1,2,3,4] written in collaboration with J.M. Gambaudo (Nice), J.E. Los (Nice), M.V. Otero-Espinar (Santiago de Compostella), D.A. Rand (Warwick) and C. Tresser (Tuscon).

DIFFERENTIAL EQUATIONS AND MAPS

Consider differential equations of the form

$$dx/dt = ax + P(x,y,z;\mu,\nu)$$
$$dy/dt = -by + Q(x,y,z;\mu,\nu)$$
$$dz/dt = -cz + R(x,y,z;\mu,\nu)$$

Measures of Complexity and Chaos
Edited by N.B. Abraham *et al.*
Plenum Press, New York

where the functions P, Q and R represent nonlinear terms in x,y and z and μ and ν are real parameters. We shall assume further that

(A1) $b > c > a > 0$;

(A2) for $\nu=0$ and μ small the differential equation has a homoclinic orbit, $\Gamma_0(t;\mu)$, biasymptotic to the origin, which is a stationary point of the flow, such that $\pi_x\Gamma_0(t;\mu)$ tends to 0 from below as t tends to $-\infty$, $\pi_z\Gamma_0(t;\mu)$ tends to 0 from above as t tends to $+\infty$ (here, $\pi_x\Gamma_0(t;\mu)$ simply denotes the x-coordinate of $\Gamma_0(t;\mu)$, etc....) and $\Gamma_0(t;\mu)$ is tangent to the z-axis as t tends to $+\infty$;

(A3) for $\mu=0$ and ν small the differential equation has a homoclinic orbit, $\Gamma_1(t;\nu)$, biasymptotic to the origin, such that $\pi_x\Gamma_1(t;\nu)$ tends to 0 from above as t tends to $-\infty$, $\pi_z\Gamma_1(t;\nu)$ tends to 0 from above as t tends to $+\infty$ and $\Gamma_1(t;\nu)$ is tangent to the z-axis as t tends to $+\infty$.

Note that if (A1) is replaced by $b > c > a > 0$ this is the same situation as in the Lorenz equations.

After some messy analysis[1] it can be shown rigorously that for generic systems of this type there exists an open neighbourhood of $(\mu,\nu)=(0,0)$ and a small tubular neighbourhood of the homoclinic orbits at $(\mu,\nu)=(0,0)$ for which trajectories are modelled by a one-dimensional map. (Strictly speaking we show that there is a return plane on which the flows define a stable foliation.) To leading order this map is given by

$$F_{\mu\nu}(x) = \begin{cases} -\mu + Ax^q & \text{if } x > 0 \\ \\ \nu + B|x|^q & \text{if } x < 0 \end{cases}$$

where $|x|$ is small (so that the slope of the map is less than one in modulus) and $q = c/a > 1$. The map is undefined at $x=0$, corresponding to trajectories which tend directly to the origin, and an iteration in $x>0$ represents a turn of the trajectory near $\Gamma_1(t;0)$ whilst an iteration in $x<0$ represents a turn of the trajectory near $\Gamma_0(t;0)$. The signs of the constants A and B give information about the geometry of the flow near the homoclinic orbits. Here we shall concentrate on the case $A<0$, $B<0$; other combinations of the signs of A and B can be found elsewhere.

From earlier results[2] we know that, in the range of validity of this return map, the o.d.e.s they model can have at most two periodic orbits and that any periodic orbits which do exist (in the small neighbourhoods of validity of these arguments) are stable. Also, if orbits are labelled by sequences of 0s and 1s corresponding to iterations in $x<0$ and $x>0$ respectively, that the codes obtained for these orbits must have a very particular (rotation-compatible) form. However, these results do not tell us which orbits are realised in a given class of system, so it is this problem that we move on to, obtaining bifurcation diagrams near the codimension two point $(\mu,\nu)=(0,0)$ for the case modelled by the return map with $A<0$, $B<0$.

THE LOCAL STRUCTURE[1]

The procedure described above leaves us with a two-parameter family of maps $F_{\mu\nu}$ which are increasing in $x < 0$ and decreasing in $x > 0$ and, in the region of validity of the analysis, with slopes of modulus less than one. This slope condition implies that any periodic orbit which exists is stable, but which periodic orbits can exist?

Theorem

Let V be a sufficiently small neighbourhood of $(\mu,\nu)=(0,0)$ in parameter space and U a tubular neighbourhood of the homoclinic orbits which exists at $(\mu,\nu)=(0,0)$. Then V can be divided into non-intersecting open regions N_n (n=0,1,2,...) and M_m (m=0,1,2,...) with $V = (Ucl(N_n)) \cup (Ucl(M_m))$ such that the differential equation modelled by $F_{\mu\nu}$ with A,B<0 has

a single periodic orbit in U with code 10^{n-1} if $(\mu,\nu) \in N_n$, $n\neq0,1$;

two periodic orbits in U, one with code 10^{n-1}, the other with code 10^n, if $(\mu,\nu) \in M_m$, $m\neq0,1$;

a single periodic orbit in U with code 1 if $(\mu,\nu) \in N_1$;

a single periodic orbit in U with code 0 if $(\mu,\nu) \in N_0$;

two periodic orbits in U, one with code 1, the other with code 10, if $(\mu,\nu) \in M_1$;

two periodic orbits in U, one with code 1, the other with code 0, if $(\mu,\nu) \in M_0$.

On the boundaries between these regions the o.d.e.s have more complicated homoclinic orbits, which create or destroy the relevant periodic orbits.

This situation is illustrated in Fig. 1a, Fig. 1b shows the situation in the case A < 0, B > 0[1]. The proof of this theorem is relatively straightforward once the messy analysis alluded to in section 2 has been done, since it then becomes possible to work with the maps rather than the flows. The sequence of bifurcations described in the introduction from period p to period p+1 should now be obvious: simply take a path with $\mu > 0$, constant and let ν decrease from N_2 through to N_0. From Fig. 1a we see that this path in parameter space passes through an infinite sequence of bifurcations accumulating on the μ-axis, and with no chaos.

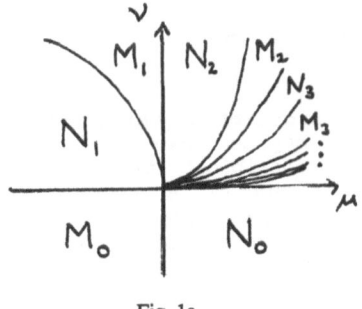

Fig. 1a

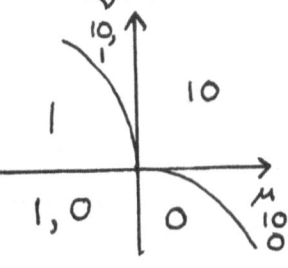

Fig. 1b

ROUTES TO CHAOS

Now let us suppose (without rigorous justification) that the stable foliation of section two, which allows the rigorous reduction to the one-dimensional maps, persists under larger perturbations of the parameters away from zero and for larger values of x. Thus we are led to consider the same families of one-dimensional maps but now we allow the slopes to become greater than one in modulus outside some neighbourhood of x=0. A full description of the transitions to chaos which arise in such maps is beyond the scope of this note[3,4], so we shall concentrate one one particular aspect: the point on the boundary of chaos on the ν-axis (i.e. with $\mu=0$). This is clearly a very degenerate point to consider since at $\mu=0$ the system has a homoclinic orbit, but it will serve to illustrate a number of points.

Consider moving up the ν-axis in Fig. 1a. We begin with a periodic orbit with code 0 and the homoclinic orbit with code 1 and at $(\mu,\nu)=(0,0)$ the periodic orbit with code 0 becomes homoclinic, giving us a pair of homoclinic orbits, with codes (1,0). Just above (0,0) we have lost the orbit with code 0 and have an orbit with code 10. After a little thought we find that the next interesting thing to happen is that the third iterate of 0 approached from below equals 0, i.e. we have a pair of homoclinic orbits for the flow again, but this time with codes (1,010). Consider the induced map $G_{\mu\nu}(x) = F_{\mu\nu}(x)$ if x>0 or $F^3_{\mu\nu}(x)$ if x<0. For some small neighbourhood of x=0 this induced map is like $F_{\mu\nu}(x)$ with A<0 and B>0, so we obtain part of the picture in Fig. 1b, but with the symbol 0 replaced by the symbols 010.

Hence just below the point at which the homoclinic pair (1,010) exist there is a periodic orbit with code 1010 (presumably obtained from the orbit 10 by period-doubling) and just above this point there is an orbit with code 010. Moving further up we find that this orbit period-doubles, giving an orbit with code 010010, which then becomes homoclinic, giving a pair of homoclinic orbits with codes (1,010010) at the parameter value $(0,\nu_1)$. Now consider the induced map defined on a neighbourhood of x=0 by $H_{\mu\nu}(x) = F_{\mu\nu}(x)$ if x > 0 or $F^6_{\mu\nu}(x)$ if x < 0. It is easy to see that the left hand branch of this map is orientation preserving, so at $(0,\nu_1)$ we are back in the same position as we were at (0,0) except with the induced map $H_{\mu\nu}$. Thus the whole process repeats infinitely often, giving the sequence of bifurcations described in the introduction, accumulating at some point $(0,\nu_\infty)$. Above this point the map is chaotic, in the sense of having positive topological entropy.

The precise sequence described here is clearly very unlikely to be observed in numerical simulations or experiment, but the point $(0,\nu_\infty)$ is on a point of codimension two on the boundary of chaos where intermediate bifurcations are complicated by the considerations of the previous section (note that all the local structure described in section three repeats off the axis near $(0,\nu_1)$, ...).

This discussion was not intended as a proof of anything, only to suggest certain new types of behaviour which may be observable. A full discussion of the maps described above is only possible by taking all the maps (with various combinations of the signs of A and B) into consideration[3,4,5,6]. For further details of the different types of transitions to chaos together with the proofs of the statements above the reader is referred to the literature.

REFERENCES

1. J.M. Gambaudo, P. Glendinning, D.A. Rand and C. Tresser (1987) preprint. (See also J.M. Gambaudo (1987), Thesis, Université de Nice, and P. Glendinning (1985), Fellowship dissertation, University of Cambridge.)
2. J.M. Gambaudo, P. Glendinning and C. Tresser (1988) Nonlinearity, 1, 203.
3. P. Glendinning, J.E. Los, M.V. Otero-Espinar and C. Tresser (1989) A new concept in renormalisation group theory for dynamical systems, preprint.
4. P. Glendinning, J.E. Los, M.V. Otero-Espinar and C. Tresser (1989), in preparation
5. C. Tresser (1984) C.R. Acad. Sci. Serie 1, 299, 253.
6. I. Procaccia, S. Thomae and C. Tresser (1987) Phys. Rev. A, 35, 1884.

CHARACTERIZATION OF SHILNIKOV CHAOS

IN A CO$_2$ LASER CONTAINING A SATURABLE ABSORBER

D. HENNEQUIN, M. LEFRANC, A. BEKKALI , D.DANGOISSE and P. GLORIEUX

Laboratoire de Spectroscopie Hertzienne, associé au C.N.R.S.

Université des Sciences et Techniques de Lille

F-59655 Villeneuve d'Ascq Cédex (France)

Studying instabilities on lasers presents several interests: (i)   lasers are used in many devices and it is important to be able to predict their dynamics. (ii) the time scale of experiments on most lasers is typically in the microsecond range and allows easy and fast exploration of wide ranges of control parameters. (iii) the modelization of lasers has been the purpose of many works for a long time and provides a good basis for theoretical studies. The Laser containing a Saturable Absorber (LSA) has recently regained interest because this laser presents a very large variety of dynamical behaviors. Two years ago, it was shown that its instability usually called Passive Q-Switching (PQS) was in fact made of different dynamical behaviors. The laser destabilizes either through a supercritical Hopf bifurcation and period-doubles to eventually reach an erratic behavior, or it exhibits undamped undulations which can be explained using mechanisms first analyzed by Shil'nikov in the 1960's. In fact the situation in the LSA is complicated by the existence of bistability between steady-states and time dependent regimes[1] and consequently a special attention has to be paid to the hysteresis (i) in the experiments where it can be detected by making large sweeps of the control parameters and (ii) in the theoretical simulations where the limitations of the linear stability analysis should be kept in mind.

The Shil'nikov dynamics occur when a system has an unstable steady-state which has the properties of a saddle-focus. Shil'nikov proved that if e.g. the diverging rate λ of the saddle exceeds the attracting rate ρ of the focus ($|\rho/\lambda|<1$), chaos occurs if there exists an homoclinic orbit. The particular scenario of these dynamics has been studied in detail[2] and resembles very much to that observed in the experiments[3]. However there remains differences between the Shil'nikov case and the laser studied here: (i) because of bistability, the point around which the trajectory is spiraling is stable, (ii) in our laser there is another unstable steady-state that can perturb the dynamics and compete with Shil'nikov's saddle-focus, (iii) our system is subjected to technical noise. A

Measures of Complexity and Chaos
Edited by N.B. Abraham *et al.*
Plenum Press, New York

quantitative characterization of chaos is thus required to check whether its dynamical behavior may be considered as Shil'nikov chaos or not.

In that purpose, we have applied to experimental signals and numerical simulations of the LSA a method recently proposed by Argoul *et al.* [4] These authors have shown that it is possible to build a multibranched 1-D map to describe spiral-type or screw-type homoclinic chaos. The procedure is the following one: rebuild the attractor in the phase space, draw a 2-D Poincaré section by intersecting the chaotic attractor with a plane transverse to the trajectory in the reinjection loop, parametrize the distance along that curve by a coordinate X and eventually define a 1-D map by plotting X_{i+1} versus X_i. They introduce a simple coding of spiral-type homoclinic chaos which consists in counting the number of turns of the trajectory around the saddle-focus and the return map of the X_i's should display a series of branches, each corresponding to a particular number of turns.

This method has first been checked on the experimental results on the LSA which was described in details elsewhere[3]. Because of noise it has not been possible to explore the region very close to homoclinicity. The closest situation that could be reached is that in which the laser operates in a chaotic regime coming from a two-undulation periodic regime and leading to the three-undulation periodic regime. Such a chaotic regime is made of an erratic sequence of pulses with less than three undulations. Figure 1 represents the reconstructed attractor, the Poincaré section and the corresponding 1-D return map which displays a set of three branches corresponding to 2, 3 or 4 turns. These results look strikingly similar to those obtained by Argoul *et al.* on the

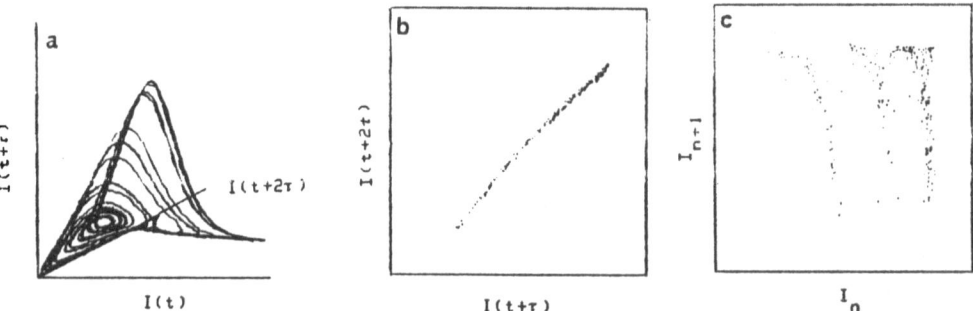

Figure 1   Illustration of the dynamics of the LSA in a chaotic regime of hesitation type.(a) reconstructed attractor from experimental time series(b)Poincaré section (c)return map. All quantities are in arbitrary units.

Belousov-Zhabotinski reaction[4.] The existence of such a mapping indicates the possibility of erratic but deterministic motion composed of orbits with different numbers of turns. The last branch is blurred by noise and does not allow to determine accurately the convergence law.

Numerical simulations of this experiment have been performed on the basis of the model of the LSA which has been shown to be in very good agreement with the experimental findings. It is basically Tachikawa's model for the LSA[5] in which it has been assumed that the absorber is very fast. The corresponding laser equations write as follows:

$$
\begin{aligned}
\dot{I} &= I(U-1) - \frac{I\overline{A}}{1+aI} \\
\dot{U} &= \varepsilon(W - U(1+I)) \\
\dot{W} &= \varepsilon(A + bU - W)
\end{aligned}
$$

where I is the laser intensity, U is the scaled population inversion in the active medium with associated damping rate ε and W is a variable representing the source term in the three-level model. In this fast absorber model, the passive cell only introduces intensity dependent losses represented by a scaled absorption $\overline{A}$ and a relative saturability a. A is the laser gain, b depends on the ratio of relaxation parameters and is fixed hereafter at a value of 0.85 as in [6]. Dots refer to derivatives with respect to time in units of the cavity damping time.

Numerical solutions of these equations with a set of parameters corresponding to our experiments (A=1.9, $\overline{A}$ =2.16, a=4.17, ε=0.137 and b=0.85) provide the results illustrated on Figure 2 for the attractor, the Poincaré section and the 1-D map . They are in excellent agreement with those of Figure 1 and we

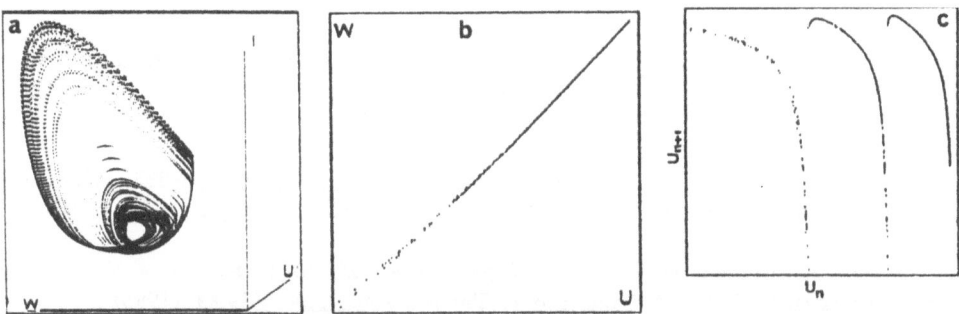

Figure 2   Results of the numerical simulation of the LSA with parameters corresponding to Figure 1.(a) attractor in the (I,U,W) phase space (b)Poincaré section (c)return map.

have explored other sets of parameters to reach situations closer to homo or heteroclinicity, as illustrated on Figure 3 where quantities similar to those of previous figures are reported for the same set of parameters of fig. 2 except for A=1.997. In the two cases, there is bistability between the cw regime and the chaotic one, because of the presence around the stable focus of an unstable cycle. This could be suppressed by a small shift of the system in the parameter space leading to the coincidence of the Hopf and the homoclinic bifurcation, which ensures the exact Shil'nikov case. Otherwise, it exists a saddle point corresponding to zero intensity. In the case of fig. 3,, we find that the return map displays a large number of branches which converge geometrically as the number of turns increases.   Both our experimental results and our simulations indicate that the dynamics of the LSA, though not in a pure Shil'nikov case, have properties which are very similar to those predicted in that case. This is consistent with Tresser's result on the heteroclinic situation in which he

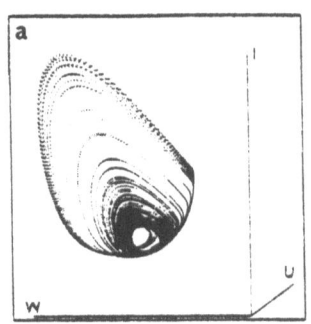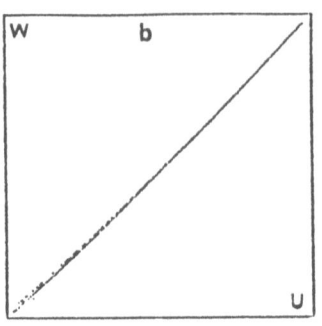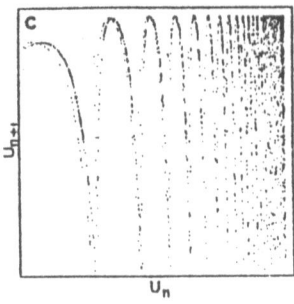

Figure 3   Results of the numerical simulation of the LSA close to heteroclinicity. (a) attractor in the (I,U,W) phase space(b)Poincaré section (c)return map.

showed that the dynamics result essentially the same if the product of the divergence rates exceeds that of the convergent rates of the two unstable points connected by the heteroclinic orbit[7]. Recently Arecchi et al proposed to draw the return map of return times[8]; this method is similar in its principle to that used here and further checks in our experiments and numerical simulations would be useful to confirm that the dynamics of our laser is also of the Shil'nikov type according to this criterium.

References

1. A. Jacques and P. Glorieux, Opt. Comm. 40:455 (1982) E. Arimondo, F. Casagrande, L.A. Lugiato and P. Glorieux, Appl. Phys. B30:57 (1983)
2. P. Glenndinning and C. Sparrow, J. Stat. Phys. 35:645 (1984) P. Gaspard, R. Kapral, and G. Nicolis, J. Stat. Phys. 35:697 (1984)
3. D. Dangoisse, A. Bekkali, F. Papoff and P. Glorieux, Eur. Phys. Lett. 6:335, (1988) D. Hennequin, F. De Tomasi, B. Zambon and E. Arimondo, Phys. Rev. A 37:2243 (1988)
4. F. Argoul, A. Arneodo and P. Richetti, Phys. Lett.A 120:269 (1988)
5. M. Tachikawa, K. Tanii, M. Tajita and T. Shimizu, Appl. Phys B39:83 (1986)
6. A. Bekkali, Thèse de Doctorat, Université de Lille (1989)
7. C. Tresser, Ann. Inst. H. Poincaré 40:441 (1984)
8. F.T. Arecchi, W. Gadomski, A. Lapucci, H. Mancini, R. Meucci and J.A. Roversi, J. Opt. Soc. Am.B5:1153 (1988)

SYMMETRY BREAKING HOMOCLINIC CHAOS

Basil Nicolaenko   Zhen-Su She

Center for Nonlinear Studies   Applied Mathematics Program
Los Alamos National Laboratory   Fine Hall
Los Alamos, NM   87545   Princeton University
     and   Princeton, NJ   08544
Department of Mathematics
Arizona State University
Tempe, AZ   85287

The essentially spatio-temporal dynamics of extended physical systems is still poorly understood, Indeed, in spite of the success of dynamical systems theory to explain the *deterministic chaos* occurring in confined situations, this approach is unable to handle the complex behaviors appearing when the spatial structure is not frozen. From this point of view, *turbulence* may be seen as the ultimate complex system, and it seems interesting to study models of extended systems somewhat "simpler" than the Navier-Stokes equations but nevertheless retaining some of the essential physics of the problem. In such models (one-dimensional P.D.E.'s, cellular automata), a transition to turbulence via *spatio-temporal intermittency* is often observed [1,2,3]. Such regimes are characterized by the coexistence of clusters of laminar cells and groups of coherent cells within a sea of turbulent patches. Now, a most intriguing problem in the theory of hydrodynamic turbulence is the formation of large-scale, coherent structures (C.S.) in a flow performing random turbulent motion of small scales. The C.S. are important in transport of heat, mass and momentum, and they are responsible for the intermittent nature of turbulence in many shear and boundary layer flows. Is it possible that the dynamics of C.S. are related to spatio-temporal intermittences observed in simpler, lower-dimensional systems?

Perhaps the simplest Navier-Stokes flow vested with intermittent C.S. in turbulent regimes is the Kolmogorov flow [4]; this is the two-dimensional flow of a viscous liquid induced by a unidirectional force field periodic in one of the coordinates:

$$\frac{\partial u}{\partial t} + u.\nabla u + \nabla p = \frac{1}{Re}\nabla u + F,$$

$$\text{div } u = 0, \quad 0 \le x, y \le L, \quad F = \gamma(\cos k_c y, \, 0),$$

(1)

Measures of Complexity and Chaos
Edited by N.B. Abraham *et al.*
Plenum Press, New York

together with periodic boundary conditions. For this flow, the generalized system of small-scale eddies turns out to be unstable to long-wave instabilities. The Kolmogorov flow is closely related to special A-B-C (Arnold-Beltrami-Childress) flows:

$$\frac{\partial u}{\partial t} + u.\nabla u + \nabla p = \frac{1}{Re}\nabla u + \mathbf{F},$$

$$\text{div } u = 0, \quad 0 \leq x, y, z \leq L, \quad \mathbf{F} = \gamma(0, \cos k_c x, \sin k_c y).$$

$$(2)$$

The basic laminar flows are shear Beltrami flows: $u = Re/k_c^2 \gamma(0, \cos k_c x, \sin k_c y)$, which are exact solutions of the Euler Equations. Frisch et al. [5] have demonstrated that the 3-D flow (2) possesses the very same linear dispension law as the 2-D Kolmogorov flows. Another key properties of such flows are their groups of symmetries; for the 2-D Kolmogorov flow:

(i) 0(2) equivariance in x, that is, arbitrary shift invariance and invariance of the stream function under $x \to -x$.

(ii) discrete 0(2) equivariance in y (discrete modulo $2\pi/k_c$ shifts in y). We denote this group by $0_D(2)$.

We suspect that *symmetry breaking* (especially parity breaking) might play an important role in the dynamics of the C.S.

We have systematically investigated the dynamics of Kolmogorov flows, for forcing up to $k_c = 16$, and Reynolds Numbers up to Re = 100, with 256×256 spectral resolution. Our principal results evidence purely dynamical systems mechanisms for the intermittent coherent structures:

Proposition. *For the above regimes of the Kolmogorov flow, intermittent C.S. within homogeneous turbulence are generated by persistent (in Re) homoclinic cycles. These cycles are the direct sum of heteroclinic connections between multiple hyperbolic states (fixed points, toris). These hyperbolic states are mapped into each other by some discrete subgroups $\Gamma \subset 0(2) \times 0_D(2)$. The corresponding homoclinic cycles corresponds to symmetry (parity) breaking of the Kolmogorov flow. The homoclinic turbulent pulses are associated with strong large-scale eddy pairs or triplets, with localized vorticity explosion.*

Figure 1 highlights the typical temporal dynamics of the Fourier modes. Figure 2 reflects the corresponding spatio-temporal organization in the physical space. In Figure 1, the evolution of the imaginary component of the mode $(k_x, k_y) = (4, 0)$ is traced for Re = 50, $k_c = 8$. It is important to notice that the mode $(4, 0)$ is not a subharmonic of the forcing mode $(0, 8)$. The dynamics follow a long plateau, then undergo a strong turbulent (chaotic) explosion; then seem to settle down to a plateau at the same level as before; then another explosion follows. Intervals between explosions are not *constant* and fluctuate randomly around some mean value. Careful study of other modes evidence that:

304

- the dynamics of the "plateau" states are in fact chaotic, but on spatio-temporal scales different from the explosive regimes.

- successive "plateaus" *do not correspond to identical dynamical states*, but rather to sequences of states S_1, S_2, S_3 ... equivalent under some subgroup action Γ. This is demonstrated by a phase analysis of the complex Fourier coefficients.

- Total vorticity during the homoclinic "explosion" increases by $O(10^2)$ at least.

Figure 2 shows a typical spatial configuration of turbulence *during a homoclinic explosion*. Contour levels are isovorticity contours. Strongly coupled localized peaks of vorticity are observed. The homoclinic explosion corresponds to temporally intermittent, spatially localized concentration of vorticity. This mirrors eddy pairs or triplets of the same polarity, with strong shearing and intense lateral vorticity transport between the closely packed large-scale eddies. This is well-known experimentally [6], but hitherto unexplained in terms of chaotic dynamical mechanisms. To our knowledge, this is the first time that intermittent large-scale eddy pairing is related to homoclinic cycles generating chaos.

The homoclinic cycles generate two turbulent regimes: first, the trajectories remain topologically close to the hyperbolic states S_i (yet are chaotic on their own scale). Second, the explosions ("pulses", "excursions") from S_i to S_{i+1} reflect a strong enstrophy pulse toward small scales, followed by inverse cascade of energy. They mirror intermittent strong dissipation toward small scales.

Mathematically, we construct the homoclinic cycles as sums of heteroclinic connections between: $S_1 \rightarrow S_2 = \Gamma S_1 \rightarrow S_3 = \Gamma S_2 \dots S_1 = \Gamma S_{n-1}$, where Γ is some discrete subgroup of $0(2) \times 0_D(2)$, $\Gamma^n = I$. We show that S_2 and S_1 lie in some invariant subspace of the Navier-Stokes dynamical systems, $S_1, S_2 \subset A_{1,2}$. Then, under action of the group Γ, S_2 and S_3 line in the image invariant subspace $A_{2,3} = \Gamma A_{1,2}$. The fundamental property of S_1 is:

$$W_u(S_1) \subset A_{1,2}, \tag{3}$$

where W_u denotes the unstable manifold of the hyperbolic state S_1. Then $W_u(S_2) = \Gamma W_u(S_1) \subset A_{2,3}$ and $W_u(S_2) \cap A_{1,2} = \emptyset$. Hence, for the dynamics restricted to $A_{1,2}$, S_1 is hyperbolic, but S_2 is an attractor! We establish the existence of an heteroclinic connection between S_1 and S_2 in $A_{1,2}$. Similarly, S_2 is hyperbolic in $A_{2,3} = \Gamma A_{1,2}$, but $W_u(S_3) = \Gamma W_u(S_2) \subset A_{3,4}$ and S_3 is an attractor in $A_{2,3}$. The homoclinic cycle is completed as a sequence of heteroclinic connections. Remarkably enough, equivariance under $0(2) \times 0_D(2)$ and symmetry breaking via the discrete subgroup, Γ are primarily responsible for the intermittent dynamics of Coherent Large Scale Eddies. Some preliminary results along the same direction for the A.B.C. flow (2) will also be presented.

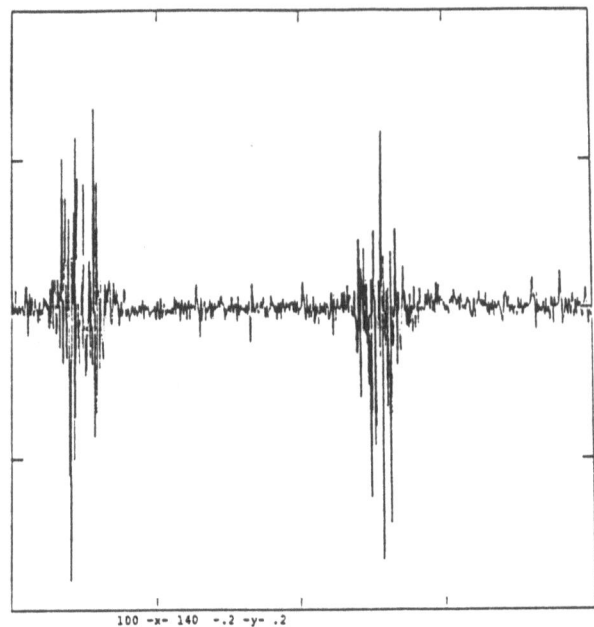

100 -x- 140 -.2 -y- .2

Fig. 1. Time evolution of the imaginary component of the Fourier mode $k_x = 4$, $k_y = 0$, for the forcing $k_c = 8$ and Re = 50.

Fig. 2. Isovorticity contours in the $x - y$ plane, for the same case, during the homoclinic explosion.

REFERENCES

[1]   H. Chaté and P. Manneville, Transition to Turbulence via Spatio-Temporal
      Intermittency, Phys. Rev. Lett. 58:112 (1987).

[2]   H. Chaté and P. Manneville, Spatio-Temporal Intermittency: a Review, Proc.
      Cargese Conf. on Recent Developments in Dynamical Systems, to appear,
      Plenum Press (1989).

[3]   H. Chaté and B. Nicolaenko, Phase Turbulence, Spatiotemporal Intermittency
      and Coherent Structures, Conf. on Recent Developments in Dynamical Systems,
      to appear, Plenum Press (1989).

[4]   V. I. Arnold and L. D. Meshalkin, Unspekki Mat. Nawk 15:247 (1960).

[5]   D. Galloway and V. Frisch, A Note on the Stability of a Family of Space–Periodic
      Beltrami Flows, J. Fluid Mech. 180:557–564 (1987).

[6]   M. Lesieur, "Turbulence in Fluids," Martinus–Nijhoff Kluwer Academic Publ.,
      Dortrecht (1987).

TIME RETURN MAPS AND DISTRIBUTIONS FOR THE LASER

WITH SATURABLE ABSORBER

F. Papoff[*], A. Fioretti and E. Arimondo
Dipartimento di Fisica, Università di Pisa, Pisa, Italy
[*] also Scuola Normale Superiore, Pisa, Italy

N. B. Abraham
Department of Physics, Bryn Mawr College
Bryn Mawr, PA

A single mode infrared CO_2 laser containing a low pressure intracavity saturable absorber is a non-linear system presenting a variety of unstable and chaotic regimes.
Fig 1a reports experimental results for the time dependence of the laser output power in an unstable operation of the 10P(30) CO_2 laser containing SF_6 as saturable absorber.

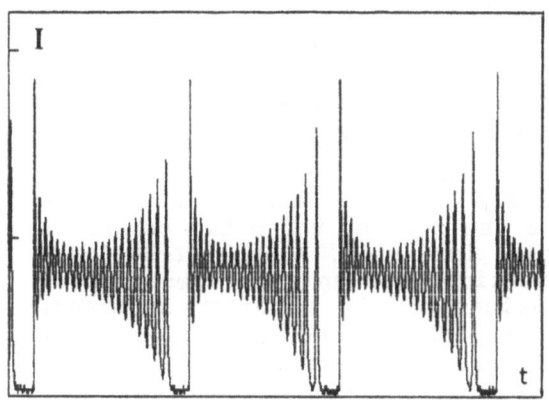
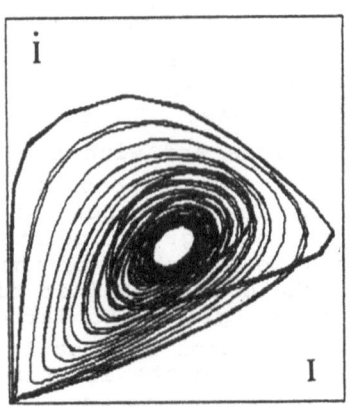

Fig. 1a. Laser output intensity
I versus time t on the 10P(30)
CO_2 laser line with 0.094 mbar
SF_6 pressure at 4.8 mA discharge
current. Timebase 1.0 ms/div.

Fig. 1b. Phase portrait
from data of Fig. 1a.

Fig 2 reports the time response of the laser output power in a chaotic regime for 10P(32) CO_2 laser line operation with CH_3I:He mixture as saturable absorber.

Measures of Complexity and Chaos
Edited by N.B. Abraham *et al.*
Plenum Press, New York

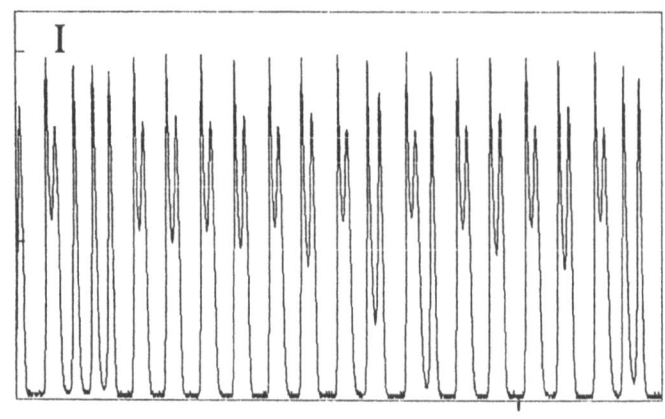

Fig.2 Laser intensity I versus time t (.2 ms/div) on the 10P(32) CO_2 line with CH_3I:He, 0.17 mbar total pressure, in 1:10 ratio at 13.2 mA discharge current.

A rate equation model for the laser with saturable absorber (LSA) has shown that the CO_2 internal dynamics plays a key role in the occurrence of the chaotic response (de Tomasi et al., 1989). By using a single two-level model for the absorbing medium, the LSA operation is described through an autonomous set of four first-order differential equations, involving the laser intensity, two population differences in the amplifier and one population difference in the absorber.

The overall behavior of the LSA has been described through orbits in the phase space revolving around the two LSA fixed points, corresponding to the non-lasing I_0 solution and to the lasing I_+ solution. I_0 is a saddle-point with one attractive eigenvalue very large as compared to all other LSA eigenvalues. I_+ is a saddle-focus attracting along two directions and with complex $\rho+i\omega$ eigenvalues in the unstable manifold, corresponding to orbits spiralling out of the saddle focus. The phase-portrait of Fig. 1b, reconstructed from the pulses of Fig. 1a, points out the role of those two points on the time evolution of the laser intensity. At the first laser threshold a homoclinic bifurcation with an orbit centered on I_0 takes place. At larger values of the pumping parameter, near the Hopf bifurcation point, homoclinicity to a periodic orbit around the I_+ points occur. The phase-portrait of Fig. 1a is an example of such an orbit.

The close resemblance of the LSA phase space evolution to the instabilites and chaos in the Belousov-Zhabotinski (BZ) reaction (Argoul et al., 1987; Richetti et al., 1987) and in the thermokinetic model of hydrocarbon oxidation (Gaspard and Wang, 1987) has suggested that LSA presents a homoclinic chaos. In particular regimes of mixed-mode oscillations, as reported in Fig. 2a, should be generated by the homoclinicity.

Properly chosen 1-Dim maps and circle maps have been used in both the BZ and thermokinetic reactions to characterize the presence of homoclinicity in those systems. We have not applied such an approach to the LSA. On the contrary our analysis was based on iteration maps of the return time to a Poincaré section of constant intensity. This approach has been used by Arecchi et al. (1988) to characterize the so-called Shil'nikov chaos in a laser with feedback. In effect quasi-homoclinic orbits in phase space present large fluctuations in the return times associated with the sensitivity of the trajectory on the rejection to the unstable point. On a more general ground, if iteration maps of the intensity or the return time constitute alternative representation of the phase space evolution, one may be more convenient than the other to have an immediate view of the general behavior.

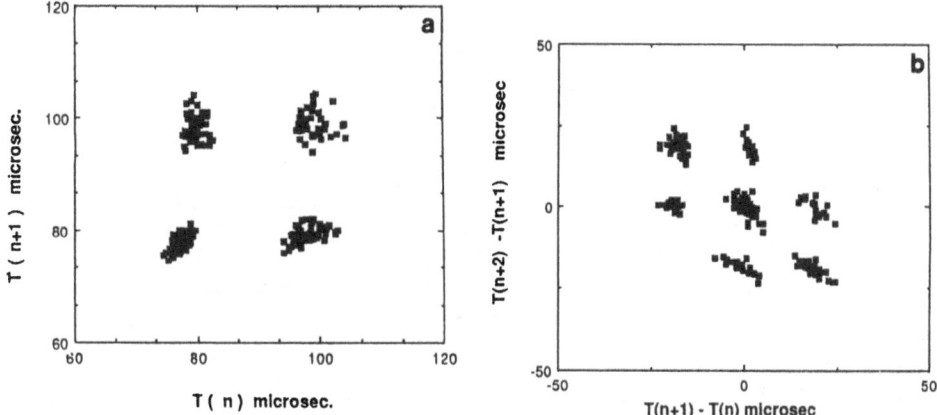

Fig. 3 Experimental results for the mixed-mode oscillations between two regimes on 10P(30) CO_2 laser line with 0.16 mbar SF_6 absorber at 6.1 mA discharge current. In a) iteration map of the return times; in b) the difference between two successive return times is plotted against the same difference one pulse earlier.

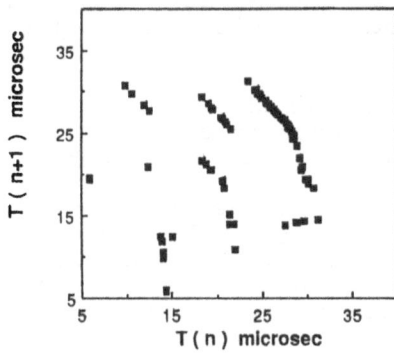

Fig. 4 Experimental results for the iteration map of the return times in the mixed-mode oscillations of Fig. 2.

In the LSA operation, with orbits visiting the I_0 and I_+ points, we have measured the time spent around each of these points and the return time. Fig. 3a represents an experimental iteration map of the return time in an LSA mixed-mode oscillation between trajectories with either one or two revolutions around I_+. The T(n+1) time measured on the (n+1)[th]-pulse has been plotted versus the time T(n) measured on the nth pulse. The observed structure appears to be regular, with a broadening produced by the noise, as corresponding to a stochastic mixture between two different regimes. In Fig. 3b an iteration map of the interperiod interval correlation has been reported, as introduced by Babloyantz and Destexhe (1988) to analyze the beat of a human heart.

Fig. 4 reports the iteration map of the return times for a LSA operation with CH_3I:He absorber, in operating conditions similar to those of Fig.2. The iteration map of Fig. 4 results quite different from the regular one of Fig. 3a. Even within the limited number of collected data (around 300), a structure with several branches, similar to that obtained in the experiments in the laser with feedback (Arecchi et al. 1987) may be recognized. Thus we believe that that iteration map is evidence of homoclinic chaos in the LSA operation. It remains to be explained why the LSA mixed-mode operation with SF_6 absorber, with the map of Fig. 3, corresponds to a stochastic mixture, while the CH_3I:He case appears to be chaotic. A large difference in the LSA phase space evolution may be connected to that different behaviour. The SF_6 orbits spends a large part of the pulse in the invariant I=0 plane containing the I_0 non-lasing solution, while the CH_3I orbits remain far from that plane. The stabilizing action of the nonlasing solution basin of attraction may preclude the occurrence of homoclinic chaos. Confirmation by the theoretical analysis is under investigation.

This research has been financially supported by the European Economy Community through the Stimulation Action.

REFERENCES

Arecchi F.T., Gadomski W., Lapucci A., Mancini H., Meucci R., Roversi J.A., Laser with feedback: an optical implementation of competing instabilities, Shi'nikov chaos, and transient fluctuation enhancement, JOSA B5, 1153 (1988)

Argoul F., Arneodo A., Richetti P. and Roux J.C., Experimental evidence for homoclinic chaos in the Belousov-Zhabotinski reaction, J. Chem. Phys. 86, 3325 (1987)

Babloyantz A. and Destexhe A., Is the Normal Heart a Periodic Oscillator?, Biol. Cybern 58, 203 (1988)
de Tomasi F., Hennequin D.,Zambon B. and Arimondo E.,Instabilities and chaos in an infrared laser with saturable absorber: experiments and vibrorotational model, JOSA B6, 45 (1989)

Gaspard P., and Wang X.-J., Homoclinic Orbits and Mixed-Mode Oscillations in Far-from -Equilibrium Systems, J. Stat. Phys. 48, 151 (1987)

Richetti P., Roux J.C., Argoul F. and Arneodo A., From quasiperiodicity to chaos in the Belousov-Zhabotinskii reaction. II. Modeling and theory, J. Chem. Phys. 86, 3339 (1987)

UNFOLDING COMPLEXITY IN NONLINEAR DYNAMICAL SYSTEMS

R. Badii

Fakultät für Physik
Universität Konstanz
7750 Constance, West Germany

ABSTRACT

Nonlinear dynamical systems produce complex temporal or spatial patterns by stretching and folding regions of phase-space in an iterative way. The topological and metric properties of this process can be extracted from the chaotic signal by using symbolic dynamics. The underlying "grammatical rules" are systematically detected and arranged on a logic tree. Predictions on the set of possible outcomes of the system are made and compared with the observation. The discrepancy between the two, evaluated through a generalization of the information gain, characterizes the complexity of the source. As a result of this unfolding procedure, the dynamics is described as a sequence of deterministic paths (blocks of symbols) which appear at random in time, with given transition probabilities. Fast hierarchical evaluations of invariant measures, dimensions, entropies and Lyapunov exponents are obtained from the logic tree, considering lower and lower levels (i.e., increasingly long symbol-sequences). Power spectra are accurately reproduced with a limited number of short orbits. The analysis applies to any system: dissipative or conservative, hyperbolic or not, invertible or not.

1 INTRODUCTION

A large variety of physical systems exhibits seemingly disordered (turbulent, chaotic) spatio-temporal behaviour which, behind its apparent irregularity, hides a high degree of organization. These phenomena, in general, cannot be simply described in terms of interactions among many modes (as it is possible, e.g., for a fluid system) and their randomness is not originated by the presence of stochastic fluctuations.

The observation that low-dimensional nonlinear dynamical systems are able to generate aperiodic bounded solutions [1] gave rise to an increasing interest for the study of chaotic behaviour, which led to the definition of strange attractors [2]. These objects have been geometrically described in terms of fractal dimensions [3] and, dynamically, by means of Lyapunov exponents and metric entropies [4]. The most striking property of such systems is that their future time evolution can be predicted only for finite (relatively short) times, although they are fully deterministic, because of the exponential amplification of the uncertainty on the initial conditions. The degree of instability of the trajectories with respect to infinitesimal perturbations is measured by the Lyapunov exponents, while the ability of the system to produce information (or, equivalently, to loose memory of the initial conditions) is estimated by the metric entropy [4,5]. Chaotic systems are characterized by positive values of these quantities.

Measures of Complexity and Chaos
Edited by N.B. Abraham *et al.*
Plenum Press, New York

A much wider class of physical phenomena, however, displays a behaviour which is clearly distinct either from simple periodicity or from complete randomness. Highly structured patterns can be produced, which are not necessarily related to chaotic motion. Examples of "complex" behaviour are provided by biological systems, hydrodynamic flows, spin glasses, neural networks, fractals, cellular automata and nonlinear dynamical systems [6,7].

The existence of some basic iterative scheme, acting on an initial seed, is a common mark of most of these phenomena. In order to understand them, it is necessary to discover the detailed generating rules of this mechanism. The concept of complexity can then be associated with those systems which are governed by a large (possibly infinite) number of rules. It is easy to show [8,9] that entropy and Lyapunov exponents are not useful indicators for the characterization of complexity. Complete randomness (as obtained, e.g., by casting dice or tossing coins) exhibits maximal entropy but is absolutely structureless. On the other hand, systems like cellular automata display complex behaviour (due to their non-invertibility), even though they may have arbitrarily low metric entropy.

Various definitions of complexity have been proposed, all of which, however, present some unsatisfactory property. The algorithmic [10] and Lempel-Ziv [11] complexities just reduce, in most cases, to the metric entropy (and the former one is, in general, not computable). Other definitions are not applicable to dynamical systems and, finally, the set and forecasting complexities (see ref. [9] for a review) are non-zero during the period-doubling cascade and diverge at the accumulation point (where the symbolic dynamics is, instead, simple [12]).

The main feature of self-generated complexity is the presence of an iterative mechanism which transforms the information contained in the initial conditions in a deterministic way. In this sense, it is possible to view complexity as **elaborated simplicity**. The aim of the present work is to discuss how to extract both the basic equations of motion and the initial seed from the observation of a signal (or, more in general, of some physical phenomenon). This is a step towards a solution of the general etiological problem as it appears in modern physics: search and investigation of the **causes** (differential equation, discrete map, natural evolutionary process) and of the **origin** (initial conditions) of the observed pattern. It should be noticed, in fact, that not all initial conditions yield a complex result, even when they are processed by a mechanism which is, in principle, able to produce a highly organized output.

The analysis of such systems can be performed most efficiently by dividing them into sub-processes which can be hierarchically allocated on suitable "logic trees" [8] and linked to each other by "parental" relations. This approach allows one to give both a local and a global characterization of complexity and to obtain fast converging approximations of the true dynamics, along with an estimate of the errors (arising from the finiteness of the available data and from the limited accuracy of the measurements). The method demands for a discretization of the input signal, in the form of a sequence of symbols, and achieves the infinite resolution limit by considering increasingly long subsequences. In some cases, the data are already in symbolic form (DNA sequences, texts written in some language, spin configurations). In general, however, experiments deal with continuous observables (e.g., recorded in hydrodynamic flows, electronic signals, lasers, nonlinear dynamical systems) and a translation to a symbolic language is needed before applying these techniques. Up to now, there is no general algorithm for the extraction of accurate symbolic dynamics from experimental data, especially if high-dimensional; this problem will be discussed in section 6.

The present work is mainly focused on low-dimensional dynamical systems, for which either an exact or an approximate symbolic dynamical representation can be obtained (section 2). This will allow the construction of a logic tree which will contain all the "grammatical rules" of the system (section 3). Accordingly, it will be possible to make predictions about the future evolution and to compare them with the observation. The discrepancy between the two, evaluated through a generalization of the information gain, characterizes the complexity of the source (section 4). This procedure, describing the way in which the dynamics folds (or cuts and displaces) phase-space, "unfolds" the iterative complexity-producing mechanism which is being studied. In addition to this, the phase-space is subdivided in a hierarchical way, so that the initial condition of a given symbolic sequence can be identified within a controllable accuracy. As a result, the signal is interpreted as a sequence of deterministic paths (**primitive blocks of symbols**) which appear at random in time, according to measurable transition probabilities. This decomposition is particularly useful for a hierarchical reconstruction of power spectra, which can be accurately reproduced with a limited number of short orbits. Moreover, the criteria for the construction of the tree can be adapted to the specific problem one is considering, in such a way that very quickly converging estimates of all interesting quantities (e.g., the metric entropy) are obtained (section 5). The aim is to compute more efficiently not only **thermodynamic** averages [13,14,15] (macroscopic description), but also **scaling functions** [16] (microscopic description), exploiting all detailed parental relations on the logic tree. The possibility of applying these techniques to the study of experimental systems and of developing general algorithms which are able to perform accurate predictions is briefly discussed in section 6, together with modifications of the algorithm which allow handling of noise-corrupted data or of non-self-generated complexity.

2 ELABORATION OF INFORMATION AND SYMBOLIC DYNAMICS

One of the main sources of difficulties with such disparate physical systems as those exhibiting complex features is that they are constructed according to different (and usually unknown) iteration schemes. For example, given an initial configuration composed of a few symbols, the successive one may be obtained by duplicating the original set and inserting new symbols between the two copies; alternatively, the system might perform a symmetry operation (e.g., a reflection) and then attach a new string to an extremum of the initial sequence, and so on. Other examples are provided by hierarchical and inflationary constructions, both occurring in nonlinear dynamical systems. The main obstacle is to establish the nature of the underlying generation mechanism. Once this is removed, the detailed rules governing the behaviour of any system can be discovered by properly arranging all relevant subsequences on a "logic" tree. As an illustration of these difficulties, consider the initial string $S_3 = 101$ and perform the substitution $(1 \rightarrow 101, 0 \rightarrow 000)$ $k - 1$ times. The output string S_{3^k} (an approximation of the ternary Cantor set) will "look" complex if analyzed in a sequential way, while it is very simple if regarded as the result of a hierarchical process. Analogously, approximating the golden mean $0.618034\ldots = (\sqrt{5} - 1)/2$ by means of continued fraction expansions yields a very simple, but slowly converging, sequence; if, however, we had used square roots, the convergence would have been immediate. In general, in absence of some physical evidence in favour of a specific scheme, it is virtually impossible to guess the appropriate strategy to adopt. Therefore, any measure of complexity can only represent a **relative** estimate of our ability to understand the system's hidden order. No absolute definition of complexity can be given [17].

In the following, for simplicity, the generation mechanism is assumed to be sequential, as it is appropriate for the study of nonlinear dynamical systems, although the method applies to any other generation procedure. Moreover, we consider the discrete-time case, whereas a discussion of continuous flows is postponed to section 6. The (unknown) equations of motion can be written as

$$\mathbf{x}_{n+1} = \mathbf{F}(\mathbf{x}_n), \tag{1}$$

where $\mathbf{x}_n$ is the position vector at time n in the d-dimensional phase-space and the function $\mathbf{F}$ describes the time evolution. In order to perform the infinite-time (-resolution) limit, which is necessary for any asymptotic estimation, a **covering** of phase-space must be chosen and each element must receive a label. In this way, a symbolic sequence $S_N = s_1 s_2 \cdots s_N$ can be associated to each trajectory $\{\mathbf{x}_1, \ldots, \mathbf{x}_N\}$ (s_i is the label of the partition element containing $\mathbf{x}_i$). In particular, we need a **generating partition** [4], which has the property of yielding a one-to-one correspondence between points (initial conditions) in phase-space and infinitely long symbol-sequences. A finite sequence S_n corresponds to a finite subset of phase-space whose volume shrinks to zero for $n \rightarrow \infty$.

Nonlinear chaotic systems are characterized by two mechanisms acting on phase-space: **stretching** (which is responsible for the exponential separation of nearby orbits [4], is essentially of local nature and is also present in linear systems) and **folding** (for **non-hyperbolic** systems) or cutting and displacing (for **hyperbolic** systems). The second process is global and typical of nonlinear phenomena. Symbol s_i contains information on the fold (or cut branch) on which point x_i lies (it has been conjectured [18] that, in non-hyperbolic systems, a generating partition can be obtained by joining "adjacent" homoclinic tangencies together). Therefore, from the knowledge of a symbolic orbit, it is possible to gain information on the set of preimages of the last observed point and on the sequence of deformations (stretching and folding) undergone by the surrounding region during the past time evolution. For $n \to \infty$, all the available information can be recovered if the partition is generating. The discovery and organization of the rules of the symbolic dynamics is particularly difficult (and interesting) when folding is not complete: in such a case, some sequences never occur and the trees present a complex topological shape since the number of rules (e.g., "after 00, there must be a 1", "after 001, there must be a 0", ...) increases very quickly with their length. In addition to giving a **topological** characterization of the system, it is possible to study its **metric** properties as well. Since dynamical systems generally yield stationary signals and admit an invariant (natural) measure [4], we can assign a probability $P(S_n)$ to each sequence S_n. Accordingly, we can investigate the probabilities of compound sequences in terms of those of their parts. In the case of non-hyperbolic transformations, memory effects will show up, which are related to singularities in the invariant measure, and will contribute to yield a positive metric complexity [8].

3   LOGIC TREES

In this section, the procedure for the construction of a logic tree is illustrated, focusing on the reproduction of the signal as a succession of elementary blocks of symbols, following each other according to specific rules (intrinsic to the system), with the goal of obtaining accurate estimates of metric entropies and of power spectra. The object under investigation is a symbolic sequence $S_N = s_1 s_2 \cdots s_N$ (with $N \approx 10^8$) produced by some nonlinear map in a sequential way: given a substring $S_n \equiv s_1 s_2 \cdots s_n$, its next extension is obtained by adding one or more symbols to the end. To each sequence S_n we may also associate a quantity Q of interest which may be the corresponding local multiplier $\bar{\mu}(S_n) \equiv \exp[n\bar{\lambda}(S_n)]$ [15,19], the size $\bar{\varepsilon}(S_n)$ of the phase-space element labelled by S_n, an expression involving several terms (such as $P(S_n)^q/\varepsilon(S_n)^r$) and so on. Usually, this quantity is of multiplicative nature: i.e., $Q(S_n) \sim a^n$ (with a being a constant), implying that $Q(S_{2n}) \approx Q^2(S_n)$. In this section, for the reasons mentioned above, we only consider the probability $P(S_n)$ of each subsequence (orbit) S_n, with length $n \in [1, n_{max}]$ (with $n_{max} \approx 20 \div 22$), which is computed as one usually does when evaluating the metric entropy $K_1 \equiv -\lim_{n\to\infty} \sum P(S_n) \ln P(S_n)/n$. The tree is constructed by allocating exclusively **allowed** sequences (i.e., with non-zero probability). Furthermore, it is important to impose a second, more restrictive, condition before accepting a sequence. This will enable us to select the relevant symbolic paths and to discover the intrinsic rules of the system in a more efficient way. In this section, we require sequences to be periodic. This choice is particularly useful for the estimation of metric entropies (related to the number of unstable periodic orbits) and for the reconstruction of power spectra. In general, the second condition should be chosen in order to optimize the evaluation of the quantities of interest (for example, we might accept only sequences with anomalously high probability for an estimation of the lowest local dimension on the strange attractor); this point will be explicitly addressed in section 5.

   If an allowed orbit is non-periodic, some rule must exist which prevents it from repeating indefinitely many times. For example, if symbol 0 appears at most twice consecutively before a 1, the rule "000 is forbidden" is found. A string S_n is considered periodic if the probability of its longest available iterate is non-zero (with $S_n = 011$ and $n_{max} = 7$, the iterate is 0110110). The aim is to decompose the original symbolic signal as a combination of **primitive** blocks of symbols, obeying all discovered rules. Such elementary strings are those sequences which cannot be subdivided into shorter strings satisfying both acceptance conditions (existence and periodicity). To be more specific, consider the roof map

$$
x_{k+1} = \begin{cases} a + 2(1-a)x_k & \text{if } x_k < 1/2 \quad (s_k = 0) , \\[2mm] 2(1-x_k) & \text{if } x_k \geq 1/2 \quad (s_k = 1) , \end{cases} \tag{2}
$$

for $a = (3 - \sqrt{3})/4$, where a Markov partition exists (the maximum belongs to an unstable period-5 orbit). The unit interval can be divided into three subsets, labelled by the sequences 1, 01 (left preimage of element 1) and 001 (left preimage of 01). The symbolic dynamics yields all possible combinations of these three strings, except for the forbidden orbit 0011. The tree is constructed by considering all sequences S_n in order of increasing length $n = 1, \ldots, n_{max}$. Symbol 1 is periodic and is allocated at the first level (see Fig. 1).

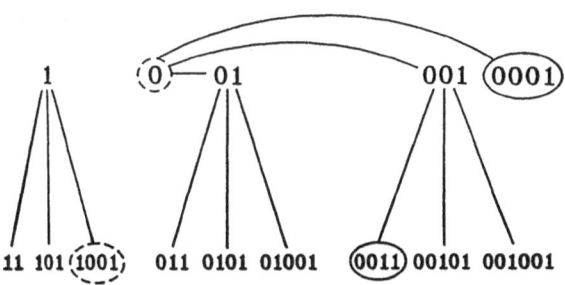

Figure 1. Logic tree for the roof map at $a = (3 - \sqrt{3})/4$. Phantom sequences are encircled with a dashed line, dead branches by a solid line. Only the first two levels are displayed.

Symbol 0 has a non-zero probability but is non-periodic. Therefore, it is only virtually placed on the tree (encircled with a dashed line, in Fig. 1) and will reappear as part of longer strings containing one or more 1's (i.e., it is a phantom). With the present knowledge, a prediction of all period-2 orbits is made. Both symbols 0 and 1 are taken as possible heads while 1 is the only admissible tail (the iterates of 0 are non-periodic). Hence, the sequences 01 and 11 are formed and allocated at the first and second level, respectively. The group 01 (as well as 1) cannot be decomposed in periodic substrings and is, therefore, called primitive. All primitive orbits are allocated at the first level. A generic sequence S_n of length n is composed by joining a head H (of length m), consisting of any allocated sequence or phantom, with a tail T (of length $n - m$), chosen from the primitive orbits only: the latter are, in fact, the elementary components of all sequences generated by the system. A phantom is not an admissible tail, since its iterate is non-periodic. The predicted orbits of length 3 are 001 (primitive), 011, 101 and 111. The first interesting information is discovered for $n = 4$. In addition to the sequence 0101 (simple iterate of 01), the procedure yields (at levels 1 and 2) 0001, 0011 and 1001: the former two have zero probability (forbidden sequences) and, therefore, constitute a surprise. Orbit 1001 is a phantom: it only exists as part of sequences of the type $1(001)^k01$, $\forall k$. Its daughters are 100101 and 1001001: the former one is periodic and is allocated at level $l = 2$ as 1001 (which is only virtually present); the latter one is a phantom (also placed at level 2) and will generate, in turn, a periodic sequence and a phantom. At this point, the only two rules of the system have been detected and all subsequent predictions will be exact. For example, the sequence $10011 = 1001 + 1$ can be excluded a priori, since it contains the forbidden substring 0011. A newly formed sequence S_n is an admissible prediction if its initial and final substrings of length $n - 1$ have non-zero probability. The first two levels for the example discussed above are displayed in Fig. 1. The lower ones will be gradually filled as $n \to \infty$. The sum of the probabilities at each level always equals 1 (here, $1 = P(0) + P(1) = P(01) + P(001) + P(1)$). The left part of phase-space (interval $[0, 1/2]$) has been refined, since symbol 0 has no special relevance if taken alone: it is the analogue of a consonant in a natural language.

Notice that the splitting of phase-space into primitives is simpler than that obtained from the Markov partition which, in this case, is composed of four intervals. In general, however, Markov partitions do not exist[20] and the successive levels of the tree yield increasingly better "Markov" approximations of the true dynamical behaviour.

4 COMPLEXITY

Having prescribed how to construct the tree and how to make predictions, the complexity of the system is defined in terms of the information gain during the process of learning. To each predicted sequence $S_n = H_m T_{n-m}$ we assign a prior (expected) probability $P_0(S_n) \equiv P(H_m)P(T_{n-m})$: i.e., we assume that probabilities approximately factorize. A refinement of this estimate, although possible, is not necessary for the purpose of defining complexity and comparing different systems with each other. Actually, the error is, in all cases investigated numerically, of the order of 1%. The (metric) complexity C_1 of the system is then given by

$$C_1 = \lim_{l \to \infty} \lim_{n \to \infty} \frac{1}{\ln N_0(l, n)} \sum_j^{N(l,n)} P(S_j) \ln \frac{P(S_j)}{P_0(S_j)} , \qquad (3)$$

where $N(l, n)$ is the number of sequences allocated at level l, after consideration of all n-periodic orbits and $N_0(l, n)$ is the number of orbits predicted at level l, based on the knowledge of all orbits up to length $n - 1$. If the predictions match perfectly ($P = P_0$, $\forall S_j$), the metric complexity C_1 is identically zero, otherwise it is a positive quantity. The normalization is here obtained through the division by $\ln N_0(l, n)$. In fact, the information gain is generally maximal when nothing is known about the system: this corresponds to assuming that $P_0 = 1/N_0$, for any sequence. If, in addition, all of the actual probabilities P happen to be zero except one, the sum in eq.(3) saturates to $\ln N_0(l, n)$ and $C_1 = 1$. This situation is expected to be the worst possible in physical systems and is very unlikely to occur. However, for "non-self-generated" complexity (i.e., for patterns prepared by changing the rules externally during the generation process) it is possible to have predictions that are systematically less accurate than the case of complete ignorance: C_1 can then be larger than 1 or even diverge. Other normalizations can be adopted, depending on the characteristics of the class of systems one is studying. In the case of dynamical systems, the ratio $P(HT)/[P(H)P(T)]$ behaves (when the lengths $L(H)$ and $L(T)$ of H and T are much larger than 1) as $\exp[-L(HT)\kappa(HT) + L(H)\kappa(H) + L(T)\kappa(T)]$, where $\kappa(S)$ is the metric entropy of the sequence S. If $\kappa(HT) \neq \kappa(H) \neq \kappa(T)$, $P(HT)/[P(H)P(T)] \sim \exp[-L(HT)\bar{\kappa}]$ (where $\bar{\kappa}$ is some average metric entropy for the three sequences HT, H and T). Therefore, the expectation value $\langle P/P_0 \rangle$ in eq. (3) should be normalized by dividing it by some typical sequence-length, at each level. The number of predictions $N_0(l, n)$ grows at most like $N_p^l(n)$, where $N_p(n)$ is the number of primitives after allocation of all sequences up to length n. Instead of dividing by $\ln N_0(l, n) \leq l \ln N_p(n)$, in eq. (3), it is also possible to divide by l: this choice would yield larger values of the complexity for trees with many primitives[8]. Here, we follow normalization (3). Notice that C_1 is an average of local, "microscopic" information gains $\ln[P/P_0]$, at variance with the effective measure complexity (EMC)[9], which describes the convergence of the block entropies[21] to their asymptotic value. The EMC, however, diverges in some cases (for example, at the period-doubling accumulation point) where, instead, the system is simple[8].

Table 1. Estimated metric (C_1) and topological (C_0) complexities for various maps. The computations were performed with sequences up to length $n = 22$ for the logistic map and $n = 20$ for the Hénon map.

Map	C_1	C_0
Logistic $a = 1.85$	0.007 ± 0.001	> 0
Logistic $a = 1.748$	0.015 ± 0.002	0
Logistic $a = 1.7905$	0.002 ± 0.002	0
Hénon $a = 1.4, b = 0.3$	0.007 ± 0.002	> 0

In order to give a complete characterization of complexity, it is necessary to generalize eq.(3) in analogy to what is usually done for dimensions and entropies [4]. The information gain can be extended to a function of a parameter q [22], thus yielding the following definition for the generalized complexity C_q:

$$C_q \equiv \lim_{l \to \infty} \lim_{n \to \infty} \frac{1}{\ln N_0(l,n)} \frac{1}{q-1} \ln \sum_{j=1}^{N(l,n)} P(S_j) \left[\frac{P(S_j)}{P_0(S_j)} \right]^{q-1} , \qquad (4)$$

for $q \geq 0$ (eq.(3) is recovered for $q \to 1$). However, it is evident that the limit $q \to 0$ does not yield a purely topological quantity, in contrast to the case of dimensions and entropies. To overcome this difficulty, definition (4) can be modified to

$$C_q \equiv \lim_{l \to \infty} \lim_{n \to \infty} \frac{1}{\ln N_0(l,n)} \frac{1}{q-1} \ln \left[\frac{(\sum P^q)(\sum P_0^q)^{q-1}}{(\sum PP_0^{q-1})^q} \right] , \qquad (5)$$

by substituting P and P_0 in eq.(4) with $PP_0^{q-1}/\sum PP_0^{q-1}$ and $P_0^q/\sum P_0^q$, respectively. The new definition (other equivalent expressions may be found) agrees with (3) and yields the topological complexity C_0 as

$$C_0 = \lim_{l \to \infty} \lim_{n \to \infty} \frac{\ln[N_0(l,n)/N(l,n)]}{\ln N_0(l,n)} . \qquad (6)$$

This is the same expression that one would obtain from eq.(4) if all P_0's were equal to $1/N_0$. The topological complexity is identically zero if all predicted orbits exist. Systems which are described by complete trees (topologically simple) with factorizing probabilities have $C_q = 0$, identically. This is the case of the generalized baker map [23] and of the logistic map at crisis (where the set complexity [9] is, instead, finite). Notice that in the latter case the metric entropy is maximal. Asymmetric critical maps are metrically complex (with $C_0 = 0$). For the example in Fig. (1) $C_0(l,n)$ becomes exactly zero starting from level $l = 3$. At generic parameter values, map (2) can be topologically complex, while its linearity implies that C_1 always vanishes. The function C_q, for $q > 0$, is therefore a measure of the nonlinearity of the system.

In addition to the previous simple cases, the generalized complexity C_q has been evaluated for the logistic map ($x' = 1 - ax^2$) at $a = 1.85$, $a = 1.748$ (close to the period-three intermittency threshold $a_0 = 1.75$) and $a = 1.7905$ (just above the crisis point [24] of the same period). The values of C_1, estimated for sequences up to length $n = 22$, are listed in table 1.

Since the topological complexity C_0 could not be evaluated with comparable precision, we only indicate the cases in which it is strictly positive. The rather small numerical values of the complexities are due to the large number of predictions generated by the phantoms at levels $l > 1$. In fact, all daughters of a phantom are allocated at the same position as the phantom itself, thus yielding a very large value of $N_0(l,n)$. This undesirable effect can be avoided by allocating all phantoms at levels lower than the first one as if they were normal periodic orbits. Such a procedure corresponds to the requirement that only the primitives are periodic sequences (with the additional property of being indecomposable in shorter periodic strings) and was followed in ref. [8]. The three examples in table 1 are chosen from systems exhibiting worse convergence properties for $l, n \rightarrow \infty$. In the second case, the precision is limited by the presence of period-3 oscillations appearing in the curves $C_1(l,n)$ vs. l, $\forall n$ (the dynamics is governed by the third iterate $x' = f^{(3)}(x)$ of the map). In the third example, the measured value is consistent with $C_1 \approx 0$ since $f^{(3)}(x)$ can be approximately rescaled to the logistic map at crisis ($a = 2$). From the observation of the curves $C_1(l,n)$ versus l, $\forall n \in [1, n_{max}]$, it is expected that analyses with sequences of length up to $n_{max} = 24 \div 25$ (requiring large computer memory) will considerably improve the convergence. The topological complexity C_0 converges more slowly than C_1, when the levels contain a large number of orbits, since every sequence receives the same weight. The estimated values of C_0 and C_1 for the Hénon map are also displayed in the table: 9 primitives were found (1, 01, 0011101, 0011111, 00111101, 00011101, 00011111, 000111101, 0011110011101), with $n_{max} = 20$, at standard parameter values $a = 1.4$, $b = 0.3$ [26], where the grammar is infinite. The logistic map close to intermittency is more complex metrically, whereas larger topological complexity is found for the Hénon map.

The analysis can also be performed when no periodic sequences exist. This situation occurs, e.g., at the accumulation point of the period-doubling cascade. A tree of phantom sequences is then constructed and the predictions are made in the usual way; the measured values of C_q converge to zero, as expected, since the symbolic dynamics is simple [12] (although it is more easily described in terms of an inflationary process rather than in terms of a sequential mechanism). For periodic chaos an alternation of real and phantom levels is obtained. It is also possible to decide to accept all symbols as primitive even when some of them are non-periodic: the resulting levels are filled more rapidly but contain less asymptotic terms and the overall convergence is not improved. Worse results are generally obtained with trees in which constant-length orbits are placed at the same level [27,25,19], since there is no special reason for which the system should refine the whole phase-space in the same way at each iteration. The right interpretation scheme requires, in general, coverings the elements of which are of different size, contain different mass and are obtained by intersecting different-order images and preimages of the generating partition, as illustrated by the examples discussed above. Complete trees (e.g., binary, for the baker map and the logistic map at crisis) are an exception. The logic trees described in this section take into account natural, system-induced, refinements in those regions of phase-space where a higher resolution is necessary. The roof map (2), at the special parameter value considered in Fig. 1, represents a typical example of this fact: the system describes either the deterministic path 01 or 001 as a whole, before jumping (at random, depending on the specific initial condition and according to the proper transition probability) onto either path 1 or 01 or 001, and so on. The symbolic dynamics appears, therefore, as an alternation of deterministic and random events. In particular cases, such as the Bernoulli shift, we only observe a random, completely uncorrelated, succession of symbols (and the tree is binary, complete). In addition to being maximally compact, trees constructed with the above discussed procedure also display a very low redundancy: in fact, many irrelevant permutations of sequences are not present (in Fig. 1, for example, the sequences 10, 110, 1010, 1110, etc. do not appear).

5 HIERARCHICAL APPROXIMATIONS

The method proposed in the previous section for the construction of the logic tree is just a member of a general class of organization schemes, each being particularly effective for the solution of a specific problem. In fact, allowed sequences (i.e., those with non-zero probability) were further distinguished according to their periodicity: non-periodic sequences, called phantoms, were just used to perform predictions, while periodic ones were regularly allocated on the tree. The periodicity condition is especially useful when dealing with dynamical systems. In this case, in fact, there is an asymptotic correspondence between symbolic and real periodic orbits and the latter ones are dense on the strange set. For hyperbolic systems, the study of periodic orbits is, for many purposes, equivalent to that of generic orbits [28,29]; for non-hyperbolic systems, differences emerge [30]. This procedure yields improved (i.e., more rapidly converging and more asymptotic) estimates of the metric entropy in the following form:

$$K_1 = - \lim_{l \to \infty} \sum [P(S) \ln P(S)/L(S)] , \qquad (7)$$

where the sum extends over all sequences at level l and $L(S)$ is the length of sequence S. Notice that the metric entropy is an average of local entropies $- \ln P(S)/L(S)$: therefore, it is necessary to evaluate them as accurately as possible. This is achieved by refining the partition (in the example of Fig. 1, symbol 0 is replaced by 01 and 001, since 000 does not exist) until the periodicity condition is met: that is, until a symbolic orbit is found which may repeat itself, in principle, indefinitely many times. No "transient" orbits (such as 0 or 00, in the same example) are considered, since they are not asymptotic enough. Similar arguments were exploited in ref. [32] to obtain the spectrum of local entropies from the looped graphs describing the symbolic dynamics of the logistic equation. The logic tree obtained with the periodicity condition can also been employed to resolve the structure of power spectra. The reconstruction procedure and the numerical results are discussed in ref. [31].

In general, the second acceptance condition for the strings to be allocated on the tree can be different from periodicity; moreover, one or more quantities can be attached to each sequence, in addition to its probability, as mentioned in section 3. Therefore, the orbits may be classified according to various properties: e.g., one can group sequences with large multiplier and small probability, or corresponding to a small phase-space element, or which contain a large amount of 0's, and so on. The choice of the conditions must be justified by some special need. For example, the invariant measure of the logistic equation at generic parameter values exhibits square-root singularities. Therefore, the dimension function $D(q)$ can be evaluated by covering the segment $[-1, 1]$ with boxes of size ε, as follows [30]:

$$\sum_{j=1}^{1/\varepsilon} P_j^q(\varepsilon) \sim \varepsilon^{(q-1)D(q)} \sim \varepsilon^{q-1} + c \cdot \varepsilon^{q/2} , \quad \text{for} \quad \varepsilon \to \infty , \qquad (8)$$

where c is a constant. Asymptotically, either the first or the second term on the r.h.s. will survive, depending on the value of q [33]:

$$D(q) = \begin{cases} 1 & \text{if } q \leq 2 , \\ q/[2(q-1)] & \text{if } q > 2 . \end{cases} \qquad (9)$$

It is clear, however, that the corner-point in $D(q)$ will not be sharply reproduced until ε becomes very small (depending on the value of q). This slowing-down effect, due to the non-asymptoticity of eq.(8), can be compensated (if not completely eliminated) by keeping in sum (8) only the relevant terms, for each q-value. This improvement is equivalent to having many more (smaller) boxes around the peaks of the distribution than around the generic points (where the mass scales as $P(\varepsilon) \sim \varepsilon$), when $q > 2$, and viceversa, when $q \leq 2$: it can be achieved by constructing a logic tree where sequences with anomalously large (small) probability are allocated, for $q > 2$ ($q \leq 2$), and by performing the sum in eq.(8) over some level l of the tree. The limit $\varepsilon \to 0$ corresponds to $l \to \infty$. Of course, the transition value q_c at which non-analytic behaviour of $D(q)$ occurs is, in general, unknown [30] and must be determined by some iterative self-consistency procedure. The construction of the tree with some q_c-dependent rule and the evaluation of a sum like the one in eq.(8) are rather fast operations (about 4 minutes of CPU time on a Microvax II), so that the procedure can be easily iterated using each time the last estimate for q_c. The time-consuming part is, in most cases, the calculation of all probabilities $P(S_n)$, for $n = 1, \ldots, n_{max}$ (with $n_{max} = 20, 22$). At the parameter value $a = 2$ (fully developed chaos) there are only two points where $P(\varepsilon) \sim \sqrt{\varepsilon}$. In such a case, no refinement of the interior of the segment $(-1, 1)$ (e.g., obtained by considering sequences 010, 011, 110, 111, instead of 01 and 11) will ever yield a large value of $P(S_n)$, for any $n \leq n_{max}$. Hence, the l-th level of the tree will contain, for $q \gg 2$, only the sequences 0^l and $1(0)^{l-1}$ (all other sequences would be phantoms), for which $P(S_l) \sim 2^{-l}$ and $\varepsilon \sim 4^{-l}$, thus yielding $D(\infty) = 1/2$. Notice that the resulting partition, in generic cases, is composed of neither constant-size, nor constant-mass nor constant-n elements. No a priori assumptions are made: the optimal approximation is approached by following, as closely as possible, the system's behaviour.

Hierarchical estimations of other quantities are also possible. For example, the invariant measure can be approximated by means of piecewise constant densities (solutions of the Frobenius-Pérron operator [34]), since any logic tree corresponds to a piecewise linear model of the true dynamics. Another application of the method, concerning relations between dimensions and Lyapunov exponents, is discussed in [19].

Finally, it should be noticed that also a microscopic description of the dynamics can be obtained from the tree. In fact, the mother-to-daughter ratio $P(HT)/P(H)$ is a **scaling function** [16] for the probabilities and can be plotted, for each level l, versus an appropriate independent variable $t \in [0, 1]$, which orders the sequences at that level. Scaling functions can be evaluated for any interesting quantity, such as lengths ε, local Lyapunov exponents λ, expressions like P^q/ε^τ, and so on. The main difficulty is finding the proper t-parametrization of the sequences at each level. So far, this has only been done, apart from very simple examples, for the phenomenon of period-doubling [35]. When the predictions are performed for a set of quantities $\vec{Q} \equiv \{Q_1, \ldots, Q_M\}$, different from the probability P, it is possible to characterize the difficulty of this task by means of modifications of eq.(3). They will be in the form of expectation values of ratios like $Q_j(HT)/Q_j(H)Q_j(T)$ (these expressions are not semi-positive defined, in general). The prediction scheme adopted in section 3, in fact, is based on a self-similarity assumption, improved at each step by using the most recent available information: i.e., the value of the actual probability $P(H)$ of the head. Further improvements to this algorithm can be devised, in order to approach the ideal limit of optimal predictions. For example, if the head H is given by the concatenation $H'T'$, we may evaluate $P_0(HT)$ as $P^2(H)/P(H')$, and not as $P(H)P(T)$. For the purpose of defining (relative) complexity, however, the factorization assumption is very satisfactory, especially in virtue of its simplicity. Definitions requiring an optimal approximation scheme, such as the algorithmic complexity, are not operationally useful.

6 EXPERIMENTAL SYSTEMS

The application of symbolic dynamical techniques to the characterization of experimental time series, although rather problematic, is very promising and challenging since it would provide a precious guidance for the constructions of global models from the knowledge of the logic tree. In addition to this, the evaluation of complexity, scaling functions, dimensions and entropies would be greatly facilitated.

The models could be utilized to interpret the unknown physical mechanism and to formulate phenomenological theories. Moreover, they may be helpful in forecasting the future dynamical evolution. Some attempts towards the problem of performing reliable predictions have been recently made [36]. They are all based on local methods (see also [37] which produce short-time (infinitesimal, in principle, because of the exponential divergency of the errors) predictions using differential approximate mappings. The main limitation of these approaches is that they do not furnish any interpretation of the underlying dynamics. Therefore, they cannot be used to infer a physical model. Another important difference with the symbolic approach is that this one deals with finite-time predictions and has the scope of reproducing a time series (by means of a global model) which is statistically equivalent to the original one: no exact forecasting is explicitly searched, at variance with local methods. Indeed, each symbol corresponds to a return onto the Poincaré section and a symbolic sequence labels a finite region of it. Therefore, the prediction of a sequence of n symbols involves a finite time (n returns); on the other hand, it is not spatially very accurate (unless n is large), since the corresponding phase-space element is not infinitesimal. Local and global methods concern different kinds of predictions: exact, fine-grained (in phase-space) and short-time the former ones, and optimal (i.e., which take into account memory effects and statistics), coarse-grained and long-time the latter ones. It would be desirable to devise algorithms which take advantage from both approaches.

The encoding of an experimental signal into a symbolic sequence requires the preliminary identification of a generating partition on a suitable Poincaré surface. The task of the construction of a generating partition, for which there is no general procedure even when the equations of motion are given, is rendered extremely difficult by the need of embedding the signal in a (usually high-dimensional) space. Problems of statistics and precision should also be taken into account, together with noise and discretization effects. An alternative to the localization of the homoclinic tangencies [18] consists of detecting all unstable periodic orbits up to a given length n_{max} [28] and of choosing a Poincaré surface which intersects all "period-1" orbits [19], in such a way that they yield the same topological return time (the lengths of periodic orbits in a continuous-time system are in general different from one another and unknown). All period-1 points are then numbered, their neighbourhoods are taken as elements of the partition, and all points belonging to longer cycles are labelled with the index of the neighbourhood in which they lie. Distinct periodic orbits might at this point still be identified by the same symbolic sequence, since the true generating partition has elements that do not contain simple fixed points, but only points belonging to longer cycles. The knowledge of such cycles allows one to further refine the partition, until a meaningful symbolic dynamical encoding can be obtained. A more detailed discussion of this problem is found in ref. [19].

The analysis proposed in this work can then be applied and the dynamics can be hierarchically approximated by means of piecewise linear Markov models. The method does not exclude the possibility of constructing fully nonlinear maps which are described by the same logic tree as the real system. Possible additional applications of these techniques include the reduction of noise and the analysis of data produced in a non-sequential way or which are not arising from dynamical systems, such as DNA sequences. Modifications of the acceptance conditions for the symbolic strings must be introduced, according to the specific problem, and "elastic" criteria may be adopted to treat signals which are corrupted by noise at a symbolic level (e.g., mutations in DNA) and may be viewed as governed by a "fuzzy" logic.

ACKNOWLEDGEMENTS

This research has been partly supported by a grant of the Von Humboldt Foundation. I wish to acknowledge very useful discussions with D. Auerbach, G. Broggi, P. Grassberger, A. Politi, I. Procaccia and M.A. Sepúlveda.

REFERENCES

[1] E.N. Lorenz, J.Atmos.Sci. **20**, 130 (1963).

[2] D. Ruelle and F. Takens, Comm.Math.Phys. **20**, 167 (1971).

[3] B.B. Mandelbrot, *"The Fractal Geometry of Nature"*, Freeman, San Francisco (1982).

[4] J.P. Eckmann and D. Ruelle, Rev.Mod.Phys. **57**, 617 (1985).

[5] P. Grassberger, in proc. conf. on *"Chaos in Astrophysics"*, Palm Coast, Florida, 1984, J. Perdang et al. editors, Reidl, Dortrecht (1985).

[6] *"Chaos and Complexity"*, R. Livi, S. Ruffo, S. Ciliberto and M. Buiatti Eds., World Scientific, Singapore, 1988.

[7] Proceedings of the conference on *Complex Systems*, Gwatt, Switzerland, 1988.

[8] R. Badii, *"Quantitative Characterization of Complexity and Predictability"*, submitted for publication.

[9] P. Grassberger, in ref.[7] and Wuppertal preprint B 9, 1988;
D. Zambella and P. Grassberger, Wuppertal preprint B 11, 1988.

[10] A.N. Kolmogorov, Probl.Inform.Transm. 1, 1 (1965);
G. Chaitin, J. Assoc.Comp.Math. 13, 547 (1966).

[11] A. Lempel and J. Ziv, IEEE Trans.Inform.Theory 22, 75 (1976).

[12] I. Procaccia, S. Thomae and C. Tresser, Phys.Rev. A35, 1884 (1987).

[13] M.J. Feigenbaum, M.H. Jensen and I. Procaccia, Phys.Rev.Lett. 57, 1503 (1986).

[14] M.H. Jensen, L.P. Kadanoff and I. Procaccia, Phys.Rev. A36, 1409 (1987).

[15] R. Badii, Riv. Nuovo Cim. 12, N° 3, 1 (1989).

[16] M.J. Feigenbaum, J.Stat.Phys. 52, 527 (1988).

[17] D.R. Hofstadter, *"Gödel, Escher, Bach: an Eternal Golden Braid"*, Vintage Books, New York (1980).

[18] P. Grassberger and H. Kantz, Phys.Lett. 113A, 235 (1985).

[19] R. Badii and G. Broggi, *"Hierarchies of Relations between Partial Dimensions and Local Expansion Rates in Strange Attractors"*, this issue.

[20] V.M. Alekseev and M.V. Yakobson, Phys.Rep. 75, 290 (1981).

[21] G. Györgyi and P. Szépfalusy, Phys.Rev. A31, 3477 (1985).

[22] A. Renyi, *"Probability Theory"*, North-Holland, Amsterdam (1970).

[23] J.D. Farmer, E. Ott and J.A. Yorke, Physica 7D, 153 (1983).

[24] C. Grebogi, E. Ott and J.A. Yorke, Phys.Rev.Lett. 48, 1507 (1982).

[25] R. Badii, unpublished.

[26] M. Hénon, Comm.Math.Phys. 50, 69 (1976).

[27] P. Cvitanović, G. Gunaratne and I. Procaccia, Phys.Rev. A38, 1503 (1988).

[28] D. Auerbach, P. Cvitanović, J.P. Eckmann, G.H. Gunaratne and I. Procaccia, Phys.Rev.Lett. 58, 2387 (1987).

[29] C. Grebogi, E. Ott and J.A. Yorke, Phys.Rev. A37, 1711 (1988).

[30] P. Grassberger, R. Badii and A. Politi, J.Stat.Phys. 51, 135 (1988).

[31] M.A. Sepúlveda and R. Badii *"Symbolic Dynamical Resolution of Power Spectra"*, this issue.

[32] A. Politi, Phys.Lett. A136, 374 (1989).

[33] E. Ott, W. Withers and J.A. Yorke, J.Stat.Phys. 36, 687 (1984).

[34] P. Collet and J.P. Eckmann, *"Iterated Maps on the Interval as Dynamical Systems"*, Birkhauser, Cambridge, MA (1980).

[35] M.J. Feigenbaum, J.Stat.Phys. **46**, 919 and 925 (1987).

[36] J.D. Farmer and J.J. Sidorowich, Phys.Rev.Lett, **59**, 845 (1987);
J.D. Farmer and J.J. Sidorowich, in *"Evolution, Learning and Cognition"*, Ed. Y.C.
Lee, World Scientific, Singapore (1989);
J.P. Crutchfield and B.S. McNamara, Complex Systems **1**, 417 (1987);
J. Cremers and A. Hübler, Z.Naturforsch. **42a**, 797 (1987);
M. Casdagli, Physica **D**, to appear (1989).

[37] M. Sano and Y. Sawada, Phys.Rev.Lett. **55**, 1082 (1985);
J.P. Eckmann, S. Oliffson Kamphorst, D. Ruelle and S. Ciliberto, Phys.Rev. **34A**,
4971 (1986).

INFERRING THE DYNAMIC,
QUANTIFYING PHYSICAL COMPLEXITY*

James P. Crutchfield

Physics Department

University of California

Berkeley, California 94720 USA

chaos@gojira.berkeley.edu

SUMMARY

Through its formalization of inductive inference, computational learning theory provides a foundation for the inverse problem of chaotic data analysis: inferring the deterministic equations of motion underlying observed random behavior in physical systems. Integrating the geometric and statistical techniques of dynamical systems with learning theory provides a framework for consistently, although not absolutely, distinguishing between deterministic chaos and extrinsic fluctuations at a given level of computational resources. Two approaches to the inverse problem, estimating symbolic equations of motion and reconstructing minimal automata from chaotic data series, are reviewed from this point of view. With an inferred model dynamic the dynamical entropies and dimensions can be estimated. More interestingly, its structural properties give a measure of the intrinsic computational complexity of the underlying process.

INTRODUCTION

During the last decade or so a number of mechanisms have been investigated by which physical processes can generate complex behavior. The most widely studied of these, deterministic *chaos*, while *globally* stable to perturbations produces noisy behavior by exponential *local* amplification of microscopic fluctuations.[1] The central problem in applying the theory of dynamical systems to the sciences is identifying the deterministic, and possibly chaotic, component in a set of observations and distinguishing it from behavior due to ever-present measurement uncertainty, extrinsic noise, and uncontrolled degrees of freedom. This is the inverse problem in nonlinear dynamics: inferring the deterministic equations of motion, if any, underlying observed random behavior in physical systems.

Over a similar historical interval, computational learning theory has formalized a range of learning paradigms for inductive inference. In this it provides a language and a collection of complexity theoretic methods appropriate to the inverse problem. When integrated with the geometric and statistical techniques of dynamical systems theory, the result is

---

* Portions of this essay were distributed as "Learning the Dynamic", an extended abstract submitted 15 April 1989 to the Conference on Computational Learning Theory to be held 31 July - 2 August 1989, University of California, Santa Cruz. This work was supported by ONR contract N00014-86-K-0154.

Measures of Complexity and Chaos
Edited by N.B. Abraham *et al.*
Plenum Press, New York

327

a framework for consistently distinguishing between deterministic chaotic behavior and extrinsic information sources to which it is coupled.

This essay gives a brief overview of two complementary approaches to the inverse problem. The first estimates symbolic equations of motion using Bayesian statistical inference[2] and the second uses techniques from stochastic grammatical inference to reconstruct minimal computational models of chaotic behavior.[3,4,5] They have two common goals. The first is to quantify the observed complexity in a data stream. The second is to capture the underlying dynamics in a form that can be related to first principles and used in forecasting and numerical modeling of the behavior. The following discussion emphases their similarities as problems in learning theory. It concludes by indicating where learning theory can contribute to a number of existing problems and remarks on the limitations of existing proposals for quantifying physical complexity. The hope is not only that dynamical systems will benefit from this synthesis but that with real world, quantitative problems computational learning theory will be made more generally accessible.

INDUCTIVE INFERENCE FORMALISM

Given the inverse problem of chaotic data analysis just outlined, this section reviews the appropriate tools from inductive inference and attempts to justify the choices made along the way from a scientific and dynamical systems viewpoint.

The types of problems of interest to physicists include the onset and structure of various forms of fluid turbulence and the occurrence of noise in superconducting Josephson junctions and other nonlinear solid state devices; just to mention a very few out of hundreds of applications in physics, chemistry, and biology. These systems generate very complex behavior and no one expects to exactly describe their detailed behavior. Nonlinear dynamicists are (or should be) satisfied with with a theory that *systematically* approximates some phenomenon. This means that with a given level of experimental limitations $\dot{I}$† and fixed data reduction (computational) resources C, observations can be explained quantitatively to within an error level δ. Furthermore, the error must vanish exponentially fast with increasing $\dot{I}$ and C.

This is a form of *identification in the limit*[6] with the additional requirement of estimating the *rate* of convergence of the inference method on an ensemble of data sets. Here, the inductive process continues indefinitely and its asymptotic behavior is used as the success criterion. The identification error is to decrease on larger and more accurate example sets.

The inverse problem can be considered also as a sequence inference problem since the example presentations are ordered by time. Typically, there are two problems in sequence inference. The first is to infer a model, such as the graph of the dynamic; this is the identification problem. The second is to forecast future observations knowing past ones from the sequence; this is the prediction problem. The formal restrictions that distinguish these subproblems in general sequence inference are not appropriate here. For the inverse problem identification and prediction are the same; the identified model can be used for forecasting.

To specify the inverse problem as a problem in inductive inference several components need to be defined. An inference method M is a computable process implemented (say) on a Turing machine. The latter is augmented to read in example data and output hypothesized models. To formalize this we must specify a success criterion C and a data representation D for the examples. The space of inference methods is then the set of all triplets (M,

---

† Specified in terms of sampling resolution and frequency as a information acquisition rate.

C, D). The basic components of this inductive inference problem are a rule space, an hypothesis space, an example set, an inference method, and its success criterion. These are defined as follows.

1. The *rule space*: This is the set of all mappings $\vec{F} : M \rightarrow M$ from the state space M into itself.

2. The *hypothesis space*: This is the space $H = \left\{ D : D = \left(\vec{F}, \vec{P}, M \right) \right\}$ of noisy discrete time dynamical systems

$$\vec{x}_{n+1} = \vec{F}(\vec{x}_n) + \vec{\xi}_n$$

   where $\vec{F}$ is a real vector-valued function, the *dynamic*; $\vec{\xi}_n$ is a realization of the real vector-valued random process $\vec{P}$; and $\vec{x}_n \in M$, the manifold of states or *state space*.

3. The *example set*: The data with which the inference method works is obtained from a single realization $\{\vec{x}_n : n = 0, 1, 2, \ldots, N-1\}$ or *orbit* of the noisy dynamical system by way of a measurement partition $P_\epsilon = \{e_i : e_i \subset M\}$. By coarse graining the state space into size ϵ cells, the partition maps the sequence of states along an orbit into a sequence of symbols, $s \in A$ in the measurement *alphabet* $A = \{i : i = 0, \ldots, k-1; k = f(\epsilon^{-1})\}$. These symbols label the partition elements in which the orbit is found at the time of measurement. The measurement sequence obtained this way will be referred to as the *measurement language* L_{D,P_ϵ}. A set of length n subsequences is derived from a sliding window of length n onto the single long sequence. The length n measurement sequences $s^n = \{s_0 \cdots s_{n-1} : s_i \in A\}$ are called *n-cylinders* in reference to the bundle of orbits that lead to the same set of n measurements.[7] The examples will be restricted to only those measurement sequences that are observed. That is, the inference method will use only *positive presentations*. An *admissible presentation* P of the rule, then, will be a set $P = \{^i s^n : i = 0, 1, \ldots, N-n\}$ of n-cylinders that are observed and that are ordered by time. The latter is guaranteed by the sliding window construction of the individual cylinders from the single long measurement sequence. In fact, we could have taken for a presentation a sequence of examples derived from orbits with different initial conditions rather than a sequence from a single orbit. In any case, the initial conditions $\vec{x}_0$ might contain transients. In the following, though, the orbit is assumed to be a *typical* orbit governed by the asymptotic probability measure on the attractor.

4. The class of *inference methods*: An inference method M takes an admissible presentation P of the measurement language and produces a model D of a noisy dynamical system: $D = M(P)$.

5. The *success criterion*: The most appropriate criterion is that developed by Wharton[8] for approximate language identification. This requires a measure of goodness of fit of the inferred models to the measurement language. For this, we need a metric on the space L of measurement languages. The weight $w(s^n)$ of an n-cylinder is the number of occurrences of that particular sequence in the presentation. The weights are effectively normalized $\sum_{s^n \in P} w(s^n) = \|P\| = N - n$. The distance between two languages $L_1, L_2 \in L$ is given by the metric on L

$$d(L_1, L_2) = \sum_{s \in L_1 \triangle L_2} w(s)$$

where $L_1 \triangle L_2$ is the symmetric difference of the two languages. Using this metric, an inference method M δ-identifies a measurement language L if and only if M converges to a grammar G associated with the hypothesized D such that

$$d(L(G), L) < \delta \cdot N$$

It is an encouraging result that if one tolerates finite error ($\delta > 0$) then the class of all languages is δ-identifiable in the limit from positive presentations by a method that infers only grammars for finite languages.[8] Requiring that the error vanish in the limit leads to a more restricted form of identification. *Convergent* identification in the limit occurs if

$$\lim_{n \to \infty} d\left(L^n\left(G\right), L^n\right) = 0$$

where G is the grammar associated with the inferred D, L^n is the measurement language of observed n-cylinders, and $L^n\left(G\right)$ is the set of n-cylinders generated by G. This is the identification criterion of interest in the following.

There are a number of other contexts in which this type of identification is problematic. For example, with the further constraint, central to deductions concerning properties of the underlying noisy dynamical system, that the inferred grammar be minimal the resulting computational problem is NP-Hard.[9,10] As will be discussed shortly, numerical results indicate that convergent identification often works, but proofs for the general case are not available. This is to be expected due to the very complex behavior arbitrary noisy dynamical systems are capable of producing. Nonconvergence can even be taken as a useful measure of the effective computational structure of the system, as noted later on. For the case of interest here there are, thus, several open problems concerning the identifiability of noisy dynamical systems. In this brief overview we can only hope to sketch the basic problem and encourage further investigation.

SYMBOLIC EQUATIONS OF MOTION

With this introduction to inductive inference for noisy dynamical systems, this section outlines the inverse problem of estimating symbolic equations of motion. This approach to modeling is closest to conventional statistical time series analysis. The basic statistical assumption for learning a dynamical system from data is that the observed time series is the result of a deterministic system in contact with a fluctuation source. The equations of motion for a continuous time process, for example, are given by a stochastic differential equation

$$\dot{\vec{v}} = \vec{F}\left(\vec{v}\right) + \vec{\xi}(t) \, , \; \vec{v}\left(0\right) \in M$$

where the second term is a random driving force.

Using a fine measurement partition P_ϵ ($\epsilon \ll 1$),[11] the goal of estimating equations of motion is to produce a symbolic representation of the dynamic $\vec{F}$ and an estimate of the extrinsic noise level due to the fluctuating force. If the deterministic behavior is stably periodic then this problem essentially reduces to conventional linear prediction theory, as originated by Wiener and Kolmogorov. The case of general interest comes in not restricting $\vec{F}$ to periodic behavior and considering the possibility that the deterministic dynamic itself produces complex behavior. The general question is then, given a noisy data stream $\vec{v}\left(t\right)$ or a function of it, how to infer that some portion of the noise is due to the fluctuating force and how much is due to the deterministic chaos.

330

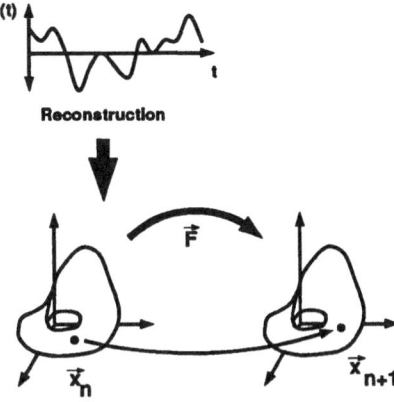

v(t)

Reconstruction

$\vec{F}$

$\vec{x}_n$

$\vec{x}_{n+1}$

Figure 1 Estimating symbolic equations of motion: The first step is to reconstruct from a time series an effective state space in which to represent the behavior.[12] In this step the continuous time series is discretely sampled in time and in value. The dynamic is the mapping $\vec{F}$ that takes the state space into itself; or equivalently, takes a single state at one time into its successor. This is the object of attention for the second step: a statistical fit then produces the coefficients of some expansion of $\vec{F}$ using a chosen function basis.

The procedure for estimating the symbolic form of the dynamic applies Bayesian statistical inference to estimate nonlinear models from reconstructed chaotic data series.[2] (Figure 1 illustrates the overall procedure.) When approached from this point of view, the inverse problem reduces to statistical quadrature: estimating the symbolic equations of motion reduces to curve fitting in a space of dimension twice that of the reconstructed state space. An *ad hoc* choice of a function basis (e.g. polynomial or Fourier functions, or local splines) is made at this point. An unsupervised learning procedure then searches the space of consistent models using an optimality criterion that trades off forecasting error against model complexity. The residual fit error then yields an estimate of the extrinsic noise level. Once a model has been estimated it can be used in an interactive simulation interpreter to reproduce behavior in the same chaotic class as the data.

The choice of a finite function basis is a coordinatization of the hypothesis space H: the coordinates are fit parameters. The finite-dimensional reduction of H this affords determines the model space $\mathcal{M}$. The observed data is a realization of some "true" distribution $p_{D'}$ on H. A model D is a good estimate then if the differences between p_D, the distribution specified by D, and $p_{D'}$ is minimized. One natural measure of the minimization is the model entropy

$$I(D, D') = \int_H dm \, p_D(m) \log \frac{p_D(m)}{p_{D'}(m)}$$

While formally appropriate to the task at hand, there are two problems with this. First, the integral is over the infinite-dimensional function space H. Second, we do not know the "true" distribution $p_{D'}$. Both of these problems can be addressed by pulling back the reconstructed data distribution on H to the finite-dimensional model space $\mathcal{M}$. In this we identify errors in the fit parameter estimation with the reconstructed distribution in H of fluctuation-induced deviations from the "true" dynamic. (This is an approximation and is not to say these are necessarily the same thing.) Both this error and the dimension of the model space play a role in finding an optimal model. By trading off these two components of the model entropy, we would like to maximize the information in the presentation P that is captured or "explained" by the estimated model. In the model space, then, we approximate p_D by the probability $p_{\mathcal{M}}(m|\mathrm{P})$ conditioned on the presentation. The model entropy is then

$$I(D, D') \approx \int_{\mathcal{M}} dm \; p_{\mathcal{M}}(m|\mathrm{P}) \log p_{\mathcal{M}}(m|\mathrm{P})$$

Using Bayes's theorem, though,

$$p_{\mathcal{M}}(m|\mathrm{P}) = \frac{p_{\mathcal{M}}(\mathrm{P}|m) p_{\mathcal{M}}(m)}{p_{\mathcal{M}}(\mathrm{P})}$$

where $p_{\mathcal{M}}(\mathrm{P}) = \int_{\mathcal{M}} dm \; p_{\mathcal{M}}(\mathrm{P}|m) p_{\mathcal{M}}(m)$ is a normalizing constant that can be dropped without affecting the minimization. $p_{\mathcal{M}}(m)$ is the probability of a particular model $m \in \mathcal{M}$ and $p_{\mathcal{M}}(\mathrm{P}|m)$ the probability that m has produced the data P. Putting this altogether, gives the following two-term approximation

$$I(D, D') \propto \int_{\mathcal{M}} dm \; p_{\mathcal{M}}(\mathrm{P}|m) \log p_{\mathcal{M}}(\mathrm{P}|m) + \int_{\mathcal{M}} dm \; p_{\mathcal{M}}(m) \log p_{\mathcal{M}}(m)$$

The first term on the right-hand side is a measure of the amount of information in the observations not explained by the model. In the case of Gaussian error distribution, this can be again approximated by the fit error $-\log \sigma$, where σ is the one-step prediction error variance and the data range has been normalized. The second term is an informational measure of the complexity of the consistent models. This term is dominated by the number of model parameters k. With these further, somewhat extreme approximations, the Bayesian version of Akaike's model identification criterion is recovered

$$I(D, D') \propto -\log \sigma + \frac{k \log N}{N}$$

where N is the number of examples in the presentation.[13]

There are two broad classes of equations of motion inference methods: global and atlas. The latter uses coordinate charts on M over which splines are fit to the graph of the dynamic. The former uses a function basis that is applied over the entire state space manifold M. The class of dynamics that can be learned with global equations of motion is clearly a subset of those inferable using atlas equations of motion. The latter, unfortunately, indicates very little about any underlying functional simplicity of the dynamic. The identification of new *laws* requires the use of both methods and so a generalized success criterion favoring global equations of motion when they are the most compact representation.

There are two problems with estimating symbolic equations of motion. The first concerns the identification of generalized *states* in the data stream. Presumably the most parsimonious use of the given data and the simplest model require detecting patterns that contain the most information and that are optimal for forecasting. This is indirectly addressed in the above procedure during the state space reconstruction step; which was given short shrift. The second problem derives from the lack of an absolute measure of a model's complexity. Here, as in all statistical curve fitting, the size of a model is measured relative to an *ad hoc* function basis. There is no general method to compare model complexities across function bases. This is essential to the search for the smallest representation for reasons of efficiency and for inferring that a given structure in the estimated model is a property of the original physical process.

Both of these problems are addressed by the second approach to the inverse problem which seeks to quantify the intrinsic computation performed by a physical process.[3] The goal is to reconstruct a minimal and unique automaton or grammar that recognizes the measurement language to within some approximation. The automata are referred to as ϵ-machines, with ϵ indicating a generic level of approximation. A reconstructed machine in principal could be based on any computation model. The Chomsky hierarchy provides a graded set useful for systematically distinguishing more from less powerful computational models. In practice an ϵ-machine is either a deterministic finite automaton (DFA) or, if a finite DFA is inconsistent, a minimal pushdown automaton (PDA). Reconstructing approximations of higher level machines from data is understandably fraught with difficulty. The computational approach reduces the inverse problem to the learning problem of stochastic grammatical inference.

Probabilistic structure in the measurement language is taken into account via a statistical mechanical formalism. One result of this is a connection between the structure of the inferred machine and traditional dynamical systems measures of the unpredictability. Additionally, this leads to a new invariant, the *complexity*, for dynamical systems based on the Rényi entropy that measures the amount of information contained in the inferred machine states. This quantity reflects the computational difficulty inherent in modeling nonlinear dynamical systems. Both repetitive and very unpredictable behavior have simple descriptions and so low complexity.[3,5]

An ϵ-machine is described by a labeled, directed graph, or *l-digraph*, that consists of a set of vertices **V** and a set of edges **E** connecting them. Its statistical structure is given by a parametrized *probabilistic connection matrix*

$$T_\alpha = \{t_{ij}\} = \sum_{s \in A} T_\alpha^{(s)}$$

that is the sum over each symbol of the state transition matrices

$$T_\alpha^{(s)} = \{t_{ij;s}^\alpha\}, \; t_{ij;s} = p(v_i|v_j; s)$$

for the vertices $\mathbf{V} = \{v_i\}$. The entries are the conditional probabilities of making a transition from state i to state j on symbol s.

For the topological case ($\alpha = 0$) the unique and minimal l-digraph and the associated connection matrix can be reconstructed from a data stream using a variant of standard grammatical inference applied to a prefix tree of unique measurement sequences.[3,5,14] For arbitrary α, probabilistic structure of the data is translated into estimates of the transition probabilities. The resulting machine in this case is only an approximation at each window length. The inferred l-digraph states, called *morphs*, represent historical templates that are optimal for forecasting.

The α-order total Rényi *entropy*, or *free information*, of a reconstructed ϵ-machine is given by

$$H_\alpha(n) = (1 - \alpha)^{-1} \log Z_\alpha(n)$$

where the *partition function* is

$$Z_\alpha(n) = \sum_{s^n \in \{s^n\}} e^{\alpha \log p(s^n)}$$

with the probabilities $p(s^n)$ defined on the observed n-cylinders $\{s^n\}$. The parameter α has two interpretations, both of interest in the present context. From the physical point of view, it plays the role of an inverse temperature in statistical mechanics which emphasizes different invariant subsets in orbit space. From the point of view of Bayesian inference α is a Lagrange multiplier specifying a maximum entropy distribution consistent with the observed cylinder probabilities.

The Rényi *specific entropy*, i.e. entropy per unit time, is approximated for n-cylinders by

$$h_\alpha(n) = n^{-1} H_\alpha(n)$$

The *metric entropy*, a measure of predictability from dynamical systems theory, is then

$$h_\mu = \lim_{n \to \infty} h_1(n)$$

The α-order *graph complexity* is defined as

$$C_\alpha = (1 - \alpha)^{-1} \log \sum_{v \in V} p_v^\alpha$$

where the probabilities p_v are defined on the machine states. It measures an ϵ-machine's information capacity in terms of the amount of information stored in the morphs. The entropies and complexities are dual in the sense that the former is determined by the maximum eigenvalue λ_α of T_α,

$$h_\alpha = \log_2 \lambda_\alpha$$

and the latter by the associated (left) eigenvector $\vec{p}_\alpha = \{p_v^\alpha : v \in V\}$ that gives the asymptotic vertex probabilities. The specific entropy is also given directly in terms of the transition probabilities

$$h_\alpha = \sum_{v \in V} \frac{p_v^\alpha}{1 - \alpha} \log \sum_{\substack{v' \in V \\ s \in A}} t_{vv';s}^\alpha$$

A complexity based on the asymptotic edge probabilities $\vec{p}_e = \{p_e : e \in E\}$ can also be defined

$$C_\alpha^e = (1 - \alpha)^{-1} \log \sum_{e \in E} p_e^\alpha$$

Not much is gained, however, since

$$C_\alpha^e = C_\alpha + h_\alpha$$

and so there are only two independent quantities for a finite ϵ-machine.

The graph complexity is also a measure of the informational fluctuations in the data. These fluctuations are most readily quantified by the total excess (α-)entropy for L-cylinders[7,15,16]

$$F_\alpha(L) = H_\alpha(L) - L h_\alpha = \sum_{n=1}^{L} [H_\alpha(n) - H_\alpha(n-1) - h_\alpha]$$

This measures the deviation of finite cylinder statistics from asymptotic. It can also be interpreted as the mutual information between the future L-cylinders and the infinite past. In the thermodynamic limit

$$C_\alpha \underset{L \to \infty}{\propto} F_\alpha(L)$$

Loosely speaking, the graph complexity measures the amount of mathematical work necessary to produce a deviation from uniform statistics. From this limit, it is seen to be the mutual information between the infinite past and the infinite future.

A measure of convergent identification in the limit can now be defined. Since the complexity is a measure of the size of the current hypothesized (DFA) model at the given level of approximation, it can be used as a diagnostic of convergence. Holding the measurement partition P_ϵ fixed, that is assuming it produces adequately representative symbols,‡ identification with longer L-cylinders is the primary concern. The identification method then converges with increasing cylinder length if the rate of change of the complexity vanishes. That is, if

$$c_\alpha = \lim_{L \to \infty} \frac{C_\alpha(L)}{L}$$

vanishes then the noisy dynamical system has been identified. If this is not the case, then c_α, a measure of the rate of divergence of the model size, is yet another measure of complexity, but at a higher level of computation.

ϵ-machines have been reconstructed from hundreds of data sets from prototype chaotic systems. The implementation is relatively fast and has been used to infer machines with several hundred states. This constructive approach to complexity has proved itself useful in elucidating the phase transition structure at the onset of chaos in systems with a control parameter.[3] Although developed in the context of reconstruction from a data series, the underlying theory provides an analytic approach to calculating entropies and complexities for a number of dynamical systems.[4] One noteworthy result is a universal description of period-doubling cascade transitions to chaos that depends only on the intrinsic computation and information processing capacity of the dynamical systems in the bifurcation sequence, and is independent of explicit nonlinearity controls.

Another application, that relies heavily on the inductive inference formalism, is a method for inferring the direction of time in a data stream and for measuring quantitatively the degree of irreversibility.[17] With this, computational learning theory sheds new light on the long-standing paradox of irreversible macroscopic processes that are described microscopically by reversible dynamics.

Finally, we note that ϵ-machine reconstruction provides a data compression technique that gives an efficient method for encoding a chaotic data stream, transmitting the compressed form, and uniquely decoding it.[18] Technically, this is referred to as data compaction; data compression allows for some error in encoding so that exact reconstruction of the original sequence is not always possible. (The chaotic data compaction method can be modified to allow for compression at a specified fidelity.) The compaction ratio r, defined as the ratio of the number output bits from the encoder to the number input, is given by

$$r = C_\alpha^e - C_\alpha$$

where log base the number of symbols is used to give a normalized ratio. This form of the data compaction ratio makes clear the dependence on branching structure in the reconstructed machine. To the extent that there is branching in the machine state transitions, bits must be passed to the output from the input. To the extent there is determinism in the transitions, the machine captures it and there is no need to output symbols on those transitions.

---

‡ That is, assuming it is a *generating* partition.[7]

335

The goal of estimating optimal models, as reflected in minimizing (say) the model entropy for selecting symbolic equations of motion, is most generally expressed by Rissanen's minimum description length (MDL) criterion.[19] This is easily explained for ϵ-machine reconstruction by recalling the preceding comments on chaotic data compaction. Briefly, the MDL criterion says to choose that computational model m that reduces the size of the model m and the length $l(P|m)$ of the data P encoded with that model. That is, the description length

$$l(P) = C_0(m) + l(P|m)$$

should be minimized. Since the reconstruction method produces the minimal and unique ϵ-machine for a complete presentation, a search procedure and so the MDL criterion are unnecessary.

CONCLUDING REMARKS

The last decade has seen tremendous progress in modeling complex natural phenomena. This has largely been accomplished through the solution of isolated problems within dynamical systems theory and its applications. These include

1. Extensive studies of nonlinear phenomenology: chaotic attractors, complex basin structures, phase transition description of the onset of chaos;
2. Data analysis techniques: reconstruction of state space from single time series, optimal coordinates, estimating equations of motion;
3. Statistical characterizations of noisy behavior: estimating the embedding and attractor dimensions as well as information production rates, a thermodynamic description of invariant measures on attractors and their orbits; and
4. Estimation of the intrinsic computational information processing within physical processes governed by noisy dynamical systems.

While some of this is unified within abstract dynamical systems theory, the practical application has been plagued by inconsistencies and unsystematic development that has produced a patchwork of theories garnered from engineering, statistics, and physics.

It is not unreasonable to explain the distance between theory and practice and the latter's difficulties as stemming from a lack of formalizing the overall goal of the enterprise: detecting and modeling deterministic structure in noisy data. This is exactly where learning theory can contribute since one of its mandates is to formalize learning paradigms. The preceding discussion has attempted to outline, however schematically, the learning paradigms appropriate to chaotic data analysis. A number of existing problems can be more clearly articulated in this context. There is much to be gained by the application of learning theory to dynamical systems.

This is all well and good for dynamical systems and the sciences it serves, but what about learning theory itself? It seems likely that it too will benefit by access to and the appreciation of the well-defined and wide-ranging class of learning problems provided by dynamical systems. One example to look forward to is a quantitative investigation of learning a parametrized class of dynamical systems that go from periodic to chaotic behavior in which the basic computational problem goes from P to NP. This is already implicit in the tension between the typical polynomial speed of machine reconstruction and the complexity theoretic results that minimal consistent DFA inference is NP-complete. The observation is that it is only noisy dynamical systems at complexity phase transitions which are hard to learn.

More generally, the type of learning problems found in dynamical systems, only some of which have been presented above, are of a different character than the symbolic AI problem traditional in learning theory. They are more akin to problems in computational geometry and in neural networks with the additional feature that dynamical systems generate a probability measure on the orbits which singles out atypical (measure zero) behavior. It is particularly important to the application of learning theory that the practical problems of approximation be taken in account from the start. In contrast, complexity theoretic approaches tend to emphasize worst case behavior in order to obtain hardness results. If these apply only to measure zero orbits or to nongeneric dynamical systems, then they might be of little practical value.

If it has not become sufficiently clear at this point let me mention before closing the underlying motivation for a computational learning theory of dynamical systems. The goal is *artificial science*. Within the limited domain of nonlinear dynamical systems subject to extrinsic fluctuations, this is the complete automation of inferring deterministic dynamics from noisy data. To the extent that this is possible, implementations will greatly accelerate experimental science and presumably scientific discovery generally. To the extent that it is not feasible, we will have an understanding of yet another limitation of scientific method.

The foregoing has briefly reviewed several approaches to consistently modeling noisy dynamical systems. It has turned on the intimate connection between complex dynamics, modeling theory, statistics, and computation. The role of modeling was emphasized, although there is an equally important scientific, rather than engineering, motivation. That is the definition of a measure of complexity that is appropriate to and *implementable* for physical processes. The framework outlined above for reconstructing ϵ-machines and for their statistical mechanical description unifies a number of proposed definitions of physical complexity, that range from the use of entropy convergence critical exponents to the relaxation rates for diffusion on hierarchical barriers.[20,7,21,22,23,24,16,25] Not only does the current framework indicate how these are related, but also how they fail to capture important structural properties of complex systems. This structure is given explicitly by the ϵ-machines.

The exclusive use of information theoretic analysis§ restricts one to the lowest level in Chomsky's hierarchy. Indeed, conventional statistical mechanics is similarly limited due to its reliance on correlation functions, that is, on $\alpha = 2$ information statistics. Such a restriction misses higher level computation that dominates at phase transitions. The general conclusion is that only systems at phase transitions, to be understood broadly as being in a "critical" state, perform computation at levels beyond information transmission and storage. Being in a critical state is a first and a minimal requirement for nontrivial computation. This is as true of ferromagnets at the Curie temperature as it is of digital or analog computers considered as physical systems and of evolving biological organisms. These systems trade their ability to store information in certain quasi-static degrees of freedom against the underlying nonlinear dynamics necessary for innovation and reliable information transmission. These needs are balanced at the borders of chaos. Computationally critical states form the substrate supporting nontrivial information processing and, presumably, are prerequisite for evolutionary development. The question remains: Why would a bowl of primordial soup spontaneously take up a computationally critical state?

§ I have in mind especially the recent resurgence of mutual information.

References

1. J. P. Crutchfield, N. H. Packard, J. D. Farmer, and R. S. Shaw. Chaos. *Sci. Am.*, 255:46, 1986.

2. J. P. Crutchfield and B. S. McNamara. Equations of motion from a data series. *Complex Systems*, 1:417, 1987.

3. J. P. Crutchfield and K. Young. Inferring statistical complexity. *Phys. Rev. Let.*, 63:10 July, 1989.

4. J. P. Crutchfield and K. Young. Computation at the onset of chaos. In W. Zurek, editor, *Entropy, Complexity, and Physics of Information*. Addison-Wesley, 1989. to appear.

5. J. P. Crutchfield and K. Young. Thermodynamics of minimal reconstructed machines. in preparation, 1989.

6. E. M. Gold. Language identification in the limit. *Info. Control*, 10:447, 1967.

7. J. P. Crutchfield and N. H. Packard. Symbolic dynamics of noisy chaos. *Physica*, 7D:201, 1983.

8. R. M. Wharton. Approximate language identification. *Info. Control*, 26:236, 1974.

9. E. M. Gold. Complexity of automaton identification from given data. *Info. Control*, 37:302, 1978.

10. D. Angluin. On the complexity of minimum inference of regular sets. *Info. Control*, 39:337, 1978.

11. J. P. Crutchfield. *Noisy Chaos*. PhD thesis, University of California, Santa Cruz, 1983. published by University Microfilms Intl, Minnesota.

12. N. H. Packard, J. P. Crutchfield, J. D. Farmer, and R. S. Shaw. Geometry from a time series. *Phys. Rev. Let.*, 45:712, 1980.

13. J. Rissanen. Modeling by shortest data description. *Automatica*, 14:462, 1978.

14. A. W. Biermann and J. A. Feldman. On the synthesis of finite-state machines from samples of their behavior. *IEEE Trans. Comp.*, C-21:592, 1972.

15. N. H. Packard. *Measurements of Chaos in the Presence of Noise*. PhD thesis, University of California, Santa Cruz, 1982.

16. P. Grassberger. Toward a quantitative theory of self-generated complexity. *Intl. J. Theo. Phys.*, 25:907, 1986.

17. J. P. Crutchfield. Time is the ultrametric of causality. in preparation, 1989.

18. J. P. Crutchfield. Compressing chaos. in preparation, 1989.

19. J. Rissanen. Universal coding, information, prediction, and estimation. *IEEE Trans. Info. Th.*, IT-30:629, 1984.

20. J. P. Crutchfield and N. H. Packard. Symbolic dynamics of one-dimensional maps: Entropies, finite precision, and noise. *Intl. J. Theo. Phys.*, 21:433, 1982.

21. S. Wolfram. Computation theory of cellular automata. *Comm. Math. Phys.*, 96:15, 1984.

22. R. Shaw. *The Dripping Faucet as a Model Chaotic System*. Aerial Press, Santa Cruz, California, 1984.

23. C. H. Bennett. On the nature and origin of complexity in discrete, homogeneous locally-interacting systems. *Found. Phys.*, 16:585, 1986.

24. C. P. Bachas and B.A. Huberman. Complexity and relaxation of hierarchical structures. *Phys. Rev. Let.*, 57:1965, 1986.

25. S. Lloyd and H. Pagels. Complexity as thermodynamic depth. *Ann. Phys.*, 188:186, 1988.

SYMBOLIC DYNAMICS FROM CHAOTIC TIME SERIES

A. Destexhe, G. Nicolis and C. Nicolis[+]

Service de Chimie Physique, Université Libre de Bruxelles
CP231 Campus Plaine, Blvd. Triomphe
B-1050 Bruxelles, Belgium

INTRODUCTION

Following the ideas of Ruelle and others [1,2], an embedding phase space can be reconstructed from experimental systems on the basis of time series data. The introduction of numerical methods for calculating dimensions, entropies, Lyapunov exponents and other related properties, has permitted extensive investigations of chaotic experimental systems these last years. However, severe restrictions about the applicability of these methods were noticed, especially for high dimensional systems [3-6].

In this paper, we use the symbolic description as an alternative approach to analyze chaotic dynamical systems, independently of any phase space reconstruction algorithms. The idea is to compress the information contained in continuous variables into a sequence of symbols which can be studied with standard statistical tools, in order to gain some quantitative knowledge about the system's dynamics. In section 1, we show from a model system that a symbolic description can reveal useful information about the regular - or irregular- aspects inherent to the chaotic dynamics. In section 2, we study two biological systems for which patterns of activity are repeated at irregular intervals. Here again, the symbolic dynamics will provide interesting information on how these intervals are interrelated.

1. SYMBOLIC DYNAMICS FROM THE ROSSLER MODEL

For the Lorenz attractor, Aizawa [7] associated the letter L (resp. R) to each orbit if the trajectory was on the left (resp. right) side of the attractor. The sequence of symbols reflects the succession of visits of the trajectory in either side of the attractor. He showed that for a given set of parameters the sequence of symbols may be a zero[th] order Markov process (Bernouilli process [8]). In other words, a deterministic system described by three coupled nonlinear differential equations of the first order may generate a sequence of totally uncorrelated symbols, as if they were generated by a stochastic process such as the coin tossing.

More generally, it has been shown [9] that certain classes of chaotic dynamical systems can be mapped into a well defined stochastic process described by a master equation. This process can be of high order.

Here, we consider the three variable Rössler model [10]:

$$\frac{dX}{dt} = -Y-Z, \qquad \frac{dY}{dt} = X+aY, \qquad \frac{dZ}{dt} = bX-cZ+XZ$$

[+] Institut d'Aeronomie Spatiale de Belgique, 1180 Bruxelles, Belgium

Measures of Complexity and Chaos
Edited by N.B. Abraham *et al.*
Plenum Press, New York

Starting from an initial condition (X=Y=Z=1.) the systems reaches a chaotic attractor (a=0.38, b=0.3, c=4.5) after a sufficient time for transients to die out (300 units of time).

The mapping into symbolic dynamics will be performed as follows [11]: once a variable crosses a prescribed threshold L, a symbol specific to that variable is produced. The sequence of these symbols will then provide the image of the various thresholds encountered successively by the dynamics. The symbols are also an image of what happens in phase space. For example, an orbit rotating around the origin near the Z=0 plane will produce the symbols **XY**. The symbol **Z** will be produced at each reinjection in the plane. The sequence obtained for thresholds L=3 reads

ZYX ZXYX ZXYX ZYX ZXYX ZYX ZYX ZX ZYX ZYX ZXYX ZYX ...

This sequence has a remarkable property: it can be entirely reformulated by introducing the hypersymbols α=**ZYX**, β=**ZXYX**, γ=**ZX**. The result reads

α β β α β α α γ α α β α ...

The second sequence is obviously less regular than the first one. In other words, we can say that the dynamics is made from the irregular repetition of three typical orbits around the origin, each one of which is preceded by a reinjection (**Z** begins all hypersymbols).

To check more quantitatively the variability of the sequence, let us consider these sequences as a Markov process of arbitrarily high order [12]. The order of the process is the number of symbols over which the system's "memory" extends and it can be evaluated as follows [11,13]. Let $X_1...X_N$ be the sequence of N symbols. The probability of observing in the sequence the word $X_1...X_L$ is $P(X_1...X_L) = P_i$ $i=1...n^L$. Here the subscript i indicates that $X_1...X_L$ is the ith word among the n^L possible words of length L (the words are sorted in alphabetic order). For a Markov process of the kth order, the conditional probability of observing X_L as the Lth symbol in the word obeys to the relation [12]

$$P(X_L|X_1...X_{L-1}) = P(X_L|X_{L-k}...X_{L-1})$$

The above two quantities can be calculated from the sequence and compared for different values of k until the values coincide for all possible words. It is also useful to compare $P(X_1...X_L)$ with $P^{(k)}(X_1...X_L)$ which is the probability of observing the sequence $X_1...X_L$, deduced by assuming a kth order Markov chain.

The values $P(X_1...X_L)$ deduced from numerical counting are compared with the computed values of $P^{(k)}(X_1...X_L)$ by using a test statistic such as

$$\chi^2 = \frac{1}{n^L} \sum_{i=1}^{n^L} \frac{\left[P_i - P_i^{(k)}\right]^2}{P_i}$$

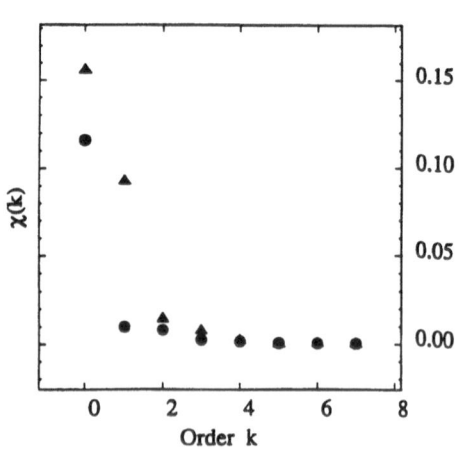

Figure 1. Test statistic χ^2 vs. order k for the Rössler model. Symbols (triangles, N=5.1 106) and hypersymbols (circles, N=2. 106) are shown. See text for other parameters.

The procedure is repeated for increasing values of k and χ^2 must converge to zero for $k \geq k_0$, where k_0 is the order of the Markov process.

Figure 1 shows χ^2 vs. k for the above described sequences. χ^2 converges to zero at a value near 5 for the sequence of symbols. A more accurate approach leads to a value of five [13], showing that the sequence of thresholds produced in the Rössler model has a memory which extends over 5 symbols. On the other hand, the sequence of hypersymbols displays saturating behavior of χ^2 at lower values of k (near 1). From these values, we can say that the Rössler model is made from the aperiodic repetition of the same characteristic orbits, following a first order Markov process.

2. SYMBOLIC DYNAMICS FROM BIOLOGICAL TIME SERIES

Figure 2 depicts the time evolution of two biological signals for which a characteristic event is produced irregularly. The first signal (Fig 2.a) represents the electrocardiogram (ECG) of a normal human heart and is measured at the level of the thorax. The second signal (Fig 2.b) shows the electrical activity of the human brain (electroencephalogram or EEG) recorded during a pathological state near deep coma. The recording of the electrical potentials and the digitization were performed using standard techniques which will not be described here (more details can be found in [14] for 2.a and in [4] for 2.b). The reconstruction of phase portraits and the evaluation of the correlation dimension [4,5,14] or Lyapunov exponents [14] all point to the same conclusion, namely that the electrical activity of the heart as well as the brain waves may follow chaotic dynamics. We will show here that the symbolic description provides an alternative approach which can be very useful to the analysis of this kind of systems [13].

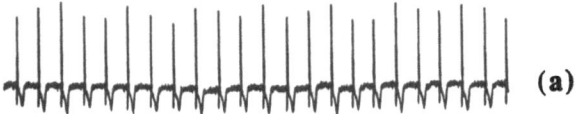 **(a)**

Figure 2. (a) Electrical activity of the heart (ECG) recorded at the level of the thorax from a normal human subject. (b) Electrical activity of the human brain (EEG) during a pathological state.

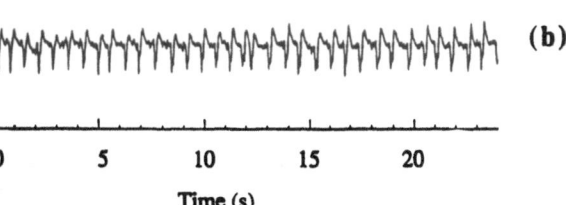 **(b)**

| 0 | 5 | 10 | 15 | 20 |

Time (s)

We will use here a different type of mapping into symbols, which takes temporal information into account (temporal mapping). From Fig.2, it is seen that both for ECG and EEG recordings, potential spikes are produced at irregular time intervals. In the case of EEG, the distribution of inter-spike time intervals may be bimodal. In this case, associating a symbol to each peak of the distribution is equivalent to studying the succession of "short" and "long" intervals. Therefore, in this case, it is natural to associate a symbol to the duration of the interval. In the following paragraphs, the intervals will be partitioned in n subsets depending on the duration and each subset will be associated with a prescribed symbol.

Figure 3 displays the values of χ^2 obtained respectively from the sequences of ECG (fig. 2a) and EEG intervals (fig. 2b). The order of the corresponding Markov process deduced from the convergence of χ^2 is of about three for ECG intervals (see fig 3c for comparison). In the case of the EEG intervals, χ^2 converges to zero at k = 6 (Fig. 3b). The convergence of χ^2 to zero does not depend significatively on the number n of different symbols used in the partition of the interval. Actually, different values of n were tested (from n=2 to n=5) and give similar results. Two results are to be retained: a) the succession of time intervals is far from being uncorrelated and b) the order of the process does not depend on the number of symbols n.

One of the most remarkable facts is that these sequences produce only few words with high frequency while some other words never appear. Among all the 2^9 possible words of length L=9 (n=2), only 279 were realized by the dynamics of ECG intervals and only 151 words were seen in EEG sequences (for comparison, the Rössler model produced only 21 sequences among 3^7 [11]). This finding shows that despite the apparent irregularity of these symbol sequences, strong "grammatical rules" may exist. These rules probably reflect the deterministic nature of the underlying dynamical system.

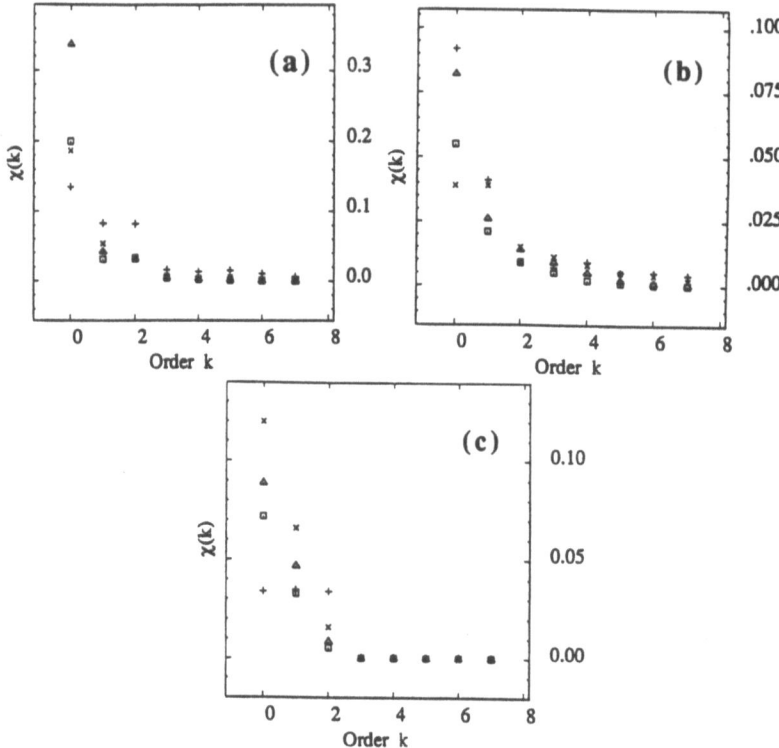

Figure 3. Test statistic χ^2 vs. order k for the sequences of symbols extracted from the signals of fig. 2 : (a) ECG (N=2132), (b) EEG (N=1756). Several different partitions of the interval in n symbols have been considered: n=2 (+), n=3 (X), n=4 (triangles) and n=5 (squares). (c) Third order Markov process obtained numerically from a random number generator

CONCLUSIONS

The mapping of the Rössler model into symbolic dynamics reveals highly correlated sequences of symbols, according to the strong regularities inherent in the structure of the attractor. From the reformulation of the sequence of symbols into hypersymbols, it appears that the dynamics reduces to a stochastic succession of three characteristic orbits.

Biological time series where the same pattern of activity is repeated at irregular time intervals can be studied through symbolic dynamics. The duration of the time interval is partitioned into several typical lengths, each corresponding to a defined symbol. The evaluation of the order of the sequence obtained from heart and brain recordings leads to unexpected high values. Despite the apparent randomness of these two rhythms, the intervals between spikes are remarkably correlated. A simple model of a noise-driven limit cycle is therefore insufficient to describe this activity. Following the previous work where chaotic dynamics were observed from the evaluation of the attractors dimensions [4,5,14,15], such strong correlations provide further evidence that the aperiodic time evolution of these important physiological states stems from chaotic dynamics.

The evaluation of the order of the Markov process obtained from the mapping of a chaotic system provides interesting insights into the system's dynamics, which are complementary to the information gained by analyzing the topological properties of the attractor. Moreover, this approach depends on relatively simple algorithms and can be used independently of any specific hypothesis concerning the reconstruction of phase space.

REFERENCES

[1] Eckmann J.P. and Ruelle D., Rev. Mod. Phys. **57** (1985), 617
[2] Packard N.H., Crutchfield J.P., Farmer J.D. and Shaw R.S.,Phys. Rev. Lett. **45** (1980), 712 ; F.Takens: in **Dynamical systems and turbulence.** Eds D.A.Rand & L.S.Young. Lect. Notes in Math. **898,** 366 (Springer, Berlin, 1981)
[3] Theiler J., Phys. Rev. A **34,** (1986) 2427, and references therein.
[4] Babloyantz A. and Destexhe A., in **Temporal disorder in human oscillatory systems** Eds. L. Rensing, U. an der Heiden and M.C. Mackey, Springer Series in Synergetics **36** ,48 (Springer, Berlin, 1987)
[5] Destexhe A., Sepulchre J.A. and Babloyantz A., Phys. Lett. A **132** (1988), 101.
[6] Eckmann J.P. and Ruelle D., preprint, (1989)
[7] Aizawa Y., Prog. Theor. Phys. **70** (1983), 1249
[8] Feller W., in: **An Introduction to Probability Theory and its Applications** (Wiley, New York, 1957)
[9] Nicolis G.and Nicolis C., Phys. Rev. A. **38** (1988), 427
[10] Rössler O.E., Ann. N.Y. Acad. Sci. **316** (1979) 376
[11] Nicolis G., Rao G.S., Rao J.S. and Nicolis C., in **Structure, Coherence and Chaos in Dynamical Systems,** Eds. Christiaensen P.L. and Parmentier R.D. (Manchester University Press, 1989)
[12] Kemeny J.G., Snell J.L. and Knapp A.W., in **Denumerable Markov Chains,** Graduate texts in Mathematics (Springer, Berlin, 1976)
[13] A. Destexhe, Phys. Lett. A, in press (1989)
[14] Babloyantz A. and Destexhe A., Biological Cybernetics **58** (1988), 203
[15] an almost exhaustive state of the art can be found in: **Dynamics of Sensory and Cognitive Processing by the Brain,** Ed. E. Basar, Springer Series in Brain Dynamics **vol.1** (1988), **vol.2** (1989).

MODELLING DYNAMICAL SYSTEMS FROM REAL-WORLD DATA

Alistair Mees

Mathematics Department
The University of Western Australia

INTRODUCTION

This summary describes some of my work on construction of dynamical system models from data, as part of a larger project to identify nonlinear dynamics and distinguish it from noise. In the space available it is only possible to look briefly at a number of different ideas and applications. The reader is referred to the bibliography for fuller details.

MODEL CONSTRUCTION

Suppose we have some data from a physical system which we believe to be well-approximated by a deterministic dynamical system. It seems reasonable to test our belief by trying to produce such a system explicitly. I have investigated a number of methods for doing so, but here I will describe only one, which I will call the tesselation method. It builds a geometric model which gives useful insights but also has the ability to make detailed numerical predictions, both of actual time series values and also of bulk properties such as Lyapunov exponents, dimensions and so on. Some alternative ways to produce a model system, such as fitting a parametrised analytic model, seem too limited, though recently in the literature there have appeared several approaches which are similar in spirit to the present one (Casgagli, 1989; Farmer and Sidorowich, 1987).

The data is first embedded if necessary; in the context of the present conference I make neither explanation nor apology for this. I shall simply assume that somehow an adequate embedding dimension, lag and so on have been identified and we are effectively given a time series of k-dimensional vectors,

$$v_0, v_1, \ldots, v_N.$$

The assumption that the data is produced by nonlinear dynamics is equivalent to

$$v_{t+1} = f(v_t)$$

Measures of Complexity and Chaos
Edited by N.B. Abraham *et al.*
Plenum Press, New York

for t=0,1,...N-1. Eventually I am going to discuss methods of approximating f from the given data. In keeping with the current dynamical systems philosophy, I assume f is smooth; in some cases it would be desirable to assume f is a diffeomorphism, but I neither make this assumption nor endeavour to make the approximation to f a diffeomorphism.

The sequence of vectors $\{v_i\}$ is often treated as merely a cloud of points in k-dimensional space; for example, fractal dimension calculations are not sensitive to the ordering in the time series. There is, however, much more information available. It is easiest to begin by looking at the geometry and then consider the dynamics.

Tesselation and geometry

Assume the points lie on an attractor. This is not necessary, and for some purposes is actually disadvantageous, but there is insufficient space to go into such matters here. The attractor will lie in some manifold, which may be part of a submanifold of the k-dimensional embedding space or may have full dimension. To identify the manifold, it is natural to triangulate it using the data points. There is a method for doing so which has certain optimality properties; this is the Delaunay triangulation. It is dual (in the points <-> lines sense) to a tiling of the space called the Dirchlet tesselation. There is one tile for each point v_i, and points in the tile are

$$\{ x : | x - v_i | < | v_i - v_j | \text{ for all } j \text{ not equal to } i \}.$$

Because of the duality, it is easy to construct the tesselation from the triangulation, or vice versa. There are several methods of constructing either the tiling or the tesselation in two dimensions, but very few in higher dimensions; I have found the method of Bowyer (1981) to be satisfactory. For present purposes, let us regard the construction as a solved problem.

Once we have constructed the simplices (that is, the generalised triangulation) we can identify whether their points appear to belong to a lower dimensional manifold than k dimensions, by examining neighbours and looking for evidence that subsimplices are approximating a tangent manifold. A simple example is a torus. If the given points are 3 dimensional and come from (say) a dense orbit on a 2-torus , we can use Bowyer's algorithm to build a triangulation consisting of 3-simplices which fill the interior of the torus as well as the hole in the middle. Even pretending we do not know what kind of object the points came from, it is relatively easy to scan through the simplices and discover that 3 of the 4 faces of each simplex may be discarded because the remaining face fits with neighbours to form an approximation to a 2-dimensional surface (Mees, 1989).

Given a triangulation of the right dimension, we now have a good image of the attractor. Already we can bring in some dynamics. To simplify the language and to give some concreteness, suppose the dimension is 2 and the triangulation really does consist of triangles. Since the embedded time series tells us the image (under iteration) of each point except v_N, we can see how the vertices of each triangle behave after one time step. Let us pretend (on the assumption of local linearity) that the interior of the triangle maps approximately to the interior of the triangle defined by the images of the vertices. Then after one step, each triangle $\{v_a, v_b, v_c\}$ is mapped into another which is identified simply as the triangle $\{v_{a+1}, v_{b+1}, v_{c+1}\}$. Call the set of triangles that results the one-step image of the triangulation. It will not usually be a triangulation any more, since the images of the triangles will often overlap, but it is still very informative.

Simple examination of the one-step image helps in identifying chaos-producing features: folds and branches are often very evident (Mees, 1989).

Moreover, we have a way of estimating Lyapunov exponents: examine the inflation and contraction of each triangle, calculate the eigenvalues, and average over all triangles.

Tesselation and dynamics

Let any point x on the attractor be written as a linear combination of the data points v_i. Some of the coefficients $p_i(x)$ in the linear combination may be zero; in the present case, only those corresponding to neighbours of x will be taken as nonzero. Here a neighbour of x is a point v_i such that, if x is inserted into the tesselation, the tiles belonging to x and to v_i are adjacent. It can be shown (see Mees, 1989 for references) that if the weight $p_i(x)$ corresponding to v_i in the linear combination is taken as pro-portional to the area of the x tile that was taken from the original v_i tile, with the sum of the weights taken to be 1, then p_i is a smooth function of x. A simple consequence of Taylor's theorem (given sufficient smoothness of f) is that the approximation to f(x) given by

$$\widehat{f(x)} = \sum_i p_i(x)f(v_i)$$

is accurate to first order. Since we know that $f(v_i) = v_{i+1}$, this defines the approximation completely, at least for any point in the convex hull H of the original data (since the triangulation only covers that convex hull).

It is now possible to define a time series starting from any point in H. This has proved remarkably successful both on artificial data (constructing the Lorenz or Henon attractors, for instance) and on real-world data including lynx population, blowfly population, sunspots, and stock market data (Mees, 1989). Dynamically interesting objects such as periodic points can be found relatively easily.

Adaptive smoothing splines

One disadvantage with the tesselation method is that the work required goes up very rapidly with k; indeed, it appears that the number of neighbours increases exponentially with k. There are also some problems with noisy data, in that once the typical triangle size is smaller than the noise scale, the effects of noise make the model behave badly. If we are prepared to forego much of the geometric information given by the triangulation, there are other methods which may involve less computation or be more robust against noise. At present I am investigating the use of adaptive smoothing splines (Friedman, 1988), which produce a smooth map from noisy data.

CONCLUSIONS

The enormous, and perhaps misguided, interest in calculating chaos indicators such as fractal dimension stems largely from a desire to discover when real-world phenomena are caused by deterministic nonlinear dynamics (and so are, in part and for a limited time, predictable) rather than being random. It is difficult to distinguish noisy simple dynamics - say, a sine wave plus noise - from noise-free, or low-noise, complex dynamics. This paper is a contribution to the program of identifying cases where the deterministic model has more to offer than the stochastic one.

REFERENCES

A. Bowyer, Computing Dirichlet tesselations, The Computer Journal, 24(2), 162-166 (1981).

M. Casdagli, Phys. Rev. Letters, to appear (1989).

J.D. Farmer and J.J. Sidorowich, Predicting chaotic time series, Phys. Rev. Letters 59(8), 845-848 (1987).

J.H. Friedman, Multivariate adaptive regression splines, Technical Report 102, Laboratory for Computational Statistics, Stanford University (1988).

A.I. Mees, Modelling Complex Systems, Proceedings of the Conference on Modelling Complex Systems, eds. L.S. Jennings, A.I. Mees and T.L. Vincent, Birkhauser-Boston (1989).

EXTRACTION OF MODELS FROM COMPLEX DATA

H.G. Schuster

Institut für Theoretische Physik und Sternwarte
Universität Kiel

In this article we are concerned with the problem of prediction and data reduction of complex dynamical systems[1]. We will try to answer the following questions:

- How does one have to choose the delaytime τ and the embedding dimension m, in order to obtain an optimal reconstruction of the strange attractor of a chaotic system from the time series of a single variable?

- How could one use unstable periodic orbits, in order to extract models that can be used for prediction?

- What is the optimal encoding of the prediction function?

I. OPTIMAL ATTRACTOR RECONSTRUCTION FROM TOPOLOGICAL
    CONSIDERATIONS[2]

According to Takens' theorem[3] the strange attractor to which the trajectory $\vec{x}_i$ in phase space of a chaotic dissipative system becomes attracted in the course of time can be reconstructed from a scalar time series $x_i = x(i \, \delta t), i = 1 \ldots N$ using delay coordinates

$$\vec{x}_i = [x_i, x_{i+\tau}, \quad \cdots \quad x_{i+(m-1)\tau}] \qquad (1)$$

where τ is a delay time and m is the embedding dimension of the attractor. While for infinitely long time series the choice of τ should be rather arbitrary this is no longer the case if the number of data points in finite.[4] We will solve the problem of finding on optimal value of τ using the idea that an embedding via delay coordinates is a topological mapping which preserves neighborhood relations. Therefore if the chosen embedding dimension is too small one obtains a projection of the attractor which violates the condition of injectivity. For the proper values of m and τ one will obtain an optimal spanning of the attractor and neighboring points will have their largest distances.

Measures of Complexity and Chaos
Edited by N.B. Abraham *et al.*
Plenum Press, New York

Fig 1 shows schematically how neighborhood relations could change for $m \to m+1$. Since distances can only increase if one goes from $m \to m+1$ we consider the ratio

$$Q(i,k) = \frac{\mathrm{dist}_{m+1}(i,j(k,m))}{d_{m+1}(i,j(k,m+1))} \geq 1 \qquad (2)$$

where the distances are defined in Fig. 1, and the equality sign holds only if the proper embedding is obtained. Since distances on strange attractors separate exponentially in time we consider $\log Q(i,k)$ and we also take the average over p neighbors of N points $\vec{x}_i$. The resulting quantity

$$\overline{W} = \frac{1}{\tau} \frac{1}{p} \frac{1}{N} \sum_{i,n} \log Q(i,k) \qquad (3)$$

in which a trivial τ dependence has been divided out should approach zero for the proper choice of m and τ.

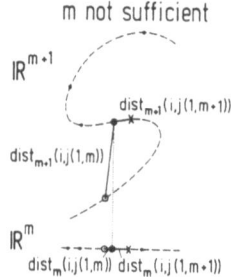

m not sufficient

Figure 1. Illustration of the delay coordinate mapping as a Projection of a (m-1)-dimensional reconstruction space onto a m-dimensional one. The change of nearest neighbor order near a reference point $\vec{x}_i$ by transition from embedding dimension m to m + 1 is visualized for not sufficient embedding dimension. The crosses mark the next neighbor in R^{m+1} and the circles the next neighbor in R^m respectively.

Fig. 2 shows $\overline{W}$ versus τ for the Rössler attractor[5] (with a = 0,15, b = 0,2, c = 10, $\delta t = \frac{\pi}{25}$) where we used 9000 data points and p = 10. The optimal values found are m = 3 and $\tau^* = 0.143\, T_c$ where T_c is the first return time of the system. It can be seen from Fig. 2b that the information dimension converges for m = 3 and τ^*. For $\tau < \tau^*$ one still finds convergence if only the embedding is extended from m = 3 to m = 4 up to 2d + 1 = 7 (where d is the true dimension of the system). However there is no convergence for $\tau^* < \tau$. This means that finding correct values of τ and m is necessary condition for proper data analysis and prediction.

II. UNSTABLE PERIODIC ORBITS AND PREDICTION[6]

It is well known that the closure of the set of unstable periodic orbits of a dynamical system defines its strange attractor.[7] The strange

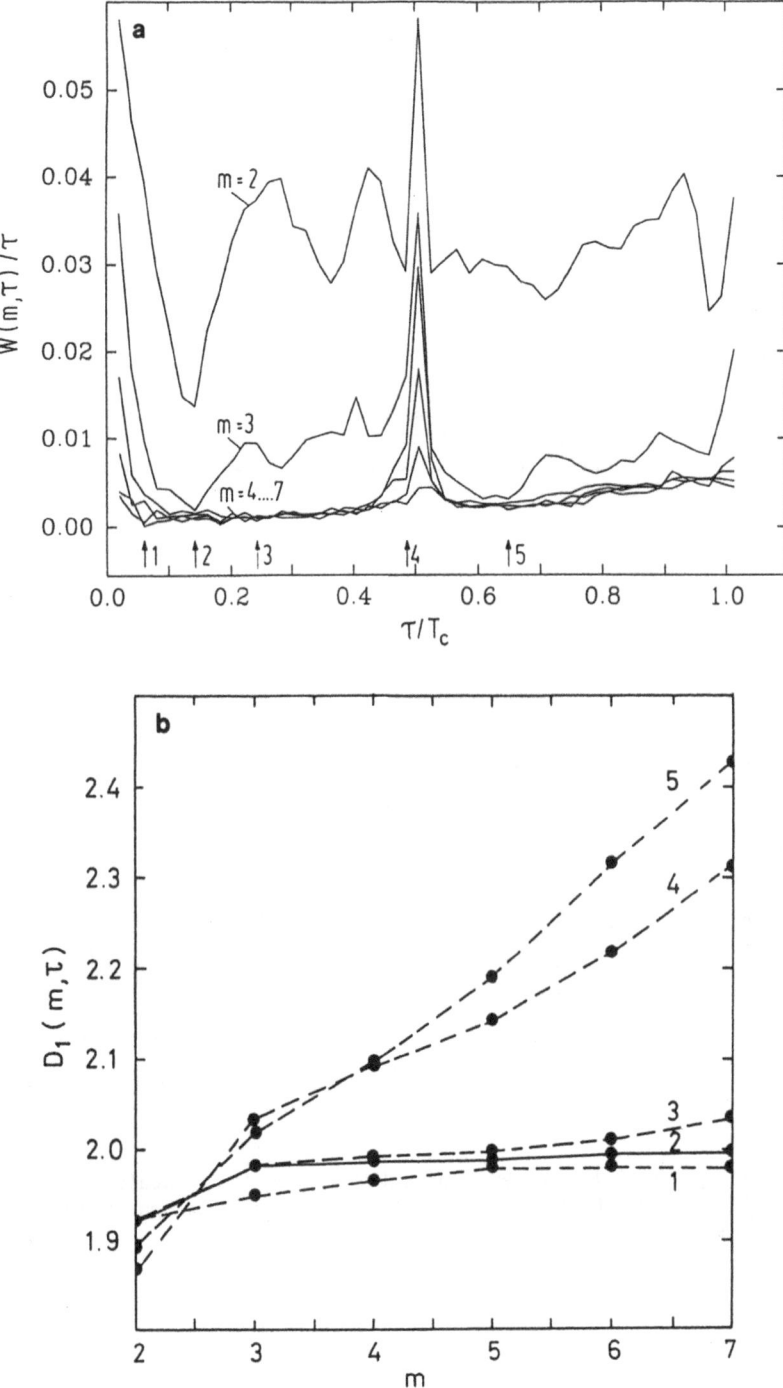

Figure 2. Numerical results for the Rössler system (see Ref. 5) with a = 0.15, b = 0.2, c = 10, δt = π/25 using N = 9000 data points.
(a) $\bar{W}$ for a range of embedding dimensions m and delay times τ in terms of the first recurrence time T_c. (b) Information Dimension versus embedding dimension m for some values of delay time τ, as marked by numbered arrows in (a).

attractor can be approximated hierarchically by unstable periodic orbits.[8] In the following we will show how these orbits can be extracted for systems which are continuous in time and how they improve modelling and predictions.

In order to extract unstable periodic orbits from the time series of a time continuous system one first has to look for the most probable return times because there are times for which there are no returns (e.g. for the Rössler system[5] the shortest return time is the order of one cycle of the system). Fig. 3 shows the inverse of $\varepsilon_p(t)$ which is the shortest distance for which there are p returns in time t for the Rössler system.[5] Large $\varepsilon_p^{-1}(t)$ signals high return probability and yields an estimate for the times for which one finds unstable periodic orbits. The value of $\varepsilon_p(t)$ itself sets the scale of accuracy in phase space within which one has to look, for return of these orbits. Having found these two numbers one can follow the methods of ref. 9 which have been developped to extract unstable periodic orbits for maps. Fig. 4 shows some projections fo unstable periodic orbits which have been obtained form 10^4 points of the Rössler system.[5]

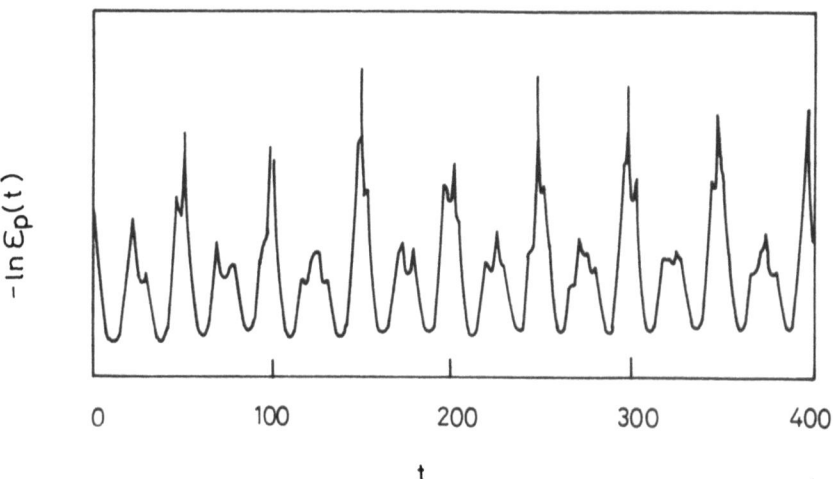

Figure 3. Log of the inverse of $\varepsilon_p(t)$ versus time obtained from 10^4 points of the Rössler system with the same parameters as in Fig. 2.

Let us now come to the question of modelling. By model we mean an approximation $\tilde{f}$ of the flow f which generates the time series. While in the true coordinates $\vec{x}_i$ the flow is a map $f : R^d \to R^d$ this changes to a map $f^\tau: R^m \to R$ if we consider delay coordinates and take time steps of size τ. The evolution

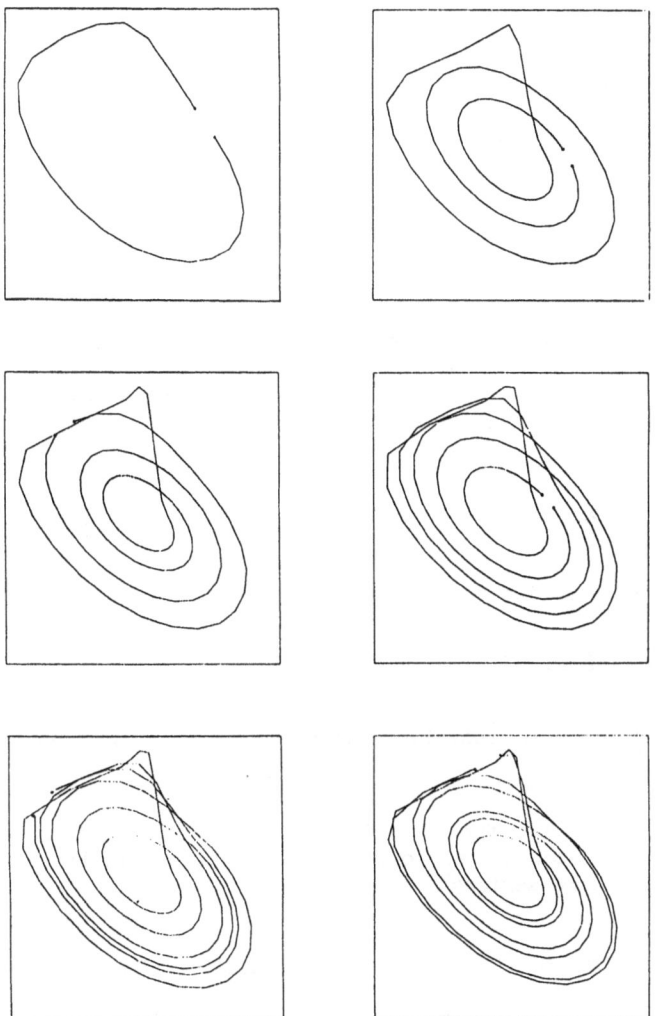

Figure 4. Some unstable periodic orbits extracted from a time series of 10^4 points of the Rössler system. The sum of these orbits tends to span the Rössler attractor.

$$\vec{x}_{i+\tau} = \vec{f}^{\tau}[\vec{x}_i] \tag{4}$$

reads explicitly in delay coordinates

$$x_{i+m\tau} = f_1^{\tau}[x_i, \ldots, x_{i+(m-1)\tau}] \tag{5}$$

$$x_{i+(m-1)\tau} = f_2^{\tau}[x_i, \ldots, x_{i+(m-1)\tau}]$$

$$x_{i+\tau} = f_m^{\tau}[x_i, \ldots, x_{i+(m-1)\tau}]$$

with m ≥ d. Next we observe that the only nontrivial map in equ. (5) is the first one

$$x_{i+m\tau} = f^{\tau}[x_i, \ldots x_{i+(m-1)\tau}] \tag{6}$$

where we dropped the index 1 .

In order to determine the function $f^{\tau}(\vec{x})$ we approximate it by a polynomial $\tilde{f}^{\tau}(\vec{x})$ and determine the free coefficients from the requirement that the prediction error of the first time step τ becomes a minimum. This means if we use N points from a time series on the attractor:

$$\frac{1}{N} \sum_i \{ x_{i+m\tau} - \tilde{f}[x_i, \ldots x_{i+(m-1)\tau}] \}^2 = \min \tag{7}$$

and if we use only periodic orbit points $\vec{x}_i^*$:

$$\sum_i \{ x_{i+m\tau}^* - \tilde{f}[x_i^*, \ldots x_{i+(m-1)\tau}^*] \}^2 \cdot e^{-\lambda_i} = \min \tag{8}$$

where $e^{-\lambda_i}$ is the sum of positive eigenvalues of the Jacobian Matrix of f at $\vec{x}_i^*$.

Fig. 5 shows that for the Rössler system[5] the use of periodic orbits leads to a much better long time prediction than the simple use of points from a time series[10].

III.   AN APPLICATION OF KOLMOGOROV'S SUPERPOSITION THEOREM[11]

In this section we want to introduce and use Kolmogorov's superposition theorem[12] to represent the "prediction function" $f^{\tau}(\vec{x})$ by means of a one dimensional function. Kolmogorov's theorem[12] states that any continuous function f in d dimensions which is defined in $x_i \in [0,1]$ i=1,...d can be written as

$$f(x_1, \ldots x_d) = \sum_{q=1}^{2d+1} g[\lambda_1 \varphi_q(x_1) + \ldots + \lambda_d \varphi_q(x_d)] \tag{9}$$

where the continuous functions φ_q are monotonically increasing in $[0,1]$ and λ_i are different positive constants with $\sum_i \lambda_i = 1$. The main point is that the $\varphi_q(x)$ can be chosen rather freely and do not depend on f

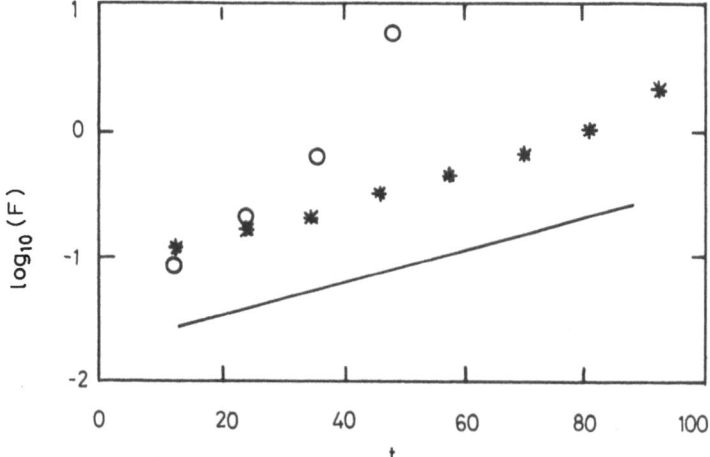

Figure 5. Log of the mean prediction error

$$F = \sum_i (x_{i+t} - \tilde{f}(\vec{x}_i))^2 / \sum_i (x_{i+t} - \bar{x})^2 , \text{ where } \bar{x} \text{ ist the time}$$

average of x_i, as a function of the prediction time t for
the Rössler system using 10^4 points. The circles and stars
are predictions using points from a time series and unstable
periodic orbits (extracted from these points) respectively.
The line whose slope is the K-entropy indicates the smallest
prediction possible error, in this chaotic system.

such that the d-dimensional function $f(x_1, \ldots x_d)$ is essentially
represented by and reduced to the one dimensional continuous (but not
necessarily differentiable) function $g(z)$. While proofs for this theorem
are given in references 13, 14 it seems from our search in the
literature that actual calculations of $g(z)$ are still lacking. Therefore
we will derive below an integral equation for $g(z)$ and demonstrate its
solution by way of examples.

For simplicity we consider only the case d = 2 and write equ. (9)
as

$$f(x,y) = \sum_{q=1}^{5} \int_0^1 dz' \; g(z') \; \delta[z_q'(x,y)-z'] \qquad (10)$$

where $z_q(x,y) = \lambda_q \varphi_q(x) + \lambda_q \varphi_q(y)$. After multiplication of equ. (10)
with $\sum_{q=1}^{5} \delta(z_q(x,y)-z)$ and integration over x,y we obtain:

$$\int_0^1 dz' \; K(z,z') \; g(z') = \chi_f(z) \qquad (11)$$

where $K(z,z') = \sum_{q \; q'} \int_0^1 dx \; dy \; \delta[z_q(x,y) - z] \; \delta[z_q'(x,y)-z'] \qquad (12)$

355

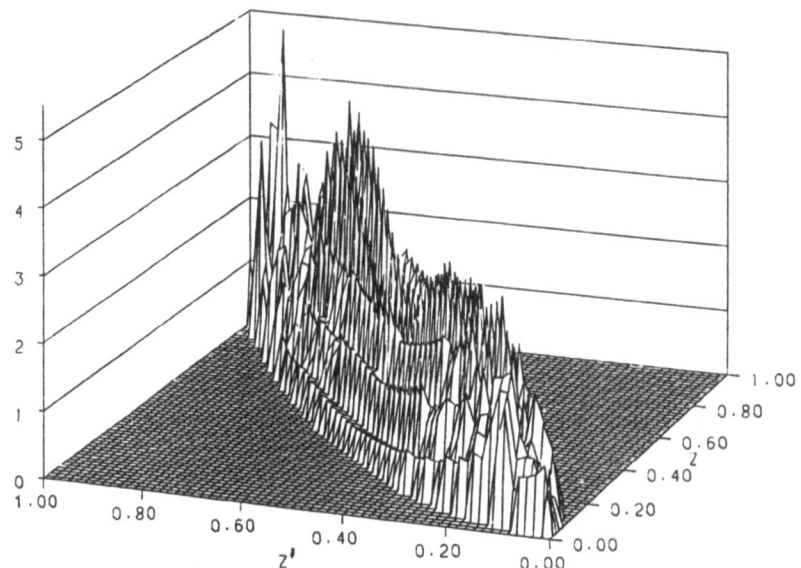

Figure 6. Kernel (K(z,z') from equ. (12) for
$z_q(x,y) = \lambda_1[(1-a_q)x^2 + a_q x] + \lambda_2[(1-a_q)y^2 + a_q y]$
($\lambda_1 = 0.67$, $\lambda_2 = 0.33$, $a_q = 0.1, 0.3, 0.5, 0.7, 0.9$)

$f(x,y) = 1 - 1.4 \, x \ast x + 0.3 \, y$

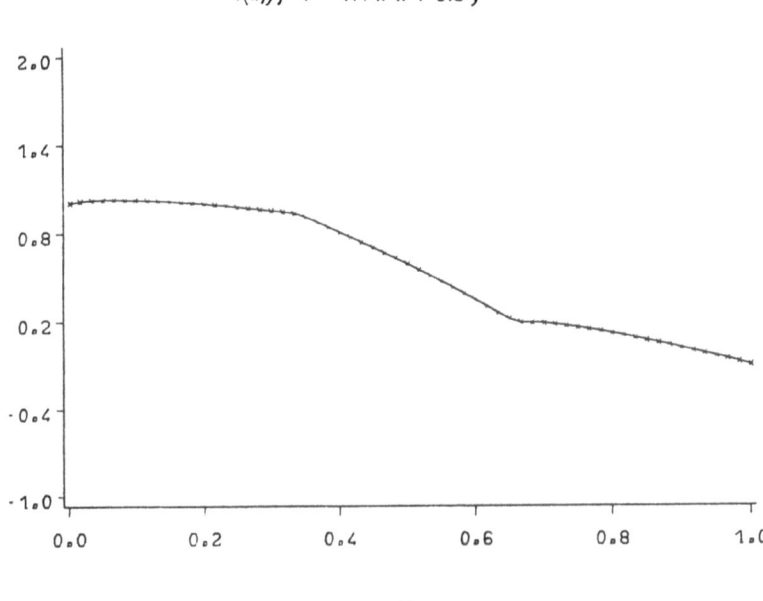

Figure 7. g(z) for the one step prediction function
$f(x,y) = 1 - ax^2 + by$ of the Hénon map (a =1.4, b= 0.3).

and $\quad \chi_f(z) = \sum_q \int_0^1 dx \quad dy \quad \delta[z_q(x,y) \quad -z] \quad f(x,y)$. $\quad$ (13)

We have solved the integral equation for $\chi_f(z) = \delta(z-\bar{z})$. The resulting Green's function $G(z,\bar{z})$ yields

$$g(z) = \int_0^1 d\bar{z} \; G(z,\bar{z}) \; \chi_f(\bar{z}) \qquad (14)$$

which reduces the computation of $g(z)$ from $f(x,y)$ to simple integrations. The Kernel $K(z,z')$ is shown in Fig. 6. We used for $\varphi_q(x)$ parabolic functions which induce integrable one over square root singularities. The example for $g(z)$ given in Fig. 7 demonstrates that the representation of the prediction function $f^\tau(\vec{x})$ via $g(z)$ offers a convenient way of data reduction. It should be noted that $g(z)$ contains all the information that is necessary to predict the behaviour of a d-dimensional nonlinear dynamical system.

Acknowledgement

It is a pleasure to thank W. Liebert, K. Pawelzik, G Radons and D. Werner for their enthusiatic cooperation and the Sonderforschungsbereich 185 for financial support.

References

1. For an introduction see e.g. H.G. Schuster, "Deterministic Chaos" second edition, VCH publishers, Weinheim (1988)

2. W. Liebert, K. Pawelzik and H.G. Schuster, to be published

3. F. Takens, in: Springer Lecture Notes in Mathematics 898, 366 (1981)

4. A.M. Fraser and H.L. Swinney Phys. Rev. 33A, 1134 (1986)

5. O.E. Rössler, Phys. Lett 57A, 397 (1976)

6. K. Pawelzik and H.G. Schuster, to be published

7. D. Ruelle, "Thermodynamic Formalism" Addison Wesley, Reading (1978)

8. P. Cvitanovic Phys. Rev. Lett 61, 2729 (1988)

9. A. Auerbach, P. Cvitanovic, J.P. Eckmann, G. Gunartne and I. Procaccia Phys. Rev. Lett 58, 2387 (1987)

10. J.D. Farmer and J.J. Sidorowich, Phys. Rev. Lett. 59, 845 (1987)

11. D. Werner, G. Radons and H.G. Schuster, to be published

12. A.N. Kolmogorov, Dokl Akad. Nauk SSSR 114 (1957); Amer. Soc. Transl. 28 55 (1963)

13   G. Lorentz Approximation of Functions Holt Rinchert and Winston
      N.Y. (1966)

14.   J.P. Kahane, Journ. Approx. Theory 13, 229 (1975)

QUANTIFYING CHAOS WITH PREDICTIVE FLOWS AND MAPS:

LOCATING UNSTABLE PERIODIC ORBITS

Leonard A. Smith

Department of Applied Mathematics and Theoretical Physics
University of Cambridge, Cambridge CB3 9EW, U.K.
las10@phx.cam.ac.uk

ABSTRACT

Several authors have suggested methods for constructing "predictors" from time series data. It is shown that using one type of predictor, constructed with radial basis functions, for some known chaotic systems, the existence of unstable periodic orbits may be established using much less data than that required by alternative methods. The general question of quantifying the error in a predictor is also addressed. Considering the fraction of the data that can be predicted as a function of the accuracy of the prediction provides a method of distinguishing different sources of error in the predictor and, in doing so, yields an estimate of the magnitude and distribution of the observational noise in the system.

INTRODUCTION

In the past, low dimensional dynamical systems have been represented by points generated directly from an observed trajectory; coverage of the relevant regions of phase space has required data records both long (in terms of the rate at which information appears in the system) and finely sampled. Recent proposals to quantify systems not with the data directly, but with reconstructed flows (Broomhead, 1988, Castagli,1989, Crutchfield and McNamara,1987, Farmer and Sidorowich, 1988, and Mees,1989) offer a new approach to predicting and quantifying nonlinear dynamical systems. While a wide variety of reconstruction techniques have been proposed, all are based on utilizing the information contained in the time ordering of the data points to construct an M-dimensional interpolation scheme. The techniques differ in the method used to form the interpolation from the known data points. Hereafter, a reconstructed flow (or map) will be referred to as a predictor. Since the information in the time ordering of the points is not used in the geometrical analysis (Grassberger and Procaccia,1983), constructing a good predictor may require less data than the direct calculation of generalized dimensions (Smith, 1988).

In this paper, predictors constructed from time series data are used to find and quantify the unstable periodic orbits of the chaotic system which generated the series. New difficulties which arise in locating the periodic orbits of continuous systems are discussed. The interpolation technique employed uses the radial basis functions (see Powell, 1985) originally applied to nonlinear time series data by Castagli(1989). When the system is known, the convergence of true and reconstructed flows is examined by comparing the unstable periodic orbits present in each system.

To quantify the quality of prediction, the fraction of the time series which can be predicted within an accuracy a is determined. The behavior of this "predictor error profile", $P_f(a)$, is related to changes in the number of points used in constructing the predictor and the time scale of the prediction. It can be used to quantify the noise in a deterministic system. The variation of $P_f(a)$ with the time scale of the prediction provides a tool for distinguishing stochasticity from determinism and suggests

Measures of Complexity and Chaos
Edited by N.B. Abraham *et al.*
Plenum Press, New York

a numerically efficient method for estimating an optimal delay time for reconstructing the system.

Predictors for several systems have been studied including the Henon map (Henon, 1976), the Moore-Spiegel system (Moore and Spiegel, 1966), the Duffing oscillator (either periodically or stochastically forced, see Stone and Holmes (1989)), an Ornstein-Ulhembeck process and experimental results of baroclinic flow (Read, 1989), boundary layer turbulence (Gaster, 1989) and an electrical oscillator (King, 1989). In addition real world data records of climate and solar activity have been considered. In general, predictors for the numerical systems give excellent results; those for low dimensional laboratory systems provide good predictions and successfully locate unstable periodic orbits. The correct interpretation of the results for the real world systems is not yet clear. This short contribution will concentrate on the numerical models with a few remarks on the laboratory experiments. A more detailed report is in preparation.

RADIAL BASIS FUNCTION PREDICTORS

An introduction to the prediction of chaotic dynamical systems using radial basis functions is given by Castagli (1989). Only a very brief summary is provided here. Consider a deterministic system with phase space dimension M_s. A trajectory, x(t), of this system may be constructed in M_s dimensions from a time series of single observable, o(t), by the method of delays (Packard et al. 1980) to yield $x(t) = (o(t), o(t + \tau_d), \ldots, o(t + (m-1)\tau_d))$. Each point on the trajectory has an image $s(t) = o(t + t_p)$, where t_p is determined either by a fixed prediction time, τ_p (i.e. $t_p = (m-1)\tau_d + \tau_p$) or through some geometric constraint (as when taking a surface of section of the trajectory).

Choosing n distinct points $(x_i, i = 1, 2, \ldots, n)$ and their n images $(s_i, i = 1, 2, \ldots, n)$, the goal is to determine a predictor $f(x_i) : R^{M_s} \to R$ such that

$$f(x_i) = s_i \tag{1}$$

Following Powell (1985), consider $f(x)$ of the form

$$f(x) = \sum_{i=1}^{n} \lambda_i \phi(||x - x_i||) \tag{2}$$

where $\phi(r)$ are radial basis functions and the λ_i are constants which are uniquely determined by equations (1) provided the matrix

$$A_{ij} = \phi(||x_i - x_j||) \tag{3}$$

is nonsingular. This is always the case (Micchelli,1985, Powell,1985) for $\phi(r)$ of the form

$$\phi(r) = (r^2 + c^2)^{-\beta} \tag{4}$$

for $\beta > -1$ and $\beta \neq 0$. The particular cases $\phi(r) = r$ and $\phi(r) = (r^2 + c^2)^{1/2}$ (referred to as the "multiquartic") will be considered here.

When the predictor is formed in delay coordinates, a single map $f(x) : R^{M_s} \to R$ is sufficient to predict the system. Prediction in singular value decomposition (SVD) coordinates (Broomhead and King, 1987), or in the true phase space, requires M_s such maps $f^j(x)$, $(j = 1, \ldots, M_s)$, one for each coordinate.

In order to both construct and test a predictor with the same data set, the trajectory is divided into two parts, a "base" portion from which the base points, x_i, are chosen, and the remaining portion on which the predictor is tested. The simplest method for choosing base points is to distribute them uniformly along the base portion of the trajectory. Since determining the λ_i involves manipulating an $n \times n$ matrix and each prediction requires computing n distances in an M_s dimensional space, the computer power required to implement this procedure increases rapidly with n; it is therefore desirable to choose the base points efficiently by sparsely sampling regions in which the flow is smooth and densely sampling the flow elsewhere. A simple way to accomplish this is to build a predictor using $\frac{n}{2}$ points and choose additional points from the base portion depending upon how well the images of these points are predicted. To do this, distribute half the base points uniformly on the base portion and then predict the base portion of the trajectory. Half the remaining number of base points (one quarter the total) are chosen from those points whose images are predicted least accurately (with the condition that no two new points are too close to each other) and the procedure is repeated.

Finally, the initial (uniformly spaced) points may be removed and the entire procedure repeated. This approach can significantly reduce the number of points required to make good predictions; it does not, of course, lessen the data requirements because the base portion must be "long enough" to explore the entire attractor. The amount of data required for a "long enough" base portion will depend on the macroscopic structure of the attractor in question (as well as its dimension). Also note that data from the transient should be included in the base portion when possible.

Two dynamical systems are used as examples below. The Henon map (Henon,1976) is given by

$$x' = 1 - ax^2 + y \tag{5}$$

$$y' = bx \tag{6}$$

with $a = 1.4$ and $b = 0.3$. The Moore-Spiegel system (Moore and Spiegel, 1966) is given by

$$\frac{d^3z}{dt^3} + \frac{d^2z}{dt^2} + (T - R + Rz^2)\frac{dz}{dt} + Tz = 0. \tag{7}$$

This system displays chaotic behavior for $R = 100.0$ and $T = 36.0$, the case considered here. A $z = 0$ surface of section of this attractor is shown in Figure 1.

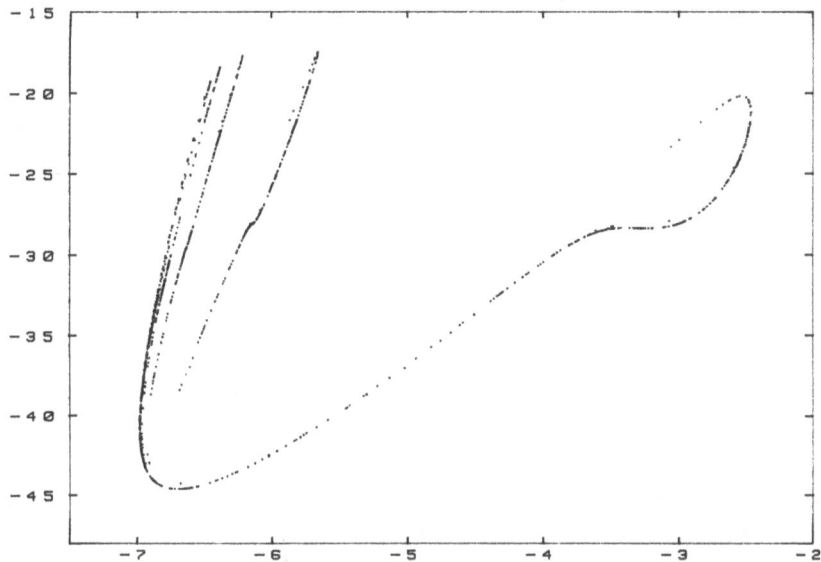

Figure 1. A $z = 0$ surface of section for the Moore-Spiegel system consisting of 2^{12} points.

EVALUATION OF A PREDICTOR

Intrinsic limits to very long term prediction of chaotic systems are imposed by the exponential divergence of nearby initial conditions, the average rate of which is quantified by the Lyapunov exponents of the system. It is misleading, however, to think of the Lyapunov exponents as limiting short term prediction at macroscopic length scales. The level of organization of the flow required to give exponential divergence (on average) is very high; there may be regions in which the flow is contracting, making prediction relatively easy. The predictability of a chaotic system should be expected to vary with location in phase space; there often exist regions of phase space in which chaotic systems are easier to predict than stochastic systems which, in the mean, diverge more slowly than exponentially and have Lyapunov exponents equal to zero.

The quality of a prediction varies for several reasons. Consider the error in predicting the image of some initial condition, x. In addition to the error introduced by the sensitive dependence to initial condition, error arises from failure of the base points to sufficiently cover the attractor; either when x is located in a region of the attractor not explored by the base portion of the trajectory or when

x is relatively close to base points, but is located in a region where the flow possesses significant fine structure. Predictors are often evaluated by computing the average difference between the predicted and observed values. This approach does not distinguish different sources of error. Comparing this average error to a "random guess" is also misleading, especially in continuous systems with persistence (or seasonality). In continuous systems, predicting the motion on a surface of section reduces the effects of persistence. To further distinguish these sources of error, consider a plot of the observed error in a one step prediction versus the distance from x to the nearest base point, d_n.

Such a plot for a predictor constructed for the Henon system with $n = 2^6$ consecutive base points is shown in Figure 2. In general, the observed error grows with increasing d_n; exceptions are the finely structured lines in regions of large error and small d_n. Points in these regions can be divided into groups which have the same nearest base point. The structures in Figure 2 are then seen to reflect the structure of the attractor near these base points; as the number of base points used in constructing the predictor increases, these structures become less apparent while the general trend of increasing predictor error with increasing d_n remains.

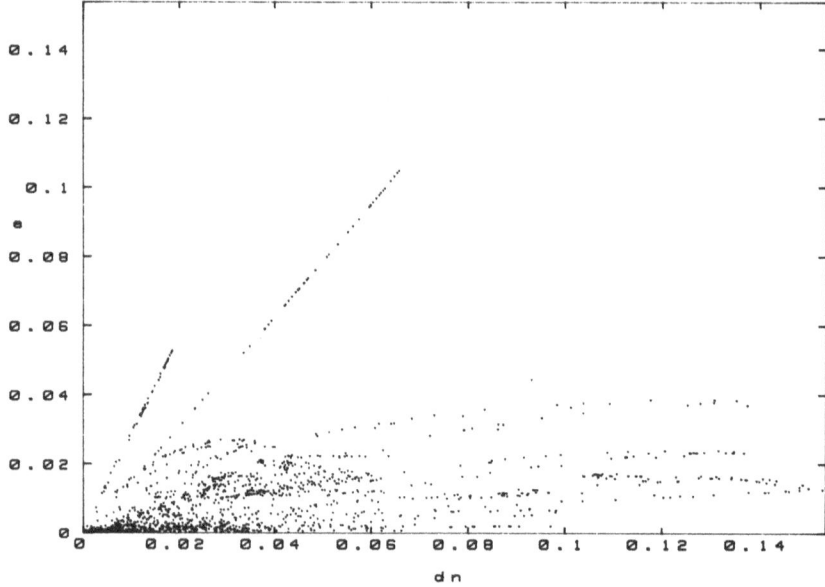

Figure 2. The observed error in the one step prediction of the Henon map versus the distance to the nearest base point.

An alternative approach to evaluating the predictor error is to consider the fraction of the data, $P_f(a)$, which can be predicted within a certain accuracy, a. Figure 3 shows this curve for predictors constructed from 2^5, 2^6, 2^7 and 2^8 base points for the Moore-Spiegel system using $\phi(r) = r$. The distribution of base points on the attractor and structure of the flow determine the shape of $P_f(a)$. As the number of base points is increased, there is improvement in the high end of the $P_f(a)$ curve (say, $P_f(a) > 0.9$) corresponding to new regions becoming predictable. The low end of the curve (say, $P_f(a) < 0.1$) is determined by the fine scale structure of the data; the details of the flow if the data is clean, the details of the noise otherwise. Given a perfect predictor of the system, predictor errors will remain due to observational effects (noise and truncation error) in the data. These effects impose a lower bound on the predictor error below which $P_f(a)$ is approximately zero. The shape and location of this transistion may be used to identify the distribution and magnitude of the observational noise. The $P_f(a)$ profiles of Gaussian, Poisson and uniformly distributed white noise are distinguishable. For well sampled attractors generated from clean experimental signals the resolution of the A/D converter is apparent as a sharp drop in $P_f(a)$. Data from numerical systems tends to show a slower transition which is dependent on the distribution of the base points; the prediction error for points located near a base point may be very low. For the case shown in Figure 3, the error in this data is off the scale, primarily arising from taking the surface of section. If noise of greater magnitude were added to the data, a sharper drop in $P_f(a)$ would be observed.

Comparing $P_f(a)$ curve for one step prediction with that for a many step prediction displays the change in predictability with increasing prediction time. When the underlying process is random, the $P_f(a)$ profile will not change (assuming persistence effects have been accounted for). For chaotic systems, the P_f profile should degrade toward some limiting case equivalent to making a random guess from an appropriate distribution. Finally, note that the quality of the predictor will vary with the details of the reconstruction. This suggests that observing the changes in $P_f(a)$ provides an efficient method for choosing the delay time, τ_d. Evidence that this is the case will be presented elsewhere.

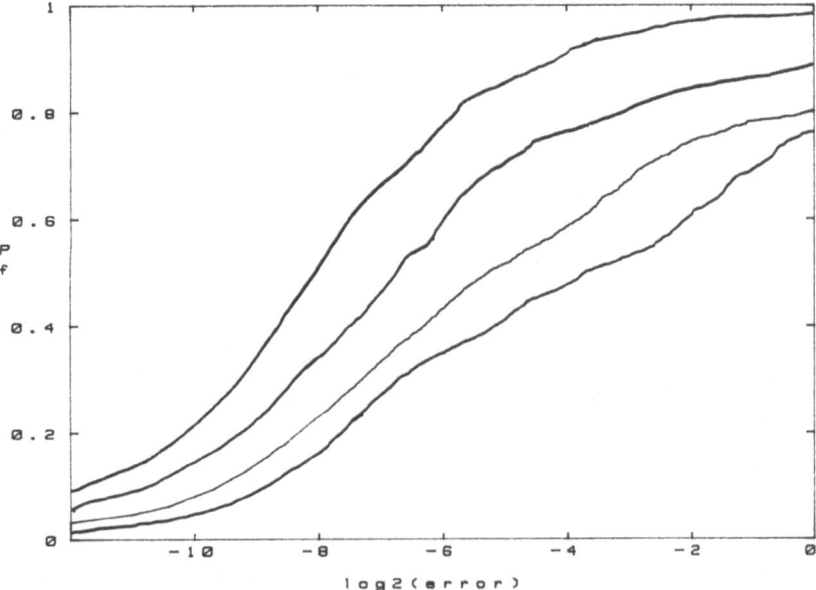

Figure 3. The predictor error profile for 4 predictors of the Moore-Spiegel system. The curves (from lowest for highest) correspond to predictors constructed with $n = 2^5, 2^6, 2^7,$ and 2^8 points.

UNSTABLE PERIODIC ORBITS

A catalogue of the unstable periodic orbits within a basin of attraction provides a useful description of chaotic attractors. These orbits may be viewed as a skeleton of the attractor; their properties (number and stability) have been successfully related to the metric properties of the attractor in the case of the Henon map (Cvitanovic et al., 1989). In this section a new method for determining these orbits is presented which consists, in short, of estimating the periodic orbits of the system by those of the predictor. The orbits and their stability are approximated by this method with modest data requirements; some orbits not accessible with other methods can be identified in a predictive flow. A detailed application to experimental data is in preparation (Smith and King, 1989).

When the system is known analytically, the original method proposed by Auerbach et al. (1987) is to select a grid of points on the basin of attraction and, starting at each point, apply a Newton-Raphson iteration scheme to search for an orbit of a given period. An alternative method is to follow a single trajectory and invoke the Newton-Raphson scheme at all near returns. This approach quickly locates the orbits near the attractor but has difficulties with orbits far from the attractor (or in rarely visited regions on the attractor). For the Henon map it finds periodic points up to period 10 in agreement with the results of Cvitanovic et al. (1989). Note however, that even in cases where the "exact" map is known, it is difficult to be certain that all the periodic orbits have been found.

In experimental flows, it is suggested that near returns be considered for each available time lag. A periodic point is then approximated by the center of mass of a ball of near returns and the eigenvalues may be estimated by assuming a linearized dynamics about this point. For the Henon map, Auerbach et al (1988) report data requirements of 2×10^5 iterations to locate orbits of period less than or equal to 10. This approach will locate only those orbits which are closely shadowed by a trajectory for at least one complete period. Using a predictive flow requires only that each segment of the orbit is visited at some point in the time series.

Considering continuous systems introduces complications not found in maps. When considering a surface of section of a continuous system, it is necessary to distinguish the number of times an orbit intersects the section from the period of the orbit. The former will be called the passage of the orbit and will, of course, depend on the section taken while the period will not. In maps, the passage and period of a periodic orbit are equal. The method of looking at a set time lag will miss periodic orbits when topologically nearby trajectories have different temporal appearances. The simplest example occurs when a periodic orbit passes close to a fixed point of the flow. The "period" (near return time) of initial conditions near such a periodic orbit will vary greatly, depending sensitively on their path near the fixed point. Two approximations of the same periodic orbit which passes through this region will not look the same when parameterized by time, although they are topologically similar. (This effect will also bias generalized dimension calculations by redistributing the probability density.) Taking a surface of section removes this problem. Points which are in the section will return, in general, after similar periods of time. Initial conditions near a periodic orbit which passes close to a stagnation point of the flow will have a wide distribution of first return times while those near other periodic orbits will have a tight distribution. This is, in fact, observed in the nonlinear electronic oscillator. Comparing the results from several surfaces of section provides an estimate of the error.

With a predictor, the methods developed for known systems may be applied to time series data. To test the stability of true orbits in the predictor, the known periodic points with periods less than or equal to 10 where used as initial conditions for a Newton Raphson style iteration procedure applied with a predictor constructed from 2^7 consecutive points on the Henon attractor with $\phi(r) = r$. In this case all periodic points of period less than 6 were present, with similar stability, to the full map. Most orbits of higher period were preserved, however their unstable eigenvalue often differed by a factor of 2 from the true orbit. Note however that the quadratic structure of the Henon map makes it a questionable test case. This is most apparent when multiquartics are used as the radial basis functions; in this case a 16 point predictor reproduces the map to astounding accuracy. Such multiquartic predictors are sensitive to noise.

A more general test case is provided by the $z = 0$ section of the Moore-Spiegel system. In this case, near returns on the section were used as initial estimates of periodic orbits in several predictors with $\phi(r) = r$. Predictors using 2^8 or fewer base points often had periodic orbits not present in Moore-Spiegel system. These spurious orbits often contain points far from the attractor. The spurious orbits are not found in a 2^9 point predictor where the location of the passage 9 points tested agree between the predictor and the full system to within 1 percent. The stability of the orbits is not well estimated, however, often differing by a factor of 2.

Before leaving this system, note that other observables may be estimated. For example, the time until the next passage through a section. Such calculations are required if the Lyapunov exponents of the reconstructed flow are to be computed from a section. Consider the series of magnifications an infinitesimal vector experiences as it is iterated along the attractor. The mean of this distribution corresponds to the largest Lyapunov exponent. For the Henon map, there is good agreement between a 2^9 point predictor and the map not only in the mean, but also in the structure of the distribution.

DISCUSSION AND CONCLUSIONS

Constructing a predictor from data when the system is not known presents new complications not considered above. These are most often clarified given "long enough" data records. For example embedding a low dimensional attractor in a large M will result in lower quality predictions. In delay coordinates, separation in the first coordinate direction(s) corresponds to distance in the true phase space at earlier times than separation in the last coordinate directions. For embedding dimensions larger than M_s, base points near the point to be predicted in the first coordinate direction(s) (but which correspond to a path which diverges from the trajectory leading to the point to be predicted) are weighted equally to those which are far away in the first coordinate direction(s) and arbitrarily close to the current position of the point in the true phase space. Conversely, embedding a high dimensional system in a low dimensional space will yield good predictions for those regions which are not saturated by projection effects. Predictions for points which lie in populated regions due to projection will be poor if the base points do not come from the same region of phase space; however the prediction of such points would still be poor in the correct embedding dimension if the region was not well sampled. This is especially true if the reconstruction is determined in SVD coordinates where the first few coordinate directions contain the majority of the signal. A method to identify flows reconstructed in $M < M_s$ needs to be developed. Periodic orbits will, of course, remain periodic in projection.

It has been shown that the predictor error profile is related to the magnitude and distribution of observational noise in a deterministic system. Observing the change of this profile as the prediction time is increased provides information on the observational noise and the chaos in the system. Predictors from reconstructed flows are capable of estimating the spectrum of unstable periodic orbits with significantly less data than other methods. Using the information contained in the time ordering of the data reduces the bulk of data required - the sampling rate and the length of the series may be shortened (nearest neighbors may be farther away) relative to the requirements of dimension calculations. It does not, of course, reduce the need to explore the entire attractor. Nonetheless, the use of reconstructed flows as discussed above should allow a much improved quantitative description of a chaotic attractor from a given series of observations.

Acknowledgments. This work has benefited from several discussions with M. Castagli and M. Powell on the use of radial basis functions. The dynamical systems insights of I. McIntosh, C. Sparrow and E. Stone have been included throughout. I would like to thank M. Gaster, G. King, and P. Read for making their data available (in a readable format). This research has been supported by the Science and Engineering Research Council.

REFERENCES

Auerbach D, et al., 1987, Exploring Chaotic Motion Through Periodic Orbits, *Phys. Rev. Lett.*, **58**:2387.

Broomhead D. and David Lowe, 1988, Multivariable Functional Interpolation and Adaptive Networks, *J. of Complex Systems*, **2**:321.

Castagli M., 1989, Nonlinear prediction of chaotic time series, to appear in *Physica D*.

Crutchfield J.P. and B.S. McNamara, 1987, Equations of Motion from a Data Series, *J. of Complex Systems*, **1**:417.

Cvitanovic P., P. Gunaratne, I. Procacia, 1989, Topological and Metric Properties of Henon Type Attractors, *Phys. Rev. A*, **38**:1503.

Farmer, J.D. and J. Sidorowich, 1987, Predicting Chaotic time series, *Phys. Rev. Lett.*, **59**:8.

Gaster, M., 1989, On the Nonlinear Phase of Wave Growth, preprint.

Grassberger, P. and I. Procaccia, 1983, Characterization of Strange Attractors,*Phys. Rev. Lett.*, **50**:346.

Henon, M., 1976, A Two-Dimensional Mapping with a Strange Attractor, *Commun. Math. Phys.*, **50**:69.

King, G., R. Jones and D.S. Broomhead, 1987, Phase portraits from a time series: a singular system approach, *Nuclear Physics B (Proc. Suppl.)*, **2**:379.

King, G, 1989, Chaos on a Catastrophy Manifold. In this volume.

Mees A. I., 1989, Modeling Complex Systems, University of Western Australia Preprint.

Moore, D.W. and E.A. Spiegel, 1966, A Thermally Excited Nonlinear Oscillator, *Astrophys. J.*, **143**:871.

Michelli, 1986. *Constr. Approx.*, **2**:11.

Packard, N.H., J.P.Cruchfield, J.D.Farmer, and R.S. Shaw, 1980, Geometry from a Time Series, *Phys. Rev. Lett.*, **45**:712.

Powell, M.J.D., 1985, Radial basis functions for multivariate interpolation: a review, Preprint, Univ of Cambridge.

Procaccia I., 1987, Exploring deterministic chaos via unstable periodic orbits, *Nuclear Physics B (Proc. Suppl.)*, 2:527.

Read, P., 1989, Chaotic regimes in rotating baroclinic flow, to be submitted to *JAFD*.

Smith L.A., 1988, Intrinsic Limits on Dimension Calculations, *Phys. Lett. A*, 133:283.

Smith L.A. and G. King, 1989, Unstable Periodic Orbits in an Experimental Chaotic Attractor, preprint.

Stone, E.F. and P. Holmes, 1989, Random perturbations of heteroclinic attractors, to appear in *SIAM*.

DEFECT-INDUCED SPATIO-TEMPORAL CHAOS

P. Coullet

Laboratoire de Physique Théorique
Université de Nice
Parc Valrose, Nice   06034 cedex FRANCE

INTRODUCTION

Systems with few degrees of freedom are well known for their propensity to develop chaos (Bergé et al., 1984). Quantitative characterization of such behavior relies upon the qualitative theory of differential equations (Arnold, 1980). Quantitative measures of chaos (Liapunov exponents, entropies, dimensions, etc.) are properties of mathematically well defined objects such as strange attractors (Guckhenheimer and Holmes, 1983).

Spatio-temporal behaviors are modelled by solutions of partial differential equations, for which the beginning of a qualitative theory is still lacking. The relation between temporal and spatio-temporal chaos is not simple. In the same way that the behavior of a N-body system and its statistical properties have very little to do with the dynamics of a three-body system, spatio-temporal complex behavior is not likely to be understood by increasing the dimension of a chaotic attractor.

In recent years models have been devised in order to study spatio-temporal chaos. These models describe spatially homogeneous physical systems, the boundaries of which have been removed to infinity. Phase turbulence (Kuramoto and Tsusuk, 1976; Pomeau and Manneville, 1979), turbulent self-focussing (Brand et al., 1986), spatio-temporal intermittency (Chaté and Manneville, 1987) and defect-mediated turbulence (Coullet and Lega, 1988; Coullet et al., 1989) are the main robust phenomena which emerge from these studies. These mechanisms are related to a collective phenomenon known as modulational instability (Benjamin and Feir, 1967; Newell, 1974).

The aim of this paper is to give a short tour of defect-induced complexity. We first briefly introduce the notion of defects and then review several mechanisms allowing their creation.

1. DEFECTS

Defects are a natural consequence of broken symmetries (Mermin, 1979; Balian et al., 1981). In the context of non-equilibrium systems defects are either *kinks* or *vortices*.

Kinks (Fig. 1) are associated with the breaking of a discrete symmetry. Domain walls in three dimensional anisotropic magnets give a typical example of such a defect. In non-equilibrium systems kinks are either associated with the breaking of the parity symmetry of x-periodic pattern ((i) $x \rightarrow -x$, where they appear as wave sinks and sources (Coullet et al., 1987; Joëts and Ribotta, 1989) (ii) $y \rightarrow -y$, where they appear as zig-zag grain boundaries (Manneville and Pomeau, 1983; Joëts, 1984; Ribotta, 1984)

Measures of Complexity and Chaos
Edited by N.B. Abraham *et al.*
Plenum Press, New York

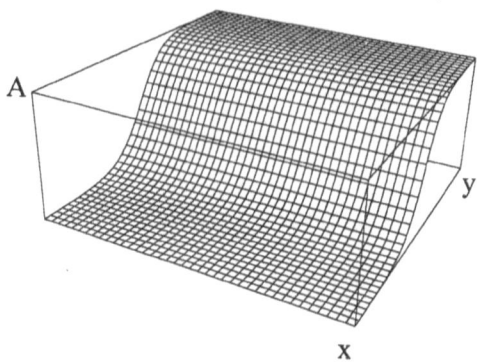

Figure 1.   Sketch of a kink.

or with an external periodic forcing which locks the phase of a natural
periodic pattern to a finite set of possible values, where there are
discommensurations (Lowe and Gollub, 1985; Coullet, 1986).

Kink interactions are short range.   Typically a kink moves with a
velocity which decreases exponentially with the distance d of a perturbation
created either by another kink or an inhomogenity, such as a wall.   ($\partial_t x = m$
$\exp(-d/d_0)$ where x is the position of the kink, m is its mobility, and d_0 is
a characteristic distance of the system).   The basic reason for such a short
range behavior has to be found in the structure of the defect, which, out of
a core with a size controlled by diffusion, reaches exponentially one of the
non symmetric possible homogeneous states.   Under some general circumstances
this exponential law is altered by a pre-factor which oscillates as a
function of the distance (Coullet and Elphick, 1987), cos (d/d_1), where d_1 is
another characteristic distance intrinsic to the system.   This oscillatory
behavior has been shown to be responsible for quasi-unidimensional spatially
chaotic structures.   In general, because of the short range character of
kink-dynamics, pinning effects are important and are likely to be at the
origin of complex stationary patterns (Coullet et al., 1987a).

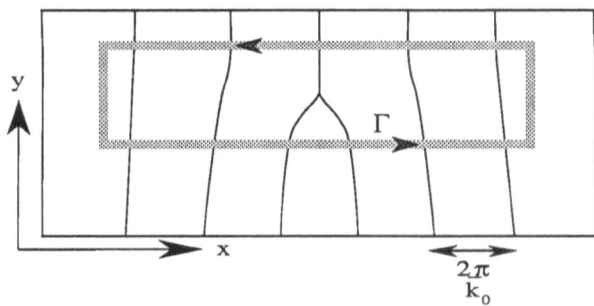

Figure 2.   Sketch of a dislocation.

Vortices are associated with the breaking of some phase symmetry.
Superfluid vortices are paradigms of such defects.   In non-equilibrium
systems vortices are associated with the breaking of translational symmetries
$(x \rightarrow x + \Sigma$ or $t \rightarrow t + \Theta$ or $x \rightarrow x + \Sigma$ and $t \rightarrow t - \Sigma/c)$.   Vortices of steady

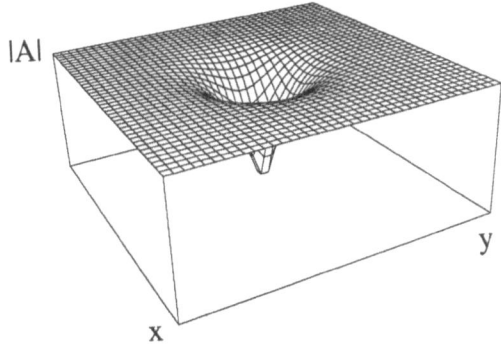

Figure 3. Modulus of the complex order parameter near a vortex.

periodic patterns are dislocations (Fig. 2) (DeGennes, 1972; Guazzelli, 1980; Siggia and Zippelius, 1981; Toner and Nelson, 1981; Dubois-Violette, 1983; Guazzelli et al., 1983; Joëts and Ribotta, 1984; Pomeau et al., 1983; Pocheau and Croquette, 1984; Tesauro and Cross, 1986; Yang et al., 1988). In the case of time periodic structures, they take the form of rotating spiral waves (Fig. 3,4,5) (Koppel and Howard, 1981; Hagan, 1982; Kuramoto, 1984; Müller et al., 1987; Bewersdorff et al., 1988). Vortices of standing wave patterns are also dislocations (Coullet et al., 1987). A standing wave pattern, which breaks independently both space and time translational symmetries supports vortex like defects (Lega, 1989) which possess a very interesting structure.

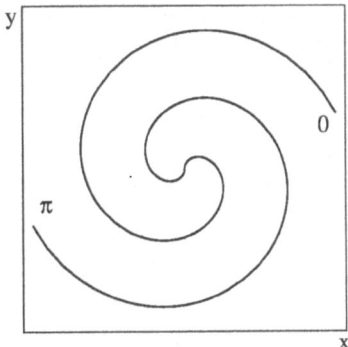

Figure 4. Equiphases π and 0 of the order parameter in the case of the spiral defect.

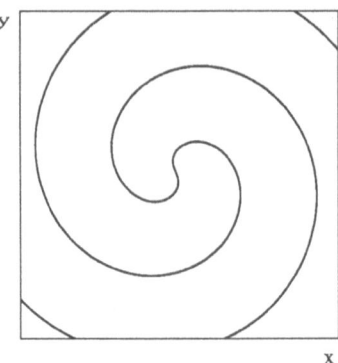

y

x

Figure 5.   Equiphases π and 0 of the electric field of an optical vortex
(Coullet et al., 1989).

2.   MECHANISMS FOR THE CREATION OF DEFECTS

        In equilibrium physics defects are produced by thermal fluctua-
tions.   In deterministic systems, several mechanisms can lead to their
formation.   In this chapter we discuss the role of inital conditions, phase
instabilities (Benjamin and Feir, 1967; Newell, 1974; Kuramoto and Tsusuki,
1976; Pomeau and Manneville, 1979) and finite amplitude transition with
symmetry changes (Coullet et al., 1989b).

        Initial states can be responsible for the existence of defects,   In
order to illustrate this first mechanism of formation, let us consider the
simple case of a physical system described by a real order parameter A
associated with the symmetry $A \rightarrow -A$.   Above the transition threshold the
system has to choose between one of the two possible states $\pm A^*$.   Below
threshold, the only stable solution is the symmetric one $A = 0$. Because of
the existence of natural fluctuations, at a given time t_0 the order parameter
is a function of x, characterized by a small amplitude.   This function is
likely to have zeroes.   In one dimensional systems these zeroes lie at
points.   They are lines and surfaces in two and three dimensions,
respectively.   Let us assume that, at this time t_0, the system is suddenly
brought above threshold.   In the regions of space where A is negative, the
natural tendency of the system is to bring A closer to $-A^*$, while in the
regions where A is positive, $A \rightarrow A^*$.   In this fast process strong gradients
appear in the vicinity of the zeroes of the initial state.   Pairs of distant
zeroes are likely to become kinks.   Two zeroes which are too close will
disappear in this transient phase, because of the interactions between
defects.   The critical distance is related to the diffusion constant and the
distance from threshold.   Because the interaction of kink defect is short
range, pinning effects are important and are likely to lead to steady
(possibly chaotic) configurations of kinks (Coullet and Elphick, 1987).

        A similar argument shows that vortices can be induced by initial
conditions, such as dislocations in large aspect ratio Rayleigh-Benard
experiments or spiral waves in oscillatory chemical reactions in large
Fresnel number lasers (Coullet et al., 1989b).   However, it is important to
stress two important differences with the previous case.   First, vortices
only exist in dimensions larger than two, then vortices have long range
interactions.   These differences come from the fact that vortices are
associated with a phase symmetry breaking transition.   In the simplest case
such a transition is described by a complex order parameter A associated with
the continuous symmetry $A \rightarrow A \exp(i\phi)$.   Again, the symmetric state $A = 0$ is
only stable below the transition threshold.   However a complex field is not
likely to have structurally stable zeroes in one dimension.   In two
dimensions the lines where the imaginary part and the real part of the field
vanish cut transversally into points.   In three dimensions zeroes of a

complex field occur generically on (closed) lines. Because the interaction is generally long-range, only distant zeroes are likely to survive.

Phase instability plays a crucial role in the creation of defects. Transitions which involve vortices, as for example the Hopf bifurcation, lead to patterns which are described in terms of a phase. These transitions are in some sense "phase breaking" transitions. Above the transition threshold a pattern is characterized by its phase ϕ. General arguments show that, in the appropriate reference frame, a phase perturbation $\phi(\mathbf{r},t)$ follows a diffusion type equation $\phi_t = D\nabla^2\phi$ (Kuramoto and Tsusuki, 1976; Pomeau and Manneville 1979). In the language of phase transition, phase modes are soft modes or Goldstone modes. More complicated patterns, as for example standing waves, require more than a single phase. In non-equilibrium system the diffusion coefficient D can change its sign, leading to a soft turbulence (Kuramoto and Tsusuki, 1976; Pomeau and Manneville, 1979) which can be described in terms of spatio-temporal fluctuations of the phase only. Actually a recent analysis (Coullet and Lega, 1987; Coullet et al., 1988b; Coullet et al., 1989) suggests that such a behavior could be only present in finite systems. The picture which emerges from this work is the following: in more than one dimension, large enough systems can sustain vortices. Phase fluctuations eventually become large enough to break, locally in space and time, this "phase only" description. In the regions where large phase gradients appear, the creation of defects is observed. Near the onset of phase instability, the presence of the defects destroys the quasi-long range order induced by the pattern. In one dimension, defects are spatio-temporal, i.e., they occur at a given spatial position, for a given time. When a phase instability occurs, from time to time the complex field A exhibits zeroes, which could also be associated with a loss of correlations (Oppo, 1989).

Lastly defects can be created in the transient process accompanying a subcritical transition (Coullet et al., 1989b). As a simple illustration, let us consider the following transition. Varying some external parameter, a one-dimensional steady periodic pattern which breaks the space translation symmetry undergoes a finite amplitude transition toward a translationally invariant pattern which breaks a discrete symmetry ($A \rightarrow -A$ for example). This scenario arises for example in the vicinity of a Liftchitz point. In this case before the transition, the pattern presents a periodic chain of zeroes. One can see it as a condensation of kink-anti-kink. When the transition occurs (melting), these zeroes disassociate to become free kinks or anti-kinks. Again distant zeroes are likely to survive. A typical experimental case where this mechanism has been identified is the transition from an undulating pattern to a chaotic array of zig-zag grain boundaries in liquid crystal experiments (Joëts and Ribotta, 1984). Similar phenomena occur for vortices or dislocations. For example, in the transition from a square pattern to a layered pattern (rolls in convective experiments), some of the zeroes located at the corners of squares are likely to transform themselves into dislocations of the layered structure.

3. CONCLUSIONS

Defects play an important role in the spatio-temporal destruction of the order induced by symmetry breaking transitions. Defects have been extensively studied in the context of the Rayleigh-Bénard instability and other hydrodynamic flows. It would be challenging to observe them in optical systems (Coullet et al., 1989b). The most efficient way to create defects in these non-equilibrium systems is related to phase instabilities. Experimental evidence of such a defect-mediated picture of turbulence exists (Rehberg et al., 1989). Let us conclude by mentioning a very recent experimental study (Nasuno et al., 1989) which also displays all the characteristic features of this type of turbulence.

ACKNOWLEDGMENT

. P. Huerre is acknowledged for a careful reading of this manuscript, and for interesting discussions. We acknowledge the financial support of the DRET (No 881453) and the EEC (No SC1-0035-C). This paper has been written while the author was visiting the Department of Aerospace Engineering of USC. The author wants to thank the members of this department for their warm hospitality. The original part of this work was done in collaboration with L. Gil and J. Lega at the University of Nice.

NOTES AND REFERENCES

Arnold V.I., 1980, Chapitres Supplémentaires de la Théorie des Equations Différentielles Ordinaires, Mir, Moscou.

Blian R., Kléman, M. and Poirier, J.P., 1981, "Physics of Defects" North-Holland, Amsdterdam.

Bejamin T.B. and Feir J.E., 1967, The Disintegration of Wave Trains on Deep Water. Part 1. Theory, J. Fluid Mech, 27: 417.

Bergé P., Pomeau Y. and Vidal Ch., 1984, "L'ordre in le Chaos, Hermann", Paris.

Bewersdorrf A., Borckmans P. and Müller S.C., 1988, "Chemical Pattern Formation in Fluid Sciences and Material Science in Space", édité par H.U. Walter, Springer-Verlag, Berlin.

Brand H.R., Lomdahl P.S. and Newell A.C., 1986, Phys. Lett 118A: 67.

Chaté H. and Manneville P., 1987, Spatio-temporal intermittency, Phys. Rev. Lett. 58:112.

Coullet P., 1986, Commensurate-incommensurate transitions in non-equilibrium systems, Phys. Rev. Lett. 56:724.

Coullet P., Elphick C., Gil L. and Lega J., 1987, Topological Defects of Wave Patterns, Phys. Rev. Lett. 59:884.

Coullet P., Elphick, C., 1987, Topological Defects Dynamics and Melnikov's Theory, Phys. Lett. 121A:233.

Coullet P., Elphick C. and Repaux D., 1987a, Nature of Spatial Chaos, Phys. Rev. Lett. 58:431.

Coullet P. and Lega J., 1988, Defect-mediated Turbulence in Wave Patterns, Europhys. Lett. 7:511.

Coullet P., Gill L. and Lega J., 1988, A form of Turbulence Associated with Defects, to appear in Physica D.

Coullet P., Gil L. and Lega J., 1989, Defect-mediated Turbulence, Phys. Rev. Lett. 62:1619.

Coullet P., Gil L. and Rocca F., 1989a, Optical Vortices, to appear in Optics Communications.

Coullet P., Lega J. and Walgraef D., 1989b, Defects at a Liftchitz point, preprint.

De Gennes P.G., 1972, Dislocation coin dans un Smectique A, C.R.A.S.B. 275:939.

Dubois-Violette E., Guazzelli E. and Prost J., 1983, Dislocation Motion in Layered Structures, Phil. Mag. A 48:727.

Gil L. and Lega J, 1988, "Defects in Waves, in Propagation in Systems far from Equilibrium", edited by J.E. Wesfreid, H.R. Brand, P. Manneville, G. Albinand and N.Boccara, Springer-Verlag, Heidelberg, p. 164.

Guckhenheimer, J. and Holmes P., 1983, "Nonlinear Oscillations, Dynamical Systems and Birfurcations of Vector Fields", Springer-Verlag, Berlin.

Guazzelli E., 1980, Nucléation Homogéne d'une Paire de Défauts in une Structure Convective Périodique, C.R.A.S., B 291:9.

Guazzelli E., Guyon E. and Wesfreid J.E., 1983, Dislocations in a roll Hydrodynamic Instability in Nematics: Static Limit, Phil. Mag. A 48:709.

Hagan P.S., 1982, Spiral Waves in Reaction-diffusion Equations, SIAM J. Appl. Math. 42:762.

Joëts A. and Ribotta R., 1984, Electro-hydrodynamical Convective Structures and Transitions to Chaos in a Liquid Crystal, in "Cellular Structures in Instabilities", edited by J.E. Wesfreid and S. Zaleski, Lecture Notes in Physics 210, Springer-Verlag, Berlin, p. 294.

Joëts A. and Ribotta R., 1989, Spatio-temporal grain boundaries, preprint, .

Koppel N and Howard L., 1981, Targand Pattern and Spiral Solutions to Reaction-Diffusion Equations with more than One Space Dimension, Adv. in Appl. Math. 2:417.

Kawasaki K. and Ohta T., 1983, Kink Dynamics in One-dimensional Nonlinear Systems, *Physica* 116A:573.

Kawasaki K, 1984, Topological Defects and Non-equilibrium, *Prog. Theor. Phys. Suppl.* 79:161.

Kuramoto Y. and Tsuzuki T., 1976, Persistent Propagation of Concentration Waves in Dissipative Media Far from Thermal Equilibrium, *Prog. Theor. Phys.* 55:356.

Kuramoto Y., 1984, "Chemical Oscillations, Waves and Turbulence", Springer Series in Synergetics, Vol 19, Springer, Berlin.

Lega J., 1989, Forme Spirale de la Dislocation des Ondes Stationnaires, submitted to "Comptes Rendus de l'Académie des Sciences".

Lowe M and Gollub J., 1983, *Phys. Rev.* A31:607.

Mermin N.D., 1979, The Topological Theory of Defects in Ordered Media, *Rev. Mod. Phys.* 51:591.

Müller S.C., Plesser T. and Hess B., 1987, Two-dimensional Spectrophotomandry of Spiral Wave Propagation in the Belousov-Zhabotinskii Reaction, *Physica* 24D:87.

Nasuno S., Sano M. and Sawada K., 1989, Phase waves in rectangular convective structure of nematic liquid crystal, to appear in *J. of Phys. of Japan*.

Newell A.C., 1974, Envelope Equations, "Lectures in Applied Mathematics 15", 157, Am. Math. Society, Providence.

Oppo, J.L., these proceedings, 1989.

Pocheau A. and Croquette V., 1984, Dislocation Motion: A Wavenumber selection Mechanism in Rayleigh-Bénard Convection, *J. Physique* 45:35.

Pomeau Y., Zaleski S. and Manneville P., 1983, Dislocation Motion in Cellular Structures, *Phys. Rev. A* 27:2710.

Pomeau Y. and Manneville P., 1979, Stability and Fluctuations of a Spatialy Periodic Convective Flow, *J. Phys. Lett.* 40:609.

Ribotta R. and Joëts A., 1984, Defects and Interactions with the Structue in EHD Convection in Nematic Liquid Crystals, in "Cellular Structures in Instabilities", edited by J.E. Wesfreid and S. Zaleski, Lecture Notes in Physics 210, Springer-Verlag, Berlin, p. 249.

Rehberg I., Resenat S. and Steinberg V., 1989, Traveling Waves and Defect-Initiated Turbulence in Electroconvecting Nematics, *Phys. Rev. Lett.* 62:756.

Siggia E.D. and Zippelius A., 1981, Dynamics of Defects in Rayleigh-Bénard Convection, *Phys Rev. A* 24:1036.

Tesauro G and Cross M.C., 1986, Climbing of Dislocations in Nonequilibrium Patterns, *Phys. Rev. A* 34:1363.

Tesauro G. and Cross M.C., 1987, Grain Boundaries in Models of Convective Patterns, *Phil. Mag. A* 56:703.

Toner J. and Nelson D.R., 1981, Smectic, Cholesteric, and Rayleigh-Bénard Order in two Dimensions, *Phys. Rev. B* 23:316.

Wesfreid J.E. and Zaleski S., 1984, éditeurs, "Cellular Structures in Instabilities", Lecture Notes in Physics 210, Springer-Verlag, Berlin, 1984.

Yang X.D., Joëts A. and Ribotta R., 1988, Localized Instabilities and Nucleation of Dislocations in Convective Rolls, in "Propagation in Systems far from Equilibrium", edited by J.E. Wesfreid, H.R. Brand, P. Manneville, G. Albinand and N. Boccara, Springer-Verlag, Heidelberg, p. 194.

LYAPUNOV EXPONENTS, DIMENSION AND ENTROPY IN COUPLED LATTICE MAPS

R. M. Everson

Center for Fluid Mechanics, Turbulence and Computation
Brown University
Providence, Rhode Island 02912

I INTRODUCTION

Recent years have seen advances in understanding of complex dynamics and chaos displayed by rather simple systems. Some routes to chaos are now well understood and have been observed experimentally in diverse fields. Quantitative measures of the chaos, principally Lyapunov exponents, dimensions and entropies[12] have been developed and used to connect theory and experiment.

A common feature of these systems is that they are <u>closed</u>. The imposition of closed boundary conditions means that the system "interacts with itself at all times",[13] so the effects of a perturbation to one location cannot escape, but influence the dynamics of the entire system. The primary instability in such systems is <u>absolute</u> and information on the dynamics of spatially distributed, but closed, systems may be obtained by studying the temporal behavior at a single location.[15] Indeed, the vast majority of experimental investigations to date have tacitly assumed that the dynamical systems under scrutiny have finite spatial extent, so they may be regarded as lumped systems and no information flows across their boundaries.

In <u>open</u> systems, where information may enter and leave the system, it is not always possible to adequately describe the dynamics by the time series at a single, fixed location. In contrast to closed systems, open systems frequently possess downstream propagating primary instabilities. A small disturbance to a steady equilibrium state, e.g. laminar flow in a pipe, may decay in a stationary frame of reference, but grow as it is advected downstream. A stationary observer reports stability; an observer moving with the disturbance reports instability; usually called <u>convective instability</u>.[10] In this case the spatial development of the flow may depend crucially on external forcing, often by low amplitude noise.[4]

Little attention has been paid to analyzing the dynamics of open systems. In this report we summarize work using coupled lattice maps to simulate open, spatially extended systems. In an effort to characterize their complexity we apply common diagnostic tools drawn from closed dynamical systems to trajectories viewed in frames moving with the flow. A more complete account is given by Everson.[7]

II COUPLED LATTICE MAPS

Coupled lattice maps are dynamical systems discrete in both time and space.[3] With one spatial dimension the maps are comprised of a chain of sites, the field variable at each site at particular time being obtained as a function of the variables at that site and its immediate neighbors at the previous time. Open systems of any required size, L, are easily simulated by regarding a segment of L consecutive sites as the system. Information is free to flow into and out of the system boundaries to and from the neighboring sites.

Measures of Complexity and Chaos
Edited by N.B. Abraham *et al.*
Plenum Press, New York

375

We concentrate on two particular coupled lattice maps, both based on the quadratic logistic map. The first map, introduced by Deissler and Kaneko,[5] is described by the equations

$$X_{n+1}^i = (1-d)f(X_n^i) + d\{(1-b)f(X_n^{i-1}) + bf(X_n^{i+1})\} + \xi_n^i,$$ (1)

where

$$f(x) = \mu x(1-x), \qquad \mu = 3.6547.$$ (2)

Here the subscript n refers to the timestep and i specifies the lattice site. The kernel, $f(x)$, is the familiar logistic map [2] with the rate parameter μ chosen to produce a chaotic trajectory with a Lyapunov exponent of 0.273. The parameter d ($0 \leq d \leq 1$) measures the spatial coupling between a lattice site and its nearest neighbors and b ($0 \leq b \leq 1$) measures its asymmetry. Thus $b = 0$ implies no downstream influence, whilst $b = 1$ implies no upstream influence. We refer to this map as the "asymmetric coupled lattice map".

Small amplitude noise, ξ_n^i, may be added to each lattice site at each time step, resulting in noise sustained structures. At time $n = 0$ the initial conditions are set as $X_0^i = x^* = f(x^*)$, the unstable fixed point of the underlying map. The upstream boundary condition is $X_n^0 = x^*$ for all n.

When the coupling between neighboring sites is sufficiently strong and asymmetric the fixed point solution $X_n^i = x^* = f(x^*)$ for all i is convectively unstable. Small amplitude noise ($|\xi_n^i| \leq 5 \times 10^{-10}$) at the upstream boundary is amplified as it moves downstream. Kinks or dislocations are formed, the trajectory consisting of segments alternately occupying the intervals $[0.288, 0.685]$ and $[0.75, 0.914]$. The dislocations move downstream at well defined speeds (dependent on the strength of the coupling) and there are a number of other speeds associated with the propagation of various spatial structures.[1] The trajectory within each segment appears chaotic and external noise is necessary for the production of dislocations.

Decreasing the coupling to neighboring sites (by decreasing d) or increasing the symmetry of the coupling (by increasing b towards $1/2$) renders the fixed point solution absolutely unstable. In this case dislocations are not formed and the trajectory is robust to small amplitude noise.

The second map, which we call the "mean lattice map" is

$$X_{n+1}^i = f(\overline{X}_n^i) + \xi_n^i$$ (3a)

$$\overline{X}_n^i = \frac{1}{3}\{(1+\sigma)X_n^{i-1} + X_n^i + (1-\sigma)X_n^{i+1}\}.$$ (3b)

Here too, $f(x)$ is the logistic map, with $\mu = 4.0$. $\overline{X}_n^i$ is a weighted average of X_n^i and its neighboring sites; σ ($-1 \leq \sigma \leq 1$) controls the symmetry of the average. When $\sigma = 0$ the symmetrical map is that examined by Grassberger.[9] In both cases we present, the symmetrical case $\sigma = 0$, and $\sigma = 1/2$ the fixed point solution $X_n^i = x^* = f(x^*)$ for all i is absolutely unstable, though we find that disturbances travel with specific velocities.

III LYAPUNOV EXPONENTS IN MOVING FRAMES

The rate of growth of infinitesimal volumes and hence the chaotic quality of the trajectory viewed in a moving frame is quantified by the Lyapunov exponents measured in that frame. Here we compute the spectrum of exponents in moving frames and infer the Lyapunov dimension and Kolmogorov–Sinai entropy from it.

A system consisting of L lattice sites has L Lyapunov exponents and the calculation of all of them requires that the evolution of L Lyapunov vectors $w_n^{(j)} \in R^L$ be followed in tangent space. Here $w_n^{(j)}$ denotes the jth Lyapunov vector at time n. In the stationary frame Lyapunov vectors evolve according to

$$w_{n+1}^{(j)} = DF(X_n)\,w_n^{(j)} - \sum_{k=1}^{j-1} w_n^{(k)}$$ (4)

where

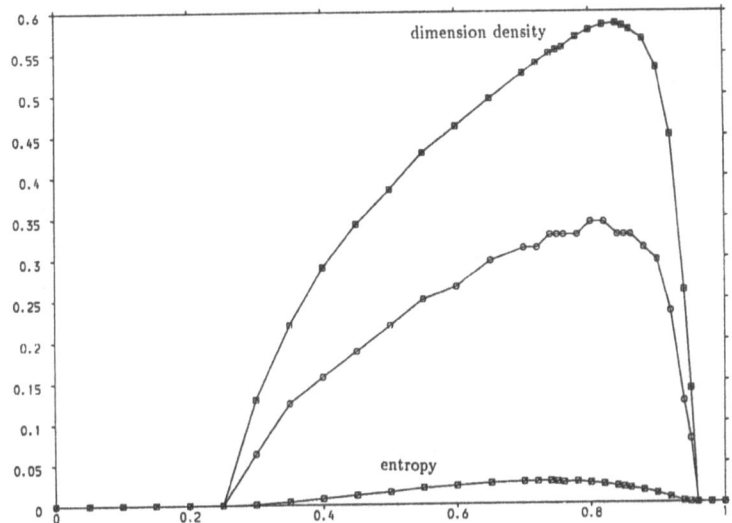

Fig. 1. Dimension density, fraction of Lyapunov exponents that are positive
and entropy density versus reference frame speed for the asymmetric coupled
lattice map in an convectively unstable regime, $d = 0.8$, $b = 0.1$.

$$DF(X_n) = \frac{\partial X_{n+1}^i}{\partial X_n^j} \qquad (5)$$

is the linearization the lattice map F about a point on the fiducial trajectory, X_n.
Lyapunov exponents are then the average logarithmic growth rate of the Lyapunov
vectors:

$$\lambda_j = \lim_{n \to \infty} \frac{1}{n} \log \frac{\|w_n^{(j)}\|}{\|w_0^{(j)}\|}. \qquad (6)$$

We use natural logarithms in the work reported here.

To calculate Lyapunov exponents in a moving frame we think of a test
section of L lattice sites embedded in a larger lattice and moving with a velocity v.
Lyapunov vectors evolve according to the linearization about the moving test
section;[5] thus in place of $DF(X_n)$ in equation 4 we use

$$DF(X_n^{[vn]}) = \frac{\partial X_n^{[v(n+1)]+j}}{\partial X_{n+1}^{[vn]+i}} \qquad 1 \le i,j \le L \qquad (7)$$

where square brackets denote the integer part of. To avoid the influence of the
spatially developing trajectory close to the upstream boundary the test section was
started several hundred lattice sites downstream and the lattice map equations
iterated until the trajectory was statistically stationary before beginning the
exponent calculations

In all moving frames and for large enough L the Lyapunov exponents scale
with the length of the test section according to the laws

$$\lambda_j \approx \Lambda_v(j/L) \qquad (8)$$

where Λ_v describes the shape of the spectrum in a frame moving with speed v.

This result lends support to increasing theoretical[8,14] and numerical
evidence[9,11] that in spatially extended systems the number of excited degrees of
freedom is proportional to its volume, V. A consequence is that the dimension, D,
and metric entropy, h, are expected to be <u>extensive</u> quantities. Accordingly,
Grassberger[9] defines the dimension density, δ, and metric entropy density, η, by

$$\delta = \lim_{V \to \infty} \frac{D}{V} \qquad \text{and} \qquad \eta = \lim_{V \to \infty} \frac{h}{V}, \qquad (9)$$

where for a one-dimensional lattice $V = L$.

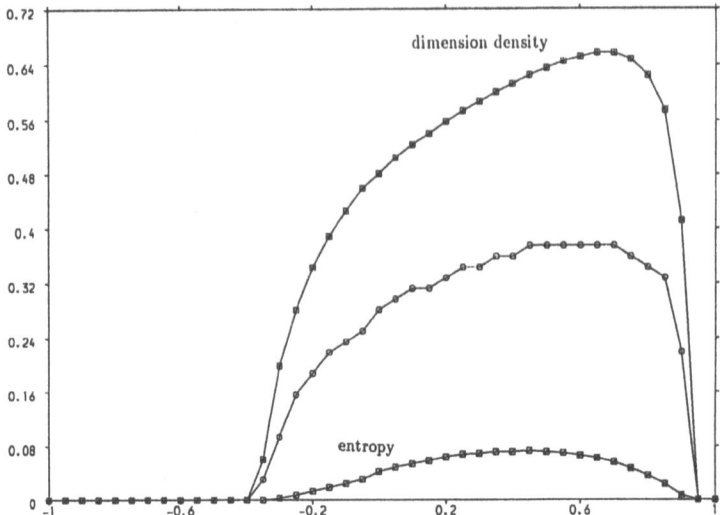

Fig. 2. Dimension density, fraction of Lyapunov exponents that are positive and entropy density versus reference frame speed for the mean lattice map, $\sigma = 0.5$.

IV DIMENSION AND ENTROPY DENSITY

The dimension density and metric entropy density are estimated from the Lyapunov spectrum via the Pesin and Kaplan-Yorke formulae:

$$h = L \int_0^\infty \Lambda_v(x)\Theta(\Lambda_v(x))\,dx \qquad (10)$$

is the sum of the positive exponents and

$$D = L\delta \qquad\qquad \int_0^\delta \Lambda(x)\,dx = 0 \qquad (11)$$

is the dimension of the (infinitesimal) phase-space volume that is neither expanded or contracted by the trajectory.

Figure 1 shows the Lyapunov dimension density, the fraction of exponents that are positive, $\rho(v)$, and the metric entropy density as a function of frame speed for the asymmetric map in the convectively unstable regime ($d = 0.8$, $b = 0.1$). Viewed in the stationary frame, there are no positive exponents and a stationary observer reports stability: growing disturbances are advected out of his frame. An observer moving with velocity between 0.297 and 0.945 sites/iteration, however, sees positive exponents, metric entropy and a dimension density as large as 0.585. Even for a moderately sized lattice of 64 sites this implies a dimension of 38, in marked contrast to the stationary observation. Observers traveling faster than 0.945 lattice sites per iteration outstrip growing disturbances: nearby trajectories appear to converge and the apparent dimension is again zero.

The asymmetric coupled lattice map with parameters $d = 0.2$, $b = 0$ is absolutely unstable. The dimension density is very high ($\delta = 0.992$) in the stationary frame, as might be expected because coupling between lattice sites is weak and the system acts almost as L independent logistic maps. In all moving frames the dimension and $\rho(v)$ is smaller than in the stationary frame, but the entropy density rises to a maximum of 0.048 at velocity 0.1 sites/iteration, falling to zero at 0.42 sites/iteration.

When $b = 0$ information cannot be propagated from downstream to upstream in the asymmetric coupled lattice map. Consequently $\delta(v)$, $\eta(v)$ and $\rho(v)$ are all zero in frames traveling upstream. The mean map with $\sigma = 0.5$ is also absolutely unstable, but information can propagate upstream and downstream. As shown in figure 2, observers traveling both upstream and downstream report chaos when moving sufficiently slowly.

Note that although the map is absolutely unstable $\delta(v)$, $\eta(v)$ and $\rho(v)$ are all larger in some moving frame that at rest. Setting $\sigma = 0$, so that the map is symmetrical, makes the graphs of δ, η and ρ symmetrical about $v = 0$, with the maxima all at rest.

Lyapunov exponents with different indices achieve their maxima at different speeds. Consequently $\delta(v)$, $\eta(v)$ and $\rho(v)$ may all be maximized at different speeds. It is unclear which speed, dimension and entropy best characterize the complexity of the dynamical system. However, we report that the speed, v_h, that maximizes the entropy density is the speed with which dislocations travel in the asymmetric lattice map, that is, the striation velocity. The mutual information $I(X_n^i; X_n^{[vn]+i})$ is also maximized at $v = v_h$. (See Everson[6]). Intuitively, one thinks of the entropy being the rate of production of information. A propagating instability transmits information, so an observer moving with the disturbance will receive more information than one who loses information as it is advected past him by the instability.

References

1. J. Brindley & R.M. Everson, "Disturbance propagation in coupled lattice maps," Phys. Lett. A 134 (1989), 229-236.

2. P. Coullet & J.-P. Eckmann, Iterated maps on the interval as dynamical systems, Progress in Physics, Birkhauser, Basel, 1980.

3. J.P. Crutchfield & K. Kaneko, "Phenomenology of Spatio-Temporal Chaos," in Directions in Chaos, H. Bai-lin., ed., World Scientific Publishing Co., Singapore, 1987.

4. R.J. Deissler, "Turbulent Bursts, Spots and Slugs in a Generalised Ginzburg-Landau Equation," Phys. Lett. A 120A (1987), 334-340.

5. R.J. Deissler & K. Kaneko, "Velocity-Dependent Liapunov Exponents as a Measure of Chaos in Open Systems," Los Alamos National Labs., Los Alamos Report, 1985.

6. R.M. Everson, "Detection and description of deterministic chaotic systems," Leeds Univ., PhD Thesis, Leeds, 1988.

7. R.M. Everson, "Lyapunov Exponents, Dimension and Entropy in Coupled Lattice Maps," Phys. Rev. A (3) (1989), (Submitted).

8. C. Fois, O.P. Manley, R. Temam & Y.M. Treve, "Asymptotic analysis of the Navier-Stokes equations," Phys. D 9D (1983), 157-188.

9. P. Grassberger, "Information content and predictability of lumped and distributed dynamical systems," University of Wuppertal Report, 1987.

10. P. Huerre, in Instabilities and Nonequilibrium Structures, E. Tirapequi & D. Villaroel, eds., Reidel, Dordrecht, 1987, 141.

11. P. Manneville, "Liapunov exponents for the Kuramoto-Sivashinsky model," in Macroscopic modelling in Turbulent Flows, H. Araki, J. Ehlers, K. Hepp, R. Kippenhahn, H.A. Weidenmuller & J. Zittartz, eds., Lect. Notes in Phys. #230, Springer-Verlag, New York-Heidelberg-Berlin, 1985, 390.

12. G. Mayer-Kress, Dimensions and Entropies in Chaotic Systems, Springer-Verlag, New York-Heidelberg-Berlin, 1986.

13. M.V. Morkovin, "Recent insights into instability and transition to turbulence in open-flow systems," Langley Research Center, NASA, ICASE , Hampton, Virginia, 1988.

14. D. Ruelle, "Large Volume Limit of the Distribution of Characteristic Exponents in Turbulence," Commun. Math. Phys. 87 (1982), 287-302.

15. F. Takens, "Detecting Strange Attractors in Turbulence," in Proc. Dynamical Systems and Turbulence, D.A. Rand & L-S. Young, eds., Springer-Verlag, New York-Heidelberg-Berlin, 1980, 366-381.

PHASE DYNAMICS, PHASE RESETTING, CORRELATION FUNCTIONS AND COUPLED MAP LATTICES

Raymond Kapral, Merk-Na Chee and Stuart G. Whittington

Chemical Physics Theory Group
Department of Chemistry
University of Toronto
Toronto, Ontario M5S 1A1, Canada

and

Gian-Luca Oppo

Department of Physics
Drexel University
Philadelphia, PA 19104, USA

INTRODUCTION

Some of the richest dynamical phenomena occur in both space and time. In view of the substantial developments that have taken place in the understanding of the onset of chaos in spatially homogeneous systems, which are described by systems of ordinary differential equations, there is some hope that parallel developments may take place in the study of spatially distributed systems that are usually described by partial differential equations. It is safe to say that this goal has not yet been achieved. The problem is twofold: not only are the phenomena diverse so that a set of organizing features like the major routes to chaos in spatially homogeneous systems has yet to be discovered, but some of the most useful nonlinear dynamics techniques, like surface of section plots, fractal dimensions and Lyapunov numbers, lose some of their utility for these high-dimensional systems. While all of the above tools can and have been brought to bear on the problem of spatio-temporal structure there is no one definitive diagnostic method.

There is another aspect of such spatio-temporal structures that cannot be overlooked: their inherent statistical character. Typically the initial state of an evolving inhomogeneous system is specified only in a statistical sense. For a fixed set of system parameters the spatial patterns that are formed may vary for different realizations of the evolution process. In addition, in physical systems one must inquire about the role of noise or internal fluctuations in influencing the nature of the spatio-temporal structures that are formed. Depending on the system such fluctuations may be important either in the initial

Measures of Complexity and Chaos
Edited by N.B. Abraham *et al.*
Plenum Press, New York

stages of the evolution or throughout the pattern formation process. In either case the structures that are formed may possess certain general characteristics in spite of detailed differences.

The statistical element of these spatio-temporal states means that some consideration must be given to the following two points: The manner in which the system state is described must incorporate an average over realizations of the evolution process and an estimate of deviations from the mean evolution. The second point is practical. If theoretical models are to be constructed for the description of such processes they must be complex enough to capture the essential features of the pattern formation process and simple enough to be able to carry out the statistical averages mentioned above.

The difficulty in dealing with partial differential equations has prompted the study of other models that are also capable of forming spatio-temporal patterns, yet possess features that make them either easier to analyse or simulate. Cellular automata and coupled map lattices constitute two such dynamical models.

Two themes related to pattern formation processes as modelled by coupled map lattices will be considered. The first concerns the modeling of specific physical systems by coupled map lattices. The second theme concerns coupled map lattices as dynamical systems in their own right and centers on whether such models can be used to explore the mechanisms for the onset of complex spatio-temporal structures.

We first present results for the phase separation of two coexisting stable states in order to point out the importance of phase boundary motion in determining the nature of the nonequilibrium patterns and to demonstrate that coupled map lattices can capture the essential features of the evolution process. Next, we examine systems that possess oscillating states and describe some of the interesting phenomena that can occur in these systems. In particular we study the nature of wave propagation in a chemical model and the spatio-temporal correlations that arise in the course of the evolution. We also consider a spatial version of phase resetting dynamics in this system. We then examine a discrete coupled map model that can exhibit the major wave propagation processes and comment on its range of validity. Finally, we briefly mention some features of coupled map lattices when no attempt is made to mimic the behavior of a specific partial diferential equation model.

COUPLED MAP LATTICES AND PHASE EVOLUTION

A simple illustration of how phase boundaries or line defects can determine the nature of pattern evolution is provided by domain growth in a system with a nonconserved order parameter. To be specific, consider a far-from-equilibrium system which when homogeneous possesses two stable steady states. In the inhomogeneous case diffusion can couple spatially distinct parts of the system and competition for territory between the two stable phases takes place. Systems showing such behavior range from chemical to magnetic or crystalline phase transformations[1]. The standard equation for the description of such phenomena is the time-dependent Ginzburg-Landau equation for an order parameter field, $x(\mathbf{r}, t)$:

$$\frac{\partial x(\mathbf{r}, t)}{\partial t} = -Ax(\mathbf{r}, t)^3 + Bx(\mathbf{r}, t) + C + D\nabla^2 x(\mathbf{r}, t), \qquad (1)$$

where A, B and C are parameters and D is the diffusion coefficient. It is not difficult to construct a coupled map lattice that faithfully reproduces the relevant dynamics, especially as far as its gross features are concerned[2]. In this case a simple Euler-like discretization suffices provided the appropriate parameter ranges are chosen for study; these may be determined easily from a stability analysis of the homogeneous states to inhomogeneous perturbations[2]. The discrete space and time coupled map model takes the form

$$x(i, t+1) = -ax(i, t)^3 + bx(i, t) + c + \gamma \sum_j{}' (x(j, t) - x(i, t)),$$ (2)

where i is a discrete position on a suitable lattice and the prime on the sum on j indicates a sum over neighboring sites. It is possible to construct map models whose dynamics is always invertable like the underlying partial differential equation[3]; both models show similar behavior in the physically relevant parameter regions. Other relations between partial differential equations and coupled map lattices have also been discussed[4].

Phase separation ensues when the system is prepared in the bistable region with spatially inhomogeneous initial conditions. Two characteristic types of behavior are observed depending on the system state. The standard critical quench case obtains when the system parameters are such that there is competition between two degenerate bistable states and the system is prepared with a narrow distribution of order parameter values about the unstable state. An initially disordered system will generate order on arbitrary length scales as the system evolves in time and domains of one phase will grow in the other. There is no external force acting on the system that favors one phase over the other; the average order parameter is zero, yet larger and larger domains form. Fig. 1(a) shows the results of the evolution from a somewhat different initial state that leads to similar domain formation. Here the initial state was generated by uniformly and randomly assigning each lattice site to one of the two degenerate states with equal probabilities.

The phenomenological theory of such domain growth is well known and was given by Allen and Cahn[5]. The principle effect that governs the domain growth is the curvature of the boundaries separating the two phases. (Note that one-dimensional models behave differently since planar interfaces are stable.) The spatial correlations in the nonequilibrium system can be characterized by the time-dependent correlation function $C(j, t)$:

$$C(j, t) = < N^{-1} \sum_i x(i, t)x(i + j, t) >,$$ (3)

where the angle brackets indicate an average over realizations and the sum is over the N lattice sites. The dynamic structure factor is the Fourier transform of this quantity:

$$S(k, t) = \sum_j exp(ik \cdot j)C(j, t)$$ (4)

The fact that the structure of the spatial patterns is governed by boundary curvature effects implies that the average domain size $R(t)$ grows like $R(t) \sim t^{1/2}$. Assuming that the domain size is the only characteristic length in the system, we have the scaling relation

$$S(k, t) \sim t\mathcal{F}(kt^{1/2})$$ (5)

where $\mathcal{F}$ is a scaling function. Hence, the apparently complicated spatial patterns possess simple universal features in this case. The validity of the scaling theory for the coupled map lattice is shown in Fig. 2.

These considerations are not restricted to this specific example but are quite general[6]. For instance, a coupled quadratic map lattice in parameter ranges where the homogeneous

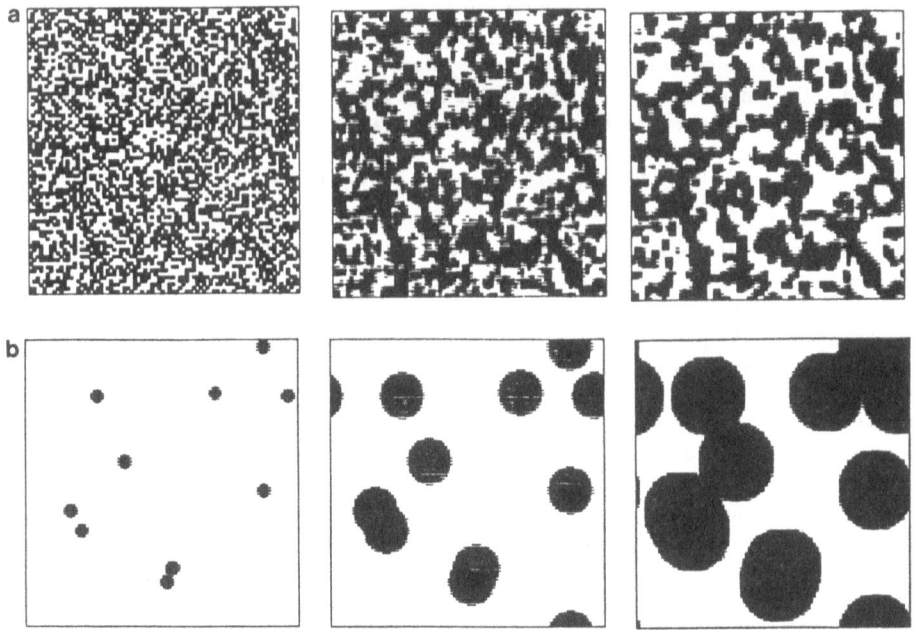

Figure 1. (a) Phase separation of equally stable states: a=1.0, b=0.3, c=0.0, and γ=0.05, (b) nucleation growth of a more-stable state into a less-stable state: a=3.0, b=0.3, c=0.035, and γ=0.05.

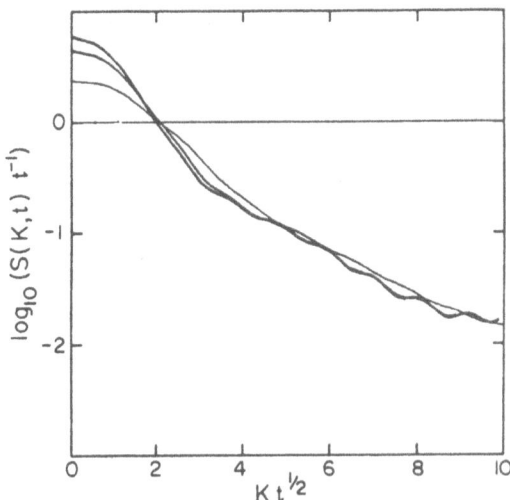

Figure 2. Plot of $log S(k,t)t^{-1}$ versus $kt^{1/2}$ for three values of t. The parameter values are: a=1.0, b=0.2, c=0.0, and γ=0.15, and the system evolved from the unstable state.

Figure 3. Period-2 phase evolution for coupled quadratic maps. The parameter λ in the map function $\lambda x(1-x)$ is $\lambda=3.2$ and $\gamma=-0.05$.

states are periodic orbits can exhibit phase competition between domains of states corresponding to different points on the cycle[7] (cf. Fig. 3). The period two fixed points necessarily have the same stability so in a stroboscopic representation the domain growth is similar to that of the bistable steady state case and is observed to be governed by curvature effects.

The cubic map example above corresponds to a system with a nonconserved order parameter. For systems with a conserved order parameter the scaling relations are somewhat more controversial and have also been studied by using coupled map lattices[3,8].

If the two states do not have the same stability the growth occurs through a nucleation process[2]. The more stable phase will consume the less stable phase provided a critical nucleus size is exceeded. In this case domains will grow linearly with time. An example of disc-shaped nucleus growth is shown in Fig. 1b.

This discussion illustrated a number of aspects of pattern evolution processes in systems with competing states. The statistical character of the pattern formation process is evident and must be taken into account when attempting to characterize the spatio-temporal structures that are formed. The results presented above corresponded to random initial conditions and deterministic evolution. Noise may also affect the pattern evolution process. The long-time scaling forms are unaffected by small amplitude noise, but noise can affect the early stages of the decay from the unstable state[1]. It is simple to incorporate noise in the coupled map model that is consistent with the fluctuation-dissipation theorem.

The central role played by boundary effects or defects is also obvious. The discrete models must be capable of describing the curvature effects that determine the growth rates of the phase domains. While the coupled map lattices described here have this feature this is not always the case for simple discrete models; for example, finite-state cellular automaton models often produce structures with sharp corners that would not correctly describe the growth process.

The study of the boundary interfacial profile in the discrete coupled map lattice bears a close similarity to the analogous problem for commensurate-incommensurate phase transitions and pinned solitons[9,10].

WAVE PROPAGATION AND OSCILLATORY DYNAMICS - CONTINUOUS SPACE AND TIME

Even more interesting phenomena arise if the dynamics is oscillatory[11]. The prototypical equation of motion in this case is the complex Ginzburg-Landau equation for the (complex) amplitude $A(\mathbf{r},t)$:

$$\frac{\partial A(\mathbf{r},t)}{\partial t} = -\alpha |A(\mathbf{r},t)|^2 A(\mathbf{r},t) + \beta A(\mathbf{r},t) + d\nabla^2 A(\mathbf{r},t), \tag{6}$$

where the parameters α and d are complex. This generic form can be derived from a reaction diffusion equation with oscillatory dynamics[12]; here we focus on one such specific example example: the Brusselator. The kinetic scheme corresponding to this reaction is well known[13] and the kinetic equations in reduced concentration variables can be written as

$$\frac{\partial U(\mathbf{r},t)}{\partial t} = A - (B+1)U(\mathbf{r},t) + U(\mathbf{r},t)^2 V(\mathbf{r},t) + D_u\nabla^2 U(\mathbf{r},t),$$

$$\frac{\partial V(\mathbf{r},t)}{\partial t} = BU(\mathbf{r},t) - U(\mathbf{r},t)^2 V(\mathbf{r},t) + D_v\nabla^2 V(\mathbf{r},t), \tag{7}$$

where A and B are constants. In the sequel we take the (real) diffusion coefficients to be equal, $D_u = D_v = D$. For B greater than some critical value $B_c = A^2 + 1$ the steady state solution of the spatially homogeneous equation undergoes a supercritical Hopf bifurcation. We have studied the response of the system to inhomogeneous perturbations in this regime[14].

Chemical waves can be initiated by introducing local perturbations in the phase of the oscillation. The chemical waves have distinctive characters in the early and late stages of the evolution[15,16]. At short times, typically of the order of a period of the oscillation, trigger waves with steep concentration gradients and constant propagation velocities are formed. At longer times the waves change their character and become phase waves which spread at a rate determined by the diffusion coefficient. The wave forms in these two different time regimes are shown in Fig. 4.

As an example of the pattern evolution processes in this oscillatory medium we consider the system in one dimension perturbed in the following way: Suppose the system has a length L which is divided into n subunits of length ℓ such that $L = n\ell$. Each subunit is then assigned a phase ϕ' on the limit cycle with probability p and a phase ϕ with probability 1-p. As in the case of domain evolution in the bistable system the pattern dynamics can be monitored by considering the nonequilibrium correlation function for either the U or V variables. Letting

$$\delta U(z,t) = U(z,t) - L^{-1}\int_0^L dz' U(z',t), \tag{8}$$

we define the correlation function as

$$C_u(z,t) = < L^{-1}\int_0^L dz' \delta U(z',t)\delta U(z+z',t) >, \tag{9}$$

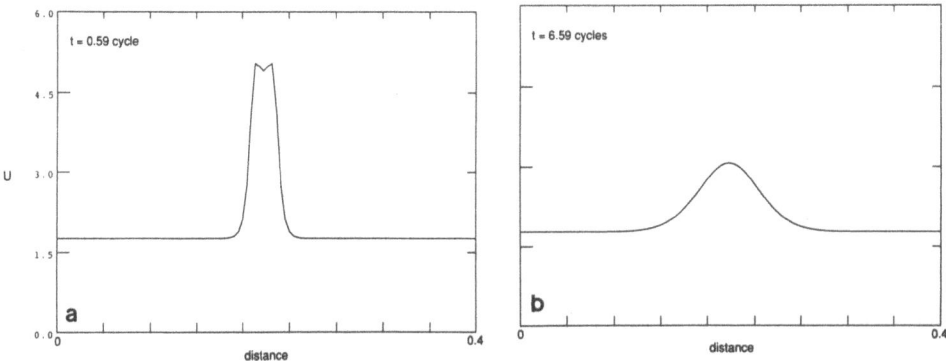

Figure 4. (a) Trigger and (b) phase waves for the Brusselator at $t = 0.59$ cycle and $t = 6.59$ cycle, respectively. The parameters are: $A = 1.0$, $B = 3.5$ and $D = 1 \times 10^{-5}$.

where $< ... >$ again signifies an average over realizations. The behavior of the correlation function for p=0.01 for several stroboscopically chosen times is shown in Fig. 5. There is a rapid (within the first cycle) build up of short-range correlations from the random initial condition due to the formation of trigger waves. The peak at the origin broadens and decreases in magnitude as the phase waves spread. An examination of the Fourier modes of this correlation function indicates that the system is evolving to a spatially homogeneous state, but with a phase different from ϕ. This feature has prompted a study of a spatial version of phase resetting dynamics.

Phase resetting dynamics

The standard phase resetting problem[17,18] consists of the application of a perturbation of given intensity to a limit cycle oscillator at a particular phase ϕ of its oscillation. The phase of the oscillator is interrogated at integral multiples of the period and the asymptotic new phase ϕ' is determined. The plot of ϕ' versus ϕ is the phase resetting map and contains a great deal of information about the dynamics of the oscillator; for example, if relaxation to the limit cycle is rapid following the perturbation the response of the system to periodic perturbations can be determined[17,19].

We have studied a spatial version of this problem[14]. The one-dimensional system of length L is again divided into n segments of length ℓ. Each subunit was assigned a particular phase ϕ on the cycle corresponding to (U_0, V_0) with probability 1-p or was perturbed away from the cycle to $(U_0 + U', V_0)$ with probability p. The system was interrogated at intervals of the period of the limit cycle and the spatially averaged phase was monitored for 30-70 cycles. The spatially averaged new phase is again termed ϕ'. We have observed that the short wavelength modes that arise from the perturbation quickly decay within this relaxation period while the long-wavelength modes decay on a very long time scale, estimated to be of order 10^3 cycles for our system size. These slowly decaying long-wavelength modes give rise to a spatially averaged limit cycle state which is a distorted version of that for a truely spatially homogeneous system; thus, the spatially averaged new phase ϕ' is characteristic of this state. This process was repeated for many realizations of the seeding (spatially inhomogeneous perturbation) process and the new phase averaged over realizations is designated $< \phi' >$. The phase resetting map for the oscillatory system with spatial degrees of freedom consists of the plot of $< \phi' >$ versus ϕ.

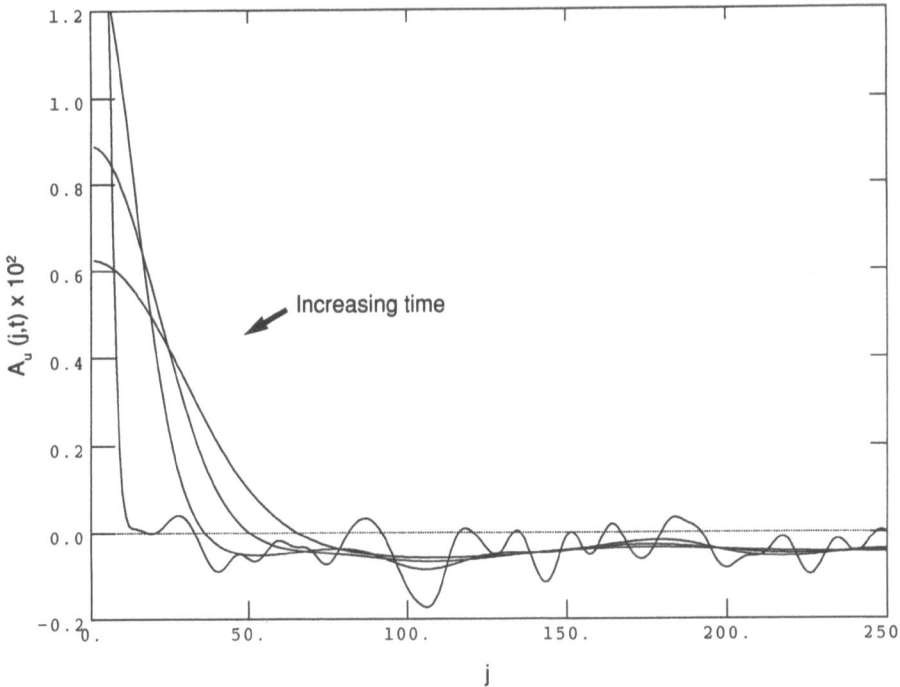

Figure 5. Correlation function for several values of the time.

Phase resetting maps typically show two types of behavior: In type 1 phase resetting the new phase ϕ' varies between 0 and 1 (mod 2π) as the original phase changes through one cycle; the average slope of the ϕ' versus ϕ plot is unity. In type 0 phase resetting ϕ' does not vary through a full cycle when ϕ varies between zero and unity; the average slope is zero[17].

The phase resetting map for fixed $U' = 10$ and several values of the seeding probability is presented in Fig. 6. The value of U' was chosen so that type 0 phase resetting was observed for the spatially homogeneous system. The figure shows that the seeding probability plays the role of an intensity variable in this case.

For very small seeding probabilities the system experiences only a small perturbation (in spite of the fact that the local perturbation may be quite strong) and type 1 phase resetting is observed. By construction type 0 phase resetting must occur for p=1. Thus, as the seeding probability increases the phase transition curve must experience a transition from type 1 to type 0. The consequences of such a transition have been discussed extensively[17,18]; for example, there must be at least one point (or region) in the $p\phi$-plane where the new phase is undefined: a phase singularity. We have observed a distinctly different type of spatio-temporal dynamics in the vicinity of this phase singularity.

Such phase singularities have been postulated to be responsible for a number of dynamical features; one example that we should take to heart is the formation of spiral waves in cardiac tissue which can lead to fibrillation and death[18]. The results outlined above indicate that the fraction of the material perturbed can play the role of an intensity parameter for the global phase resetting dynamics of the system. This observation suggests a number of extensions of the idea of phase resetting dynamics to the spatial domain; this is an interesting extension since in physical and biological systems oscillatory dynamics is

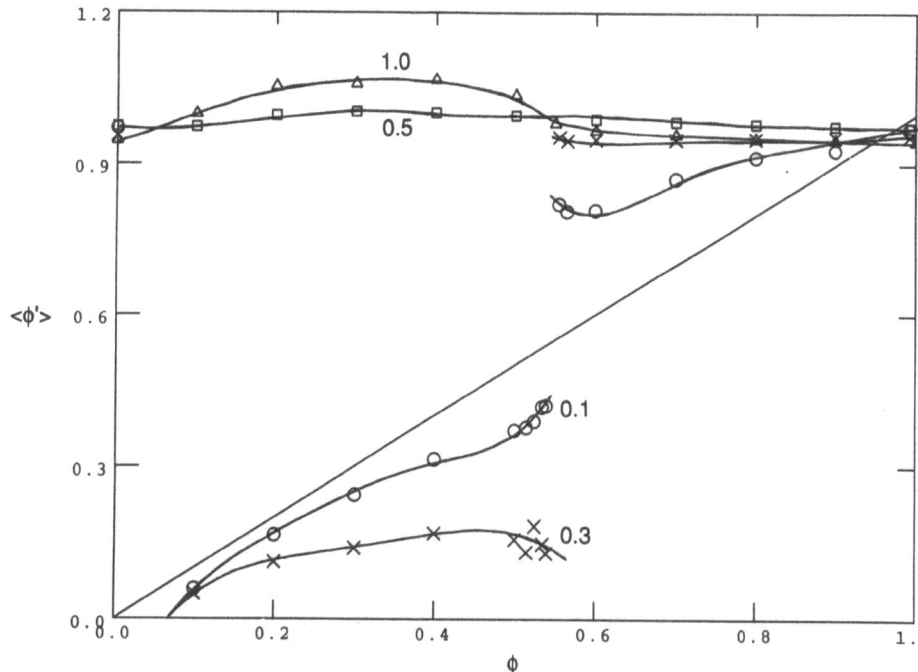

Figure 6. Phase resetting maps for several values of the seeding probability.

accompanied by wave propagation processes[18]. We have carrried out extensive studies to document such behavior[14].

WAVE PROPAGATION IN OSCILLATORY MEDIA - DISCRETE MODELS

In the preceding section we described some of the different types of wave propagation and phase resetting phenomena that can occur in oscillatory media. Two features are worth emphasizing: the chemical wave propagation processes are somewhat subtle, involving changes in character as time progresses, and, as has been noted earlier, to obtain a true picture of the spatio-temporal dynamics averages over many realizations of the evolution process must be performed. In order to extend such studies to higher space dimensions it is useful to construct discrete models for such systems.

In order to be able to describe the chemical waves that occur in the Brusselator in the relaxation oscillation regime, models that are more elaborate than those described above must be employed. The fact that the velocity of the system is a strong function of the phase of the limit cycle makes discrete modelling especially difficult. We[14] have constructed a coupled map lattice model for the Brusselator using a scheme similar to that of Ref. [3].

Consider the spatially homogeneous Brusselator, eq.(7) without the diffusion terms. This set of equations may be integrated numerically using any convenient method and yields the trajectory

$$\mathbf{W}(t + \tau) = \mathbf{F}_\tau(U(t), V(t)) , \qquad (10)$$

where the vector $\mathbf{W}^T(t) = (U(t), V(t))$. Given a suitably chosen set of discrete times

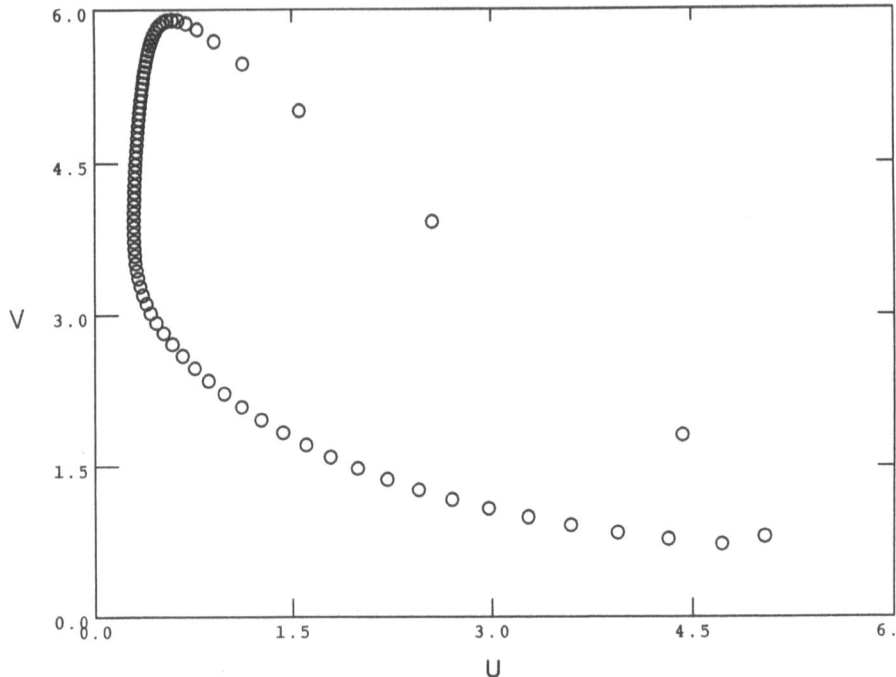

Figure 7. Discrete limit cycle for the Brusselator ($A = 1.0$ and $B = 3.5$).

separated by time intervals of τ, and a coarse-graining of the UV-plane a table of F_τ values can be generated that constitutes a numerical discrete mapping which is necessarily invertable since it is constructed from the solution of the original differential equation. For this relaxation oscillator the time interval must be chosen to be small enough to describe the fast part of the limit cycle dynamics accurately. The discrete limit cycle for $\tau = 0.01$ is shown in Fig. 7.

The coupled map model for the spatially distributed system is then written as

$$\mathbf{W}(i, t + \tau) = \mathbf{F}_\tau(\mathbf{W}(i, t)) + \gamma \sum_j{}'(\mathbf{F}_\tau(j, t) - \mathbf{F}_\tau(i, t)), \tag{11}$$

where γ is related to the space δr and time τ steps by $\gamma = D\tau/(\delta r)^2$. As before i is a discrete lattice position and the sum on j runs over neighboring sites. For sufficiently small space and time steps the solutions of this discrete equation converge to those of the underlying partial differential equation; however, its utility lies in the fact that the gross aspects of the wave propagation processes can be studied with a rather crude discretization leading to an increase of about an order of magnitude of computational speed.

We have verified that the transition from trigger to phase waves is captured by this map model. In addition the computation of statistically averaged quantities in two and three dimensions becomes feasible. Fig. 8 shows one realization of the spread of chemical waves in the oscillating medium in two dimensions. The system was prepared in the spatially homogeneous state and a small number of uniformly and randomly distributed perturbations in the phase of the oscillator were introduced. The propagation of the waves and the collisions between them are evident in the figure and we have confirmed that the spatial structure is qualitatively similar to that of the continuous space-time system.

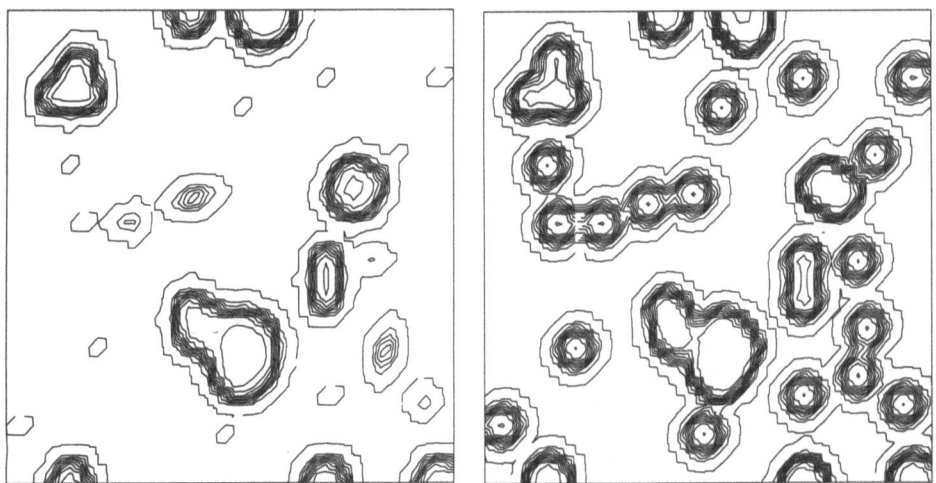

Figure 8. Wave propagation in two dimensions for the Brusselator.

Recently a coupled map lattice model similar to the one described here has been used to investigate vortex structures in the complex Ginzburg-Landau equation in two dimensions[20].

COUPLED MAP LATTICES AS DYNAMICAL SYSTEMS

If no care is taken to insure that the coupled map lattice mimics solutions of a particular partial differential equation a much richer bifurcation structure can be obtained. Viewing the discrete model in this way, as a dynamical system in its own right with no antecedents in an underlying partial differential equation, one can attempt to analyse and study how complex spatio-temporal states, like fully-developed turbulence, arise in systems with a large number of degrees of freedom. One can, for example, inquire if there are well-defined scenarios for the onset of chaos with universal features that transcend the map model. This is the spirit in which some of the earliest studies of coupled map lattices were carried out[21-23].

Numerical simulations have shown that coupled map lattices can exhibit a wide range of behavior with analogs in physical systems; for example, spatio-temporal intermittency has been studied using these discrete models[7,24-28]. Some of the different types of behavior have been cataloged in recent reviews[24,25] where an extensive set of references can be found.

There is evidence that some aspects of the spatio-temporal states in coupled map lattices possess universal scaling features similar to those that have been discovered for spatially-homogeneous systems. In particular, there are analogs of Feigenbaum's[29] scaling for the period-doubling route to chaos in a single one-dimensional map. One of the simplest such scaling relations concerns the sizes (and shapes) of the regions in the $\lambda\gamma$-plane corresponding to spatially-homogeneous period-2^n solutions for a coupled quadratic map lattice with bifurcation parameter λ and coupling strength γ. The heights (λ intervals) of these regions scale like the λ bifurcation values for an isolated map: the ratio of the height of the period-2^n region to that of the period-2^{n+1} region approaches Feigenbaum's δ as $n \to \infty$. What is more interesting is that the widths (γ intervals), which depend on the coupling strength between the maps, scale as α, Feigenbaum's orbit scaling exponent[21]. Such scaling features have not been thoroughly explored for these spatially-distributed systems.

Another interesting topic in this area is the construction of a statistical mechanics of spatially-distributed systems in the context of coupled map lattices[30]. In view of their simplicity in comparison with partial differential equation models there is some hope for progress along analytical lines on this problem.

ACKNOWLEDGEMENTS

This research was supported in part by grants from the Natural Sciences and Engineering Research Council of Canada.

REFERENCES

1. See for instance, J.D. Gunton, M. San Miguel and P.S. Sahni, in "Phase Transitions and Critical Phenomena", edited by C. Domb and J.L. Lebowitz (Academic Press, NY, 1983), Vol. 8, pg. 267.

2. R. Kapral and G.-L. Oppo, "Competition Between stable States in Spatially Distributed Systems", Physica **23D**, 455 (1986); G.-L. Oppo and R. Kapral, "Domain Growth and Nucleation in a Discrete Bistable System", Phys. Rev. A **36**, 5820 (1987). For a study of a system with two order parameters see, R. Kapral, G.-L. Oppo and D.B. Brown, "Phase Separation in a Two-Variable Discrete Model", Physica **147A**, 77 (1987); G.-L. Oppo and R. Kapral, "Evolution from an Unstable State in a Discrete-Model for Spatially-Distributed Chemical Systems", to be published.

3. Y. Oono and S. Puri, "Computationally Efficient Modeling of Ordering of Quenched Phases", Phys. Rev. Lett. **58**, 836 (1987); S. Puri and Y. Oono, "Study of Phase-Separation Dynamics by Use of Cell Dynamical System. II. Two-Dimensional Demonstrations", Phys. Rev. A **38**, 1542 (1988).

4. For a special case of a periodically driven reaction-diffusion system a coupled map model has been derived by J.D. Keeler, "An Explicit Relation Between Coupled Maps, Cellular Automata and Reaction-Diffusion Equations", UCSD preprint.

5. S.M. Allen and J.W. Cahn, "A Microscopic Theory for Antiphase Boundary Motion and its Application to Antiphase Domain Coarsening", Acta Metall. **27**, 1085 (1979).

6. A discussion of the role of boundary motion and curvature effects in determining the nature of the bifurcation structure of noisy coupled maps has been given by T. Bohr, G. Grinstein, Y. He and C. Jayapakash, "Coherence, Chaos and Broken Symmetry in a Classical Many-Body Dynamical System", Phys. Rev. Lett. **58**, 2155 (1987).

7. G.-L. Oppo and R. Kapral, "Discrete Models for the Formation and Evolution of Spatial Structure in Dissipative Systems", Phys. Rev. A **33**, 4219 (1986).

8. T.M. Rogers, K.R. Elder and R.C. Desai, "Numerical Study of the Late Stages of Spinodal Decomposition", Phys. Rev. B **37**, 9638 (1988); K.R. Elder, T.M. Rogers and R.C. Desai, "Early Stages of Spinodal Decomposition for the Cahn-Hillard-Cook Model of Phase Separation", Phys. Rev. B **38**, 4725 (1988).

9. S. Aubry, in "Bifurcation Phenomena in Mathematical Physics" edited by C. Bardis and D. Bessis (Reidel, Dordrecht, 1981), p. 163.

10. V.L. Pokrovsky, "Splitting of Commensurate-Incommensurate Phase Transitions", J. Phys. (Paris) **42**, 761 (1981); P. Bak and V.L. Pokrovsky, "Theory of Metal-Insulator Transition in Systems with Nearly Half-Filled Bands", Phys. Rev. Lett. **47**, 958 (1981); S.E. Burkov, V.L. Pokrovsky and G. Uimin, "Soliton Structures in a Discrete Chain", J. Phys. A **15**, L645 (1982).

11. See, for example, P. Ortoleva in "Oscillations and Traveling Waves in Chemical Systems", editors, R.J. Field and M. Burger (Wiley, NY, 1985).

12. Y. Kuramoto, "Chemical Oscillations, Waves and Turbulence", (Springer, Berlin, 1980).

13. G. Nicolis and I. Prigogine, "Self-Organization in Nonequilibrium Systems" (Wiley, NY, 1977).

14. M.-N. Chee, R. Kapral and S.G. Whittington, unpublished.

15. P. Ortoleva and J. Ross, "Phase Waves in Oscillatory Chemical Reactions", J. Chem. Phys. **58**, 5673 (1973); P. Ortoleva and J. Ross, "On a Variety of Wave Phenomena in Chemical Reactions", J. Chem. Phys. **60**, 5090 (1974).

16. J.M. Bodet and J. Ross, "Experiments on Phase Diffusion Waves", J. Chem. Phys. **86**, 4418 (1987); P.M. Wood and J. Ross, "A Quantitative Study of Chemical Waves in the Belousov-Zhabotinsky Reaction", J. Chem. Phys. **82**, 1924 (1985).

17. A.T. Winfree, "The Geometry of Biological Time", (Springer-Verlag, NY, 1980).

18. A.T. Winfree, "When Time Breaks Down", (Princeton University Press, Princeton, NJ, 1987).

19. L. Glass, M.R. Geuvara, J. Belair and A. Shrier, "Global Bifurcations of a Periodically Forced Biological Oscillator", Phys. Rev. A **29**, 1348 (1984); M.R. Gurvara, A. Shrier and L. Glass, "Phase Resetting of Spontaneously Beating Embryonic Ventricular Heart Cell Aggregates", Am. J. Physiol. **251** (Heart Circ. Physiol. 20), H1298 (1986).

20. T. Bohr, A.W. Pedersen, M.H. Jensen and D.A. Rand, "Vortex Dynamics in a Coupled Map Lattice" in eds., P. Coullet and P. Heurre, "New Trends in Nonlinear Dynamics and Pattern Forming Phenomena", (Plenum, NY), to be published.

21. I. Waller and R. Kapral, "Spatial and Temporal Structure in Systems of Coupled Nonlinear Oscillators", Phys. Rev. A **30**, 2047 (1984); R. Kapral, "Pattern Formation in Two-Dimensional Arrays of Coupled, Discrete-Time Oscillators", Phys. Rev. A **31**, 3868 (1985).

22. K. Kaneko, "Period-Doubling of Kink-Antikink Patterns, Quasiperiodicity in Antiferro-Like Structures and Spatial Intermittency in Coupled Logistic Lattice", Prog. Theor. Phys. **72**, 480 (1984); K. Kaneko, "Spatiotemporal Intermittency in Coupled Map Lattices", Prog. Theor. Phys. **74**, 1033 (1985).

23. R.J. Diessler, "1-d Strings, andom Fluctuations, and complex chaotic Structures", Phys. Lett. A **100**, 451 (1984).

24. J. P. Crutchfield and K. Kaneko, "Phenomenology of Spatiotemporal Chaos", in "Directions in Chaos", editor, Hao Bai-lin, (World Scientific, Singapore, 1987).

25. K. Kaneko, "Pattern Dynamics in Spatiotemporal Chaos", Physica D **34**, 1 (1989).

26. J.D. Keeler and J.D. Farmer, "Intermittency, Confinement of Turbulence, and Functional Attractors in High Dimensional Coupled Maps", Physica D **23**, 413 (1986).

27. L.A. Bunimovich, A. Lambert and R. Lima, "The Emergence of Coherent Structures in Coupled Map Lattices", preprint.

28. H. Chate and P. Manneville, "Spatio-Temporal Intermittency in Coupled Map Lattices", Physica D **32**, 409 (1988).

29. M.J. Feigenbaum, "The Universal Metric Properties of Nonlinear Transformations", J. Stat. Phys. 21, 669 (1979).

30. L.A. Bunimovich and Ya. G. Sinai, "Space-Time Chaos in the Coupled Map Lattice", Nonlinearity, to be published.

CHARACTERIZATION OF SPATIOTEMPORAL STRUCTURES IN

LASERS: A PROGRESS REPORT

Gian-Luca Oppo*, Mauro A. Pernigo, Lorenzo M. Narducci

Department of Physics,
Drexel University, Philadelphia, PA, 19104, U.S.A.

Luigi A. Lugiato

Dipartimento di Fisica, Politecnico di Torino, Torino, 10129, Italy

1. INTRODUCTION

The introduction of the transverse dependence of the field in the analysis of the dynamics of lasers has already explained the presence of temporal instabilities at low values of the pump parameter for lasers operating in the good cavity limit [1,2]. However, the relevance of these studies does not expire with the detection of chaotic motion. The interplay of spatial and temporal degrees of freedom together with the highly nonlinear character of the Maxwell-Bloch equations of motion, make lasers with large transverse section good candidates for the characterization of complex spatiotemporal patterns and perhaps also of turbulent states. Moreover, the time scale of the laser dynamics offers a clear advantage over other fields of physics where spatiotemporal phenomena have been detected, from hydrodynamics to oscillating chemical reactions.

Here we show that a laser with transverse effects displays spatiotemporal structures with different degrees of complexity. The evaluation of a suitably generalized correlation function allows one to discriminate between different structures and to detect regions of the transverse plane where the coherence of the output field can be easily lost. The existence of peculiar singular solutions in the spatiotemporal evolution of the system appears to be responsible for the loss of correlation. The temporal origin of these phenomena is also discussed.

The paper is organized as follows. In Section 1 we provide a brief introduction together with an outline of the paper (you are indeed reading it). Section 2 is devoted to the modal expansion of the Maxwell-Bloch equations for a laser with transverse dependence of the field. Spatiotemporal structures originating from the competition of different Gauss-Laguerre modes for the same population of excited atoms are shown in Section 3. In Section 4 we introduce a generalized correlation function as a characterizing parameter of spatial complexity, and discuss its relation to singular solution detected in the temporal evolution of the laser output. Conclusions, acknowledgements and disacknowledgements end the presentation.

Measures of Complexity and Chaos
Edited by N.B. Abraham *et al.*
Plenum Press, New York

2. THE MODEL

The treatment of transverse effects in laser dynamics requires the proper inclusion of (a) the diffraction due to the finite cross section of the field, (b) the curvature of the wave-front due to the spherical mirrors and (c) the transverse profile of the pump of the active medium. For a medium of homogeneously broadened two-level atoms with a transition frequency ω_A, the slowly varying envelopes of the field F and of the polarization P, and the population inversion D obey to the following generalized Maxwell-Bloch equations [1]

$$- \frac{i}{4} \nabla_\perp^2 F + \frac{\partial F}{\partial z} + \frac{1}{v} \frac{\partial F}{\partial t} = i \frac{\delta\Omega}{v} F - \alpha\Lambda P$$

$$\frac{\partial P}{\partial t} = - (FD + (1 + i\Delta)P) \tag{1}$$

$$\frac{\partial D}{\partial t} = - \gamma \left[D - \frac{1}{2} (F^*P + FP^*) - \chi(r) \right]$$

where $\gamma = \gamma_\parallel / \gamma_\perp$ is the decay rate of the population inversion normalized to that of the polarization, z is the longitudinal variable, v is the scaled phase velocity ($v = c/\Lambda\gamma_\perp$), Λ is the length of the cavity, $\delta\Omega$ is the frequency shift (normalized to $\gamma_\perp$) between the (unknown) laser frequency and the reference frequency ω_0, α is the unsaturated gain per unit length, Δ is the difference between $(\omega_A - \omega_0)/\gamma_\perp$ and $\delta\Omega$, and finally $\chi(r)$ is the pump parameter which depends explicitly on the transverse radial variable r.

Equations (1) can not be solved easily even by standard numerical techniques. Then, we introduce a supplementary set of assumptions:
a) Large separation of longitudinal modes;
b) High reflectivity of the mirrors and small gain per pass;
c) Nearly plane-plane cavity configuration (large radii of curvature of the mirrors);
d) Cylindrical symmetry;
e) Diffraction losses.
    Under these conditions the variables of the system (1) can be conveniently expanded as [1]

$$\begin{bmatrix} F(r,t) \\ P(r,t) \\ D(r,t) \end{bmatrix} = \sum_p A_p(r) \begin{bmatrix} \Psi_p(t) \\ P_p(t) \\ d_p(t) \end{bmatrix} \tag{2}$$

where $A_p(r)$ are the (empty cavity) Gauss-Laguerre modes

$$A_p(r) = 2 L_p(2r^2) \exp(-r^2) \tag{3}$$

where $L_p(.)$ are the Laguerre polynomials of index p (integer) and given argument. $A_p(r)$ constitute a complete set of modes as they satisfy the orthonormality condition

$$\int_0^\infty dr \, r \, A_p^*(r) \, A_{p'}(r) = \delta_{pp'} \tag{4}$$

Equations (1) can be rewritten as a set of ordinary differential equations for the modal amplitudes introduced in (2) because the transverse dependence of the solutions is entirely contained in the superposition of $A_p(r)$ and the longitudinal dependence has been eliminated through the uniform field limit approximation (assumption (b) above). The new equations read as

$$\frac{d\Psi_p}{dt} = -k\left[g_p\Psi_p + i(a_p - \Delta)\Psi_p + 2Cp_p\right]$$

$$\frac{dp_p}{dt} = -\sum_{p'p''}\Gamma_{pp'p''}\Psi_{p'}d_{p''} + (1 + i\Delta)p_p \tag{5}$$

$$\frac{dd_p}{dt} = -\gamma\left[d_p - \frac{1}{2}(\sum_{p'p''}\Gamma_{pp'p''}\Psi_{p'}^*p_{p''} + c.c.) - \chi_p\right]$$

where k is the cavity damping rate (normalized to $\gamma_\perp$), $g_p = (1+\beta p^4)$ describes the diffraction losses which increase for higher order modes because of the limited transverse size of the resonator, $a_p = pa_1$ is the detuning between the frequency of the Gauss-Laguerre mode order p and that of the p=0 mode, 2C is the new gain coefficient, $\Gamma_{pp'p''}$ are the mode-mode coupling coefficients defined as

$$\Gamma_{pp'p''} = \int_0^\infty dr\, r\, A_p(r)\, A_{p'}(r)\, A_{p''}(r) \tag{6}$$

and

$$\chi_p = \int_0^\infty dr\, r\, A_p(r)\, \chi(r) \tag{7}$$

It is important to note that the expansion using Gauss-Laguerre modes of the polarization and of the population inversion are not strictly necessary in order to integrate numerically the equations. Similar results than the ones presented here, can be obtained from a slightly modified version of equations (5) where a radial integral of the polarization appear on the right hand side of the equation for Ψ and there is no dependence of P and D from the modal index p.

3. SPATIOTEMPORAL STRUCTURES

A great advantage of the modal expansion of the Maxwell-Bloch equations is that it is possible to study regimes where only few modes are coupled without losing physical insight. Gauss-Laguerre modes with radial size larger than the transverse size of the resonator have large diffraction losses and do not play any role in the dynamics. It is then possible to select a proper value of β such that only a desired number of modes are dynamically coupled. For the simulation reported here, we selected the following set of parameters: k=1.0, β=0.005, γ=0.05, Δ=0.18, 2C=1.2 and a total number of modes of 7 even if the chosen value of β assures the presence of only three active modes (p=0,1,2). A useful scanning parameter is the detuning a_1 which controls the relative gain of the different active modes and can be related to experiments where the radius of curvature of a mirror is changed by an intracavity telescope [2]. Large values of a_1 correspond to single Gaussian mode operation whereas a_1=0 identifies the plane-plane cavity limit where all modes have degenerate frequencies.

In order to better characterize the changes in the dynamical behavior of the laser with decreasing a_1, we evaluated the total flux $\Phi(t)$ which is defined as

$$\Phi(t) = 2\pi\int_0^\infty dr\, r\, |F(r,t)|^2 = 2\pi\sum_{p=0}^{pmax}\Psi_p^*(t)\Psi_p(t)\int_0^\infty dr\, r\, A_p^*(r)A_p(r) = 2\pi\sum_{p=0}^{pmax}|\Psi_p(t)|^2 \tag{8}$$

Fig.1 shows two regions of instability alternating with cooperative frequency locking states. While the first region (large values of a_1) shows purely oscillatory behavior enclosed by a

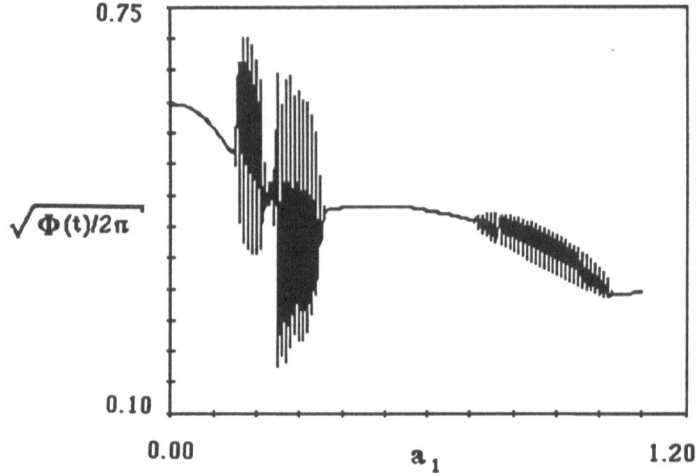

Fig. 1. Evolution of the total flux versus a_1. For each value of the control parameter the maximum and minimum of the temporal evolution are reported.

direct and a reverse Hopf bifurcation, in the second instability region (counting right to left) the system evolves to quasiperiodic and chaotic states as well [3]. We focus our attention on the range of values of the control parameter close to the onset of aperiodic regimes where there are three active modes. In Fig.2 we report three time evolution of the transverse section of the field for different values of a_1. One can be convinced rather easily that the third plot shows a spatiotemporal structure with a higher degree of complexity than the first diagram of Fig.2. However, we need a complexity index which is capable of discriminating between similar patterns such as that shown in Fig.2 a and b. The question we try to address here has a relevant application in the field of communications where pulses and signals generated by semiconductor lasers operating in a multi transverse mode configuration have been used to propagate information through fibers. How small changes of the operating parameters can affect the spatiotemporal coherence of such signals ? What kind of measurements can be performed in order to avoid spatially uncorrelated states ?

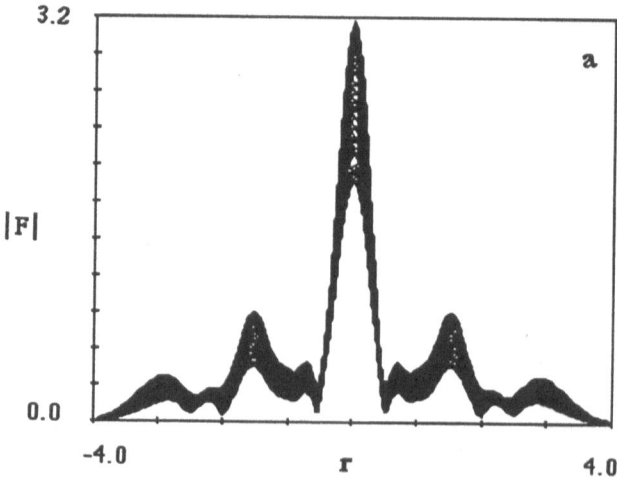

Fig. 2. Spatiotemporal evolution of the module of the total field for (a) $a_1=0.34$, (b) $a_1=0.32$ and (c) $a_1=0.25$.

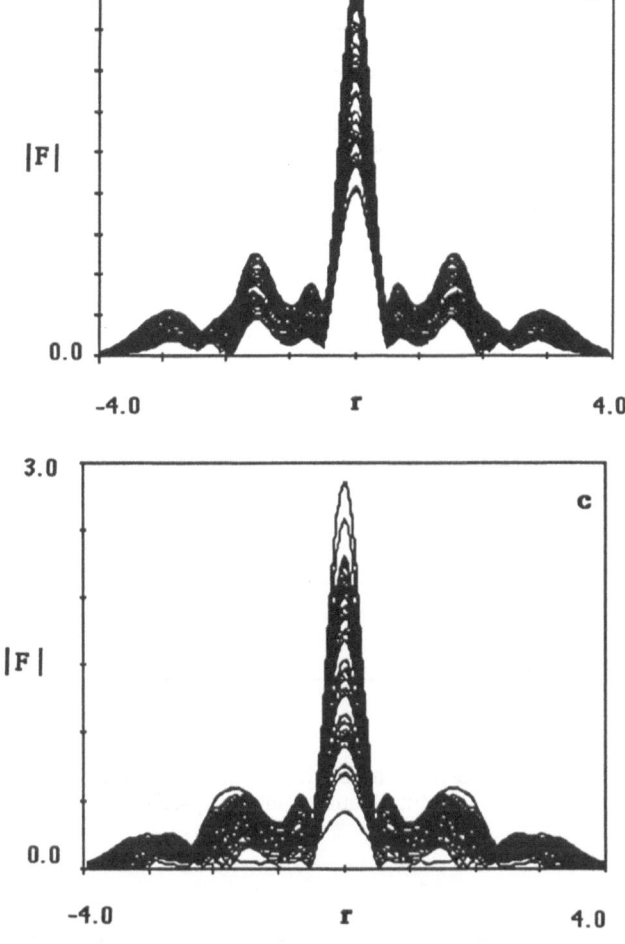

Fig. 2. Spatiotemporal evolution of the module of the total field for
(a) $a_1=0.34$, (b) $a_1=0.32$ and (c) $a_1=0.25$.

4. CORRELATION FUNCTIONS

We characterize spatiotemporal patterns with a complexity index defined as the maximum of the intensity correlation function with respect to the delay time τ

$$C_{MAX}(0,r) = Max_{\forall \tau} \, C(0,r,\tau)$$

$$C(0,r,\tau) = \frac{<I(0,t) \, I(r,t+\tau)> - <I(0,t)><I(r,t)>}{\sigma(I(0,t)) \, \sigma(I(r,t))} \tag{9}$$

$$\sigma(I) = \sqrt{<I^2> - <I>^2}$$

where the $<.>$ represents the temporal average and $I(r,t)$ is the field intensity at the point r.

$C_{MAX}(0,r)$ offers some advantages over other complexity indexes previously introduced: (a) it is a well defined quantity over the interval (0,1), (b) it is equal to 1 for signals at different places which are merely dephased, (c) it is equal to 1 for oscillatory signals generated by beat notes, and (d) it is measurable experimentally [4]. It is interesting to compare the usual correlation function $C(d,\tau=0)$ averaged over the starting spatial point (see Fig.3) with $C_{MAX}(0,r)$ (see Fig.4a) for $a_1=0.34$. The presence of propagating waves in the system induces dephasing of the signals at different values of r. Such an effect is smeared out by using the correlation function as defined in (9). Fig.4 shows $C_{MAX}(0,r)$ for the three spatiotemporal structures of Fig. 2.

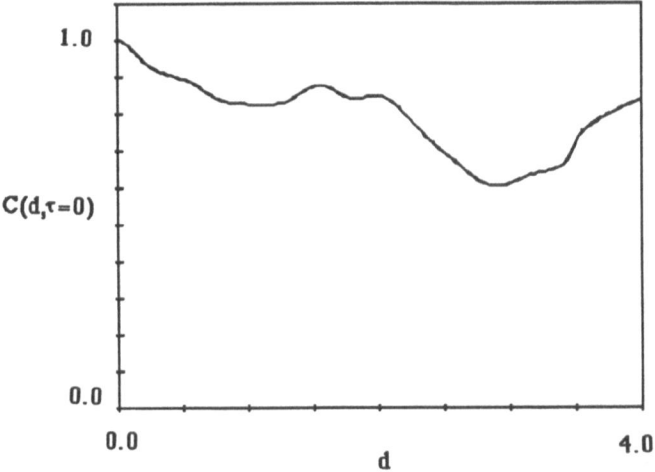

Fig. 3. The correlation function $C(d,\tau=0)$ averaged over the starting spatial point for $a_1=0.34$.

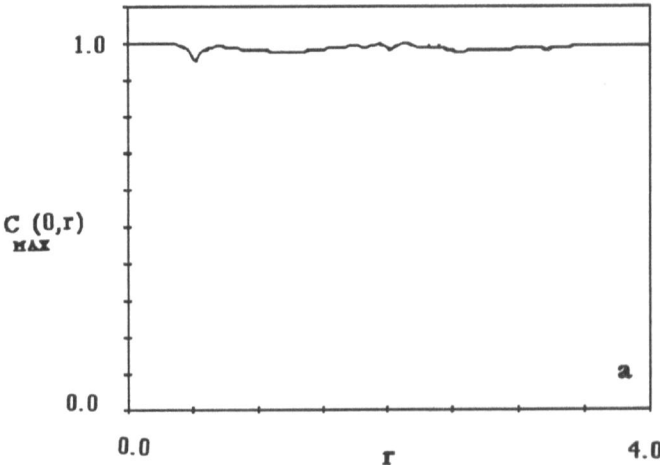

Fig. 4. The maximum of the correlation function as defined by Eq. (9) for (a) $a_1=0.34$, (b) $a_1=0.32$ and (c) $a_1=0.25$.

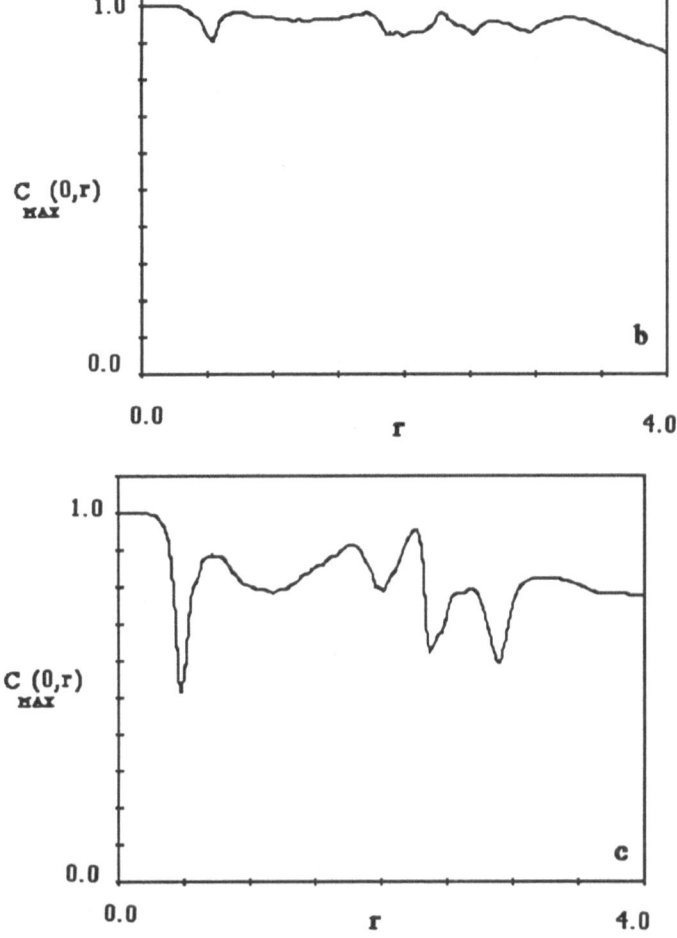

Fig. 4. The maximum of the correlation function as defined by Eq. (9) for (a) $a_1=0.34$, (b) $a_1=0.32$ and (c) $a_1=0.25$.

By decreasing the control parameter a_1, the correlation over the radial direction decreases progressively. More important, dips in the correlation function develops in correspondence to some critical regions. To better understand this behavior, we try to link the phenomenon here observed, with a recent conjecture of Coullet, Gil and Lega [5] about defect mediated turbulence. The presence of peculiar singular solutions during the evolution of the spatial structure can be claimed to be responsible for the loss of spatiotemporal correlation. For example, the crossing of the surface identified by F=0 (which is an invariant unstable surface) can induce discontinuous jumps of the phase when moving along the radial direction. Fig.5 shows the presence of such singular solutions (simultaneous crossing of the zero by the real and imaginary part of the total field F) in the spatiotemporal evolution obtained for $a_1=0.32$ and $a_1=0.25$. It is important to notice that no singular points have been detected for the perfectly correlated structure at $a_1=0.34$ while their presence affected any evolution for a_1 between 0.33 and 0.24. Another important feature is that the regions where the singular points appear correspond to the regions where the correlation function shows dips. The effect of these singularities on the phase of the total field is presently under intensive investigation.

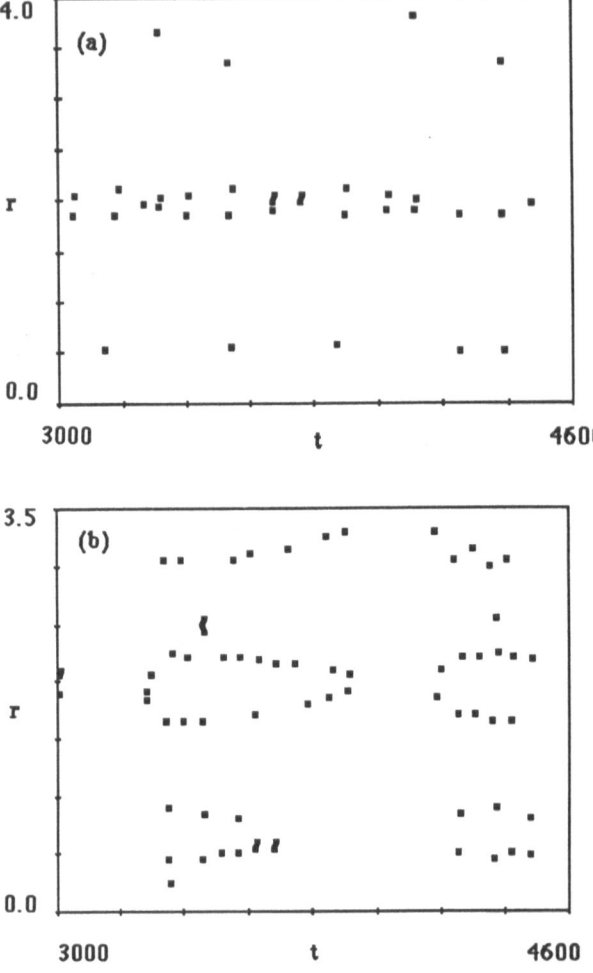

Fig. 5. Time evolution of the singular solutions Re(F)=Im(F)=0
for (a) a_1=0.32 and (b) a_1=0.25

A careful analysis is needed before we can claim that these solutions corresponds to defects. Our simulations have been performed with a very low number of Gauss-Laguerre modes (seven for the pictures showed in this work) which appears to be in contrast with the large aspect ratio requirement for the detection of defects. However, there is numerical evidence that the presence and especially the motion of the singular points happen in conjunction with the loss of spatiotemporal correlation. Another difference between our analysis and simulations performed on partial differential equations regards the origin of the loss of the spatiotemporal correlation. The system of equations (5) involves the amplitudes of the different modes while their spatial dependence is entirely contained in the Gauss-Laguerre functions $A_p(r)$. We have then evaluated the correlation between the time evolution of the different modal intensities by computing

$$C_{MAX}(p,p') = Max_{\forall \tau}\, C(p,p',\tau)$$

$$C(p,p',\tau) = \frac{\langle I_p(t) \, I_{p'}(t+\tau)\rangle - \langle I_p(t)\rangle\langle I_{p'}(t)\rangle}{\sigma(I_p(t)) \, \sigma(I_{p'}(t))} \qquad (10)$$

$$\sigma(I) = \sqrt{\langle I^2\rangle - \langle I\rangle^2}$$

where p and p' are two generic modal indexes and <.> indicates again the temporal average. The results are displayed in Fig.6 where specific correlations between the three active modes are shown. The agreement with the spatiotemporal functions of Fig.3 clarify the temporal nature of the loss of spatial coherence due to partial factorization of the equation of motion. In other words the presence of singular solutions seems to be related to the progressive loss of correlation between the different modes. Fig.6 also shows that in the region of investigation there are two different behavior where the leading loss of correlation is exchanged between the couple (p=1,p'=2) and (p=0,p'=2). Moreover a sudden crisis of the attractor at $a_1=0.24$ restores the spatiotemporal coherence.

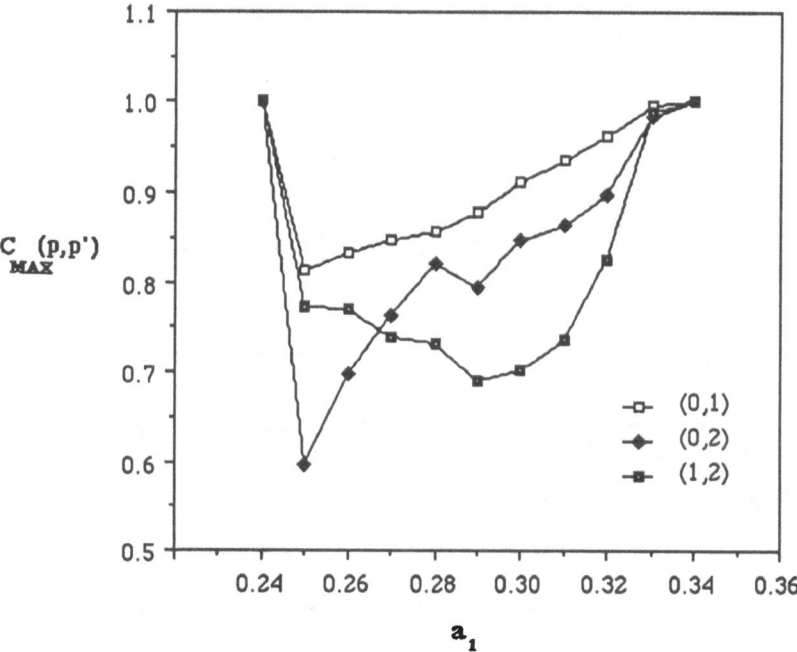

Fig. 6. The mode-mode temporal correlation function as defined by Eq.(10) for decreasing values of a_1. The symbols correspond to couples of modes (p,p') as shown in the legend.

5. CONCLUSIONS

We have shown and discussed the existence of complex spatiotemporal patterns in the dynamics of lasers. Dips in the correlation function along the radial direction detects regions where the coherence of the laser source build by quantum mechanical effects can be easily destroyed by purely classical phenomena. Present studies focus on the relation of the singular solutions detected in the dynamics of the output laser intensity and the defects found in hydrodynamical and chemical systems. The inclusion of the angular dependence [6] of the transverse field as well as multi-mode simulations of the dynamics will clarify to what extent it is possible to talk about optical turbulence nowadays.

ACKNOWLEDGEMENTS

We like to express our special thanks to Jorge Tredicce for sharing with us his ideas, insights and experience about the subject of this paper. Unfortunately, we also regret that he did not accept to add his name to the author list of this communication. Useful discussions with Pierre Coullet, Antonio Politi, Charles Green, and B. Gabriel Mindlin are also gratefully acknowledged.

DISACKNOWLEDGEMENTS

One of us (G.-L.O.) owes **no** thanks to the Head of the Physics Department Dr. R.D. Haracz, to the Dean of the College of Science Dr. J.W. Burley, and to the particle physics group of Drexel University.

* Present Address: Department of Physics and Applied Physics, University of Strathclyde, John Anderson Building, Glasgow G4 0NG, Scotland, U.K.

References:

[1] L.M. Narducci, J.R. Tredicce, L.A. Lugiato, N.B. Abraham, and D.K. Bandy, Phys. Rev. A 33, 1842 (1986);
L.A. Lugiato, G.-L. Oppo, M.A. Pernigo, J.R. Tredicce, L.M. Narducci, and D.K. Bandy, Opt. Comm. 68, 63 (1988);
L.M. Narducci, G.-L. Oppo, J.R. Tredicce, and L.A. Lugiato, in preparation.

[2] J.R. Tredicce, E.J. Quel, A.M. Ghazzawi, C. Green, M.A. Pernigo, L.M. Narducci, and L.A. Lugiato, Phys. Rev. Lett. 62, 1274 (1989).

[3] For smaller values of k, no chaotic solutions have been detected, in agreement with the experimental results of [2].

[4] J.R. Tredicce very private communication.

[5] P. Coullet, L. Gil, and J. Lega, Phys. Rev. Lett. 62, 1619 (1989).

[6] L.A. Lugiato, F. Prati, L.M. Narducci, and G.-L. Oppo, Opt. Comm. 69, 387 (1989).

AMPLITUDE EQUATIONS FOR HEXAGONAL PATTERNS OF CONVECTION IN NON-BOUSSINESQ FLUIDS

C. Pérez-García (*)(‡), E. Pampaloni (**) and
S. Ciliberto (**)

(*) Dept. Física, Univ. Autònoma Barcelona,
    08193 Bellaterra, Catalonia, Spain
(**) Istituto Nazionale di Ottica, Largo
     Enrico Fermi 6, 50125 Firenze, Italy

Hexagonal patterns can be observed in a fluid heated from below, mainly in three situations: 1) when the upper surface has a temperature-dependent surface tension (Bénard-Marangoni convection)[1], 2) when its transport coefficients vary with the temperature (non-Boussinesq convection)[2] and 3) under a modulated heating.[3] Convective patterns can be described by means of the so-called **amplitude equations**, that are obtained either from the hydrodynamic equations[4] or simply from symmetry arguments.[5] This system of equations is simpler than the hydrodynamic nonlinear equation and it is easily simulated in computers.[6]

A hexagonal pattern are made by the superposition of three sets of straigth rolls characterized by wavenumbers that obey $\Sigma k_i = 0$, $|k_i \cdot k_j| = k^2/2$ $(i \neq j)$ and $i = 1,2,3$. The corresponding amplitude equations for the three modes read as

$$\tau_0 \frac{\partial A_i}{\partial t} = \left\{ \varepsilon + \xi_0^2 \left[\frac{\partial}{\partial x_i} - \frac{i}{2k_c} \frac{\partial^2}{\partial y_i^2} \right]^2 \right\} A_i + a\, A_{i+1}^* A_{i+2}^* -$$

$$- b \left[\sum_{i \neq j} |A_j|^2 \right] A_i - c\, |A_i|^2 A_i \qquad (i = i \bmod 3) \qquad (1)$$

Here $\varepsilon = (\Delta T - \Delta T_c)/\Delta T_c$ denotes the temperature difference above threshold. The relaxation time τ_0, the correlation length ξ_0 and the critical Rayleigh number R_c and wavenumber k_c are obtained from a linear stability analysis of the full hydrodynamic equations of the system. The superscript * denotes the complex conjugate and $\partial/\partial x_i$ and $\partial/\partial y_i$ are the spatial derivatives parallel and perpendicular to the vector k_i, respectively. The physical meaning of the nonlinear coefficients a, b and c, will be specified later on.

Measures of Complexity and Chaos
Edited by N.B. Abraham *et al.*
Plenum Press, New York

These equations can be written in a variational form

$$\tau_0 \frac{\partial A_1}{\partial t} = - \frac{\partial \mathcal{F}}{\partial A_1^*} \tag{2}$$

where $\mathcal{F}$ is a functional, analogous to a potential, defined as

$$\mathcal{F} = \int dxdy \left\{ \sum_{i=1}^{3} \left[-\varepsilon |A_i|^2 + \left| \xi_0 \left(\frac{\partial}{\partial x_i} - \frac{1}{2k_c} \frac{\partial^2}{\partial y_i^2} \right) A_i \right|^2 \right] - a \prod_{i=1}^{3} |A_i| \right.$$

$$\left. + \frac{1}{2} \sum_{i=1}^{3} \left[\sum_{j \neq i} |A_j|^2 |A_i|^2 + \frac{c}{2} |A_i|^4 \right] \right\} + f(\varepsilon) \tag{3}$$

provided that on the boundaries $A_1 = 0$ or $\partial A_1/\partial x_1 = \partial A_1/\partial y_1 = 0$. The function $f(\varepsilon)$ is added in order to ensure that stationary solutions give $\mathcal{F} = 0$. The term $\prod A_i^*$, simply gives $\prod |A_i|$, because the hexagonal symmetry implies that the sum of the phases of the three system of rolls must vanish. The existence of solutions in the weakly nonlinear regime imposes that the quartic positive term must be greater than the quadratic negative one in (3). Therefore, $\mathcal{F}$ is a positive-definited functional, whose temporal derivative gives

$$\frac{d\mathcal{F}}{dt} = - 2\tau_0 \int dxdy \sum_{i=1}^{3} \left| \frac{\partial A_1}{\partial t} \right|^2 \leq 0 \tag{4}$$

Therefore, $\mathcal{F}$ is a Lyapunov functional. As a consequence, the evolution of the an initial to the stationary state is monotonous. (For a discussion of the limitations of such a kind of variational formulation see ref.[7]).

Stationary solutions are obtained by the relation $\delta \mathcal{F}/\delta A_1^* = 0$. In the homogeneous case these are a) a pattern of rolls $|A_1| \neq 0$, $|A_2| = |A_3| = 0$ and b) a hexagonal pattern $|A_1| = |A_2| = |A_3|$. The linear stability of these solutions can be determined by the matrix $\delta^2 \mathcal{F}/\delta A_i \delta A_j^*$ linearized around the stationary solutions. This allows to determine that rolls are the unstable solutions for $0 \leq \varepsilon \leq \varepsilon_R$, where ε_R is given by

$$\varepsilon_R = \frac{a^2 c}{(b-c)^2} \tag{5}$$

Hexagons are stable in the interval $\varepsilon_A \leq \varepsilon \leq \varepsilon_H$ with

$$\varepsilon_A = - \frac{a^2 c}{4(2b+c)} \qquad \varepsilon_H = \frac{a^2(b+2c)}{(b-c)^2} \tag{6}$$

The Liapunov functional $\mathcal{F}$ provides an interesting tool to determine the dynamics of the system, but it is not directly accesible from experiments. Some elaborations, based on local measurements must be made in order to exploit such an analysis.[7] Moreover in some cases (convection in liquid ^{4}He) optical inspection of the cell is not possible.

It is interesting to relate the local description of amplitude equations with the global heat flow measurements which are experimentally accesible. The comparison can be made taking into account that the normalization chosen for eqs. (1) gives the following relation[8]

$$N \equiv \frac{(N-1)R}{R_c} = \frac{1}{S}\int dxdy \sum_{i=1}^{3}|A_i|^2 \qquad (7)$$

where N is the Nusselt number, the ratio between convective heat flow and the total heat flow, and S the horizontal area of the layer. For rolls the normalized Nusselt number $\mathcal{N}$ gives

$$\mathcal{N}_r = |A_r| = \varepsilon/c \qquad (8).$$

For hexagons the expression is simply

$$\mathcal{N}_h = 3|A_h| = \frac{3}{C}\left\{\frac{a^2}{4C} + \varepsilon + a\sqrt{a^2 + 4C\varepsilon}\right\} \qquad (9)$$

where C = 2b+c. As a consequence the slope of $\mathcal{N}_r(\varepsilon)$ gives a direct information about the coefficient c, while the determination of b and a from $\mathcal{N}_h$ requires a more delicate fitting of the dependence on ε and $\varepsilon^{1/2}$.

Of course, for a comparison of (8)-(9) with heat flow measurements one must take into account lateral effects, as well as the contributions of defects into the heat transport. Therefore, the aim of the present analysis is to give an approximate value of the coefficients in the amplitude equations from heat flow measurements. We restrict the analysis to the case of a non-boussinesquian fluid. When ε is small a hexagonal pattern of convection appears in this case. For a certain threshold two modes vanish and that pattern is replaced by a pattern of rolls. The threshold for the stability of these two modes have been calculated by Busse[2a] and observed in some recent experiments.[2c]

Equations (5)-(6), (8)-(9) can be compared with the corresponding equations obtained by Busse[2a]. This allows to identify the coefficients a, b, c in the amplitude equation as

$$a^2 = \frac{3\mathcal{P}^2}{\mathcal{R}_c}, \quad b = \frac{1}{2}\left(3\mathcal{R}_h - \mathcal{R}_r\right), \quad c = \mathcal{R}_r, \quad L_2 = (b-c) \qquad (10)$$

The coefficients $\mathcal{P}$, $\mathcal{R}$ and L_2 are related with the transition between hexagons and rolls. (The values of $\mathcal{R}$ are quoted in p. 641, those of $\mathcal{P}$ can be obtained from eq. (7.7) and Table II and L_2 from Table I in ref.[2a]).

In the particular case of a horizontally unbounded layer of a fluid with an infinite Prandtl number these coefficients are

$$a^2 = (1.76 \times 10^{-3})\ \mathcal{P}^2, \qquad b = 0.99, \qquad c = 0.70 \qquad (11)$$

The values of these coefficients obtained by experiments are of the same order than the theoretical ones computed from (10) in the case of convection in water[2c]. The discrepancy are probably due to finite-size effects that are not taken into account in theoretical calculations.

Acknowledgements

This work has been partially supported by a grant of the ECC, project SC1-0035-C(CD), and by an Integrated Action (n∘46) of the Spain-Italy Scientific Cooperation Program.

(‡) also at Departamento de Física, Universidad de Navarra, 31080 Pamplona, Navarra, Spain.

References

[1] H. Bénard, Rev. Gén. Sci. Pur Appl., **11**, 1261, 1309   (1900)
  A. Cloot and G. Lebon, J. Fluid Mech., **145**, 447 (1984)
  P. Cerisier, C. Pérez-García, C. Jamond and J. Pantaloni, Phys. Rev. A, **35**, 1949 (1987)
[2] F. H. Busse, J. Fluid Mech., **30**, 625 (1967)
  M. Dubois, P. Bergé and J.E. Wesfreid, J.Physique, **39**, 1253 (1978)
  S. Ciliberto, E. Pampaloni and C. Pérez-García, Phys. Rev. Lett., **61**, 1198 (1988)
[3] M.N. Roppo, S.H. Davis and S. Rosenblatt, Phys. Fluids, **27**, 796 (1984)
  P.C. Hohenberg and J.B. Swift, Phys. Rev. A, **35**, 3855 (1987)
  C.W. Meyer, D.S. Cannell, G. Ahlers, J.B. Swift and P.C. Hohenberg, Phys. Rev. Lett., **61**, 947 (1988)
[4] A.C. Newell and J.A. Whitehead, J. Fluid. Mech., **38**, 279 (1969)
  L.A. Segel, J. Fluid Mech., **38**, 203 (1969)
[5] P. Coullet, these Proceedings.
[6] M. Bestehorn and H. Haken, Z. Physik, **57**, 329 (1984)
[7] M.S. Heutmaker, P.N. Fraenkel and J.P. Gollub, Phys. Rev. Lett, **54**, 1369 (1985)
[8] G. Ahlers, M.C. Cross, P.C. Hohenberg and S. Safran, J. Fluid Mech., **110**, 297 (1981)
[9] G. Ahlers, J. Fluid Mech., **98**, 137 (1980)

FRACTAL DIMENSIONS IN COUPLED MAP LATTICES

A. Politi, G. D'Alessandro, and A. Torcini

Istituto Nazionale di Ottica

50125 Firenze, Italy

The discovery of erratic behaviour in deterministic systems opened entirely new perspectives in the comprehension of dynamical behaviour of nonlinear systems. The existence of broad-band spectra no longer necessarily requires a coupling with an external uncontrollable thermal bath. Chaos, providing an information flow from irrelevant to relevant digits, naturally transforms the indetermination on the initial condition into a seemingly stochastic behaviour in time domain [1]. New classes of indicators have been consequently introduced, which allow to distinguish between truly stochastic motion and low-dimensional chaotic behaviour: Lyapunov exponents, metric entropy and fractal dimensions are dynamical invariants which measure the degree of chaoticity [2].

Accordingly, we can nowadays say that the mechanisms underlying a low-dimensional chaotic behaviour are, up to a great extent, fairly well understood. On the contrary, the characterization of spatially extended systems with their infinite number of degrees of freedom, remains a still open problem, which goes under the generic name of spatio-temporal turbulence. The, in principle, infinite dimensionality, and infinite metric entropy indicate in this case that a straightforward extension of the standard indicators to describe spatio-temporal phenomena does not allow to distinguish between turbulent and stochastic behaviour. It is the aim of this paper to investigate in detail the fractal properties of finite portions of an initial infinite system. To make the analysis as simple as possible, and assuming that some of the relevant features of true turbulent motion are maintained, we study a one dimensional lattice of I diffusively coupled logistic maps [3].

Two main limitations are associated with such a model:

a) Discreteness in space, which prevents the reproduction of the typical features occurring at high wave vectors in real turbulent behaviour. This is not a crucial difficulty as long as one is interested in describing a "mild" turbulence. We can for instance think of the single logistic map as mimicking the local chaotic motion of a single convecting roll in Rayleigh-Bénard instability.

b) Non-invertibility of the dynamics. Such a difficulty is immediately removed by replacing the one-dimensional logistic map with an invertible 2-d recursive relation (e.g. the Henon map), thus doubling the phase space dimension ($2I$). However, we can reasonably conjecture, that the lattice of 1-d maps reproduces the main features

Measures of Complexity and Chaos
Edited by N.B. Abraham *et al.*
Plenum Press, New York

displayed by a projection of the behaviour of the more realistic lattice of invertible maps, at least when the dimension of the attractor in the I-dimensional space is strictly less than I, as it is the case in the regime we have investigated.

Bearing these limitations in mind, let us introduce the following dynamics

$$x_{n+1}^i = f(y_n^i) \quad , \tag{1}$$

where i is the spatial index ranging between 1 and I, n is the time label, and $f(y)$ is a map of the interval into itself. As already anticipated, we limit our analysis to the logistic map

$$f(y) \equiv a - y^2 \quad , \tag{2}$$

where the parameter a controls the single-site chaoticity. Finally, the argument of f in Eq. (1) is given by,

$$y_n^i = \varepsilon x_n^{i-1} + (1 - 2\varepsilon)x_n^i + \varepsilon x_n^{i+1} \quad , \tag{3}$$

to schematize a diffusive coupling. The parameter ε controlling the strength of the coupling, is bound between 0 (uncoupled case) and 1/2 (two independent space-time lattices). Periodic boundary conditions will always be chosen throughout our analysis, namely, $x_n^0 = x_n^I$, $x_n^1 = x_n^{I+1}$.

In literature, there is a growing numerical evidence and some preliminary theoretical results [4] in favour of the existence of a limit Lyapunov spectrum in systems intinitely extended in space. More precisely, by defining $\lambda(i, I)$ as the i-th largest characteristic exponents in a lattice of I maps, we have

$$\lim_{I \to \infty} \lambda(i, I) = \Lambda(i/I) \qquad 1 \le i \le I \quad . \tag{4}$$

As a consequence, the Kaplan-Yorke relation naturally suggests to introduce the concept of density of dimensions

$$\sigma^L \equiv D_L(I)/I \quad , \tag{5}$$

where $D_L(I)$ is the Lyapunov dimension. From Eq. (4), it follows that σ^L is independent of I, and is given implicitely by

$$\int_0^{\sigma^L} \Lambda(x)dx = 0 \quad . \tag{6}$$

Some numerical results are reported in Fig. 1, where the spectrum of Lyapunov exponents has been computed for a chain of $I = 120$ maps, $a = 2$ (fully developed chaos in the single map) and $\varepsilon = 1/3$. Eq. (6) provides an upper bound to the density σ of information dimension [5]. Despite only an inequality has been theoretically proved, numerical simulations suggest that the equality holds generically for invertible maps. In the case of non-invertible dynamics, instead, we may expect a finite difference, as, for instance, already found in the simple Sinai map [6]. Preliminary simulations indicate that the difference, if any, is small for coupled logistic maps [7].

However, the direct proportionality of the dimension $D(I)$ to the length of the chain, for I sufficiently large, is an indisputable fact. The meaning of such a relation is that

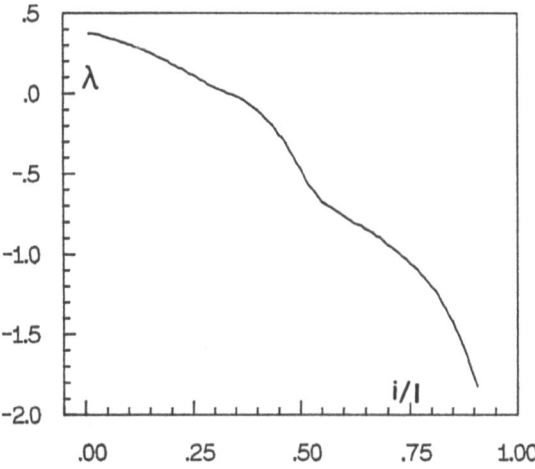

Fig. 1 - *Spectrum of Lyapunov exponents estimated from a chain of length $I = 120$. In this and all the following figures the parameter values are $a = 2$, $\varepsilon = 1/3$.*

the dynamics of a long chain can be essentially interpreted as that of many juxtaposed independent pieces of large enough length L. Accordingly, the dimension turns out to be proportional to the length of the chain. By pushing farther on this argument one is lead to conclude that, monitoring the evolution of a finite portion of a (in principle infinite) chain, a dimension smaller than the chain length is found, as if the boundary conditions were irrelevant for L large enough [8]. More precisely, starting from a generic site i, and building a vector $V_n(i) \equiv (x_n^i, x_n^{i+1}, \ldots, x_n^{i+L})$, i.e. working in a spatial embedding-space of dimension L, it is expected that all such points fill the space with a dimension equal to $\sigma_s L$ (we have added a subscript s to the σ^L defined in Eq. (5) to stress the fact that we are making a spatial embedding). In other words, as claimed in Ref. [9], the embedding technique allows to distinguish between a truly random noise (where it is expected to have $D(L) = L$) and a deterministic signal arising from infinitely many coupled maps.

The same conjecture has been extended to the time domain, where the more familiar temporal embedding can be applied. With reference to this latter case, let us notice the difference with respect to low-dimensional chaos where $D(L)$ converges, for increasing L, to some finite value which is interpreted as the dimension of an underlying attractor. In the present case instead, no saturation is expected,and but the growth rate of $D(L)$ becomes an interesting indicator to look at. In Ref. [9] an attempt has been made to express the temporal dimension density σ_t in terms of dynamical indicators.

The main result of our analysis is to show the incorrectness of such conjectures. We start with a general argument against the hypothesis of a dimension density strictly smaller than one. Assuming that the dynamics of our chain of maps is mixing (that is reasonably true in the fully turbulent regime), it follows that the distribution of values $V_n(i)$ estimated at the same time n, starting from different sites i, is the same as the distribution of values estimated in the same site at different times. Now, the distance $d_L(i, n_1, n_2)$, between two sequences of values $V_n(i)$, $V_m(i)$ in an embedding space of dimension L,

$$d_L(i, n, m) = \sqrt{\sum_{j=i}^{i+L}(x_n^j - x_m^j)^2},\qquad(7)$$

is nothing but the distance between the projections in a suitable L-dimensional subspace of the original points (i.e. with all variables taken into account). In other words, the dimension $D(L)$ of a portion of length L of a chain with $I \gg L$ maps, is the dimension of the projection of a suitable $(\sigma_s I)$-dimensional subset onto a (generic) subspace of dimension L. As $\sigma_s I > L$, we must in general expect $D(L) = L$, except for very exceptional cases.

Our conclusion has been rigorsously proved in a recent paper by Bunimovich and Sinai [10] for everywhere expanding maps, with a different coupling. With this idea in mind we have carefully checked the outcome of numerical experiments. We started with a "rough" simulation using the Grassberger-Procaccia algorithm [11] with the same amount of data points used in Ref. [8] (30,000). Rather than plotting directly the correlation integral, we show in Fig. 2 the behaviour of its derivative (coarse grained dimension - CGD) as a function of the decimal logarithm of the distance. The wild fluctuations occurring at small distances are due to lack of sufficient statistics and prevent us to draw any definite conclusion on the convergence of the coarse grained dimension to a precise limit value. We can only observe the big difference between the estimated CGD values and the dimension of the embedding space.

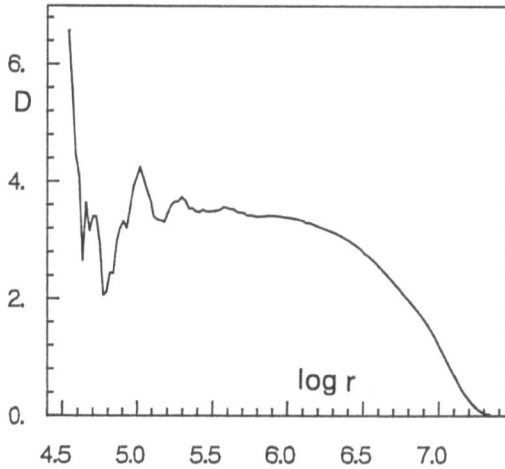

Fig. 2 - *Coarse grained dimension (Grassberger Procaccia method) versus the decimal logarithm of interpoint distance r. The chain length is $I = 30,000$, and the spatial embedding dimension is $L = 6$.*

In order to improve the quality of the simulations and to better control the error on the CGD we have changed algorithm and introduced a number of "tricks". We have chosen the nearest-neighbour method [12], which is based on the comparison between a randomly chosen reference point and an increasing number n of points (also randomly chosen). The logarithm of the distance of the k-th nearest-neighbour for different n-values is then averaged over m different reference points to decrease as much as possible the statistical error. Therefore, the number n of points controls the length scale (larger n values yield smaller distances), while the number m of reference points controls the statistical error on the distance $\delta(k, n)$. From the behaviour of the distance $(\delta(k, n) \simeq (k/n)^{-1/D}$ [12], where D is the information dimension) it follows that $-(d \ln n / d \ln \delta)$ can be interpreted as a CGD, and plotting it versus $\ln n$, we can check its dependence on the observational resolution as done in Fig. 2 for the Grassberger-Procaccia algorithm.

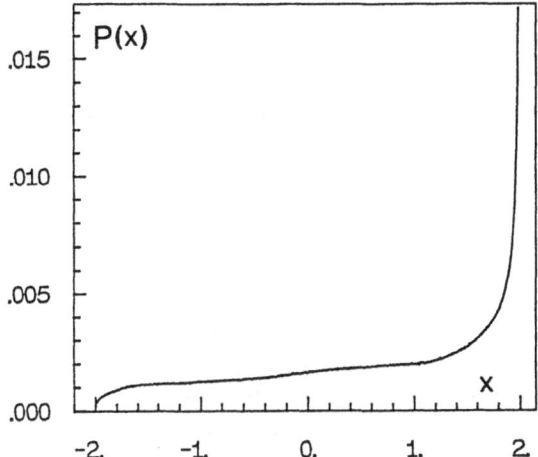

Fig. 3 - *Invariant measure $P(x)$ computed with 3,000,000 data points.*

Let us go to the specific points regarding the numerical simulations, which all refer to the case $\varepsilon = 1/3$ and $a = 2$. At first we have looked at the invariant single site distribution $P(x)$. We recall that in the uncoupled case ($\varepsilon = 0$),

$$P(x) = \frac{1}{2\pi\sqrt{4 - x^2}} \quad , \tag{8}$$

with two square root singularities at the extrema of the invariant interval $[-2, 2]$. The distribution for $\varepsilon = 1/3$, reported in Fig. 3, exhibits only one singularity on the right which derives from the existence of a quadratic maximum in the map. Indeed, assuming a flat distribution around the preimage of the maximum of Eq. (2), a singularity with exponent $\beta_1 = -1/2$ is necessarily generated around $x = 2$. A careful numerical analysis around the minimum value $x = -2$ shows that $P(x)$ goes instead to zero with a scaling exponent $\beta_2 \simeq 1/3$. This remarkable difference with respect to the single-map case (where $\beta_2 = -1/2$ is found) can be qualitatively explained. In fact, the minimum value $x = -2$ is reached at time $(n+1)$ iff a sequence of three maxima is observed at time n. Assuming x_n^i to be independent of x_n^{i+1}, and using the local linearity of the map around $x = 2$ with an expansion coefficient k (i.e. a segment δx around $x = 2$ is mapped onto a segment of length $k\delta x$ around $x = -2$), it follows that

$$(\delta x)^{1+\beta_2'} \equiv P(x = -2)\delta x = \left(P(x = 2)(\frac{\delta x}{k})\right)^3 \equiv (\frac{\delta x}{k})^{3(1+\beta_1)} \quad , \qquad (9)$$

$$\Rightarrow \quad \beta_2' = 3(1 + \beta_1) - 1$$

from which an exponent $\beta_2' = 1/2$ is found which explains the convergence to zero of $P(x)$. The difference between β_2' and the actual value $\beta_2 \simeq 1/3$ is to be attributed to correlations between nearby sites.

The relevant point emerging from the analysis of the invariant measure is the occurrence of a spike around $x = 2$. Such a singularity does not affect any generalized dimension D_q with q strictly smaller than 2, as shown in Ref. [13]. In particular it does not affect the correlation exponent which results from the Grassberger-Procaccia algorithm, and it does not affect the information dimension extracted with the nearest-neighbour algorithm. However, this is rigorously true only in the limit of infinite resolution (i.e. zero distance). Since numerical simulations allow to investigate only a limited range of distances, and such a range becomes increasingly small for increasing dimension, it becomes crucial to get rid as much as possible of the "concentrations" of mass which can heavily affect the CGD on larger distances. Therefore, we have performed a smooth change of coordinates (with a singularity at the upper edge) from x to y, such that the y-probability density $Q(y)$ were constant. Such a change of variable again does not affect D_q for $q < 2$.

A second improvement of the numerical results can be achieved by reducing the edge effects. Let us assume to have a generic distribution which covers only a finite region S of phase space. The density around a point distant δ from the S-edge, if estimated from the mass in a ball of radius r, turns out to be obviously underestimated as long as $r > \delta$. This, in turn, leads to an underestimation of the dimension. This effect becomes more and more relevant for increasing dimensions as already discussed in Ref. [14]. In order to reduce it, we have changed the topology of the phase space. Let us explain the idea in the case of a sequence of L identically, independently, and uniformly distributed x-variables. They fill uniformly an L-dimensional hypercube, so that the density in any ball, chosen sufficiently close to an edge, is affected by the external empty regions. However, if we identify opposite edges, transforming the hypercube into a torus T^L, all empty regions disappear and the edge effect is completely removed. Such a trick can be easily implemented on a computer especially when working with integer numbers. It turns out that using the nearest neighbour algorithm, with 27,000 points in a 6-dimensional space (with an average over 3,000 reference points, and looking at the 20-th nn), we obtain $D = 6$ with an error around 0.2%, to be compared with $D \simeq 5.47$ determined from the original data points.

Obviously, whereas the previous trick proves to be very effective in the case of independent variables, since all empty regions are removed, it is expected to be less powerful in the case of coupled maps, because of correlations. The difficulty is confirmed by the analysis of the two dimensional distribution $P_2(x^i, x^{i+1})$. Its support covers only a portion of the allowed square, as shown in Fig. 4. In order to increase again the accuracy of our simulations for fixed amount of data points, we have performed a further change of variables which expanded the covered region to the full square. To this end we have first generated 300 millions points to estimate numerically the border of the allowed region. Such borders have been then rectified to get rid of statistical fluctuations and, finally, a uniform vertical expansion has been introduced. Obviously, we still expect edge effects to arise because of higher order correlations yielding empty regions in higher dimensional spaces.

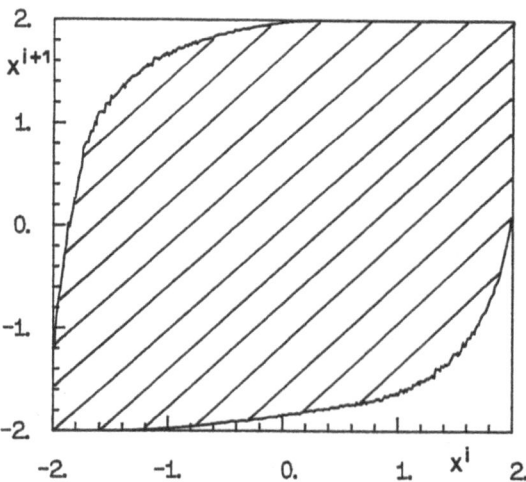

Fig. 4 - *Support of the spatial two dimensional distribution $P_2(x^i, x^{i+1})$ computed from 300,000,000 data points.*

After having performed all such changes of variables, we have finally used the nearest-neighbour algorithm with 2.7 million points and averaging over 30,000 reference points, monitoring simultaneously the 4-th, 10-th and 30-th nearest neighbour (nn). The simulations required 14 days of CPU time on a HP835 computer. The results for the CGD values are always larger for smaller order k of nn (whenever the number of points is insufficient to yield saturation) since smaller distances are reached. However if, on one side, the results for the 4-th nn are more asymptotic, on the other hand, for increasing k, the fluctuations decrease and more consistent curves are obtained. For this reason we present in Fig. 5 the results in the intermediate case $k = 10$. Embedding dimensions $L = 3$, 4 and 5 indicate that a saturation of the CGD to a value coinciding with the space dimension is essentially reached. In the case $L = 6$ the CGD arrives at most at 5.5, but no evidence of saturation is detected, so that this value definitely appears as an underestimation. A comparison with the starting simulation reported in Fig. 2 is instructive (we recall that high resolutions corresponds to points lying on the left of Fig. 2, and on the right of Fig. 5). From a value around 3.7 reached in the first case, we pass to 5.5 which is still not saturated. Part of the difference is due to the fact that now we are computing the information dimension which is expected to be larger than the correlation exponent. However, multifractality cannot explain the difference as confirmed by a more robust simulation with the Grassberger-Procaccia algorithm with 50,000 points, where a constant increasing trend of the CGD is found, as well. Therefore, we can reasonably assume that the asymptotic value of D is 6 in the case $L = 6$. Figure 5e confirms the increasing difficulty in reaching the asymptotic value: the distance from the embedding dimension is already around 2 for $L = 9$, but again no evidence of saturation is exhibited.

Therefore, our conclusions point in the direction of suggesting that the equality $D(L) = L$ holds, indicating that the results shown in Refs. [8,9] are strongly affected by a lack of sufficient data points. A further question now arises whether the underestimation is essentially due to the trivial fact that the number of points must

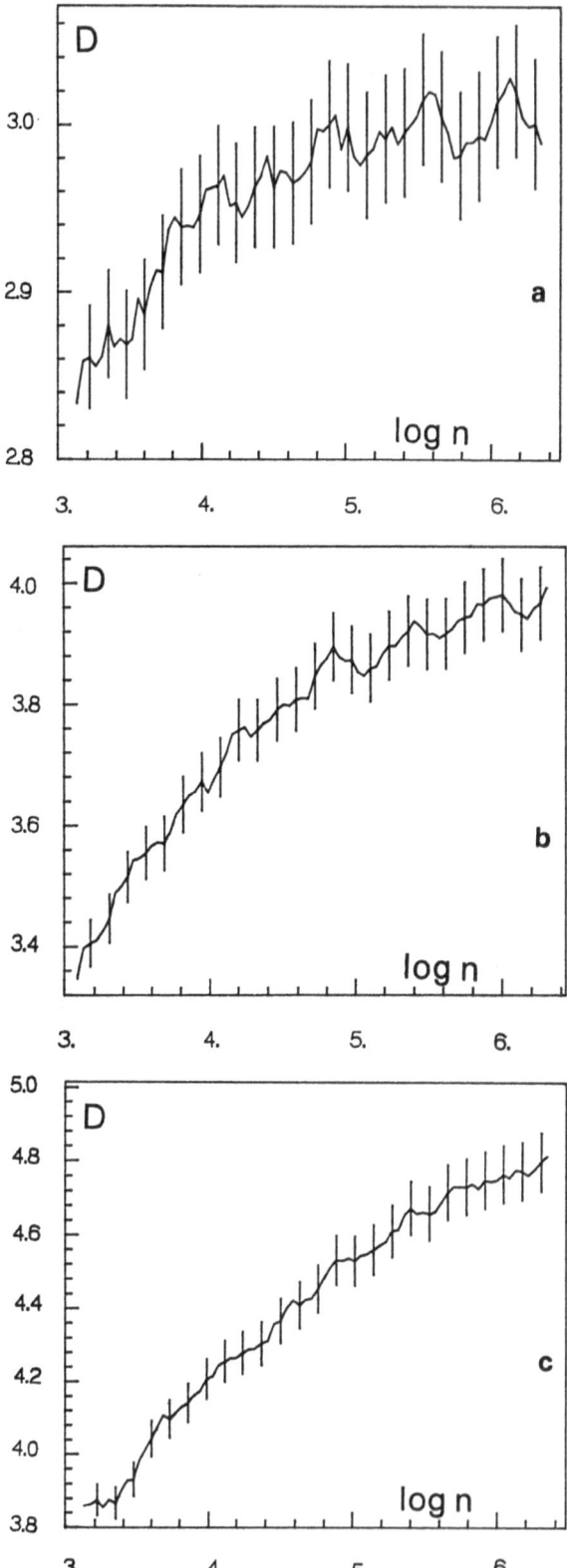

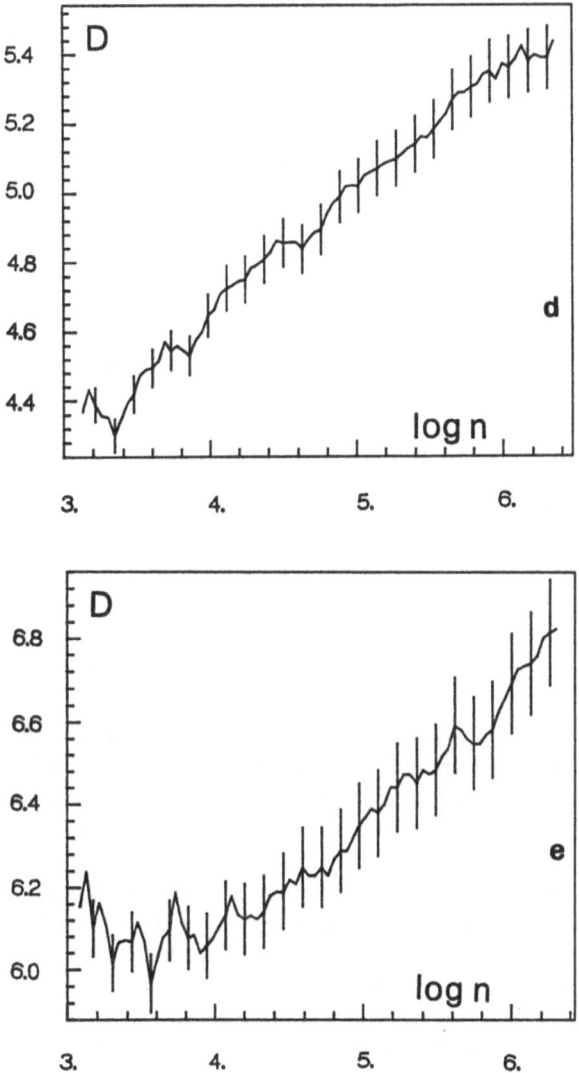

Fig. 5 - *Coarse grained dimension (nearest neighbour method) versus the decimal logarithm of the number of points examined. Spatial embedding on 2,700,000 data points with an average over 30,000 reference points. The figures a,b,c,d,e refer to embedding dimensions 3,4,5,6,9 respectively, and nearest neighbour of order 10.*

be increased when looking at sets of increasing dimension, or to a somehow "strange" geometric structure of the set. To answer this question we proceed empiricaly assuming the following dependence of CGD on $\log n$,

$$D(L, \log n) = L - a \exp[-b \log n] \quad , \tag{10}$$

with a, b being free parameters. A power law convergence in n (exponential convergence in $\ln n$) is the most natural hypothesis in the case of a fractal structure. Objectively, we may expect many exponential correction terms to be simultaneously present, so that, as long as $D(L)$ is very close to saturation, Eq. (10) reproduces very well the expected D-dependence, whereas less reliable results should be found for less asymptotic values. Starting from Eq. (10), we have fitted all the curves shown in Fig. 5, to determine a and b. This allows to extrapolate the minimum number of points n^* needed to observe $D = (1 - f)L$.

$$n^* = \left(\frac{a}{fL} \right)^{1/b} . \tag{11}$$

In order to have a more objective indicator (the number of points depends on the number of nn considered in the analysis), we have translated n^* in terms of distances. From Eq. (10), recalling that

$$D = -\frac{1}{< \log \delta >'} \quad , \tag{12}$$

where $< \ldots >$ means average over different reference points, and the prime indicates the derivative with respect to $\log n$, we find

$$< \log \delta > = -\frac{\log n}{L} - \frac{\log[L - a \exp(-b \log n)]}{Lb} + C \quad , \tag{13}$$

where C is the integration constant to be determined, and the first term in the r.h.s. yields the expected asymptotic behaviour. By determing C with a further fit of the results of the direct simulations, we are able to convert n^* determined from Eq. (11) into a $< \log \delta^* >$. The distance δ^* corresponds to the maximum distance over which the asymptotic dimension is observed. In Fig. 6 we have plotted $< \log \delta^* >$ versus L, for all the three nn used in our simulations. Although a striking decrease of the distances is shown (when passing from $L = 4$ to $L = 9$, δ^* is reduced by a factor 10), we cannot conclude anything about the scaling behaviour, which appears to be roughly exponential in L, because of the large uncertainty at $L = 8, 9$, which is indicated by the relative difference among the three curves. Anyhow, the decrease of the distance is rather unequivocal. This remains true even after considering the edge effects. We have tried to make them as small as possible, but they are still there, and it is important to estimate their scaling behaviour. We present a calculation in the case of a hypersphere, being rather confident that the scaling behaviour remains unchanged for different (not too strange) geometries. Let us call R the radius of the sphere, and let r be the radius of the generic ball used to estimate the local probability density, which we assume to be uniform, for simplicity. The fraction p of balls (randomly chosen), whose mass is affected by the edges, is obviously given by the fractional volume of the sphere contained between radii $(R - r)$ and R. For $r << R$ we obtain

$$p = L(r/R) \tag{14}$$

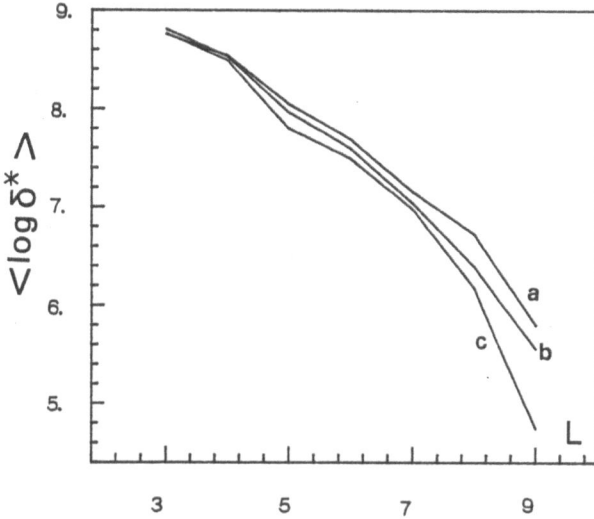

Fig. 6 - *Logarithm of minimal resolution for different spatial embedding dimensions; curves a, b, and c refer to the 4-th, 10-th, and 30-th nearest neighbour, respectively.*

The fraction p has to be kept constantly small for increasing L values, if we want to keep the underestimation error under control. As a result, the range of distance we have to reach to guarantee such a condition goes to zero as (p/L). Hence, contrary to what is usually believed, we see that already in this naive case, with uniform distribution, the number of points must grow faster than exponentially in L, namely

$$N \simeq \left(\frac{L}{p}\right)^L .$$

$$(15)$$

Let us recall the standard argument used to determine the minimum number of points required to estimate the dimension. Assuming that the distance among points along any direction has to be smaller than a given fraction g of the actual size of the attractor, we must at least generate $N = g^{-L}$ points to fulfill such a condition along all directions. Ruelle and Eckmann have been able to prove that such a limit can be reduced to $N = g^{-L/2}$ in the case of Grassberger-Procaccia algorithm, noticing the peculiarity of that method [15]. Indeed, the counting procedure is formally equivalent to computing all distances from a single fictitious reference point around which the distributions seen from any reference point, are rigidly shifted and overlapped. As a result, we have an "equivalent" number of points around the fictitious reference one which is half of the square of the number of effective avalaible points, thus accounting for the square root factor found in Ref. [15]. However, Eq. (15) indicates that a more dangerous problem can arise from the border of the support of the probability density. Coming back to our simulations, the factor 10 increase of the minimal resolution, when passing from $L = 4$, to $L = 9$, cannot be entirely explained by the edge effect which accounts only for a factor 2.25 (see Eq. (14)). A plausible explanation is that the thickness of the invariant set (projected onto an L-dimensional space) becomes smaller and smaller along some directions for increasing L. Therefore, besides the difficulty of estimating the dimension of moderately high-dimensional sets expressed

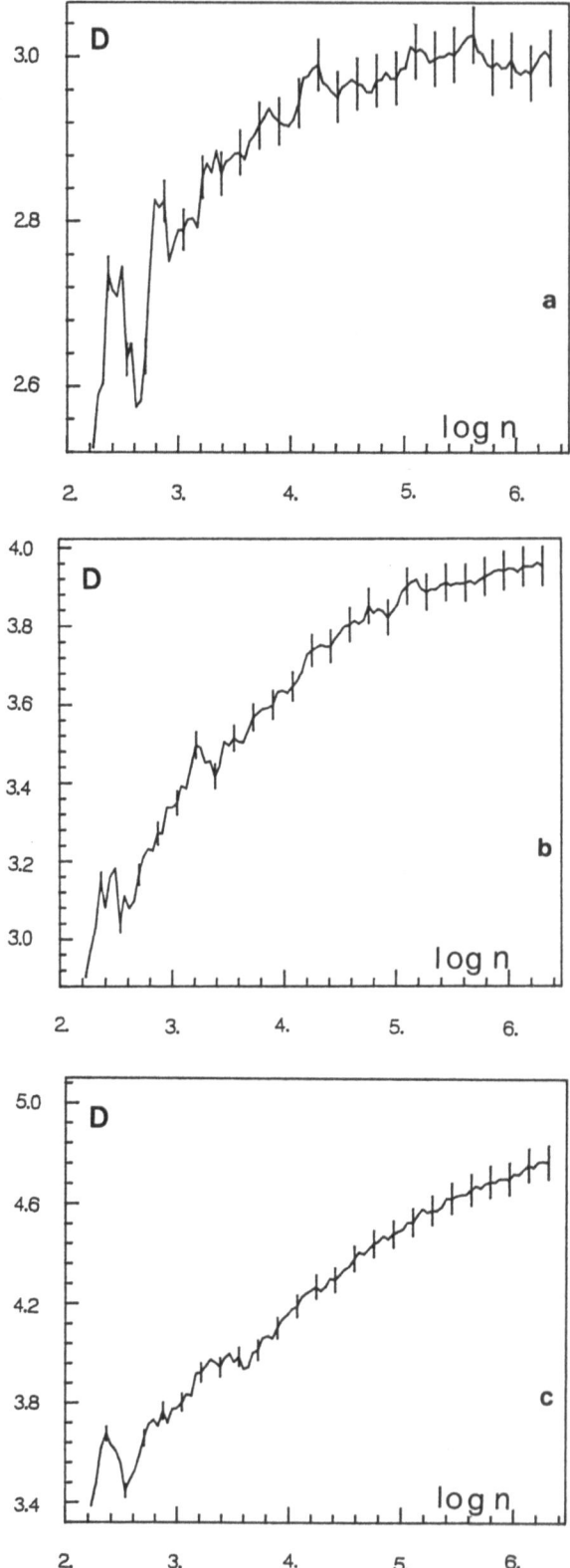

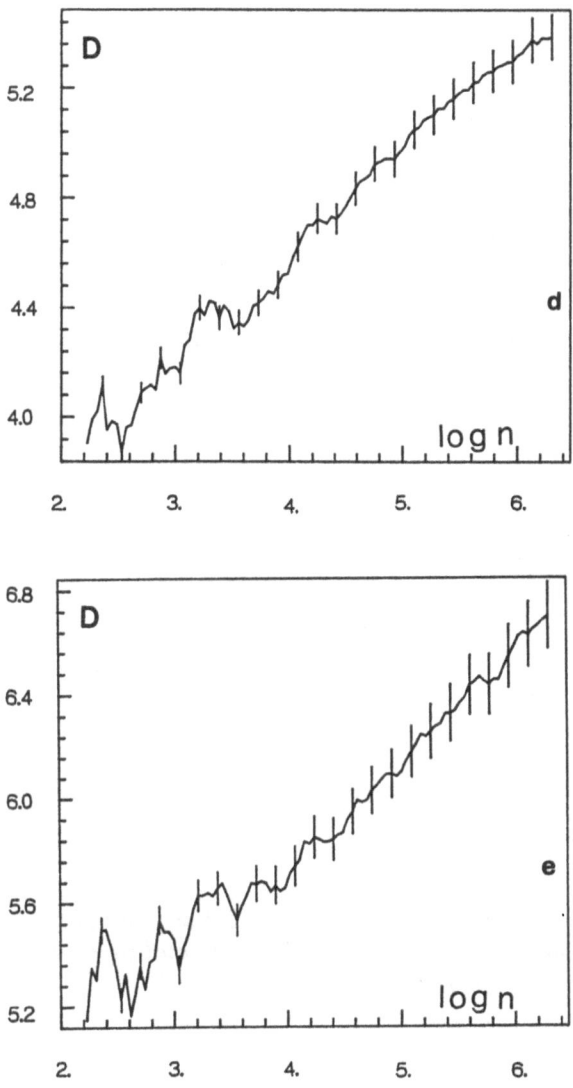

Fig. 8 - *Coarse grained dimension (nearest neighbour method) versus the decimal logarithm of the number of points examined. Temporal embedding on 2,700,000 data points versus 60,000 reference points. The figures a,b,c,d,e refer to embedding dimensions 3,4,5,6,9 respectively, and nearest neighbour of order 10.*

by Eq. (15), here we must add the nonuniform filling of phase space. As a consequence it becomes practically impossible to use any direct algorithm to compute dimensions in case of spatio-temporal chaos. To give an idea, from our extrapolations it follows that the number of points needed to find $D = 9$ with 3% accuracy in the case $L = 9$ turns out to be around 10^{22}!! However, at least from a theoretical point of view, an interesting question is opened on the scaling behaviour of the minimal resolution for increasing L.

For completeness, we have investigated the temporal embedding case as well, proceeding exactly in the same way as for the spatial case. Namely, we have changed the variable x to make the single-site invariant measure uniform. Then we have extended the two-dimensional domain to the whole allowable square and changed the space topology. At variance with the spatial case, it is possible to find an analytical "envelope" for the support of the two-dimensional temporal distribution $P_2(x_n, x_{n+1})$. If the values of the variable x_n^i are uncorrelated at different sites, then the variable y_n^i defined in Eq. (3) is limited to the interval

$$I_y \equiv [y_m, y_M] = [(1 - 2\varepsilon)x - 4\varepsilon, (1 - 2\varepsilon)x + 4\varepsilon] \quad . \tag{16}$$

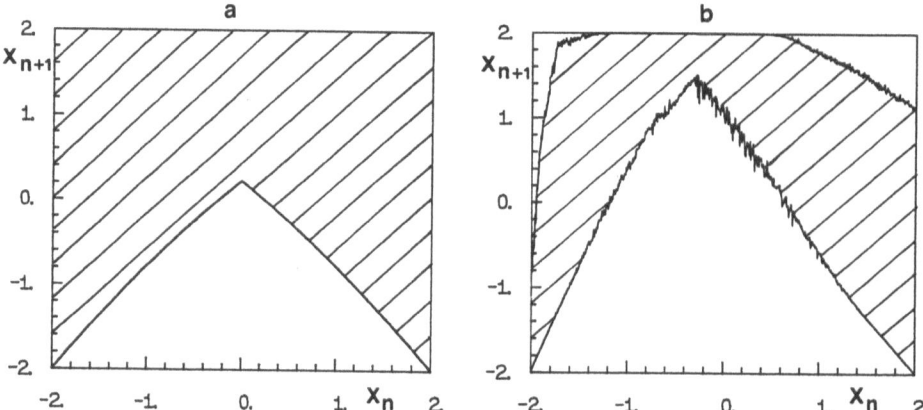

Fig. 7 - *Estimate of the support of the temporal 2-d distribution, $P_2(x_n, x_{n+1})$. (a) is the theoretical result under the hypothesis of uncorrelated sites; (b) is the numerical result computed with 3,000,000 data points.*

I_y is stretched and folded according to Eq (2). Depending wether $0 \in I_y$, its image is of the type $[x_m, 2]$ or $[x_m, x_M]$, where $x_m = \min[f(y_m), f(y_M)]$ and $x_M = \max[f(y_m), f(y_M)]$. For $\varepsilon > 1/4$, $x = 0$ always belongs to I_y. The theoretical estimate and the numerical results of the support boundaries are shown in Fig. 7. The relative difference must be attributed to correlations between x^i's which restrict the allowed region. The final results are shown in Fig. 8. They are very similar to the spatial case displayed in Fig. 5. This is further confirmed by the dependence of the minimal resolution on the embedding dimension L plotted in Fig. 9.

At variance with the previous case, the application of the Grassberger-Procaccia algorithm gave contradictory results: for large L-values (larger than 6) the CGD reaches a maximum for a given resolution, to drop to small values at smaller distances. So far we do not have a reasonable explation for these strange results. However we can conclude asserting that we have presented enough numerical evidence in favour of the following statement: the fractal dimension of the coupled map lattice estimated in an embedding space of dimension L coincides with L itself.

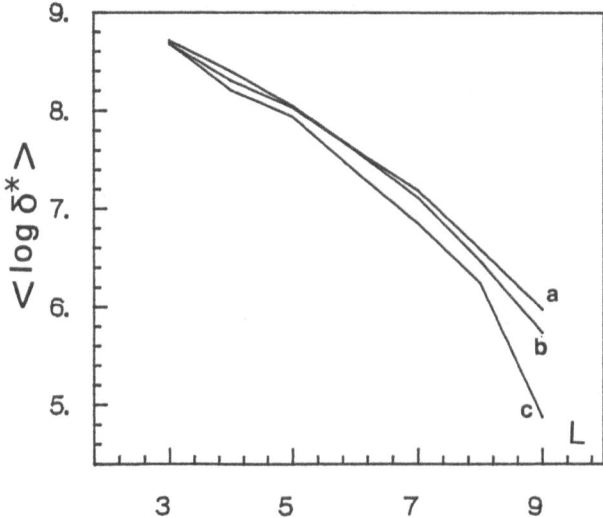

Fig. 9   *Logarithm of minimal resolution for different temporal embedding dimensions; curves a, b, and c refer to the 4-th, 10-th, and 30-th nearest neighbour, respectively.*

Acknowledgements

One of us (AP) benefitted from discussions with P. Grassberger.

References

[1] J.D. Farmer, Z. Naturforsch. **37A**, 1304 (1983)

[2] J.-P. Eckmann and D. Ruelle, Rev. Mod. Phys. **57**, 617 (1985); R. Badii and A. Politi, in **Instabilities and Chaos in Quantum Optics II**, N.B. Abraham, F.T. Arecchi, and L.A. Lugiato eds. (Plenum, New York, 1988), p. 335

[3] K. Kaneko, Prog. Theor. Phys. **72**, 480 (1984)

[4] Y. Pomeau, A. Pumir, and P. Pelce, J. Stat. Phys. **37**, 39, (1984); J,-P. Eckmann and C.E. Wayne, J. Stat. Phys. **50**, 853 (1988)

[5] F. Ledrappier, Commun. Math. Phys. **81**, 229 (1981)

[6] R. Badii and A. Politi in **Fractals in Physics**, L. Pietronero and E. Tosatti eds., (Elsevier, Amsterdam 1986), p. 453

[7] A. Torcini, G. D'Alessandro, and A.Politi, in preparation

[8] P. Grassberger, "Information content and predictability of lumped and distributed dynamical systems", preprint, Wuppertal WB 87-8

[9] G. Mayer-Kress and K. Kaneko, J. Stat. Phys. **54**, 1489 (1989)

[10] L.A. Bunimovich and Ya G. Sinai, Nonlinearity **1**, 491 (1989)

[11] P. Grassberger and I. Procaccia, Phys. Rev. Lett. **50**, 346 (1983)

[12] R. Badii and A. Politi, Phys. Rev. Lett. **52**, 1661 (1984)

[13] E. Ott., W. Withers, and J.A. Yorke, J. Stat. Phys. **36**, 687 (1984)

[14] G. Mayer-Kress, in **Directions in Chaos**, Hao Bai-Lin ed. (World Scientific, Singapore, 1987)

[15] J.-P. Eckmann and D. Ruelle, "Fundamental limitations for estimating dimensions and Lyapunov exponents in dynamical systems", preprint

WEAK TURBULENCE AND THE DYNAMICS OF TOPOLOGICAL DEFECTS

Itamar Procaccia

Department of Chemical Physics
Weizmann Institute of Science
Rehovot 76100, Israel

Low dimensional chaos appears in physical systems whose spatial extent is small. In large aspect-ratio systems chaos sets in concurrently with the loss of spatial coherence[1], leading to a state known as weak turbulence, whose description with low dimensional dynamical systems theory seems at present impossible. Typically, the spatial correlations are destroyed by the spontaneous generation of defects in the macroscopic patterns, and the dynamics of these defects is responsible for much of the chaotic motion in such systems.

In this short note I wish to report briefly on a joint experimental-theoretical effort at the Weizmann Institute to elucidate some of the characteristics relating to this interesting phenomenon, and to use this example to raise a list of questions that seem to me relevant for a reasonable understanding of weak turbulence in general. A fuller description of our results is available upon request[2,3,4].

The experiment, conducted by S. Rasenat and V. Steinberg[4] employed the nematic liquid crystal MBBA, confined in a cell of dimensions $15\mu \times 1cm \times 3cm$, and driven by AC electric field across the layer, to create a perfectly ordered Williams domain of convective rolls, which is destroyed at higher fields by the appearance of topological defects and weak turbulence. The cell supports about 2000 rolls, and is

Measures of Complexity and Chaos
Edited by N.B. Abraham *et al.*
Plenum Press, New York

425

therefore an example of a very large aspect ratio limit. When weak turbulence sets in, one sees spontaneous creation of topological defects which move about following complicated trajectories, where every now and then two defects of opposite topological charge annihilate each other, just to be compensated by a generation of another pair elsewhere[5]. A set of relevant questions seems to be the following:

1. What is the appropriate dynamical description of the state close to the onset of weak turbulence?

2. What are the mechanisms for the creation of defects? Are they "generic" or system specific?

3. Can the defects be treated as localized objects? Can one write equations of motion for the defects themselves?

4. If the last question can be answered in the affirmative, what are the forces acting between defects, and between the defect and the background fields?

5. Can the problem be reduced to that of a gas of defects, with creation and annihilation included?

6. Can one introduce statistical mechanics into the game? Response functions? Fluctuation dissipative theorems?

It is commonly expected that the description of fluctuations around the ordered state before the appearance of defects is adequately given by "amplitude" or "envelope" equations which blame the space and time dependence on the amplitude of the modes that create the cellular state (like convection rolls[6]). With defects[7], the situation is less clear. For example, in the theoretical work done in the context of the nematic electroconvection (together with G. Goren) we concluded that a gauge field has to be added to the theory, turning the commonly used amplitude equations into a classical gauge field theory. It would not surprise me if it would be seen that methods of gauge field theory[8,9] turned out to be very useful in weak turbulence.

The affirmative answers to questions 3 and 4 in this (the nematic) example seems to me at present the main theoretical advance that we can report. It turned out that the defects are localized having an "influence length" λ above which the influence of the defect decays exponentially. The forces between the defects were found, their interaction with the background and their mobility were evaluated analytically. Trajectories of pairs of defects on their ways to annihilation were calculated theoretically and compared with experiments, yielding excellent (parameter free) comparisons[2-4]. We do not know however whether these conclusions are specific to the system at hand, or general.

At present we have no answers to questions 2, 5 and 6. It is my judgement that the answers are not known for any system of this type.

To gain intuition that would lead to progress in understanding this important precursor to developed turbulence, we need more experiments in diverse physical systems. Weak turbulence should in principle be observable in any physical system that showed temporal chaos in the small aspect ratio limit. Accumulating experimental data will allow us to separate the generic from the specific, and to anchor our thoughts in the right direction.

References

1. G. Ahlers and R.P. Behringer, Phys.Rev.Lett. 40:712 (1978); J.P. Gollub and R.K. Steinman, Phys.Rev.Lett. 47:505 (1981); G. Ahlers, D.S. Cannell and V. Steinberg, Phys.Rev.Lett. 54:1373 (1985); A. Pocheau, V. Croquette and O. LeGal, Phys.Rev.Lett. 55:1099 (1985); P. Coullet, L. Gil and J. Lega, Phys.Rev.Lett. 62:1619 (1989).

2. G. Goren, I. Procaccia, S. Rasenat and V. Steinberg, Phys.Rev.Lett., in press.

3. G. Goren and I. Procaccia, submitted to Nonlinearity.

4. S. Rasenat and V. Steinberg, in preparation.

5. Similar observations of weak turbulence in nematics were reported before. See for example R. Ribota and A. Joets in Cellular Structures and Instabilities, eds. J.E. Westfreid and S. Zaleski (Springer, Berlin 1984).

6. A.L. Nowel and J.A. Whitehead, J.Fluids Mech. 38:279 (1969); L.A. Segel, J.Fluid Mech. 38:203 (1969).

7. a) E.D. Siggia and A. Zippelius, Phys.Rev.A 24:1036 (1981); b) A. Zippelius and B. Siggia, Phys.Fluids 26:2906 (1983); c) Y. Pomeau, S. Zaleski and P. Manneville, Phys.Rev.A 27:2710 (1983); d) G. Tesauro and M.C. Cross, Phys.Rev.A 34:1363 (1986).

8. I.E. Dzyaloshinskii, in "Physics of Defects", eds. R. Balian, M. Kleman and J.-P. Poirier (North Holland, Amsterdam 1981).

9. K. Kawasaki, Prog.Theor.Phys.Suppl. 79:161 (1984).

PATTERN CARDINALITY AS A CHARACTERIZATION OF DYNAMICAL COMPLEXITY

John Ringland and Mark Schell

Department of Chemistry

Southern Methodist University, Dallas, TX 75275, USA

In this note we pursue an idea of Lozi[1] in order to characterize the complexity of the behavior of rhythmic systems where a dichotomy of pulses can be naturally imposed.

As a model of such systems, we consider the class of one-dimensional maps of the interval such as the one shown in Fig. 1, which are continuous and monotone except for a single discontinuity. (Maps with discontinuity can model continuous dissipative dynamical systems where a region of strong separation of trajectories exists, e.g. at a fold of a slow manifold[2]). Iterates can be classified as lying to the right (R) or the left (L) of the point of discontinuity. Let us write an itinerary as $R^{M_1}L^{N_1}R^{M_2}L^{N_2}...$, and call each distinct group of the form $R^M L^N$ a *pattern*[1].

The parameter plane has an extremely intricate structure with respect to asymptotic dynamics.[3] Yet it is relatively straightforward to divide it into regions according to the number of patterns from which an itinerary can be constituted. Figure 2 depicts maps in 2- and 3-pattern regions. In Fig. 3, we show the division of the plane for a piecewise linear case. Regions of given pattern cardinality form jagged bands across the parameter plane. The picture for any piecewise monotone map with typical parametrization is qualitatively similar.

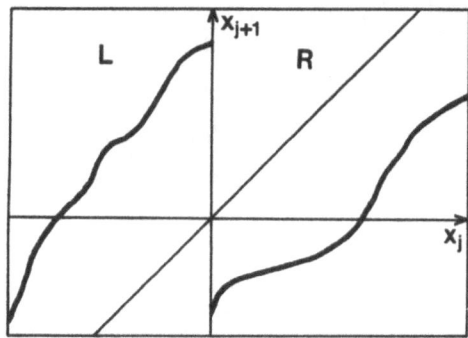

Fig. 1. A map of the interval everywhere continuous and monotone except for a single discontinuity.

Measures of Complexity and Chaos
Edited by N.B. Abraham *et al.*
Plenum Press, New York

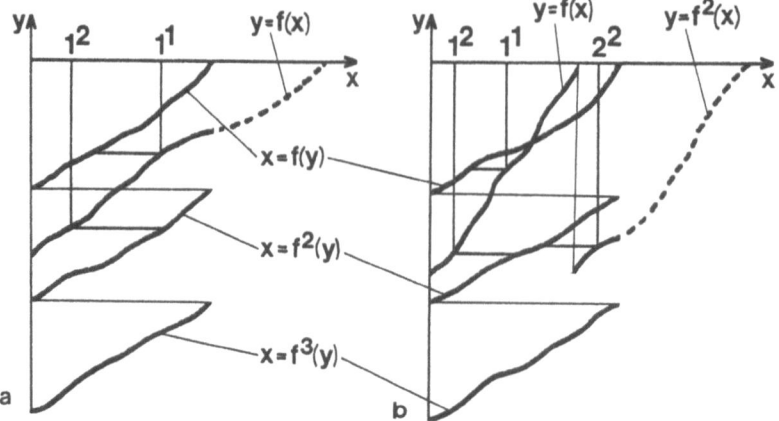

Fig. 2. (a) A map, f, in a 2-pattern region. Only the patterns 1^1 and 1^2 can possibly appear, where for compactness we are writing the pattern $R^M L^N$ as M^N. For visualization, multiple compositions of the map are shown on the appropriate sub-intervals, and the L part is graphed sideways. (b) In a 3-pattern region where only 1^1, 1^2 and 2^2 can appear.

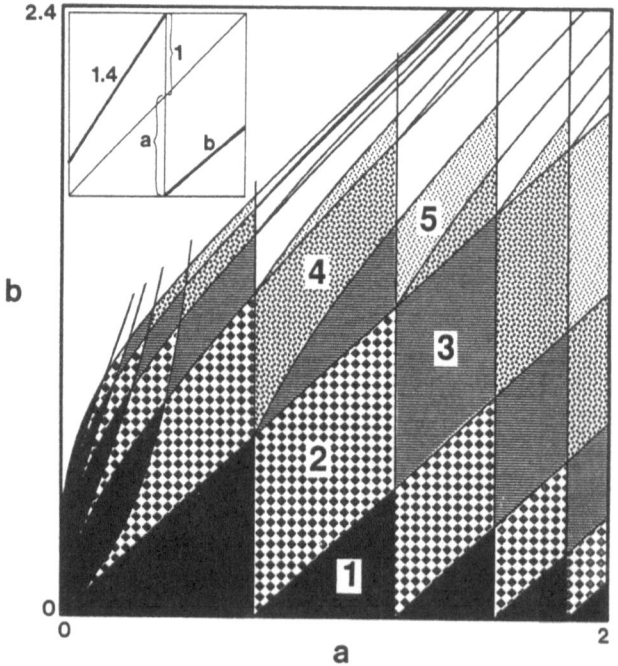

Fig. 3. Division according to pattern cardinality of the parameter plane of the map defined in the inset. In the checkered regions for example, all itineraries, including all but the earliest transients, are constituted of at most two patterns.

We propose that the pattern cardinality is a useful operational measure of the complexity of the dynamical behavior. It is a quantity which does not vary wildly from point to point (the boundaries are ordinary curves). It does not require precise quantitative measurement of time-series for its determination, and it is a property of transients as well as of asymptotic states. For all of these reasons it is amenable to rapid experimental survey.

Moreover, in practical applications, a quantity like

$$\rho = \sum_{i=1}^{t} \frac{M_i}{M_i + N_i}$$

(non-infinite t) may be of primary significance. When the location in parameter space is imprecisely known or perhaps somewhat variable, or in the presence of noise, the pattern cardinality is much more strongly related to the uncertainty of a quantity like ρ than say the Liapunov number; whether a particular parameter space point has periodic or chaotic asymptotic behavior is of little relevance.

To our knowledge, the first and only experiments along these lines are those of Maselko and Swinney[4] on the Belousov-Zhabotinskii reaction, where 1-, 2-, 3- and 4-pattern regions were located. Since systems with dichotomous pulses are very common, more experiments of this kind could be performed profitably.

References

1. R. Lozi, Sur un modèle mathématique de suite de bifurcations de motifs dans la réaction Belousov-Zhabotinsky, C.R. Acad. Sci. Paris, I-294, 21 (1982).
2. D. Barkley, Near-critical behavior for one-parameter families of circle maps, Phys. Lett. A, 129, 219 (1988).
3. J.P. Keener, Chaotic behavior in piecewise continuous difference equations, Trans. Am. Math. Soc., 261, 589 (1980).
4. J. Maselko and H.L. Swinney, Complex periodic oscillations and Farey arithmetic in the Belousov-Zhabotinskii reaction, J. Chem. Phys., 85, 6430 (1986).

CHARACTERIZING SPATIO-TEMPORAL CHAOS

IN ELECTRODEPOSITION EXPERIMENTS

F. Argoul, A. Arneodo, J. Elezgaray and G. Grasseau

Centre de Recherche Paul Pascal, Chateau Brivazac
33600 Pessac Cedex, France

INTRODUCTION

Pattern formation in systems far from equilibrium is a subject of considerable current interest[1-5]. Recently, much effort has been directed toward the study of fractal growth phenomena in physical, chemical and biological systems. Unfortunately, the understanding of phenomena like viscous fingering in Hele-Shaw cells[6] and electrochemical deposition[7] is hampered by the mathematical complexity of the problem. Highly ramified structures generally are produced in the zero surface tension limit. In this limit both processes are equivalent to the Stefan problem[8]: a diffusion problem for the pressure or the electrochemical potential, with boundary values specified on the moving interface, whose local velocity is in turn determined by the normal gradient of the Laplace field. This highly nonlinear problem is not readily amenable even to modern numerical simulations. When solving the Stefan problem by direct means, the interface develops unphysical cusps in a finite time[6]. One is thus led to introduce some short-distance cutoff which in some sense mimics surface-tension[9,10]. Thus far no computer simulations of the equations of motion achieve the necessary size to make definite conclusions about the deterministic character of the fractal patterns observed in the experiments[1-5].

An alternative to solving the Stefan problem consists of simulating the diffusion-limited aggregation (DLA) model introduced by Witten and Sander[11] in 1981. In this model, an aggregate is grown by the successive addition of random walkers to the perimeter sites of the cluster. Extensive computer investigations[11,12] have shown that complex, apparently randomly branched fractals are produced (Fig. 1a). The structure of these aggregates has been analyzed by computational[11-18] and analytical[19-22] methods and shows a strong resemblance to the arborescent patterns observed in electrodeposition experiments[7,23-26] (Fig. 1b). From the visual inspection of the objects formed, many people have been tempted to identify the electrodeposition mechanism, and more generally, the diffusion-limited aggregation mechanism to a random growth process[27-29]. However, despite the apparent simplicity of the DLA model, there is still no rigorous theory for diffusion-limited growth processes. Many important theoretical questions remain unanswered; in particular, it is still an open question whether the fractal geometry of DLA clusters is a product of the randomness in the growth process[27-29] or the result of a proliferation of deterministic tip-splitting instabilities[7,9,10,30]. In this communication, we report on the first experimental evidence for spatio-temporal chaos in electrochemical deposition[31]. This experimental discovery, together with the results of a numerical analysis of screening effects in DLA growth processes[32], suggest that the fractal geometry of diffusion-limited aggregates is the signature of a chaotic dynamics which displays sensitivity to initial conditions[33-40]. These results also show that the

Measures of Complexity and Chaos
Edited by N.B. Abraham *et al.*
Plenum Press, New York

433

permanent external noise due to random arrival of particles does not play any crucial role in fractal growth phenomena.

SELF-SIMILARITY OF DIFFUSION-LIMITED AGGREGATES AND ELECTRO-DEPOSITION CLUSTERS

Thus far, there has been no convincing experimental demonstration that fractal aggregates like viscous fingers and electrodeposition clusters actually belong to the DLA universality class. In fact, only a little is known in quantitative terms about the structure of DLA clusters. Most of the numerical analysis of growing patterns has focused on the determination of the fractal dimension[1-3,11,13]. However, the fractal dimension is a global property which does not provide deep insight into the geometrical complexity of the aggregate. Only very recently[14,41] more attention has been paid to the computation of the generalized fractal dimensions[42-45] D_q, which are closely related to the $f(\alpha)$ spectrum of singularities[46-50] of strength α (α is also called the local scaling exponent). In a comparative study of DLA clusters and two-dimensional electrochemical deposits grown in either a strip[51] or a radial[52] geometry, we have recently elucidated the conjectured self-similarity of these fractal aggregates. Using box-counting and fixed-mass algorithms, we have shown that small-mass on-lattice DLA clusters are statistically self-similar with dimensions $D_q = 1.60 \pm 0.02$, independently of q. Similarly, zinc and copper electrodeposition clusters in the limit of small ionic concentration and small voltage (or small current) were also found to be self-similar with $D_q = 1.63 \pm 0.03$. This study has been further assisted by a wavelet transform analysis of the local scaling properties of these clusters[52,53]. For both DLA and electrodeposition clusters, the local scaling exponent $\alpha(\vec{x})$ was found to take on a unique value over the entire aggregate. The exponent $\alpha = 1.60$ for DLA clusters (Fig. 1c), and $\alpha = 1.63$ for electrodeposition clusters (Fig. 1d), are in fact the same within the experimental uncertainty. This result strongly suggests that the electrodeposition growth mechanism is very likely to be governed by the two-dimensional DLA process[51,52].

The wavelet transform[54] is a mathematical technique which has been recently introduced for characterizing the local self-similarity of fractal objects[55-57]. The wavelet transform can be seen as a mathematical microscope which is well suited to resolve local scaling properties through the determination of the local pointwise dimension $\alpha(\vec{x})$ (Fig. 1). The ability of this microscope to identify the construction rules of fractals has been recently exploited to analyze fractals arising at the critical points of nonequilibrium phase transitions[58]. In the context of critical phenomena[59,60], the wavelet transform turns out to be of fundamental interest since it naturally reveals the renormalization operation[61] which is essential to the theoretical understanding of the universal properties of these transitions.

As seen through the wavelet microscope, the self-similarity of both DLA and electrodeposition clusters seems to be strongly related to the existence of a characteristic angle θ between branches of two successive generations[52,53]. Although our statistical sample is not large enough to allow us to produce a reliable estimate of this angle, our average value is quite compatible with a rough pentagonal symmetry[19-21,62,63] $< \theta > \sim 36°$. There exist, however, fluctuations in the sequences of screening angles $\theta_1, \theta_2, \ldots, \theta_n, \ldots$ between branches of successive generations. These fluctuations are at the origin of the nonperiodic oscillations observed in the scaling law behavior of the wavelet transform (Figs 1b and 1d). This observation raises the fundamental question whether the departure from periodicity is due to noise or deterministic chaos. Noisy oscillations would mean that diffusion-limited aggregates are random fractals generated by a nondeterministic construction rule[27-29]. Conversely, chaotic oscillations would indicate that these clusters are chaotic fractals constructed according to a deterministic recursive process[7,9,10,30]. When progressively increasing the magnification, the wavelet microscope clearly reveals the existence of spatial correlations between branches. For example, when the screening angle θ_n is small at a given stage, the screening angle θ_{n+1} at the next stage is very likely to be large. The specific property that distinguishes growth phenomena from other non-equilibrium spatio-temporal organization phenomena (e. g. pure reaction-diffusion processes or hydrodynamic turbulence), is that their spatial structure (the fossilized aggregate) retains the full memory of their temporal evolution. Thus, one can reasonably suspect that the screening effects irreversibly influence the dynamics of the growth process. The fractal geometry of DLA and electrodeposition clusters is very likely the result of the instability of the interface coupled to a nonlinear chaotic competition process between branches[52,53].

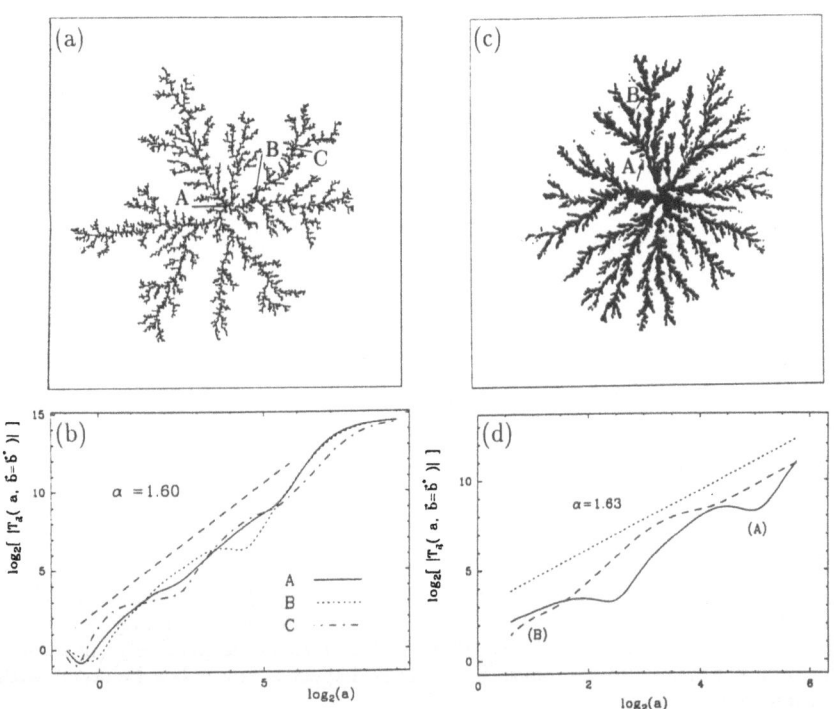

Fig. 1 Wavelet transform analysis[52,53] of (a) a DLA cluster of mass $M = 13000$ and (c) a copper electrodeposition cluster in the limit of small ionic concentration and low current. When focussing the wavelet microscope to a given point $\vec{x}$ of the aggregate and increasing the magnification factor a^{-1}, the wavelet transform $T_g(a, \vec{x})$ behaves as a power law with exponent $\alpha(\vec{x})$. The estimate of the local scaling exponent (pointwise dimension) at different points arbitrarily chosen on the clusters is shown in (b) for the DLA cluster: $\alpha(\vec{x}) = 1.60$, and (d) for the electrochemical deposit: $\alpha(\vec{x}) = 1.63$.

Unfortunately our wavelet transform analysis faces computational and storage difficulties. Only a few generations can be analyzed with DLA clusters of reasonable mass. The study of high mass ($M = 10^6$) off-lattice clusters is currently in progress. The photographs of the electrodeposition clusters are digitized with a resolution 512×512 which prevents the investigation of a wide range of scales. Because the complexity of the geometry is intrincately connected with the dynamical evolution, it may be worthwhile instead to analyze the temporal behavior of the growth process in the electrochemical deposition experiments with the tools from dynamical systems theory[33-40] (phase portraits, Poincaré maps, 1-D maps, etc . . .).

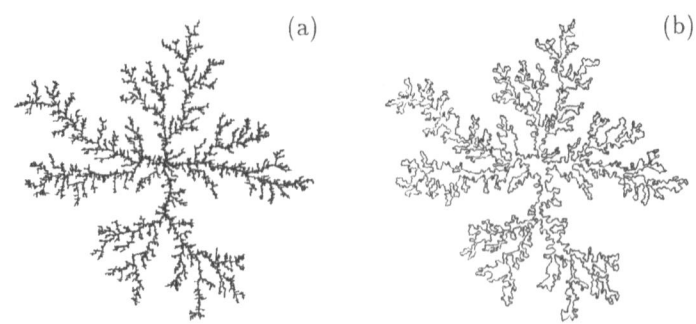

Fig. 2 (a) An on-lattice DLA cluster of mass $M = 13000$ computed with the random-walker model of Witten and Sander[11]. (b) The active zone of the cluster boundary[32].

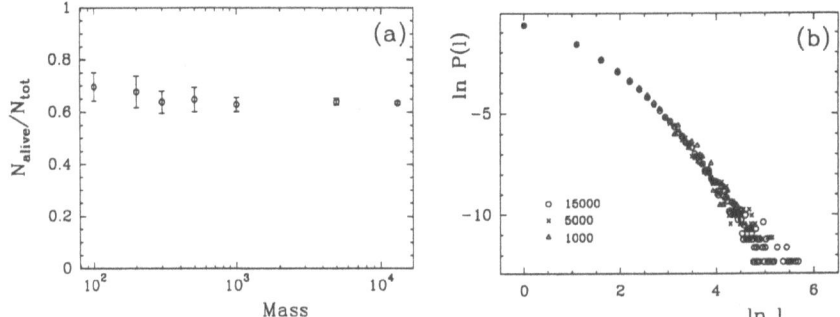

Fig. 3 (a) The proportion of active sites of the cluster boundary versus the mass of the DLA clusters[32]. (b) Histograms of the lengths of the dead boundary zones[32] of three DLA clusters of mass $M = 1000, 5000$ and 15000 respectively (in log-log scales). For ℓ large enough, these histograms display a power-law behavior with exponent $\delta = -3.1 \pm 0.3$

SCREENING EFFECTS AND LACUNARITY OF THE GROWTH PROBABILITY DISTRIBUTION IN DIFFUSION-LIMITED AGGREGATION

Recently a large number of numerical[10,64−69] and experimental[70−72] studies have investigated the growth probability distribution of structures whose surface is fractal. These studies have revealed that the growth probability measure (perimeter occupancy probability) has a finite range of scaling indices with an $f(\alpha)$-spectrum which is a convex function very similar to those found for multifractal measures[49,50]. The sticking probability distribution is thus a nonhomogeneous distribution over the cluster perimeter; the physical basis for this fact is that the "hot" tips of the main branches correspond to active regions of growth, while the growth of many "cold" tips and fjords is invisibly small. The possibility that some fjords are nearly completely screened (e.g. with exponentially small field) causes a phase transition in the corresponding $f(\alpha)$-spectrum[62,73−75] ($\alpha_{max} \to +\infty$). In Figs 2 and 3 we show the first numerical evidence that regions of the DLA cluster perimeter are not only inactive but completely screened by the hot tips[32]. These fossilized regions correspond to sites that are no longer accessible by the random walking particles because of the finite size of the sticking length (physical interaction length below which the diffusing particles cannot penetrate each other) that characterizes the aggregation mechanism. These regions are thus dead forever with a sticking probability which is strictly zero.

In Fig. 2, we compare a DLA cluster of mass $M = 13000$ with its active boundary sites as computed by a specially written numerical algorithm[32]. By simple visual inspection it is clear that only a finite subset of the boundary sites are active. In Fig. 3a, we measure the proportion of active sites with respect to the total number of boundary sites for a collection of DLA clusters of different mass. Surprisingly, this proportion does not seem to be sensitive to $M(> 500)$: about 63% ($\pm 3\%$) of the perimeter sites are alive, while 37% have stopped growing. Moreover, as seen in Fig. 3b, there exist extinct fjords of every size from the lattice mesh size up to the size of the entire cluster. The histogram of lengths ℓ of these totally screened regions behaves apparently as a power law ℓ^δ at large ℓ values, with an exponent $\delta \simeq -3.1 \pm 0.3$ which does not depend upon the mass of the aggregate. This remarkable result[32] strongly suggests that the active part of the DLA cluster boundary is a fat fractal[76] with nonzero measure. This fat fractal has the same fractal dimension as the underlying space (i.e. the cluster boundary) $D_F = 1.60$. Since the dimension of fat fractals does not provide an appropiate means of characterization, several scaling exponents have been introduced for this purpose[77-80]. The exponent δ extracted from the power-law behavior of the histogram in Fig. 3b is directly related to these exponents. An immediate consequence of this result is that the support of the growth probability distribution is lacunar[32]; this observation suggests that the DLA cluster boundary is not locally connected.

These results differ from previous analysis of the multifractal properties of the growth probability distribution in fractal growth phenomena[10,62,64-75]. In particular, they address the fundamental question of the validity of identifying the sticking probability distribution with the harmonic measure of the aggregate[81]. Without a doubt the demonstration that the DLA cluster boundary is not locally connected makes the application of conformal mappings[21,62,73] quite questionable. One may still pursue the analogy developed in refs 62,73 between aggregates generated by probabilistic diffusion and deterministically generated Julia sets[82], provided one considers non-locally connected Julia sets[83]. One may also try to circumvent the difficulty by redefining the cluster boundary from the active sites only.

DETERMINISTIC CHAOS IN ELECTRODEPOSITION EXPERIMENTS

Among the various experimental illustrations of fractal pattern formation, electrochemical deposition is currently considered the paradigm for theoretical studies of diffusion-limited aggregation[7,30]. In fact, by varying the concentration of metal ions and the applied constant current, one can explore different morphologies like dense radial aggregates[84], dendritic patterns[26,85] and fractal aggregates[23-26,51,52]. Fractal patterns are usually obtained in the double limit of small ionic concentrations and low current.

To quantify this deposition problem better than the simple characterization of the fractal geometry of electrochemical deposits, we have chosen to study the potential signal[31,86]. This signal provides a measurement of the conductivity of the medium (electrodes + electrolyte) when a constant current is applied. Since the conductivity is expected to fluctuate as the structures grow on the surface of the cathode, the temporal evolution of the voltage across the cell provides interesting information about the screening effects and selection processes that govern electrochemical deposition.

Fig. 4 shows several photographs of a zinc electrodeposition experiment in a strip geometry at the early stages of the growth[31]. At the beginning, many trees emerge from the cathode. The significant feature is that very quickly these trees start competing. Some of these trees stop because of shadowing effects. Cooperative phenomena (simultaneous extinction of several trees) are observed depending on the experimental conditions. This screening-induced selection process is apparent in Fig. 4 where only a few branches survive this struggle for life, 15 min after the beginning of the growth.

When recording the cell voltage between the two electrodes, an induction regime is observed when turning on the applied current. This induction regime corresponds to

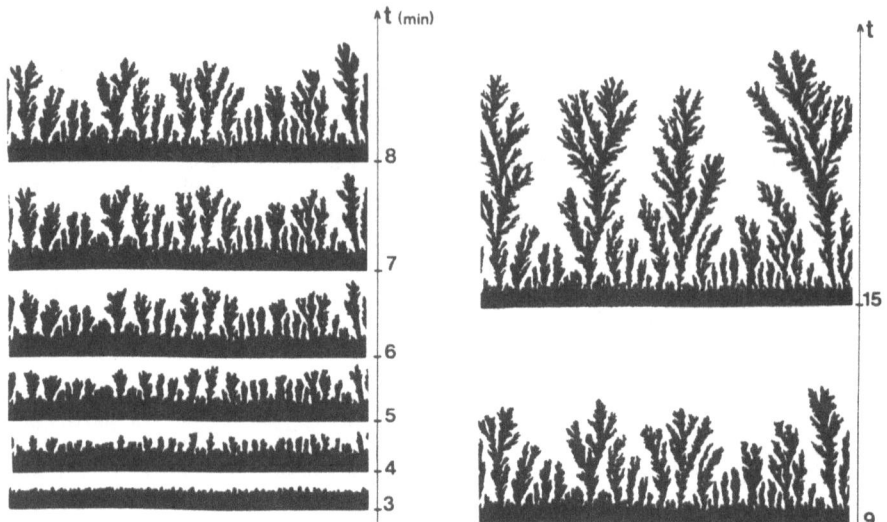

Fig. 4 Zinc-metal trees (about 15 mm long after 15 min of growth) photographed during the early stages of the chemical deposition process[31]. The experimental set-up is similar to the one described in ref. 51. These zinc trees are grown from 0.1 M $ZnSO_4$ (aq) with an applied current of 0.5 mA.

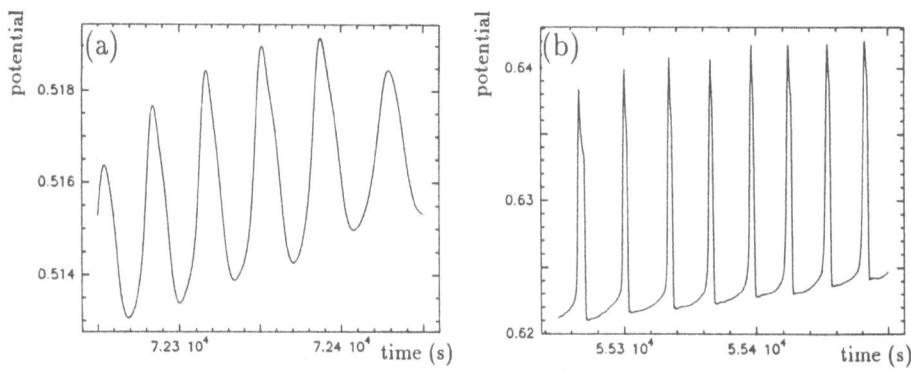

Fig. 5 Two periodic regimes extracted from potential measurements in an electrodeposition experiment. (a) "Quasi"-sinusoidal oscillations: $ZnSO_4$=0.15 M, I=0.9 mA. (b) Relaxation oscillations: $ZnSO_4$ = 0.25 M, I=1.5 mA. The drift in the signal is inherent to the growth process (moving interface).

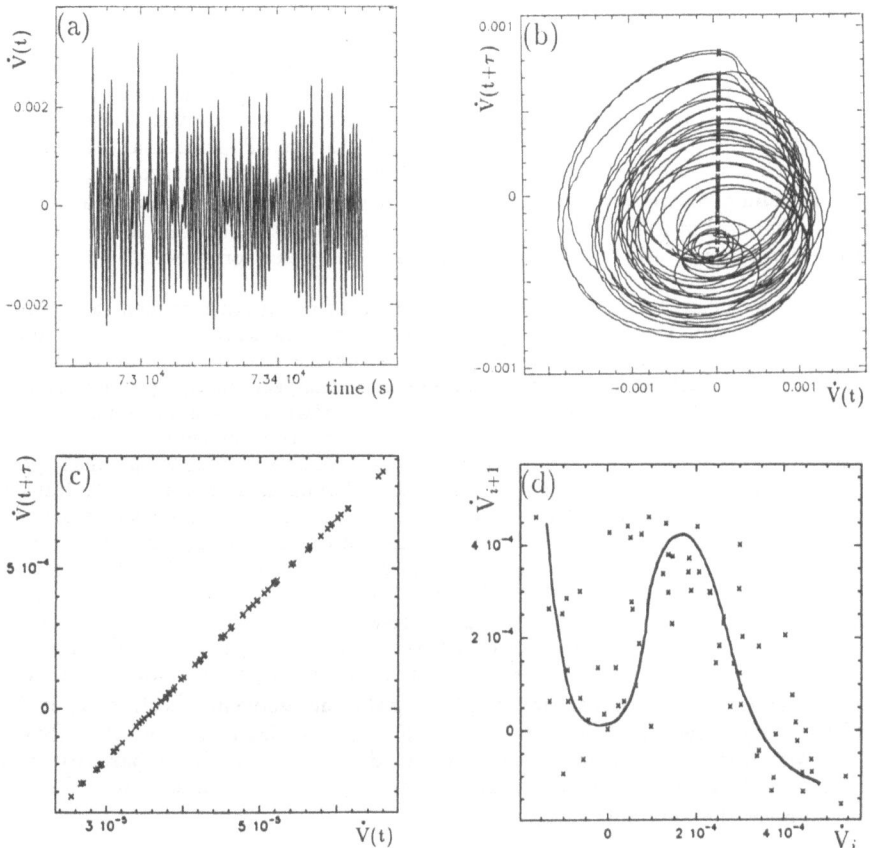

Fig. 6 A chaotic regime extracted from potential measurements in an electrodeposition
experiment[31]: $ZnSO_4 = 0.1$ M, I=0.5 mA. (a) The derivative of the potential $\dot{V}(t)$
vs time. (b) A two-dimensional projection of a three-dimensional phase-portrait
$(\dot{V}(t), \dot{V}(t+\tau), \dot{V}(t+2\tau))$ reconstructed from the time-series in (a) using the time-
delay method[93,94] with $\tau = 2.5\,s \simeq 1/4$ of the main period of oscillation. (c) A
Poincaré section constructed by the intersection of positively directed trajectories
with the plane (normal to the paper) passing through the line in (b). (d) A one-
dimensional map obtained by plotting as ordered pairs $(\dot{V}_i, \dot{V}_{i+1})$, the successive
values of the ordinate of trajectories when they cross the crossed line in (b); the
hand-drawn curve suggests the existence of a 1-D map with two extrema.

a slow drift of the voltage. When the voltage exceeds some critical value, one detects oscillations[31,86]. These oscillations are observed simultaneously in the selection process shown in Fig. 4. They appear when the instability of the cathode develops into a forest of zinc trees. The initial number of trees is found to depend essentially upon the concentration of the zinc sulfate, the applied current intensity and the ratio between the width of the cell and the thickness of the layer between the two glass plates. The oscillations almost disappear when the selection process between trees is over. In fact oscillations of different amplitude are still present in the data; these oscillations characterize screening effects between branches of a same tree.

When adjusting the control parameters, one can find regions in the parameter space where the recorded oscillations appear to be coherent over many cycles (Fig. 5). Once the drift inherent to growth processes with moving interface is removed, these oscillations turn out to be periodic[31]. We have observed both "quasi"-sinusoidal oscillations (Fig. 5a) which resemble the oscillations of small amplitude that emerge from a Hopf bifurcation[87], and relaxation oscillations (Fig. 5b) which are current phenomena in non-equilibrium chemical systems[88−92]. A strong correlation exists between the coherence time of the oscillations and the recursive character of the selection process.

In Fig. 6, we describe a chaotic regime observed at small ionic concentration and low current[31]. The zinc trees which are involved in the selection process have a ramified structure very similar to the arborescent fractal structure of diffusion-limited aggregates. The nonperiodicity of the recorded time series is analyzed using a phase-portrait, Poincaré map and one-dimensional map[33−40]. Fig. 6b shows a two-dimensional projection of a three-dimensional phase portrait constructed from the time-series of Fig. 6a using the time delay method. This phase portrait looks strikingly similar to the phase portrait of strange attractors displayed by low-dimensional dynamical systems[33−40]. Rather than analyze phase portrait directly, it is easier to look at the Poincaré section formed by the intersections of the trajectories with a plane approximately normal to the orbits. As shown in Fig. 6c, the points on the Poincaré section lie to a good approximation along a smooth curve. The fact that the Poincaré section is not a scatter of points rules out any suspicious stochastic interpretation of these complex oscillations; moreover, it demonstrates the low-dimensional nature of this chaotic state (the orbits lie approximately on a two-dimensional sheet). Further insight into the dynamics can be achieved by constructing a one-dimensional map by plotting the successive intersections $\dot{V}_i$ of the ordinate $\dot{V}(t + \tau)$ of the orbits with the line in Fig. 6b. A plot of $\dot{V}_{i+1}$ vs $\dot{V}_i$ is shown in Fig. 6d; within the experimental resolution, the data appear to fall apparently on a smooth curve, indicating that the dynamics is governed by a deterministic law: for any $\dot{V}_n$, the map gives the next value $\dot{V}_{n+1}$.

The results in Fig. 6 bring the first inescapable evidence for the existence of low-dimensional deterministic chaos in fractal growth phenomena[31]. Similar chaotic oscillations also exist in the late stages of the growth where screening effects mainly result from the competition between internal branches of the surviving trees. These observations strongly indicate that the fractal aspect of diffusion-limited aggregates is the spatial footprint of a chaotic dynamics governed by screening effects. The low-dimensionality of the recorded chaos suggests that this apparently complicated spatio-temporal evolution can be understood in terms of the nonlinear interaction of a small number of elementary instabilities. A systematic investigation of the routes to chaos in electrochemical deposition is currently in progress. This study generalizes our approach to the different morphologies encountered in electrodeposition experiments, i. e. fractal aggregates, dendritic patterns and dense radial aggregates. The results of this study will be detailed in a forthcoming publication.

We would like to thank E. Kostelich for a careful reading of the manuscript. This work was supported by a Grant ATP-CNRS, CNES (Sciences Physiques en Microgravité).

REFERENCES

1. H. E. Stanley and N. Ostrowsky, eds., "On Growth and Form: Fractal and Non-Fractal Patterns in Physics", Martinus Nijhof, Dordrecht (1986), and references quoted therein.
2. L. Pietronero and E. Tosati, eds., "Fractal in Physics", North-Holland, Amster-dam (1986), and references quoted therein.
3. H. E. Stanley, ed., "Statphys 16", North-Holland, Amsterdam (1986), and references quoted therein.
4. W. Guttinger and D. Dangelmayr, eds., "The Physics of Structure Formation", Springer-Verlag, Berlin (1987), and references queted therein.
5. H. E. Stanley and N. Ostrowsky, eds., "Random Fluctuations and Pattern Growth", Kluwer Academic Publisher, Dordrecht (1988), and references quoted therein.
6. D. Bensimon, L. P. Kadanoff, S. Liang, B. I. Shraiman and Chao Tang, Rev. Mod. Phys. 58:977 (1986), and references quoted therein.
7. L. M. Sander, in ref. 4., p. 257.
8. L. Rubinstein, "The Stefan Problem", A.M.S., Providence (1971).
9. L. M. Sander, P. Ramanlal and E. Ben Jacob, Phys. Rev. A 32:3160 (1985).
10. P. Ramanlal and L. M. Sander, J. Phys. A 21:L995 (1988).
11. T. Witten and L. M. Sander, Phys. Rev. Lett. 47:1400 (1981); Phys. Rev. B 27:5686 (1983).
12. P. Meakin, in "Phase Transitions and Critical Phenomena", Vol. 12, C. Domb and J. L. Lebowitz, eds., Academic Press, Orlando (1988), and references quoted therein.
13. P. Meakin, Phys. Rev. A 27:1495 (1983).
14. P. Meakin and Z. R. Wasserman, Chem. Phys. 91:391 (1984).
15. P. Meakin and L. M. Sander, Phys. Rev. Lett. 54:2053 (1985).
16. R. C. Ball and R. M. Brady, J. Phys. A 18:L809 (1985).
17. P. Meakin, Phys. Rev. A 33:3371 (1986).
18. P. Meakin, R. C. Ball, P. Ramanlal and L. M. Sander, Phys. Rev. A 35:5233 (1987).
19. L. A. Turkevitch and H. Scher, Phys. Rev. Lett. 55:1026 (1985).
20. R. C. Ball, Physica A 140:62 (1986).
21. T. C. Halsey, P. Meakin and I. Procaccia, Phys. Rev. Lett. 56:854 (1986).
22. M. Matsushita, K. Konda, H. Toyoki, Y. Hayakawa and H. Kondo, J. Phys. Soc. Jpn. 55:2618 (1986).
23. R. M. Brady and R. C. Ball, Nature 309:225 (1984).
24. M. Matsushita, M. Sano, Y. Hayakawa, H. Hongo and Y. Sawada, Phys. Rev. Lett. 53:286 (1984).
25. M. Matsushita, Y. Hayakawa and Y. Sawada, Phys. Rev. A 32:3814 (1985).
26. D. Grier, E. Ben-Jacob, R. Clarke and L. M. Sander, Phys. Rev. Lett. 56:1264 (1986).
27. H. E. Stanley, in ref. 4, p. 210.
28. J. Nittmann and H. E. Stanley, Nature 321:663 (1986).
29. J. Nittmann and H. E. Stanley, J. Phys. A 20:L1185 (1987).
30. L. M. Sander, in ref. 2, p. 241.
31. F. Argoul and A. Arneodo, Deterministic chaos in two-dimensional electrodepo-sition processes, preprint (July 1989), to be published.
32. F. Argoul, A. Arneodo, J. Elezgaray and G. Grasseau, Lacunarity of the growth probability distribution in diffusion-limited aggregation processes, preprint (June 1989), to be published.
33. J. Guckenheimer and P. Holmes, "Nonlinear Oscillations, Dynamical Systems and Bifurcations of Vector Fields", Springer-Verlag, Berlin (1983), and references quoted therein.
34. P. Cvitanovic, ed., "Universality in Chaos", Hilger, Bristol (1984), and references quoted therein.

35. Bai-Lin Hao, ed., "Chaos", World Scientific, Singapore (1984), and references quoted therein.

36. H. G. Schuster, "Deterministic Chaos", Physik-Verlag, Weinheim (1984), and references quoted therein.

37. A. V. Holden, ed., "Chaos", Manchester University Press, Manchester (1986), and references quoted therein.

38. P. Bergé, Y. Pomeau and C. Vidal, "Order within Chaos", Wiley, New-York (1986), and references quoted therein.

39. H. B. Stewart, "Nonlinear Dynamics and Chaos", Wiley, New-York (1986) and references quoted therein.

40. P. Bergé, ed., "Le Chaos: Théorie et Expériences", Collection du C.E.A., Eyrolles (1988), and references quoted therein.

41. P. Meakin and S. Havlin, Phys. Rev. A 36:4428 (1987).

42. A. Renyi, "Probability Theory", North-Holland, Amsterdam (1970).

43. P. Grassberger, Phys. Lett. A 97:227 (1983).

44. H. G. Hentschel and I. Procaccia, Physica D 8:435 (1983).

45. P. Grassberger and I. Procaccia, Physica D 13:34 (1984).

46. E. B. Vul, Ya. G. Sinai and K. M. Khanin, Usp. Mat. Nauk. 39:3 (1984); J. Russ. Math. Surv. 39:1 (1984).

47. R. Benzi, G. Paladin, G. Parisi and A. J. Vulpiani, J. Phys. A 17:3521 (1984).

48. G. Parisi, in U. Frisch, Fully developped turbulence and intermittency, in "Turbulence and Predictability in Geophysical Fluid Dynamics and Climate Dynamics", M. Ghill, R. Benzi and G. Parisi, eds, North-Holland, Amsterdam (1985) p. 84.

49. T. C. Halsey, M. H. Jensen, L. P. Kadanoff, I. Procaccia and B. I. Shraiman, Phys. Rev. A 33:1141 (1986).

50. P. Collet, J. Lebowitz and A. Porzio, J. Stat. Phys. 47:609 (1987).

51. F. Argoul, A. Arneodo, G. Grasseau and H. L. Swinney, Phys. Rev. Lett. 61:2558 (1988).

52. F. Argoul, A. Arneodo, J. Elezgaray, G. Grasseau and R. Murenzi, Wavelet transform analysis of the self-similarity of diffusion-limited aggregates and electrodeposition clusters, preprint (June 1989), to be published.

53. F. Argoul, A. Arneodo, J. Elezgaray, G. Grasseau and R. Murenzi, Phys. Lett. A 135:327 (1989).

54. J. M. Combes, A. Grossmann and P. Tchamitchian, eds, "Wavelets, Time-Frequency Methods and Phase Space", Springer-Verlag, Berlin (1989), and references quoted therein.

55. M. Holschneider, J. Stat. Phys. 50:963 (1988).

56. A. Arneodo, G. Grasseau and M. Holschneider, Phys. Rev. Lett. 61:2281 (1988).

57. A. Arneodo, G. Grasseau and M. Holschneider, in ref. 54, p. 182.

58. A. Arneodo, F. Argoul, J. Elezgaray and G. Grasseau, in "Nonlinear Dynamics", G. Turchetti, ed., World Scientific, Singapore (1989) p. 130.

59. S. K. Ma, "Modern Theory of Critical Phenomena", Benjamin Reading, Mass. (1976).

60. D. Amit, "Field Theory, The Renormalisation Group and Critical Phenomena", Mc Graw-Hill, New-York (1978).

61. K. G. Wilson, Rev. Mod. Phys. 55:583 (1983).

62. I. Procaccia and R. Zeitak, Phys. Rev. Lett. 60:2511 (1988).

63. A. Arneodo, Y. Couder, G. Grasseau, V. Hakim and M. Rabaud, Uncovering the analytical Saffman-Taylor finger in unstable viscous fingering and diffusion-limited aggregation, preprint (April 1989), to be published.

64. P. Meakin, Phys. Rev. A 34:710 (1986); 35:2234 (1987).

65. C. Amitrano, A. Coniglio and F. diLiberto, Phys. Rev. Lett. 57:1016 (1986).

66. P. Meakin, A. Coniglio, H. E. Stanley and T. A. Witten, Phys. Rev. A 34:3325 (1986).

67. Y. Hayakawa, S. Sato and M. Matsushita, Phys. Rev. A 36:1963 (1987).

68. T. Nagatani, Phys. Rev. A 36:5812 (1987).

69. R. Ball and M. Blunt, Phys. Rev. A 39:3591 (1989).

70. J. Nittmann, H. E. Stanley, E. Touboul and G. Daccord, Phys. Rev. Lett. 58:619 (1987).

71. S. Ohta and H. Hongo, Phys. Rev. Lett. 60:611 (1988).

72. M. Blunt and P. King, Phys. Rev. A 37:3935 (1988).

73. T. Bohr, P. Cvitanovic and M. H. Jensen, Europhys. Lett. 6:445 (1988).

74. J. Lee and H. E. Stanley, Phys. Rev. Lett. 61:2945 (1988).

75. R. Blumenfeld and A. Aharony, Phys. Rev. Lett. 62:2977 (1989).

76. J. D. Farmer, in "Fluctuations and Sensitivity in Nonequilibrium Systems", W. Horstemke and D. Kondepudi, eds, Springer-Verlag, New-York (1984), p. 172.

77. D. K. Umberger and J. D. Farmer, Phys. Rev. Lett. 55:661 (1985).

78. C. Grebogi, S. W. McDonald, E. Ott and J. A. Yorke, Phys. Lett. A 110:1 (1985).

79. K. Nakamura, Y. Okazaki and A. R. Bishop, Phys. Lett. A 117:459 (1986).

80. R. Eykholt and D. K. Umberger, Phys. Rev. Lett. 57:2333 (1986); Physica D 30:43 (1988).

81. N. G. Makarov, Proc. London Math. Soc. 51:369 (1985).

82. A. Douady and J. H. Hubbard, C. R. Acad. Sci. (Paris) 294:123 (1982).

83. A. Douady, private communication.

84. D. G. Grier, D. A. Kessler and L. M. Sander, Phys. Rev. Lett. 59:2315 (1987).

85. Y. Sawada, A. Dougherty and J. P. Gollub, Phys. Rev. Lett. 56:1260 (1986).

86. R. M. Suter and P. Wong, Nonlinear oscillations in electrochemical dendritic growth, preprint (July 1987).

87. J. E. Marsden and M. McCracken, "Hopf Bifurcation and its Applications", Applied Math. Sci. 19, Springer, New-York (1976).

88. C. Vidal and A. Pacault, eds, "Nonlinear Phenomena in Chemical Dynamics", Springer-Verlag, Berlin (1981) and references quoted therein.

89. C. Vidal and A. Pacault, eds, "Nonequilibrium Dynamics in Chemical Systems", Springer-Verlag, Berlin (1984) and references quoted therein.

90. J. C. Roux, P. Richetti, A. Arneodo and F. Argoul, J. Mec. Th. & Appl., numéro spécial: "Bifurcations et Comportements Chaotiques" (1984) p. 77.

91. A. Arneodo, F. Argoul, P. Richetti and J. C. Roux, in "Dynamical Systems and Environmental Models", H. G. Bothe, W. Ebeling, A. M. Zurzhanski and M. Peschel, Akademie-Verlag, Berlin (1987) p. 122.

92. F. Argoul, A. Arneodo, P. Richetti, J. C. Roux and H. L. F. Swinney, Acc. Chem. Res. 20:436 (1987).

93. N. H. Packard, J. P. Crutchfield, J. D. Farmer and R. S. Shaw, Phys. Rev. Lett. 45:712 (1980).

94. F. Takens, Lect. Notes in Maths. 898:366 (1981).

CHARACTERIZING SPACE-TIME CHAOS IN AN EXPERIMENT

OF THERMAL CONVECTION

S.Ciliberto

Istituto Nazionale Ottica

Largo E.Fermi 6-50125 Firenze-Italy

1) Introduction

In the last decade, dynamical system theory has provided several tools to analyse quantitatively the transition to low dimensional chaos[1] observed in many experiments[2]. In contrast a general characterization of spatiotemporal chaos is still an unsolved problem, that has recently received a lot of interest both from a theoretical [3-10] and experimental point of view[11-13]. In fluid dynamics the main goal of this research is to understand the relationship between chaos and the turbulent regimes, where the fluid motion presents a chaotic evolution both in space and time. Indeed in spatially extended systems, such as a fluid, the transition to low dimensional chaos is associated with relevant spatial effects, but the unpredictable time evolution does not influence the spatial order. In other words the correlation length is comparable with the size of the system. So, in order to give more insight into the problem of the transition to turbulence it is very important to study the role of the spatial degrees of freedom in temporal chaotic regimes and the reasons why the spatial coherence is lost. The simplest mathematical models, in which the features of the transition to spatiotemporal chaos may be analysed, are systems of coupled maps[4-6], one dimensional partial differential equations[3,6,7,8] and cellular automata[9]. One of the typical regimes, presented by these models, is spatiotemporal intermittency,that consists of a fluctuating mixture of laminar and turbulent domains with well defined boundaries. Such a phenomenon has a physical relevance because it is similar to those observed in Rayleigh-Benard convection [13,14] and in boundary layer flows [15].

In this paper we descibe an experiment in which the space time evolution of Rayleigh-Benard convection has been studied in order to investigate how the spatial order is lost in temporal chaotic regimes and to test methods to characterize the transition. The behaviours of our system are compared with those of the above mentioned mathematical model. We have also studied the statistical properties of the onset of spatiotemporal chaos. Our results display features typical of phase transitions similar to those obtained in numerical simulations[6-7].

The general properties of Rayleigh-Benard convection,that is thermal convection in a horizontal fluid layer heated from below, may be found in standard text books and in review papers[16], thus we briefly remind, in section 2), only the main features of this instability. The rest of the paper is organized as follows. In section 3) the experimental apparatus is described. In section 4) the different space time dynamics observed as function of the control parameter are discussed. In section 5) the results concerning the transition to turbulence are reported and compared with those observed in numerical simulations. In section 6) we describe the statistical properties of some global quantities such as the energy and the entropy, that may

Measures of Complexity and Chaos
Edited by N.B. Abraham *et al.*
Plenum Press, New York

be useful to describe the transition to turbulence. Finally conclusions are presented in section 7).

2) Rayleigh-Benard Convection

To illustrate the general features of Rayleigh-Benard convection let us consider a fluid layer confined between two horizontal solid plates and heated from below. The most relevant parameter of this instability is the Rayleigh number $Ra = \beta g \Delta T d^3/(\nu\chi)$, where β is the volumetric expantion coefficient, g the acceleration of gravity , ν the kinematic viscosity, χ the thermal diffusion coefficient, d the depth of the layer and ΔT the difference of temperture between the two horizontal plates. When Ra exceeds the threshold value Rc a steady convective flow arises, producing a pattern of parallel rolls with a well defined wavenumber q. The roll pattern is parallel to the shortest side of the cell containing the fluid. The values of Rc and q depend on the sizes of the cell and on the nature of boundary conditions, for example in the case of an infinite layer and perfect conduting plates $Rc = 1708$ and $q = 3.11/d$. Thus, from an experimental point of view, it is very important to define two other parameters, that are the aspect ratios $\Gamma_x = L_x/d$ and $\Gamma_y = L_y/d$, where L_x and L_y are the two horizontal lengths of the cell. The time dependent regimes of Rayleigh Benard convection observed at $Ra >> Rc$ are strongly influenced by the aspect ratios and also by the Prandtl number $Pr = \nu/\chi$. Indeed in the experiments in which the transition to low dimensional chaos has been studied[2] Γ was of the order of $2\pi/q$.

In the experiment that we describe in this paper the cell containing the working fluid has an annular geometry. Indeed with this geometry and a suitable choise of the radial aspect ratio, it is possible to construct a pattern that is almost a one dimensional chain of radial rolls(roll axis along radial directions, see also Fig.2) with periodic boundary conditions. These features of the spatial pattern are very useful in order to compare the results of our experiment, with those obtained in the above mentioned mathematical models.

3) Experimental apparatus

A schematic cross section of the cell is reported in Fig.1. The lateral walls of the cell are made of plexiglass. The outer and inner diameters of the annulus are 8 Cm and 6 Cm respectively. The depth of the layer d is 1 Cm. With these dimensions the radial aspect ratio is 1, whereas the aspect ratio along the circle, of diameter $2r_o = 7cm$, is 21.99. The bottom plate of the cell is made with a copper plate whose upper surface is finished to a mirror quality and is protected with a film of nickel to prevent oxidation. The plate is heated with an electrical resistor R_1. The upper plate is made of a sapphire window SW whose top is cooled by the water circulation Wa, that is confined on the other side by the glass window GW. This arrangement allows an optical investigation of the convective motion. The cell is inside a temperature stabilized box that reduces the thermal fluctuations of the enviroment. The temperture of the cooling water Wa is stabilized by a thermal bath. The long term stability of ΔT is $\pm 0.001°C$. The working fluid is silicon oil with $Pr = 30$. The critical difference of temperature , computed with $Rc = 1708$,is $\Delta T_c = 0.06°C$.

The qualitative features of the patterns are determined by a shadowgraph technique[17]. An optical technique, based on the deflection of a laser beam that sweeps the fluid layer[18], enable us to obtain quantitative global and local characteristics of the pattern. The shadowgraph and laser beam deflection techniques are not perturbative and rely upon the changes of the index of refraction induced by the temperature field. The principle of the sweeping technique has been described elsewhere[18]. The actual set up provides the possibility of measuring on the circle of radius $r_o = 3.5cm$ (that is on the circle of mean diameter), with a twelve-bit

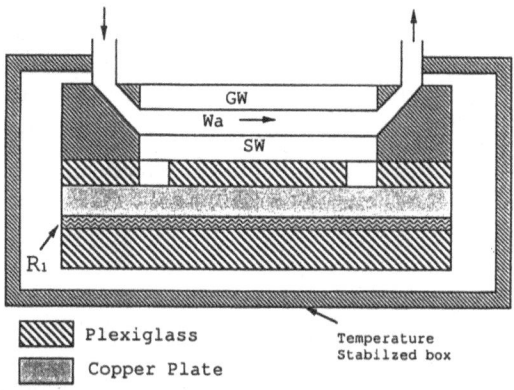

▨▨▨ Plexiglass

▨▨▨ Copper Plate

Temperature
Stabilzed box

Figure 1. Schematic diagram of the cell: R_1 heating resistor, SW sapphire window, GW glass window, WA cooling water

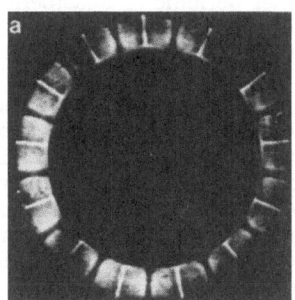

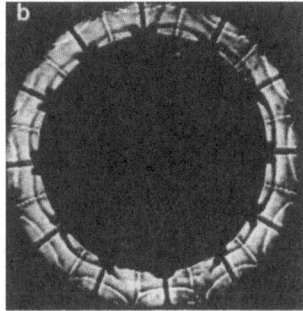

Figure 2. Shadowgraphs of typical spatial patterns. White and dark regions correspond to cold and hot currents respectively. a)Stationary spatial pattern at $\eta = 13$. b) Snapshot of the spatial pattern at $\eta = 190$ in a time dependent regime.

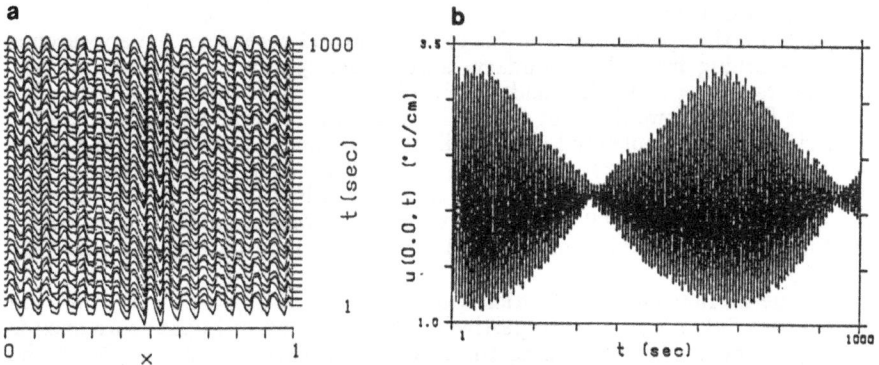

Figure 3.a) Space time evolution of u(x,t) at $\eta = 164$;b) Corresponding time evolution of the point x=0. The vertical scale has been amplified in b) because the time dependent modulation slightly perturbes the spatial pattern shown in a),where the maximum amplitude is roughly $4°C/cm$

resolution, the two components of the thermal gradient averaged along the vertical direction, in the polar coordinate reference frame r, θ. The accuracy of the measurement is about 7%, the sensitivity $0.01°C/cm$ and the spatial resolution about 1 mm. In time dependent regimes only the component of the gradient perpendicular to the roll axis has been recorded. This component will be called u(x,t), with $x = \theta/(2\pi)$. The function u(x,t) is sampled at 128 points in space. In time dependent regimes u(x,t) is recorded for at least 5000 times at interval of 1 sec. that is roughly 1/10 of the main oscillation period of the system.

3) Spatial patterns

Analysing the fluid behaviour as a function of $\eta = \Delta T/\Delta T_c$, we observe, that for η around 1, the spatial structure has about 22 rolls. This number increases with η and reaches 38 at η around 200. A detailed analysis of the wavenumber selection process has been reported elsewhere[19]. In Fig.2a) we show the shadowgraph of the spatial pattern at $\eta = 13$. Dark region correspond to the hot currents rising up and white regions to the cold ones, going down. We observe that our geometry constrains the spatial structure to an almost one dimensional chains of rolls.

The spatial structure remains stationary for $\eta < 164$ where a subcritical bifurcation to the time dependent regime takes place. For $\eta > 164$ the time evolution is chaotic but, reducing η, the system presents either periodic or quasiperiodic oscillations, and at $\eta = 149$ it is again stationary. In the range $149 < \eta < 200$ the time dependence consists of rather localized fluctuations that slightly modulate the convective structure, which maintains its periodicity. This is clearly seen in Figs.2b) where a snapshot of the spatial structures at $\eta = 190$ is reported. The presence of hot and cold currents transverse to the main set of rolls merit a special comment. Such a two dimensional effect certainly influence the dynamics. However considering that the ratio between the length and the width of the annulus is roughly 22 we realise that the system can be considered almost one dimensional for what concerns the propagation time of thermal fluctuations along the circle, because the two time scales are very well separated. Besides, we also observe that the time dependent fluid motion is still very correlated along the radius.

The space time evolution of u(x,t) and the corresponding time evolution of the point x=0 at $\eta = 164$ are shown in Fig.3a and Fig.3b. In looking at Fig.3b we clearly see that the time evolution is quasiperiodic. However this time dependent modulation is hardly seen in Fig.3a, because it slightly perturbes the spatial pattern that mantain its original periodic structure. Increasing η the time evolution becomes chaotic but the spatial order is still mantained. The fractal dimension and the orthogonal decomposition [20] indicate that the number of degrees of freedom involved in the dynamics is around 3.

At higher η the spatial order begins to be destroyed because of the appearance of bursts, detaching from the boundary layer. This spatiotemporal intermittent regime appears at $\eta = 200$. The shadographs of typical spatial patterns at $\eta = 230$ are shown in Fig.4) at two different times. They present, several domains where the spatial periodicity is completely lost (we will refer to them as turbulent) and other regions (that we call laminar) where the spatial coherence is still mantained. The space time evolution of u(x,t) at $\eta = 216$ is shown in Fig.5a),5b) at two different times. We notice that for $1000 < t < 1040$ there are strong oscillations that locally destroy the spatial order whereas for $1500 < t < 1540$ the pattern is again very regular.

The time averaged spatial Fourier spectra at $\eta = 164$, $\eta = 216$, $\eta = 347$ are reported in Figs.6a),6b), and 6c) respectively. The spectrum of Fig.6a) corresponds to a quasiperiodic regime and being the spatial structure still very ordered the spectrum presents well defined peaks. In contrast Fig.6b), corresponding to a value of η that is very close to the threshold for spatiotemporal intermittency presents a broadened third harmonic. This indicate that the most important length scales for this transition are the shortest ones. Finally in Fig.6c) the spectrum, corresponding to a value of η far above the transition point, is totally broadened because the

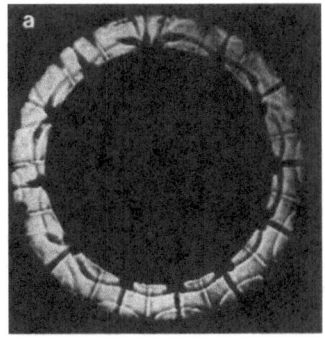
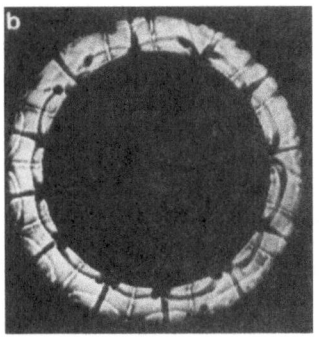

Figure 4). Snapshots of the spatial patterns at $\eta = 230$. The time interval between a),b) is 30 sec.

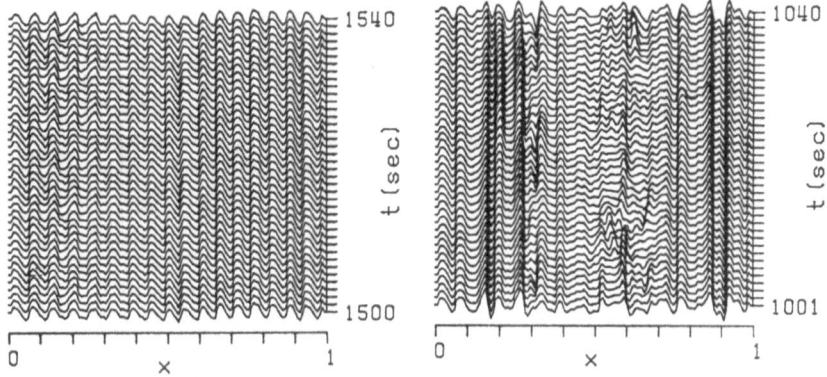

Figure 5). Space time evolutions of u(x,t) at $\eta = 216$ at two different time intervals of 40 sec each.

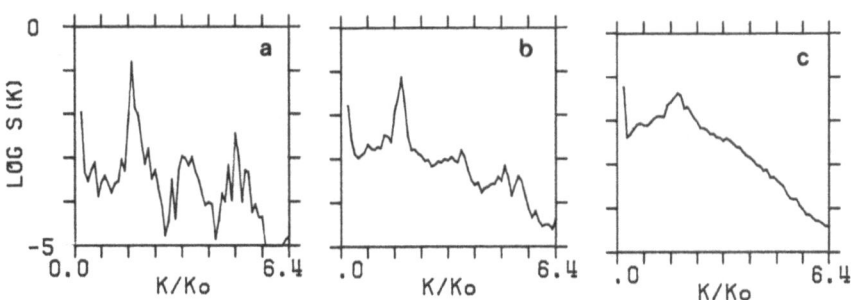

Figure 6). Spatial power spectra at different values of η:a) $\eta = 164$;b) $\eta = 216$;c)$\eta = 348$.

spatial order has been destroyed. Notice the exponential decay at high k and the flat region for the small ones. These features are rather similar to those observed in the turbulent regimes of the Kuramoto-Shivanshinsky equation[8].

4) The weak turbulence

As we discussed in the previous section, the space time evolution of u(x,t) shows that in the turbulent domains the time evolution is characterised by the appearence of large oscillatory bursts. Instead in laminar regions the oscillations remain very weak. Thus the two regions can be identified by measuring the local peak to peak amplitude u_{pp}, for a time interval δ comparable with the mean period of the oscillation, that is:

$$u_{pp}(x,t) = max[u(x,\tau)] - min[u(x,\tau)]$$

with $t < \tau < (t+\delta)$. Choosing a cutoff α, setting to 1 all the points in which $u_{pp} > \alpha$ and to 0 all the other points, the space time dynamics is reduced to a binary code in which 1 stands for "turbulent" and 0 for laminar. As an example of such a code we show the spacetime evolution of u(x,t) at $\eta = 216$, in Fig.7a), and $\eta = 248$ in Fig.7b),the black and white regions correspond to turbulent and laminar domains respectively. We remark that the qualitative features of these pictures are rather independent of the precise value of the cutoff. We can easily verify that the code catches the main properties of the dynamics by comparing Fig.7a) with Figs. 5a) and 5b). Indeed we clearly see that at the most oscillating and disordered regions of Figs.5) correspond to black points in Fig.7a) whereas the ordered and not oscillating regions are represented by white points.

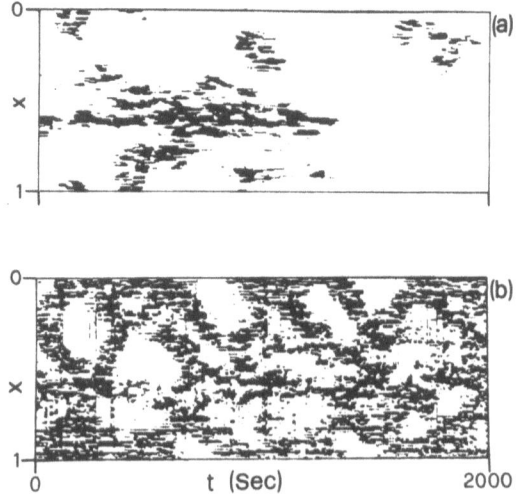

Figure 7).Binary representation, at $\alpha = 1.5°C/cm$, of the space-time evolution of u(x,t) at $\eta = 216$ a) and $\eta = 248$ b). The dark and white area correspond to turbulent and laminar domains respectively.

At $\eta = 216$,Fig.7a), a wide laminar region surrounds completely the turbulent patches that remain localized in space, after their appearence. Furthermore, the nucleation of a turbulent domain has no relationship with the relaxation of another one. In contrast, at $\eta = 248$ Fig.7b),the turbulent regions migrate and slowly invade the laminar ones. This last regim that sets in for $\eta > 245$ is very similar to those obtained in theoretical models $5 - 7$. The change from the regime of Fig.7a) to that of Fig.7b) is reminescent of a percolation 6, that, indeed, has been proposed as one of the possible mechanisms for the transition to spatiotemporal intermittency.

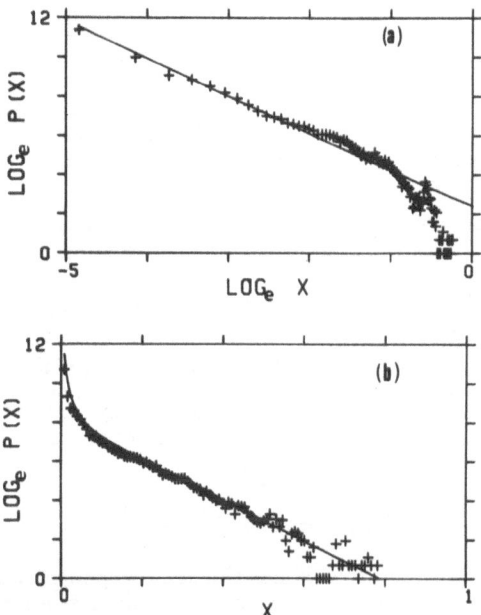

Figure 8). Distribution P(x) of the laminar regions of length x.(a) $\eta = 241$, algebraic decay with exponent 1.9; (b) $\eta = 310$ and $\alpha = 1.6°C/cm$, exponential decay with a characteristic length $1/m = 0.10$. The solid lines are obtained from Eq.3).

Following a method also used in numerical models [5-7], we quantitatively characterize such a behaviour by computing, over a time interval of $10^4 sec$, the distibution P(x) of the the laminar domains of length x. For $\eta < 248$ P(x) decays with a power law.The exponent does not depend within our accuracy, either on α or on η . Its average value is $\mu = 1.9 \pm 0.1$. On the other hand, for $\eta > 248$,the decay of $P(x)$ for $x > 0.1$ is exponential with a characteristic length $1/m$.The existence of two different regimes is clearly seen in Figs.8a),8b) which display P(x) versus x at $\eta = 241$ and $\eta = 310$. Looking at Fig.8a) we clearly see that the decay of P(x) begins for a length scale that is smaller than the roll size. This rather strange result has an explanation, because ,as we remarked in section 3), the main energy contribution to the time dependent regimes is coming from the spatial high frequencies.

We find that the dependence of m on η is the following:

$$m(\alpha, \eta) = m_o(\eta) exp(-\alpha/\alpha_o) \tag{1}$$

with $\alpha_o = (0.87 \pm 0.06)°C/cm$ independent of η. The dependence of m_o versus η is reported in Fig.9. The linear best fit for $\eta > 246$ of the points of Fig.9) gives the following result:

$$m_o(\eta) = m_1(\frac{\eta}{\eta_s} - 1)^{\frac{1}{2}} \tag{2}$$

with $\eta_s = 247 \pm 1$ and $m_1 = 117 \pm 2$. This equation shows the existence of a well defined threshold η_s for the appearence of an exponential decay in P(x). Besides we see that the characteristic length $1/m_o$ diverges at $\eta = \eta_s$. In the range $200 < \eta < 400$, P(x) is very well approximated by the following equation:

$$P(x) = (Ax^{-\mu} + B)exp[-m(\alpha, \eta)x] \tag{3}$$

where $m(\alpha, \eta)$ is given by 1) and μ has the previous determined value. A,B are instead free parameters that can be very easily determined. It is possible to fit our experimental P(x),in the range $0.4°C/cm < \alpha < 3°C/cm$, with $A = 10$ $B \simeq 4 \cdot 10^3$ for $\eta > \eta_s$ and $B = 0$ for $\eta < \eta_s$.

The features of P(x) displayed by equations 2),3)are typical of phase transitions . Therefore, being the transition point η_s very close to the point where the behaviour like that of fig.8b) sets in, we conclude that the transition to this behaviour may be a phase transition[21] The main features of P(x) and m for $\eta > \eta_s$ qualitatively agree with those obtained in coupled maps [5,6], partial differential equations [6,7], and in a phenomenological cellular automaton model[9] in spatiotemporal intermittent regimes.

The transition may also be characterized by measuring p_o that is the probability of finding a laminar point[9]. If we suppose that alaminar site is generated at a certain time with space-time independent probability p_o, the probability of finding a laminar region of length x is given by $P(x) \propto exp[xlog(p_o)/l_o]$, where l_o is a suitable characteristic length. We can verify this hypothesis by computing directly p_o on the experimental data. By following the same procedure, used to compute m as a funtion of α, we find that $\log p_o$ extrapolated at $\alpha = 0$ has the following dependence on η :

$$|\log p_o| = const.(\frac{\eta}{\eta_c} - 1)^{\frac{1}{2}} \tag{4}$$

with $\eta_c = 216$. So we conclude that it has the same exponent but different critical threshold. This means that the appearence of a laminar site may be considered a statistical independent process for $\eta >> \eta_s$ and that a certain correlation exists between laminar and turbulent sites near the critical value η.

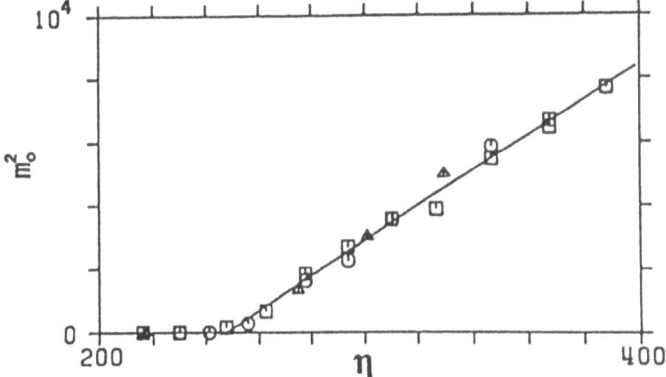

Figure 9). Dependence of m_o^2 on η ,the different symbols pertain to different sets of measurements done either increasing or decreasing η. The solid line is obtained from Eq.2).

The presence of a power law decay of P(x) for $\eta_c < \eta < \eta_s$ may be due to finite size effects [18]. Indeed the cellular automaton model of this transition [8], presents the same features when the number of cell is reduced.

6) Statistical properties of the energy and spectral entropy

In the previous section we have clearly demonstrated that the transition to space-time chaos has the properties of a phase transition. This result has been obtained by studying the statistic of the sizes of laminar regions.

In this section we show that this transition may be also characterized by analysing the statistical properties of some global quantities such as the energy $E(t)$ and the spectral entropy $Ent(t)$, that may be computed from the spatial Fourier spectra $S(k,t)$[22]. Indeed in section 4) we have shown that the time averaged spatial Fourier spectra changes as a function of the control parameter and they become broadened for $\eta > \eta_s$. As a consequence the Fourier spectra are good candidates to study the transition to the space time chaos. A similar approach as been recently proposed also by Hohenberg and Shraiman[23]. It has the advantage of dealing with averaged quantities such as the thermodynamic ones of a system near thermal equilibrium.

The two quantities $E(t)$ and $Ent(t)$ are defined as following:

$$E(t) = \sum_{k=0}^{n} S(k,t)$$

and

$$\sigma(t) = Ent(t)/Ent_o = -\sum_{k=0}^{n} \Phi_k(t) \cdot \log(\Phi_k(t))/Ent_o$$

with $\Phi_k(t) = S(k,t)/E(t)$ and $Ent_o = \log n$, where n is the number of Fourier modes. The quantity σ is equal to 1 when all the modes have the same energy, whereas it is 0 when only one mode is excited. The time evolution of $\sigma(t)$ for 4 values of η is shown in Fig.10). We observe that the mean value of $\sigma(t)$ grows as a function of η and that the fluctuations of $\sigma(t)$ are enhanced when η aprroaches η_s. This effect is clearly seen in the distributions of $\sigma(t)$ and $E(t)$, reported in Figs.11), for different values of η. N is the number of times a particular value of σ is found in 5000 sec.

We observe that the distributions not only change their shape as a function of the control parameter but they are considerably larger for η close to η_s. For $\eta > \eta_s$ the distributions are similar to a gaussian but a significant deviation is observed. These are very preliminar results, showing that the idea of constructing a statistical mechanics of the transition to space time chaos is probably correct. However much work has to be done in order to study the dependence on the control parameter of the above mentioned quantities and to define a generalized temperature of the system.

7) Conclusion

Rayleigh-Benard convection in an annular geometry is very useful to investigate the transition from low dimensional chaos to weak turbulence because both the regimes are found as a function of the control parameter.

The onset of spatiotemporal intermittency, in our cell, displays features of a phase transition that is reminescent of a percolation. This result has been obtained by reducing the space-time dynamics to a binary code, that catches the relevant features of the phenomenon. A cellular automaton model, whose transition probabilities have been obtained from the experiment, indicates that the power law decay

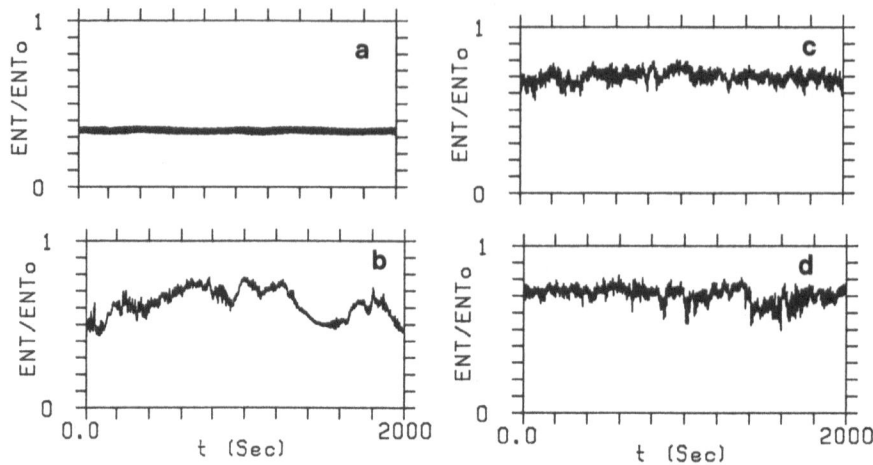

Figure 10). Dependence of the spectral entropy on time for different values of η: a) $\eta = 164$,b) $\eta = 216$, c) $\eta = 310$,d) $\eta = 348$.

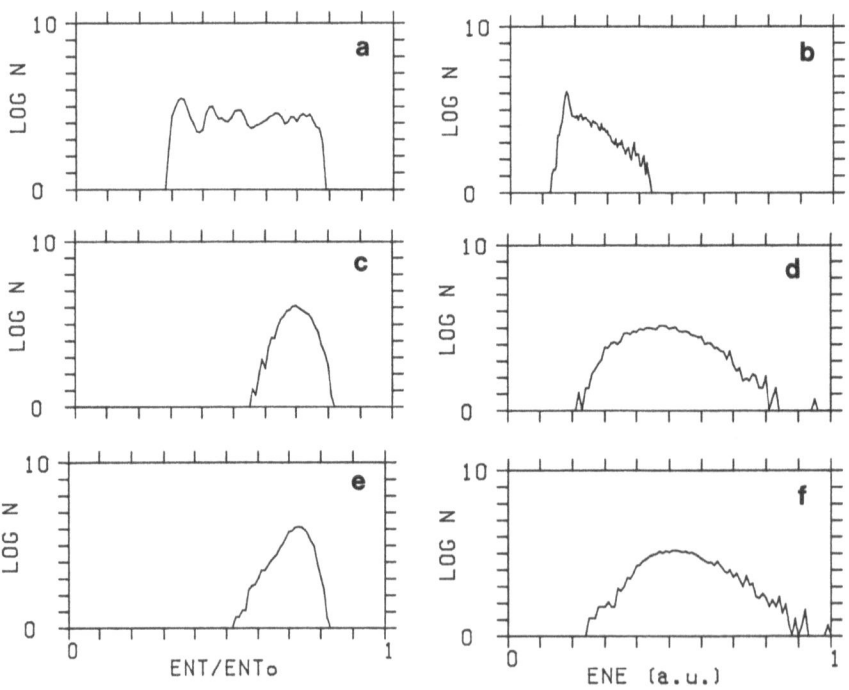

Figure 11). Distributions of the spectral entropy Ent and of the energy $E(t)$ for different values of η: a),b) $\eta = 216$;c),d) $\eta = 310$;e),f) $\eta = 348$.

of P(x), observed in a wide range of η, is probably due to finite size effects. Work is in progress to identify the universality class of the transition by measuring other critical exponents.

Another useful approach to study the transition to space time chaos is the analysis of space averaged quantities. Indeed it could allow to compare this behaviour with that of the system in thermal equilibrium.

Another important problem to solve in spatially extended systems is the estimation of the number of degrees of freedom. It is clear that the calculation of the fractal dimension is not feasible in such a complex regimes[10]. However one could try to obtain a rough estimation by using the orthogonal decomposition[20] that has also the advantage of giving some informations about the most important spatial structures involved in the dynamics. This method has been applied to the data obtained from the experiment, here described, and it gave very promising results.

As a general comment we can say that the techniques that we have used to analyse the transition to space time chaos in our experiment of thermal convection are very general and they can be succesfully applied also in many other physical systems.

Acknowledgement

This work has been partially supported by GNSM and by CEE contract number sci-0035-c(cd).

References

1. For a general review of low dimensional chaos see for example: J. P. Eckmann, D. Ruelle, Rev. Mod. Phys. 1987; P. Berge, Y. Pomeau, Ch. Vidal, L'Ordre dans le Chaos (Hermann, Paris 1984).
2. A. Libchaber, C. Laroche, S. Fauve, J. Physique Lett. 43, 221, (1982); M. Giglio, S. Musazzi, U. Perini, Phys. Rev. Lett. 53, 2402 (1984); M. Dubois, M. Rubio, P. Berge', Phys. Rev. Lett. 51, 1446 (1983); S. Ciliberto, J. P. Gollub, J. Fluid Mech. 158, 381 (1984).
3. B. Nicolaenko, in " The Physics of Chaos and System Far From Equilibrium", M. Duong-van and B. Nicolaenko, eds. (Nuclear Physics B, proceeding supplement 1988); A. R. Bishop, K. Fesser, P. S. lomdhal, W. C. Kerr, M. B. Williams, Phys. Rev. Lett. 50, 1095 (1983).
4. G. L. Oppo, R. Kapral Phys. Rev. A 3, 4219 (1986).
5. K. Kaneko, Prog. Theor. Phys. 74, 1033 (1985); J. Crutchfield K. Kaneko in "Direction in Chaos", B. L. Hao (World Scientific Singapore 1987); K. Kaneko Physica 34D, 1 (1989); R. Lima, Bunimovich preprint.
6. H. Chate', P. Manneville, Phys. Rev. Lett. 54, 112 (1987); Europhysics Letters 6,591(1988);Physica D 32, 409 (1988)
7. H. Chate', B. Nicolaenko, to be published in the proceedings of the conference: "New trends in nonlinear dynamics and pattern forming phenomena", Cargese 1988;
8. Y. Pomeau, A. Pumir and P. Pelce', J. Stat. Phys. 37, 39 (1984)
9. F. Bagnoli, S. Ciliberto, A. Francescato, R. Livi, S. Ruffo, in "Chaos and complexity" , M. Buiatti, S. Ciliberto, R. Livi, S. Ruffo eds., (World Scientific Singapore 1988); F. Bagnoli, S. Ciliberto, R. Livi, S. Ruffo , in the proceedings of the school on Cellular Automata, Les Houches (1989).
10. P. Coullett and P. Procaccia in these proceedings.
11. A. Politi these proceedings.
12. P. Kolodner, A. Passner, C. M. Surko, R. W. Walden, Phys. Rev. Lett. 56, 2621 (1986); S. Ciliberto, M. A. Rubio, Phys. Rev. Lett. 58, 25 (1987); A. Pocheau Jour. de Phys. 49, 1127 (1988)I; I. Rehberg, S. Rasenat , J. Finberg, L. de la Torre Juarez Phys. Rev. Lett. 61, 2449 (1988); N. B. Trufillaro, R. Ramshankar, J. P. Gollub Phys. Rev. Lett. 62, 422 (1989).
13. P. Berge', in " The Physics of Chaos and System Far From Equilibrium", M.Duong-van and B. Nicolaenko, eds. (Nuclear Physics B, proceedings supplement 1988); F. Daviaud, M. Dubois, P. Berg Europhysics Lett. 9, 441 (1989).

14. S.Ciliberto,P.Bigazzi,Phys.Rev.Lett. 60, 286 (1988).
15. M. Van Dyke, An Album of Fluid Motion (Parabolic Press, Stanford, 1982); D. J. Tritton, Physical Fluid Dynamics (Van Nostrand Reinold, New York, 1979), Chaps.19-22.
16. S. Chandrasekar, Hydrodynamic and Hydromagnetic Stability, Clarendon Press, Oxford 1961; F. H. Busse, Rep. Prog. Phys. 41 (1978) 1929; Ch. Normand, Y. Pomeau, M. Velarde Rev. Mod. Phys. 49, 581,(1977).
17. W. Merzkirch, Flow Visualisation, Academic Press, New York 1974.
18. S. Ciliberto, F. Francini,F. Simonelli, Opt. Commun. 54, 381 (1985).
19. S. Ciliberto, M. Caponeri, F. Bagnoli, submitted to Nuovo Cimento D.
20. S.Ciliberto,B.Nicolaenko submitted for publication; A.Newell, D. Rand , D. Russell, Physica 33d, 281 (1988).
21. H. Muller-Krumbhaar in 'Monte Carlo Methods in Statistical Physics", edited by K. Binder (Springer- Verlag,New York 1979); D. R. Nelson ,'Phase transitions and critical phenomena' , edited by C. Domb and J.L. Lebowitz (Academic Press London 1983).
22. R. Livi, M. Pettini, S. Ruffo, M. Sparpaglione, A. Vulpiani, Phys. Rev. A 31, 1039 (1985); M. A. Rubio, P. Bigazzi, L. Albavetti, S. Ciliberto. J. Fluid Mech. to be published.
23. P. C. Hohenberg, B. I. Shraiman ,preprint.

CHARACTERIZING DYNAMICAL COMPLEXITY IN INTERFACIAL WAVES

J.P. Gollub

Physics Department
Haverford College
Haverford, PA 19041 U.S.A. and

David Rittenhouse Laboratory
The University of Pennsylvania
Philadelphia, PA 19104 U.S.A.

Introduction

Waves on a fluid interface may be excited parametrically by vertical excitation of the container (Faraday, 1831). These waves illustrate the full spectrum of dynamical complexity, including stationary patterns, periodically or chaotically oscillating patterns, and spatiotemporal chaos. A variety of experiments performed at Haverford and illustrating these phenomena are reviewed, with emphasis on the development of methods for studying complex dynamics. A general review of this problem has been given by Miles & Henderson (1990).

Background

When a bounded cell containing a fluid with a free surface is subjected to a sinusoidal vertical oscillation, a standing interfacial wave may be excited if the driving amplitude Δ is larger than a frequency-dependent threshold Δ_o. The basic instability (Benjamin & Ursell, 1954) may be described as a parametric process, i.e., the gravitational acceleration g may be considered to be oscillating. The excited waves (when close to threshold) are similar to the modes of a drum whose shape is given by the boundary of the container.

For example, in a rectangular container with horizontal dimensions L_x and L_y, the modes have the form

$$Z(x,y,t) = [\cos(m\pi x/L_x)\cos(n\pi y/L_y)]g_{mn}(t) .$$

The waves oscillate at half the driving frequency ω_o, and this picks out particular modes, namely those which satisfy the capillary wave dispersion relation

$$x(k) = \omega_o/2 = \gamma k^3/\rho, \text{ where}$$

$$k^2 = (m\pi/L_x)^2 + (n\pi/L_y)^2 .$$

Similar statements may be made about waves in a cylindrical cell, in which case the linear modes are of course given by Bessel functions in the radial coordinate.

Two cases. There are two important special cases that may be conceptually distinguished. For low driving frequency, the typical wavelength λ is a significant fraction of the system size and the modes are generally well separated in frequency. In that case, low-dimensional chaos can occur either through the growth of nonlinear interactions as the driving amplitude Δ is increased (Gollub & Meyer, 1983), or through the interaction between two modes that happen to be degenerate (Ciliberto & Gollub, 1984 and 1985a; Simonelli & Gollub, 1989). An example is described below. On the other hand, when the driving frequency is sufficiently high, λ is much less than the system size, the density of modes per unit frequency is quite high, many modes can be excited, and one observes spatiotemporal chaos for amplitudes not to far above Δ_0.

Dynamical Complexity from Interacting Modes

One of the particular difficulties of characterizing complex temporal dynamics has been the difficulty of obtaining sufficient information to construct a valid phase space. The use of embedding methods provides little useful physical information. It is much more desirable to choose as the coordinates for a phase space description the amplitudes of the modes that would be used in a theoretical explanation of the dynamics. For chaos near the onset of instabilities, the amplitudes of the linear modes of the problem are the most natural phase coordinates to use. Simonelli & Gollub (1989) showed that this can be accomplished for the interaction between several wave modes in a rectangular geometry. They also showed how multiple attractors and repellers can be detected, their respective basins of attractions determined, and the bifurcations of these fixed points followed as a function of two parameters. We are not aware of other examples of such a detailed realization of the program of nonlinear dynamics in a hydrodynamic system.

The amplitudes that were measured for the construction of phase trajectories are defined by the equation

$$g_{mn}(t) = A_{mn}(t)\cos(\omega t/2) + B_{mn}(t)\sin(\omega t/2) .$$

The coefficients $A_{mn}(t)$ and $B_{mn}(t)$ give the components of the slow variation of the pattern in phase and out of phase with the driving amplitude. The actual measurement of these amplitudes is accomplished by suitably placed local probes in the shadowgraph image of the waves, in conjunction with phase sensitive detection. No embedding is required!

Particularly interesting dynamics occurs in a square or nearly square cell where two modes (m,n) and (n,m) are essentially degenerate. This allows the two modes to be present simultaneously and to interact. Simonelli & Gollub were able to detect as many as 8 stable fixed points corresponding to different symmetries of the interacting modes. Some of these correspond to "pure states" and others to "mixed modes" or nonlinear superpositions of the linear eigenmodes. Furthermore, by measuring transient trajectories, the basins of attraction and the unstable fixed points could also be detected. This information is commonly found in numerical simulations, but rarely in experimental studies.

In nonlinear dynamics, as is well known, the fixed points organize the structure of the phase space. Thus, it is extremely important to know how the fixed points evolve as a function of the parameters of the problem, in this case the driving frequency ω_0 and amplitude Δ. A detailed study of the bifurcations affecting these fixed points as a function of the parameters was in fact carried out by Simonelli & Gollub (1989).

These observations have stimulated several theoretical studies (Silber & Knobloch,1989; Guckenheimer, 1989; Umeki, 1990). Non-autonomous systems such as this one can be rather accurately described by maps corresponding to stroboscopic observation. Silber and Knobloch (1989) propose that, because of the symmetry of the system, the unfolding of the codimension-three bifurcations of an appropriate D_4-equivariant map can explain the main experimental observations. Here, we emphasize that the symmetry properties that are apparently so central to the theoretical explanation are revealed naturally by the type of modal decomposition employed in the experiment. They would not be revealed by using embedding methods.

Chaos. Chaotic dynamics occurs only when the spatial symmetry of the cell is broken, i.e., when L_x and L_y are slightly different. One of the prominent routes to chaos involves successive bifurcations of limit cycles that lengthen the period, but not by factors of 2. Though the chaotic dynamics of this system arises through interaction between only two modes, a quantitative explanation has not been proposed.

Spatiotemporal Chaos

When the driving frequency is raised sufficiently that the wavelength λ is much shorter than the size of the system (say $L/100$), spatiotemporal chaos can occur. This phenomenon was studied by Tufillaro et al. (1989), who found a well-defined order/disorder transition as a function of the dimensionless driving amplitude

$$\epsilon = (\Delta - \Delta_c)/\Delta_c .$$

For $\epsilon < 0.1$ in a square cell of size 40λ (8 cm), ordered patterns having square symmetry and very little time-dependence are found. The pattern selection leading to this state is actually not well understood. As ϵ is increased, long wavelength modulations lead to the generation of defects and the destruction of the order of the pattern. This phenomenon was first noted by Ezerskii et al. (1985). We studied the transition process quantitatively by measuring the spatial extent of translational and orientational correlations. Strikingly, the loss of long-range translational order occurs in a narrow range of ϵ. It coincides with the onset of temporal chaos and a dramatic reduction of long range orientational order. However, some orientational order persists above the transition.

Ramshankar & Gollub (1990) used a number of additional diagnostic tools for the study of spatiotemporal chaos in interfacial waves. They measured the probability distribution of the spatial Fourier amplitudes, and showed that it evolves toward a Gaussian distribution in the spatiotemporally chaotic state. This is evidence that the dynamics is dominated by many uncorrelated regions of size $\xi << L$. However, the two-dimensional spatial Fourier spectra show a very significant power law rise at low wavenumber q. The spectral density S(q) rises approximately as

$$S(q) \sim q^{-2.5} \text{ as } q \to 0 .$$

This observation and the persistence of some orientational order indicate that residual long-range correlations must be important in spatiotemporal chaos in interfacial waves.

Conclusion

Methods for phase space measurement of surface wave modal interactions have been demonstrated. The achievement of a satisfactory understanding of spatiotemporal chaos is a major challenge for the future. Surface wave dynamics is an appropriate context for the development of theoretical models and methods of experimental characterization. Statistical methods are bound to play a substantial role in studying spatiotemporal chaos.

Acknowledgement

This work was supported by the University Research Initiative program under Contract No. DARPA/ONR N00014-85-K-0759.

References

Benjamin, T.B. & Ursell, F., 1954, The stability of the plane free surface of a liquid in vertical periodic motion, *Proc. R. Soc. Lond.* A225:505.

Ciliberto, S. & Gollub, J.P., 1984, Pattern competition leads to chaos, *Phys. Rev. Lett.* 52:922.

Ciliberto, S. & Gollub, J.P., 1985, Chaotic mode competition in parametrically forced surface waves, *J. Fluid Mech.*, 158:381.

Faraday, M., 1831, On the forms and states assumed by fluids in contact with vibrating elastic surfaces, *Phil. Trans. R. Soc. Lond.*, 121:319.

Funakoshi, M. & Inoue, S., 1988, Surface waves due to resonant horizontal oscillation, J. Fluid Mech., 192:219.

Gollub, J.P. & Meyer, C.W., 1983, Symmetry-breaking instabilities on a fluid surface, *Physica D*, 6:337.

Gollub, J.P. & Ramshankar, R., 1990, Spatiotemporal chaos in interfacial waves, *in*: "New Perspectives in Turbulence," S. Orszag and L. Sirovich, ed., Springer-Verlag, Berlin.

Miles, J. & Henderson, 1990, Parametrically forced surface waves, *Ann. Rev. Fluid Mech.*, to appear.

Silber, M. & Knobloch, E., 1989, Parametrically excited surface waves in square geometry, to appear.

Simonelli, F. & Gollub, J.P., 1989, Surface wave mode interactions: effects of symmetry and degeneracy, *J. Fluid Mech.* 199:471.

Tufillaro, N.B., Ramshankar, R., & Gollub, J.P., 1989, Order-Disorder Transition in Capillary Ripples, *Phys. Rev. Lett.* 62:422.

Umeki, M., 1990, Faraday resonance in rectangular geometry, *J. Fluid Mech.*, to appear.

CHARACTERIZATION OF IRREGULAR INTERFACES: ROUGHNESS AND SELF-AFFINE
FRACTALS

Miguel A. Rubio[a,b], Andrew Dougherty[a], and Jerry P. Gollub[a,c]

[a]Physics Department, Haverford College, Haverford, PA 19041
[b]Dept. Física Fundamental, Universidad Nacional de Educación a Distancia, Aptdo. Correos 60141, Madrid 28080, Spain
[c]Physics Department, University of Pennsylvania, Philadelphia, PA 19104

1. INTRODUCTION

Many physical systems with complex spatiotemporal behavior give rise to structures with fractal geometries in phase space or real space[1,2]. The paradigm of such a fractal structure in phase space is the strange attractor appearing in the chaotic motion of a dissipative system. The structure of a strange attractor is statistically self-similar. Several techniques of evaluating the fractal dimension have been widely used[3].

In the case of real space fractals the situation is more complex. For self-similar structures the fractal dimension may be computed from the scaling properties of the radius of gyration, the mass distribution, or the two-point density-density correlation function[4]. The complication arises because self-affine fractal structures often appear[5]. These structures show different scaling properties in different directions, and the definition of fractal dimension is not unique.

Here we present a useful method of characterizing irregular interfaces. We analyze them by computing their roughness[6] (rms variation of the interface position) as a function of length scale. We have applied this method to self-affine Weierstrass-Mandelbrot (WM) curves, and to experimental water-air interfaces in a model porous medium. In both cases power law behavior has been found. The roughness exponent β is directly related to the box dimension. The higher precision in the measurement of β and the more intuitive physical interpretation makes the characterization of the structures in terms of β more convenient.

2. CHARACTERIZATION OF ROUGH INTERFACES

Self-affine (anisotropic) fractal structures are often found in systems with rough interfaces. These interfaces may be characterized by computations of their fractal dimensions. The "box" dimension, D_b, is computed by counting the number of boxes of size L needed to cover the interface, and the "divider" (or "compass") dimension, D_d, by counting the number of steps needed to cover the curve with a divider of length L. For self-affine fractals these two dimensions yield different results[5], and this fact may be used to identify their self-affine character. Moreover, the values of the "box" and "divider" dimensions are related to the distinct scaling properties in different directions. A characteristic exponent may be defined by $\Delta y \approx (\Delta x)^\beta$, where Δx and Δy represent typical fluctuations in the directions parallel and perpendicular to the interface, respectively. Then[5,2] $D_b = 2 - \beta$, and $D_d = 1/\beta$; when β approaches unity, D_b and D_d converge, and this method of characterization of self-affine fractals becomes unpractical.

To overcome this difficulty we make a direct estimation of the exponent β. We compute the "roughness" of the profiles as a function of the length scale as is done in studies of interfacial phase transitions[6] and growth models[7]. The roughness $w(L)$ is defined as the rms value of the deviations of the profile from the average perpendicular coordinate computed as a function of the longitudinal length scale L. For self-affine profiles a power law, $w(L) = AL^\beta$ is expected. We have checked the validity and usefulness of this approach for two types of interfaces: the WM self-affine curves, and rough interfaces obtained experimentally in a model porous medium system.

The WM functions are one-dimensional self-affine functions that are constructed by the following series[1,8]:

$$h(x) = \sum_{n=-\infty}^{\infty} b^{-nH} (1 - \cos b^n x)$$

where $b > 1$, and $0 < H < 1$. We have fixed $b = 2.1$ and $0.60 \le x \le 0.61$. For these curves $\beta = H$. Two curves with $H = 0.3$ and $H = 0.7$ are plotted in Fig. 1.

Rough interfaces were obtained experimentally[9] in a thin horizontal cell made of two glass plates, separated by a teflon gasket 0.15 cm deep. The porous medium consisted of tightly packed glass beads. Water containing a small amount of dye was injected at constant flow rate. The relevant control parameter is the capillary number, $Ca = U\mu/\gamma$, where U is the average velocity of the interface, μ the dynamic viscosity of water, and γ the air-water interfacial tension. We varied Ca between 10^{-5} and 10^{-3}. The motion of the interface was recorded with a CCD videocamera, digitized with a spatial resolution of 512x480 pixels and 8 bit intensity resolution, and the interfacial shape $h(x)$ was determined as in Ref. 10. Some sample interfaces obtained at $Ca = 1.38 \times 10^{-4}$ with a time interval of 30 seconds are shown in Fig. 2.

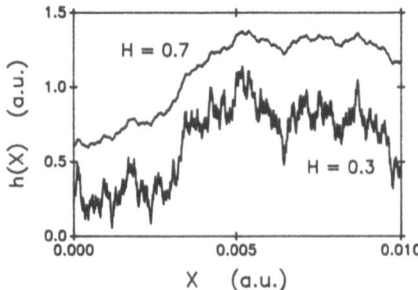

Fig.1. Weierstrass-Mandelbrot curves for $H = 0.3$ and $H = 0.7$.

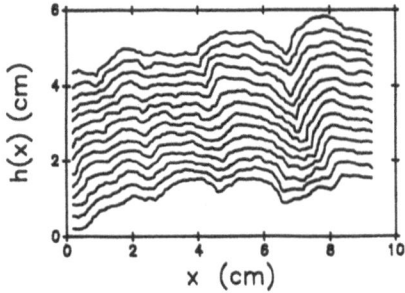

Fig.2. Plot of 14 successive interfaces at Ca = 1.38x10^{-4}. The time
interval between single interfaces is 30 seconds.

3. RESULTS

In Fig. 3 we show a log-log plot of w(L) as a function of L for the
two WM curves in Fig. 1. The power-law behavior appears for log(x)>-3.5.
The corresponding exponents are β = 0.264±0.002 and β = 0.695±0.005,
respectively. In Fig. 4 we show the analogous plot for the interfaces in
the porous medium experiment at two different values of Ca. A remarkable
feature is that in this case the exponent, β = 0.73±0.03, turns out to be
independent of Ca and the bead size b.

We have checked β against direct computations of D_b and D_d, both for
the WM curves and the porous medium interfaces (Table I). (The divider
dimension has been computed with amplification of 10^3 in the vertical
coordinate[5,11]). We note that β≈H, as expected, and D_d is close to 1/β.
However, D_b is somewhat lower than 2-β≈2-H. (This deviation has also been
observed in Ref. 7). The uncertainty of β is always lower than that of D_b
and D_d, for both types of interfaces. In other words, more reliable
estimates of dimensions can be made *indirectly* from β.

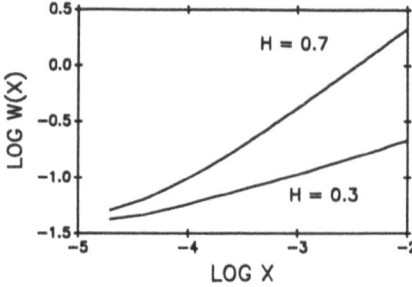

Fig.3. Roughness vs length scale for the WM curves in Fig.1.

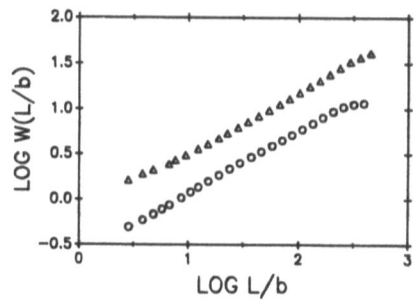

Fig.4. Roughness vs length scale for porous medium interfaces.
$((\Delta)$ Ca=6.9x10^{-5}, β=0.69; (O) Ca=2.75x10^{-4}, β=0.72).

Table I. Average values of β, D_b, and D_d.

Weierstrass-Mandelbrot Functions							
Theor. Values			Measured Values				
H	2-H	1/H	β	D_b	2-β	D_d	1/β
0.3	1.7	3.33	0.264±0.002	1.47±0.03	1.736	3.85±0.12	3.79
0.7	1.3	1.43	0.695±0.005	1.18±0.03	1.305	1.45±0.08	1.439
Porous Medium Interfaces (Measured average values)							
β		D_b		2-β		D_d	1/β
0.73±0.03		1.18±0.08		1.27±0.03		1.36±0.08	1.37±0.05

In conclusion the roughness exponent β completely quantifies the structure of self-affine fractal interfaces, allowing the fractal dimensions to be easily obtained. Moreover, the accuracy is higher than for direct measurement of the dimensions. The roughness exponent β is therefore an appealing tool for the characterization of rough interfaces.

We are indebted to Chris Edwards for his contribution to the development of the porous medium experiment. This work has been supported by the NSF Low Temperature Physics Program DMR-8503543. M.A.R. was supported in part by a NATO Fellowship.

REFERENCES

1. B. B., Mandelbrot, "The Fractal Geometry of Nature", Freeman, New York (1982).
2. J. Feder, "Fractals", Plenum Press, New York (1988).
3. G. Mayer-Kress, "Dimensions and Entropies in Chaotic Systems", Springer-Verlag, Berlin (1986).
4. P. Meakin, in: "Phase Transitions and Critical Phenomena", Vol. 12, C. Domb and J.L. Lebowitz, eds., Academic Press, London (1988).
5. B. B. Mandelbrot, Phys. Scripta, 32:257 (1985).
6. M. E. Fisher, J. Chem. Soc., Faraday Trans. 2, 82:1569 (1986).
7. F. Family, J. Phys. A, 19:L441 (1986).
8. B. Dubuc, J. F. Quiniou, C. Roques-Carmes, C. Tricot and S. W. Zucker, Phys. Rev. A, 39:1500 (1989).
9. M. A. Rubio, C. Edwards, A. Dougherty and J. P. Gollub, to be published (1989).
10. A. Dougherty and J. P. Gollub, Phys. Rev. A, 38:3043 (1988).
11. S. R. Brown, Geophys. Res. Lett., 14:1095 (1987).

THE FIELD PATTERNS OF A HYBRID MODE LASER:
DETECTING THE "HIDDEN" BISTABILITY OF THE OPTICAL PHASE PATTERN

Christian Tamm

Labor 4.42, Physikalisch-Technische Bundesanstalt
Braunschweig, D-3300 Braunschweig, FRG

It is well known that the output field of lasers
oscillating in a superposition of higher-order transverse
modes of the optical resonator typically is characterized by a
complex spatiotemporal structure /1/. The "cooperative frequency locking" /2/ of near-degenerate transverse resonator
modes is accompanied by the formation of stationary transverse
patterns of the optical field.

In two recent experiments, the intensity patterns that
arise from the frequency locking of transverse laser modes
have been analyzed /3,4/. This contribution demonstrates that
a complete characterization of transverse optical patterns may
require to study not only the transverse variation of
intensity, but also that of the optical phase.

The particular example considered here is that of a
laser oscillating in frequency-locked TEM_{10} and TEM_{01} Gauss-
Hermite optical-resonator modes where the stationary output
pattern is given by a TEM^*_{01} hybrid mode /3/, i.e., the two
constituent modes are locked in phase quadrature /5/. It is
shown in the following that the two possible phase-locking
angles of $+\pi/2$ and $-\pi/2$, although yielding identical intensity
patterns, correspond to physically distinguishable, mirror-
symmetric field patterns. A simple transverse-mode converter

is described that allows to discriminate between the two field
patterns; the device has been used to demonstrate bistability
and optically induced switching of the field patterns of a
TEM*01 hybrid mode laser /6/.

Near to a beam waist of width wo, the optical field of a
TEM*01 hybrid mode can be written as

$$\vec{E}(\vec{r},t) = \vec{E}_o\, e^{i(\omega t - kz + \phi_o)}\, e^{-(x^2+y^2)/w_o^2} \left(\tfrac{x}{w_o} + e^{i\phi}\tfrac{y}{w_o}\right) + c.c. \;,\; \phi = \pm\tfrac{\pi}{2}\,, \tag{1}$$

where ϕ is the optical phase difference between the constituent
TEM01 and TEM10 modes, ϕ_o is an arbitrary common phase, ω is
the optical frequency, $k = 2\pi/\lambda$ (λ: wavelength), and z denotes
the direction of propagation. It is easy to conclude from Eq.
(1) that the transverse intensity distribution $|\vec{E}(x,y)|^2$ is
rotationally symmetric around the optical axis and equal for
the two values of ϕ; the equiphase surfaces of the field $\vec{E}(\vec{r})$
however have a different shape for $\phi = +\pi/2$ and $\phi = -\pi/2$: They
are mirror-symmetric helices winding around the optical axis
clockwise or counterclockwise, respectively.

The mode converter used to discriminate between the two
mirror-symmetric field configurations relies on converting the
two different field-phase patterns into different intensity
patterns with the use of astigmatic imaging of optical-
resonator modes /7/: A one-dimensional beam waist is created
for the TEM*01 field in the space between two cylindrical
lenses whose axes are oriented in parallel (Fig. 1), introducing a diffractive phase shift difference $\delta\phi$ between the two
constituent modes. A suitable choice of the geometrical
arrangement (confocal parameter and beam waist location of
ingoing field, focal lengths and spacing of the cylindrical
lenses) allows to convert hybrid mode fields with $\phi = +\pi/2$
($\phi = -\pi/2$) into "pure" TEM10 (TEM01) fields, with the nodal
planes of the output fields rotated by 45° with respect to the
axes of the cylindrical lenses.

For a perfect mode conversion, it follows from comparison
with Eq. (1) that the three conditions $|\delta\phi| = |\phi| = \pi/2$, $w_x = w_y$,
and $w_x(z) = w_y(z)$ (see Fig. 1) for the outgoing field must be

met simultaneously. From the elementary imaging and propagation properties of higher-order Gaussian modes /7/, it follows that the first condition is equivalent to

$$\arctan\frac{2d_1}{b_1} + \arctan\frac{2d_2}{b_2} + \arctan\frac{2d_o}{b_o} - \arctan\frac{2(d_o+d_1+d_2)}{b_o} = \frac{\pi}{2} \quad , \tag{2}$$

with

$$b_1 = \frac{b_o}{\left(1-\frac{d_o}{f_1}\right)^2 + \left(\frac{b_o}{2f_1}\right)^2} \quad , \quad d_1 = f_1 + \frac{d_o-f_1}{\left(1-\frac{d_o}{f_1}\right)^2 + \left(\frac{b_o}{2f_1}\right)^2} \quad , \tag{3}$$

while the second condition leads to

$$b_1\left[1+\left(\frac{2d_2}{b_1}\right)^2\right] = b_o\left[1+\left(\frac{2(d_o+d_1+d_2)}{b_o}\right)^2\right] \quad . \tag{4}$$

These two conditions are satisfied by the surprisingly simple relation

$$\left(\frac{d_o}{f_1}\right)^2 + \frac{1}{4}\left(\frac{b_o}{f_1} - 2\right)^2 = 1 \quad , \quad b_o > 2f_1 \text{ if } d_o > 0 \quad . \tag{5}$$

The third condition determines the focal length of the second cylindrical lens:

$$\frac{f_2}{f_1} = \frac{1}{b_o - b_1}\left[b_1\left(d_o + d_1 + d_2\right) + d_2 b_o\right] \tag{6}$$

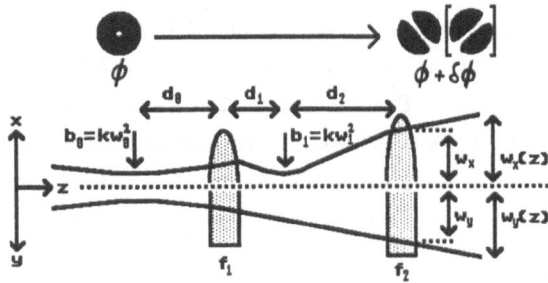

Fig. 1. Schematic of mode converter, showing the beam envelopes and the effect of the cylindrical lenses in the x-z plane (upper section) and y-z plane (lower section).

In practice, satisfactory operation of a mode converter of this type could be obtained (and the bistability of the field pattern of a TEM^*_{01} laser could be observed /6/) using the following experimental setup: The output of a Helium-Neon TEM^*_{01} laser with a symmetric, linear spherical-mirror resonator (radius of curvature, 600 mm; mirror separation, 300 mm) is refocused by a spherical lens with f=30 mm placed at a distance of 160 mm from the outcoupling mirror. A cylindrical (glass rod) lens with f_1=1.2 mm is placed at $d_0 \approx 32$ mm behind the spherical lens, the second cylindrical lens with f_2=40 mm is placed at d_1+d_2=15 mm.

It may be noted finally that only a simultaneous recording of the spatial distribution of the output intensity of a TEM^*_{01} laser <u>and</u> of the output intensity pattern behind the mode converter allows to fully characterize the spatio-temporal evolution of the system for the cases where periodic or irregular pulsing of the local output intensity is observed /3/.

ACKNOWLEDGEMENT

This work was supported by the Deutsche Forschungsgemeinschaft.

REFERENCES

/1/ J.P. Goldsborough, Appl. Opt. **3**, 267 (1964); T. Uchida, Appl. Opt. **4**, 129 (1964).

/2/ L.A. Lugiato, C. Oldano and L.M. Narducci, J. Opt. Soc. Am. **B5**, 879 (1988).

/3/ Chr. Tamm, Phys. Rev. **A38**, 5960 (1988).

/4/ J.R. Tredicce et al., Phys. Rev. Lett. **62**, 1274 (1989).

/5/ W.W. Rigrod, Appl. Phys. Lett. **2**, 51 (1963).

/6/ Chr. Tamm, Phys. Rev. Lett., submitted for publication.

/7/ H. Kogelnik and T. Li, Appl. Opt. **5**, 1550 (1966); D.C. Hanna, IEEE J. Quantum Electron. **QE-5**, 483 (1969)

CONTRIBUTORS

Abraham, N.B.   Department of Physics, Bryn Mawr College, Bryn
                       Mawr, PA 19010-2899, USA

Albano, A.M.   Department of Physics, Bryn Mawr College, Bryn
                       Mawr, PA 19010-2899, USA

Arecchi, F.T.   Istituto Nazionale di Ottica, Largo Enrico Fermi 6,
                       I-50125 Firenze, Italy

Arimondo, E.   Dipartimento di Fisica, Università di Pisa, I-56100
                       Pisa, Italy

Atmanspacher, H.   Max-Planck-Institut für Extraterrestrische Physik,
                       D-8046 Garching, Federal Republic of Germany

Auerbach, D.   Department of Chemical Physics, Weizmann Institute,
                       Rehovot 76100, Israel

Badii, R.   Fakultät für Physik, Universitat Konstanz, D-7750,
                       Konstanz, Federal Republic of Germany

Barkley, D.   Applied Mathematics, California Institute of
                       Technology, Pasadena, CA 91125, USA

Bekhali, A.   Laboratoire de Spectroscopie Hertzienne, associé au
                       CNRS, Université des Sciences et Techniques de
                       Lille, F-59655 Villeneuve d'Ascq Cédex, France

Bensen, D.   Center for Complex Systems Research, Department of
                       Physics, Beckman Institute, 405 N. Mathews Ave.,
                       Urbana, IL 61801, USA

Brock, W.A.   Department of Economics, University of Wisconsin,
                       Madison, WI 53706, USA

Byers, R.E.   Deparment of Zoology, University of Toronto, 25
                       Harbord St., Toronto, Ontario M5S 1A1, Canada

Caputo, J.G.   LESP, INSA de Rouen, BP 8, F-76131 Mont-Saint-
                       Aignan Cédex, France

Caruso-Haviland, L.   Dance Program, Bryn Mawr College, Bryn Mawr, PA
                       19010-2899, USA

Chang, K.   Center for Complex Systems Research, Department of
                       Physics, Beckman Institute, 405 N. Mathews Ave.,
                       Urbana, IL 61801, USA

Chee Merk-Ne   Chemical Physics Theory Group, Department of
                       Chemistry, University of Toronto, Toronto, Ontario
                       M5S 1A1, Canada

Coullet, P.	Laboratoire de Physique Théorique, Université de Nice, Parc Valrose, F-06034 Nice Cédex, France
Crutchfield, J.P.	Physics Department, University of California, Berkeley, CA 94720, USA
Cummings, A.	AT&T Bell Laboratories, Murray Hill, NJ 07974, USA
D'Alessandro, G.	Istituto Nazionale di Ottica, Largo Enrico Fermi 6, I-50125 Firenze, Italy
Dangoisse, D.	Laboratoire de Spectroscopie Hertzienne, associé au CNRS, Université des Sciences et Techniques de Lille, F-59655 Villeneuve d'Ascq Cédex, France
Dechert, W.D.	Department of Economics, University of Houston, Houston, TX 7704, USA
Demmel, V.	Max-Planck-Institut für Extraterrestrische Physik, D-8046 Garching, Federal Republic of Germany
Dextexhe, A.	Service de Chimie Physique, Université Libre de Bruxelles, CP-231 Campus Plaine, Boulevard du Triomphe, B-1050 Bruxelles, Belgium
Everson, R.M.	Center for Fluid Mechanics, Turbulence and Computation, Brown University, Providence, RI 02912, USA
Fang Jinqing	Center for Studies in Statistical Mechanics, University of Texas at Austin, Austin, TX 78712, USA
Farrell, M.E.	Naval Air Development Center, Warminster, PA 18974, USA
Fioretti, A.	Dipartimento di Fisica, Università di Pisa, I-56100 Pisa, Italy
Frame, M.	Department of Mathematics, Union College, Schenectady, NY 12308, USA
Frank, G.W.	Department of Applied Mathematics, University of Western Ontario, London N6A 5B9 Canada
Fraser, A.M.	Center for Nonlinear Dynamics, Department of Physics, The University of Texas at Austin, Austin, TX 78712, USA
G. Broggi	Fakultät für Physik, Universität Konstanz, D-7750, Konstanz, Federal Republic of Germany
Gaito, S.T.	Nonlinear Systems Laboratory, Mathematics Institute, University of Warwick, Coventry CV4 7AL, UK
Gilmore, R.	Department of Physics and Atmospheric Science, Drexel University, Philadelphia, PA 19104, USA
Glendinning, P.	Department of Applied Mathematics and Theoretical Physics, Cambridge University, Silver St., Cambridge CB3 9EW, UK
Glorieux, P.	Laboratoire de Spectroscopie Hertzienne, associé au CNRS, Université des Sciences et Techniques de Lille, F-59655 Villeneuve d'Ascq Cédex, France

Glover, J.N.	Department of Applied Mathematics, University of Western Australia, Nedlands 6009, Australia
Goel, A.	Deparment of Electrical Engineering, Villanova University, Villanova, PA 19085, USA
Gollub, J.P.	Department of Physics, Haverford College, Haverford, PA 19104; David Rittenhouse Laboratory, The University of Pennsylvania, Philadelphia, PA 19104, USA
Hansell, R.I.C.	Deparment of Zoology, University of Toronto, 25 Harbord St., Toronto, Ontario M5S 1A1, Canada
Hartt, K.	Department of Physics, The University of Rhode Island, Kingston, RI 02881-0817, USA
Hediger, T.	Naval Air Development Center, Warminster, PA 18974, USA
Hennequin, D.	Laboratoire de Spectroscopie Hertzienne, associé au CNRS, Université des Sciences et Techniques de Lille, F-59655 Villeneuve d'Ascq Cédex, France
Hübler, A.	Center for Complex Systems Research, Department of Physics, Beckman Institute, 405 N. Mathews Ave., Urbana, IL 61801, USA
Hübner, U.	Physikalish-Technishche Bundesanstalt, D-3300 Braunschweig, Federal Republic of Germany
Hunt, E.R.	Department of Physics and Astronomy, Ohio University, Athens, OH 45701, USA
Kahn, L.M.	Department of Physics, The University of Rhode Island, Kingston, RI 02881-0817, USA
Kapral, R.	Chemical Physics Theory Group, Department of Chemistry, University of Toronto, Toronto, Ontario M5S 1A1, Canada
King, G.P.	Nonlinear Systems Laboratory, Mathematics Institute, University of Warwick, Coventry CV4 7AL, UK
Klische, W.	Physikalish-Technishche Bundesanstalt, D-3300 Braunschweig, Federal Republic of Germany
Lange, W.	Institut für Angewandte Physik der Westfälischen Wilhelms-Universität Münster, D-4400 Münster, Federal Republic of Germany
Lapucci, A.	Istituto Nazionale di Ottica, Largo E. Fermi 6, 50125 Firenze, Italy
LeBerre, M.	Laboratoire de Photophysique Moléculaire, Bât. 213, Université Paris-Sud, 91405 Orsay, France
Lefranc, M.	Laboratoire de Spectroscopie Hertzienne, associé au CNRS, Université des Sciences et Techniques de Lille, F-59655 Villeneuve d'Ascq Cédex, France
Li Wentian	Center for Complex Systems Research, Department of Physics, Beckman Institute, 405 N. Mathews Ave., Urbana, IL 61801, USA
Lookman, T.	Department of Applied Mathematics, University of Western Ontario, London N6A 5B9, Canada

Lugiato, L.A.   Dipartimento di Fisica, Politecnico di Torino, Corso Duca degli Abruzzi 24, I-10129 Torino, Italy

Madras, N.   Department of Mathematics, York University, 4700 Keele St., Downsview, Ontario M3J 1P3, Canada

Mayer-Kress, G.   Department of Mathematics, University of California at Santa Cruz; Department of Chemical Engineering, Princeton University; Center for Nonlinear Studies, Los Alamos National Laboratory, MS-B258, Los Alamos, NM 87545, USA

Mees, A.I.   Department of Mathematics, The University of Western Australia, Nedlands, Perth 6009, Australia

Meucci, R.   Istituto Nazionale di Ottica, Largo Enrico Fermi 6, I-50125 Firenze, Italy

Meyer, T.   Center for Complex Systems Research, Department of Physics, Beckman Institute, 405 N. Mathews Ave., Urbana, IL 61801, USA

Mindlin, G.   Department of Physics and Atmospheric Science, Drexel University, Philadelphia, PA 19104, USA

Möller, M   Institut für Angewandte Physik der Westfälischen Wilhelms-Universität Münster, D-4400 Münster, Federal Republic of Germany

Morfill, G.   Max-Planck-Institut für Extraterrestrische Physik, D-8046 Garching, Federal Republic of Germany

Narducci, L.M.   Department of Physics and Atmospheric Science, Drexel University, Philadelphia, PA 19104, USA

Nerenberg, M.A.H.   Department of Applied Mathematics, University of Western Ontario, London N6A 5B9 Canada

Nicolaenko, B.   Center for NonLinear Science, MS B-258, Los Alamos National Laboratory, Los Alamos, NM 87545, USA

Nicolis, C.   Service de Chimie Physique, Université Libre de Bruxelles, CP-231 Campus Plaine, Boulevard du Triomphe, B-1050 Bruxelles, Belgium

Nicolis, G.   Service de Chimie Physique, Université Libre de Bruxelles, CP-231 Campus Plaine, Boulevard du Triomphe, B-1050 Bruxelles, Belgium

Olsen, L.F.   Institute of Biochemistry, Odensee University, Odensee, Denmark

Oppo, G.-L.   Department of Physics and Atmospheric Science, Drexel University, Philadelphia, PA 19104, USA

Packard, N.   Center for Complex Systems Research, Department of Physics, Beckman Institute, 405 N. Mathews Ave., Urbana, IL 61801, USA

Papoff, F.   Scuola Normale Superiore, I-56100 Pisa, Italy

Passamante, A.   Naval Air Development Center, Warminster, PA 18974, USA

Pernigo, M.A.   Center for Complex Systems Research, Department of Physics, Beckman Institute, 405 N. Mathews Ave., Urbana, IL 61801, USA

Politi, A.   Istituto Nazionale di Ottica, Largo Enrico Fermi 6, I-50125 Firenze, Italy

Procaccia, I.   Department of Chemical Physics, Weizmann Institute, Rehovot 76100, Israel

Rapp, P.E.   Department of Physiology and Biochemistry, The Medical College of Pennsylvania, 3300 Henry Ave., Philadelphia, PA 19129, USA

Raymer, M.G.   Department of Physics and Chemical Physics Institute, University of Oregon, Eugene, OR 97403, USA

Ressayre, E.   Laboratoire de Photophysique Moléculaire, Bât. 213, Université Paris-Sud, F-91405 Orsay, France

Ringland, J.   Department of Chemistry, Southern Methodist University, Dallas, TX 75275, USA

Rollins, R.W.   Department of Physics and Astronomy, Ohio University, Athens, OH 45701, USA

Sayres, C.   Department of Economics, University of Houston, Houston, TX 77204-5882, USA

Schaffer, W.M.   Department of Ecology and Evolutionary Biology, The University of Arizona, Tucson, AZ 85721, USA

Schell, M.   Department of Chemistry, Southern Methodist University, Dallas, TX 75275, USA

Schuster, H.G.   Institut für Theoretische Physik und Sternwarte, Universität Kiel, Federal Republic of Germany

Sepulveda, M.A.   Department of Chemical Physics, Weizmann Institute, 76100 Rehovot, Israel

She Zhen-Su   Applied Mathematics Program, Princeton University, Princeton, NJ 08544, USA

Smith, L.A.   Department of Applied Mathematics and Theoretical Physics, University of Cambridge, Cambridge CB3 9EW, UK

Solari, H.G.   Department of Physics and Atmospheric Science, Drexel University, Philadelphia, PA 19104, USA

Su, Z.   Department of Physics and Astronomy, Ohio University, Athens, OH 45701, USA

Tallet, A.   Laboratoire de Photophysique Moléculaire, Bât. 213, Université Paris-Sud, F-91405 Orsay, France

Tamm, C.   Labor. 4.42, Physikalish-Technische Bundesanstalt, D-3300 Braunschweig, Federal Republic of Germany

Theiler, J.   Institute for Nonlinear Science, University of California at San Diego and MIT Lincoln Laboratory, L-244, Lexington, MA 02173-0073, USA

Torcini, A.   Istituto Nazionale di Ottica, Largo Enrico Fermi 6, I-50125 Firenze, Italy

Tufillaro, N.B.   Department of Physics, Bryn Mawr College, Bryn Mawr, PA 19010-2899, USA

Voges, W. Max-Planck-Institut für Extraterrestrische Physik, D-8046 Garching, Federal Republic of Germany

Weiss, C.O. Labor 4.42, Physikalish-Technishche Bundesanstalt, D-3300 Braunschweig, Federal Republic of Germany

Welge, M. Center for Complex Systems Research, Department of Physics, Beckman Institute, 405 N. Mathews Ave., Urbana, IL 61801, USA

Whittington, S.G. Chemical Physics Theory Group, Department of Chemistry, University of Toronto, Toronto, Ontario M5S 1A1, Canada

Wiedenmann, G. Max-Planck-Institut für Extraterrestrische Physik, D-8046 Garching, Federal Republic of Germany

INDEX

Accuracy in Measures
  of Complexity, 79
Astronomy, 5,33
Bifurcation, 295
Biology, 51,111,187,213,341
  (see also neurophysical
signals;
    neurobiology models)
Capacity, 2
Chaos, 1,4,8
  heteroclinic, 4,5,291
  homoclinic, 4,5,11,281,
295,299,303
  on a catastrophe manifold, 235
  proof of Existence, 79
  routes to 4, 297
Chemistry, 5,147
Complexity, 8,9,63,117,203,318,327
  reduction by forcing, 253
Convection, 193,405,445
Correlation Dimension, 2,33,57,
  99,115,133,137,181,199
  accuracy, 79,137,183,199
Correlation functions,
  42,129,181,249,381
Dance, 29
Defects, 367,395,425
Dendritic growth, 5
Delay, effects of, 245
Differential Time Interval Plot,
  56
Digitizing, Effects of, 137,140
Dimensions, 2,9,10,33,57,99,
  173,375,409
  accuracy, 79,183,199
  d_q, 2,37,157
  procedures for calculating, 35
    embedding, 3,99
    window, 3,99,349
    data requirements 3
  effects of filtering, 138
  error estimates, 3,133
    effects of noise, 133,137,141
    effects of digitizing, 137,140
  local intrinsic dimension, 3
  partial dimensions, 63
  (see Kaplan-Yorke Conjecture)
  (see Correlation Dimension)
Driven Systems, (see Forced
Systems)
Duffing, 176,265

ε-machines, 333

Economics, 5,79
Electrodeposition, 433
Entropies, 2,4,9,10,99,137,
  181,333,375
  correction of entropies, 144
  influence of filtering, 137
Epidemic growth rates, 5,187
$f(\alpha)$, 2,37,43,46,191
Filtering (see Entropies;
Dimensions)
Flames, 5
Flow, Kolmogorov, 303
Fluids 5,11,193,303
Forced Systems, 209,225,229,253
Fractals, 11,41,461
  (see also Multifractals)
Fractal dimension (see Dimensions)
Hausdorff dimension, 2
Hénon, 125,176,356,361
Homoclinic chaos
  (see Chaos, Homoclinic)
Information dimension, 2,63
Interfaces 461
Interfacial waves, 457
Intermittency, 303
Kaplan-Yorke conjecture, 4
Lasers, 5,133,181,269,281,299,
  309,395,465
  (see also Nonlinear Optics)
Logistic Map, 214
Local Dimension, 3,63,125,173
Lorenz Model, 125,129,133,269
Lyapunov Exponents, 3,11,63,111,
  113,121,152,313,375,409
Maps
  discontinuous, 295
  lattice, 375,381,409
  return, 309
Models from Data, 10,345,327,354
Multifractals, 33,75
Mutual Information, 3,117,249
Neurobiology models, 111
Neurophysical signals, 5,7,8,51,
  113
Nonlinear Optics, 12
  (see also Lasers)
Nonlinear Resonance Spectroscopy,
  255
Periodic orbits 3,147,203,241,261,
  349,359
Phase measurments, 269
Phase dynamics, 381

Pointwise dimensions, 156
Power Spectra, 257
Prediction algorithms 5,10,327,
    345,349,359
Quasiperiocity, 75
Recurrence Plots, 52
Relative Rotation Rates, 261,265
Reviews, 12
Rössler Attractor, 121,176,339,352
Singular value decomposition, 3,
    125,173
Shil'nikov Chaos
    (*see* Chaos, Homoclinic)
Solitons, 254
Spatio-temporal complexity, 1,6,
    367,375,381,395,405,409,425,

429,433,445,457,461,465
    (*see also* Turbulence)
correlation functions, 6
defect-mediated, 6,425
in convection,
in electrodeposition, 433
in interfacial waves, 457
in lasers, 395,465
Symbolic Dynamics, 3,64,151,
    257,313,339,359
Tesselation, 345
Time Evolution of Complexity
    Measures, 155
Turbulence, 1,12,367,425,445
    homoclinic behavior 5